The Science, Etiology and Mechanobiology of Diabetes and its Complications

The Science, Etiology and Mechanobiology of Diabetes and its Complications

Edited by

Amit Gefen

Department of Biomedical Engineering,
Faculty of Engineering,
Tel Aviv University, Tel Aviv, Israel

Academic Press is an imprint of Elsevier
125 London Wall, London EC2Y 5AS, United Kingdom
525 B Street, Suite 1650, San Diego, CA 92101, United States
50 Hampshire Street, 5th Floor, Cambridge, MA 02139, United States
The Boulevard, Langford Lane, Kidlington, Oxford OX5 1GB, United Kingdom

Copyright © 2021 Elsevier Inc. All rights reserved.

No part of this publication may be reproduced or transmitted in any form or by any means, electronic or mechanical, including photocopying, recording, or any information storage and retrieval system, without permission in writing from the publisher. Details on how to seek permission, further information about the Publisher's permissions policies and our arrangements with organizations such as the Copyright Clearance Center and the Copyright Licensing Agency, can be found at our website: www.elsevier.com/permissions.

This book and the individual contributions contained in it are protected under copyright by the Publisher (other than as may be noted herein).

Notices

Knowledge and best practice in this field are constantly changing. As new research and experience broaden our understanding, changes in research methods, professional practices, or medical treatment may become necessary.

Practitioners and researchers must always rely on their own experience and knowledge in evaluating and using any information, methods, compounds, or experiments described herein. In using such information or methods they should be mindful of their own safety and the safety of others, including parties for whom they have a professional responsibility.

To the fullest extent of the law, neither the Publisher nor the authors, contributors, or editors, assume any liability for any injury and/or damage to persons or property as a matter of products liability, negligence or otherwise, or from any use or operation of any methods, products, instructions, or ideas contained in the material herein.

Library of Congress Cataloging-in-Publication Data
A catalog record for this book is available from the Library of Congress

British Library Cataloguing-in-Publication Data
A catalogue record for this book is available from the British Library

ISBN: 978-0-12-821070-3

For information on all Academic Press publications visit our website at https://www.elsevier.com/books-and-journals

Publisher: Mara Conner
Acquisitions Editor: Fiona Geraghty
Editorial Project Manager: Gabriela Capille
Production Project Manager: Prem Kumar Kaliamoorthi
Cover Designer: Mark Rogers

Typeset by TNQ Technologies

CONTENTS

Contributors   xiii
Preface   xix

1. Heel ulcers in patients with diabetes   1
Michael Clark

The impact of pressure ulcers	1
Changing terms and definitions of pressure ulcers?	2
Is the wound a pressure ulcer or a diabetic foot ulcer? Does this matter?	2
Is diabetes a risk factor for heel pressure ulcers?	3
Do heel ulcers in diabetes heal?	4
Do heel ulcers in diabetics heal slower than other wound types?	4
What can be inferred about the healing of heel ulcers in diabetics and nondiabetics from wound registry data?	5
Conclusions	10
References	10

2. Diabetic foot ulcers and their wound management   13
Norihiko Ohura *and* Katsuya Hisamichi

Introduction	13
Assessment of the wound	14
Wound bed preparation/debridement and amputation	21
Revascularization	23
Treatments for the promotion of wound healing	24
Reconstruction of the soft tissue	25
Summary	32
References	32

3. Computational modeling of the plantar tissue stresses induced by the clinical practice of off-loading of the diabetic foot   35
Hadar Shaulian, Amit Gefen *and* Alon Wolf

Introduction	35
Methods	36
Results	39

Discussion 41
References 41

4. Modeling effects of sustained bodyweight forces on adipose tissue microstructures and adipocytes in diabesity ... 43

Maayan Lustig, Golan Amrani, Adi Lustig, Liran Azaria, Raz Margi, Yoni Koren, Avraham Kolel, Nurit Bar-Shai, Avior Exsol, Maya Atias and Amit Gefen

Introduction 43
Methods 45
Results 53
Mechanical loads in the mesoscale of lean and obese tissues 54
Discussion 56
Mechanical loads in the mesoscale of lean and obese tissues 57
Conclusion 58
Acknowledgments 59
References 59

5. Mechanisms underlying vascular stiffening in obesity, insulin resistance, and type 2 diabetes ... 63

Michael A. Hill, Yan Yang, Zhe Sun, Liping Zhang and James R. Sowers

Abbreviations 63
Introduction 64
Extracellular matrix and vascular stiffening in obesity and diabetes 66
Cellular contributions to vascular stiffening in obesity and diabetes ... 69
Metabolic and endocrine activation of vascular stiffening in obesity, INS resistance, and diabetes 73
Conclusion 80
References 80

6. Pathomechanics of diabetic foot ulceration: revisiting plantar shear and temperature ... 89

Metin Yavuz

Biomechanics overview 89
Plantar pressure 92
Comparison with pressure injuries (ulcers) 96
Shear stress 97
Plantar callus and shear 100
Temperature increase and shear 101
The need to revisit the etiology 102

Acknowledgments	102
References	103

7. Novel technologies for detection and prevention of diabetic foot ulcers — 107

Neil D. Reeves, Bill Cassidy, Caroline A. Abbott *and* Moi Hoon Yap

Health and societal burden of diabetic foot ulcers	107
Risk factors for the development of diabetic foot ulcers	107
Foot pressures and diabetic foot ulcer development	108
The need for technology solutions for diabetic foot ulcer prevention	111
Intelligent insole system: reduced diabetic foot ulcer incidence through continuous daily foot pressure feedback	113
Acknowledgments	119
References	119

8. The role of tissue biomechanics in improving the clinical management of diabetic foot ulcers — 123

Panagiotis Chatzistergos, Roozbeh Naemi *and* Nachiappan Chockalingam

Introduction	123
Basic engineering concepts	124
The case for assessing tissue biomechanics	127
Clinically applicable methods to assess plantar soft tissue biomechanics	131
Concluding remarks	138
References	140

9. The mechanobiology of adipocytes in the context of diabetes: much more than a fat depot — 143

Shirley L. Yitzhak-David *and* Daphne Weihs

Introduction	143
From stem cells to mature adipocytes: mechanobiology of migration and differentiation	147
References	156

10. Optical Coherence Tomography to determine and visualize pathological skin structure changes caused by diabetes — 161

Raman Maiti, Roger Lewis, Daniel Parker *and* Matt J. Carré

Background and working principle of OCT	161
Capabilities for skin imaging	161
Case study 1: use of OCT for skin biomechanics measurements	163

Case study 2: use of OCT for the diagnosis of atopic dermatitis — 168
Future potential of OCT for the diagnosis and management of diabetes-related complications — 168
References — 170

11. Effects of hyperglycemia and mechanical stimulations on differentiation fate of mesenchymal stem cells — 173

Tasneem Bouzid *and* Jung Yul Lim

Introduction — 173
Effect of hyperglycemia on MSC function and fate — 174
Effect of mechanical loading cues on MSC fate — 181
Perspective — 193
Acknowledgments — 196
References — 196

12. Clinical complications of tendon tissue mechanics due to collagen cross-linking in diabetes — 201

Jennifer A. Zellers, Jeremy D. Eekhoff, Simon Y. Tang, Mary K. Hastings *and* Spencer P. Lake

Clinical problem — 201
Clinical syndromes of tendon dysfunction in individuals with DM — 202
Advanced glycation end products in collagenous tissues — 204
Collagen sliding and homeostasis — 207
Noncollagenous proteins in the tendon in diabetic conditions — 210
Tendon mechanics — 211
Summary — 218
References — 219

13. A phenomenological dashpot model for morphoelasticity for the contraction of scars — 227

F.j. Vermolen

Introduction — 227
The mathematical model — 229
Analysis and computational methodologies — 233
Numerical results — 235
Discussion and conclusions — 241
Acknowledgment — 244
References — 244

14. Mechanobiology of diabetes and its complications: from mechanisms to effective mechanotherapies 247

Chenyu Huang *and* Rei Ogawa

Mechanobiological changes in DM that affect wound healing 248
Mechanotherapy for diabetic wounds 249
References 252

15. Application of tissue mechanics to clinical management of risk in the diabetic foot 255

Daniel Parker *and* Farina Hashmi

Plantar soft tissue 255
Effect of diabetes on plantar soft tissue structure 259
Biomechanics of the plantar soft tissue 260
Effect of diabetes on plantar soft tissue biomechanics 270
Relevance for clinical management of diabetes 274
Outlook in the context of clinical practice 275
References 276

16. Bone carriers in diabetic foot osteomyelitis 283

Cristian Nicoletti

Background 283
DFO current diagnosis and treatment 283
Rationale for the use of topical antibiotic therapy 284
Classes of local antibiotic delivery systems 285
Evidences for the use of local antibiotics in diabetic foot infections 286
Summary 289
References 289

17. Vascular mechanobiology and metabolism 291

Sarah Basehore, Jonathan Garcia *and* Alisa Morss Clyne

Introduction 291
Endothelial cell metabolism 292
Endothelial cell metabolism impacts function 295
Tools to measure endothelial cell metabolism 295
Endothelial cell response to flow 297
Endothelial cell metabolism in varied flow conditions 300
Computational models of vascular mechanobiology and metabolism 300
Translational potential and future work 304
References 305

18. Effect of type 2 diabetes on bone cell behavior — 313
Rachana Vaidya, Anna Church and Lamya Karim

Introduction	313
Bone cells	315
Bone modeling and remodeling	317
Effect of T2D on bone cells	318
Conclusion	322
References	322

19. What makes a good device for the diabetic foot — 327
Evan Call, Darren F. Groberg and Nick Santamaria

Introduction (Daren Groberg, DPM)	327
Background	328
What constitutes a good device	329
Benefits a device must provide	330
Drawbacks of devices	332
Importance of interventions	333
Offloading in prevention	334
Offloading in treatment	334
Science of tissue damage	335
Specific devices	338
Heel lift boots	338
Total contact cast	339
Compression systems, wraps, and garments	339
Static stiffness index	340
Dressings for treatment	340
Dressings for prevention	341
Device life	341
Conclusions	342
References	343

20. Allostasis: a conceptual framework to better understand and prevent diabetic foot ulcers — 347
Laurel Tanner, Craig Oberg and Evan Call

Introduction	347
Conclusion	359
Conflict of interest	360
References	360

21. Footwear for persons with diabetes at high risk for foot ulceration: offloading, effectiveness, and costs — 363

Sicco A. Bus

Introduction	363
The person with diabetes at high risk	363
Offloading	364
The offloading effect of footwear	364
Effectiveness of footwear for ulcer prevention	365
Adherence to wearing diabetic footwear	367
Recommendations from international guidelines	368
Costs related to the use of footwear for ulcer prevention	369
A few considerations	370
Future research	370
Conclusions	370
References	371

22. Compounding effects of diabetes in vessel formation in microvessel fragment−based engineered constructs — 375

Omar Mourad, Blessing Nkennor and Sara S. Nunes

Introduction	375
Isolated microvessel fragments	375
Elements affecting microvessel fragment vascularization outcomes	377
Mechanisms controlling engineered vessel function and maturation	377
Impact of diabetes on engineered vessel function	380
Future directions	382
References	385

23. Dressing selection challenges in diabetic foot local treatment — 389

Paulo Alves, Tania Manuel, Nuno Mendes, Emília Ribeiro and Anabela Moura

Therapeutic approach in DFUs	389
Wound dressings	390
Exudation, frequency and comfort	397
Ulcer protection and pressure relief	401
References	404

Index — *407*

Contributors

Caroline A. Abbott
Research Centre for Musculoskeletal Science and Sports Medicine, Faculty of Science and Engineering, Manchester Metropolitan University, Manchester, United Kingdom

Paulo Alves
Universidade Católica Portuguesa, Health Sciences Institute | Wounds Research Lab, Center for Interdisciplinary Research in Health (CIIS), Portugal; Portuguese Wound Management Association (APTFeridas), Portugal

Golan Amrani
Department of Biomedical Engineering, Faculty of Engineering, Tel Aviv University, Tel Aviv, Israel

Maya Atias
Department of Biomedical Engineering, Faculty of Engineering, Tel Aviv University, Tel Aviv, Israel

Liran Azaria
Department of Biomedical Engineering, Faculty of Engineering, Tel Aviv University, Tel Aviv, Israel

Nurit Bar-Shai
Department of Biomedical Engineering, Faculty of Engineering, Tel Aviv University, Tel Aviv, Israel

Sarah Basehore
Drexel University, Philadelphia, PA, United States; University of Maryland, College Park, MD, United States

Tasneem Bouzid
Department of Mechanical and Materials Engineering, University of Nebraska-Lincoln, Lincoln, NE, United States; Nebraska Center for the Prevention of Obesity Diseases (NPOD), University of Nebraska-Lincoln, Lincoln, NE, United States

Sicco A. Bus
Associate Professor and Head Gait Lab, Amsterdam UMC, University of Amsterdam, Department of Rehabilitation Medicine, Amsterdam, Netherlands

Evan Call
Department of Microbiology, Weber State University, Ogden UT, United States

Matt J. Carré
University of Sheffield, Sheffield, United Kingdom

Bill Cassidy
Centre for Applied Computational Science, Faculty of Science and Engineering, Manchester Metropolitan University, Manchester, United Kingdom

Panagiotis Chatzistergos
Centre for Biomechanics and Rehabilitation Technologies, Staffordshire University, Stoke-on-Trent, United Kingdom

Nachiappan Chockalingam
Centre for Biomechanics and Rehabilitation Technologies, Staffordshire University, Stoke-on-Trent, United Kingdom

Anna Church
Department of Bioengineering, University of Massachusetts Dartmouth, Dartmouth, MA, United States

Michael Clark
Professor of Wound Study, Birmingham City University, Birmingham, United Kingdom; Welsh Wound Innovation Centre, Ynysmaerdy, Wales

Alisa Morss Clyne
Drexel University, Philadelphia, PA, United States; University of Maryland, College Park, MD, United States

Jeremy D. Eekhoff
Washington University in St. Louis, St. Louis, MO, United States

Avior Exsol
Department of Biomedical Engineering, Faculty of Engineering, Tel Aviv University, Tel Aviv, Israel

Jonathan Garcia
Drexel University, Philadelphia, PA, United States; University of Maryland, College Park, MD, United States

Amit Gefen
Department of Biomedical Engineering, Faculty of Engineering, Tel Aviv University, Tel Aviv, Israel

Darren F. Groberg
DPM, Utah Musculoskeletal Specialists, Salt Lake City, UT, United States

Farina Hashmi
School of Health and Society, University of Salford, Salford, United Kingdom

Mary K. Hastings
Washington University School of Medicine in St. Louis, St. Louis, MO, United States

Michael A. Hill
Dalton Cardiovascular Research Center, University of Missouri, Columbia, MO, United States; Department of Medical Pharmacology and Physiology, University of Missouri School of Medicine, Columbia, MO, United States

Katsuya Hisamichi
Shimokitazawa Hospital Chairman, Tokyo, Japan

Chenyu Huang
Department of Dermatology, Beijing Tsinghua Changgung Hospital, School of Clinical Medicine, Tsinghua University, Beijing, China

Lamya Karim
Department of Bioengineering, University of Massachusetts Dartmouth, Dartmouth, MA, United States

Avraham Kolel
Department of Biomedical Engineering, Faculty of Engineering, Tel Aviv University, Tel Aviv, Israel

Yoni Koren
Department of Biomedical Engineering, Faculty of Engineering, Tel Aviv University, Tel Aviv, Israel

Spencer P. Lake
Washington University in St. Louis, St. Louis, MO, United States

Roger Lewis
University of Sheffield, Sheffield, United Kingdom

Jung Yul Lim
Department of Mechanical and Materials Engineering, University of Nebraska-Lincoln, Lincoln, NE, United States; Nebraska Center for the Prevention of Obesity Diseases (NPOD), University of Nebraska-Lincoln, Lincoln, NE, United States

Maayan Lustig
Department of Biomedical Engineering, Faculty of Engineering, Tel Aviv University, Tel Aviv, Israel

Adi Lustig
Department of Biomedical Engineering, Faculty of Engineering, Tel Aviv University, Tel Aviv, Israel

Raman Maiti
Loughborough University, Loughborough, United Kingdom

Tania Manuel
Universidade Católica Portuguesa, Health Sciences Institute | Wounds Research Lab, Center for Interdisciplinary Research in Health (CIIS), Portugal; Portuguese Wound Management Association (APTFeridas), Portugal

Raz Margi
Department of Biomedical Engineering, Faculty of Engineering, Tel Aviv University, Tel Aviv, Israel

Nuno Mendes
Grupo Saúde Nuno Mendes, Portugal

Anabela Moura
Centro Hospitalar de São João (CHSJ), Portugal; Portuguese Wound Management Association (APTFeridas), Portugal

Omar Mourad
Toronto General Hospital Research Institute, University Health Network, Toronto, ON, Canada

Roozbeh Naemi
Centre for Biomechanics and Rehabilitation Technologies, Staffordshire University, Stoke-on-Trent, United Kingdom

Cristian Nicoletti
Diabetic Foot Unit, Pederzoli Hospital, Peschiera del Garda, Italy

Blessing Nkennor
Toronto General Hospital Research Institute, University Health Network, Toronto, ON, Canada

Sara S. Nunes
Toronto General Hospital Research Institute, University Health Network, Toronto, ON, Canada

Craig Oberg
Department of Microbiology, Weber State University, Ogden UT, United States

Rei Ogawa
Department of Plastic, Reconstructive and Aesthetic Surgery, Nippon Medical School, Tokyo, Japan

Norihiko Ohura
Kyorin University School of Medicine, Tokyo, Japan

Daniel Parker
School of Health and Society, University of Salford, Salford, United Kingdom

Neil D. Reeves
Research Centre for Musculoskeletal Science and Sports Medicine, Faculty of Science and Engineering, Manchester Metropolitan University, Manchester, United Kingdom

Emília Ribeiro
Grupo Saúde Nuno Mendes, Portugal

Nick Santamaria
Faculty of Medicine Dentistry and Health Sciences, University Of Melbourne Victoria, Melbourne, Australia; Visiting Professorial Fellow, Cardiff University, Wales, United Kingdom

Hadar Shaulian
Faculty of Mechanical Engineering, Technion-Israel Institute of Technology, Haifa, Israel

James R. Sowers
Dalton Cardiovascular Research Center, University of Missouri, Columbia, MO, United States; Department of Medical Pharmacology and Physiology, University of Missouri School of Medicine, Columbia, MO, United States; Diabetes and Cardiovascular Center, University of Missouri School of Medicine, Columbia, MO, United States; Department of Medicine, University of Missouri School of Medicine, Columbia, MO, United States

Zhe Sun
Dalton Cardiovascular Research Center, University of Missouri, Columbia, MO, United States; Department of Medical Pharmacology and Physiology, University of Missouri School of Medicine, Columbia, MO, United States

Simon Y. Tang
Washington University School of Medicine in St. Louis, St. Louis, MO, United States

Laurel Tanner
EC-Service, Centerville, UT, United States

Rachana Vaidya
Department of Bioengineering, University of Massachusetts Dartmouth, Dartmouth, MA, United States

F.j. Vermolen
Computational Mathematics Group (CMAT), Division of Mathematics and Statistics, Faculty of Sciences, University of Hasselt, Diepenbeek, Belgium

Daphne Weihs
Faculty of Biomedical Engineering, Technion-Israel Institute of Technology, Haifa, Israel

Alon Wolf
Faculty of Mechanical Engineering, Technion-Israel Institute of Technology, Haifa, Israel

Yan Yang
Dalton Cardiovascular Research Center, University of Missouri, Columbia, MO, United States

Moi Hoon Yap
Centre for Applied Computational Science, Faculty of Science and Engineering, Manchester Metropolitan University, Manchester, United Kingdom

Metin Yavuz
Department of Applied Clinical Research, Prosthetics and Orthotics Program, University of Texas Southwestern Medical Center, Dallas, TX, United States

Shirley L. Yitzhak-David
Faculty of Biomedical Engineering, Technion-Israel Institute of Technology, Haifa, Israel

Jennifer A. Zellers
Washington University School of Medicine in St. Louis, St. Louis, MO, United States

Liping Zhang
Dalton Cardiovascular Research Center, University of Missouri, Columbia, MO, United States; Department of Medical Pharmacology and Physiology, University of Missouri School of Medicine, Columbia, MO, United States

Preface

This Preface is written during the present COVID times; however, a much larger pandemic, type 2 diabetes, continues to threat humanity and ravage our civilization, taking lives, mutilating, causing suffering and agony, and consuming massive global healthcare resources daily. Diabetes starts silently and kills its victims slowly and viciously. The disease first attacks neural cells, small blood vessels, and the immune system and only then progresses to the macroscale, for example, the coronary vessels. It remarkably changes vital organs such as the liver and kidneys, irreversibly impacts the brain and the retinas, reduces blood perfusion to peripheral organs and thereby, starves them, and also compromises the function of the heart itself. In essence, diabetes, with or without obesity, systematically destroys the structure and function of almost every tissue type, leading to a slow and painful death, from cells to tissues and organs and eventually, the whole body. Mechanobiology appears to play a pivotal role in the damage inflicted by diabetes to all the body tissues and systems. This book is the first to focus on the essential roles that physical factors play via the process of mechanotransduction in the context of diabetes.

My own research work has been focusing on chronic wounds and in particular, on diabetic foot ulcers for more than two decades. These are one of the many horrific consequences of diabetes: necrosis caused by exposure to mechanical forces which spreads in the soft tissues of the feet—the foundations of the body—sometimes causing osteomyelitis and sepsis and often leading to limb amputations. I constantly felt that there are inherent gaps between disciplines and academic fields related to the study of diabetic foot ulcers which hinder technological progress to prevent the diabetes-related amputations or at least, achieve better care of these chronic wounds. More effective multidisciplinary bridges will facilitate faster translation of science and innovation to the medical device markets for protecting the feet of patients with diabetes. The case of the diabetic foot also makes a good example of a central medical problem caused by diabetes which falls between two stools: Diabetic foot ulcers are neither a pure podiatric challenge (as they are part of a systemic disease, not limited to just the foot) nor are they a pure surgical problem. In fact, the bioengineering and clinical communities need to invest the majority of efforts and resources in prevention, as opposed to the last resort of amputation. Importantly, diabetic foot ulcers are one of the most complex, multifaceted medical conditions and one of the most serious ones, as well.

The only way for us, as a professional community, to provide the state-of-the-art, cost-effective healthcare which is supported by the latest science and based on rigorous evidence for diabetes and its complications, such as diabetic foot ulcers and related amputations, is to work closely together. This is facilitated by this multidisciplinary book, which blends the top-tier science and technology in the management of diabetes and its complications with the latest bioengineering and clinical research across Europe, the United Kingdom, Israel, the United States and Canada, Japan, and other countries. All the 23 contributed chapters in this book have been cherry-picked by myself and have been written by international leaders who are pioneers in the fight against diabetes and its complications worldwide. The book describes the frontier of science and state of knowledge and further points to the numerous challenges that lie ahead as the population ages and diabetes spreads further. This book is valuable for all professionals involved in the prevention, diagnosis, and treatment of diabetes and its complications, whether from a clinical perspective or if involved in research and development, in academia, medical settings, and industry. Being the best current synthesis of research knowledge with regard to the cell and tissue damage cascades in diabetes, and since this book identifies current and emerging research directions and future technology trends in the field, it should serve all professionals and graduate students who focus on the diagnosis, prevention, early detection, management, treatment, and care of diabetes and its complications.

Finally, the joint interests that exist with academia and clinicians for delivering better care of diabetes should be translated to work hand-in-hand with the global medical industry to achieve the goal of ultimately eliminating the disease and prior to that, prevent or reverse its horrible complications. This book does just that, by providing a comprehensive and integrated source of knowledge for this researchers-industry dialogue, which makes it a cornerstone in the fight against diabetes.

Professor Amit Gefen
The Editor

CHAPTER 1

Heel ulcers in patients with diabetes

Michael Clark[1,2]
[1]Professor of Wound Study, Birmingham City University, Birmingham, United Kingdom; [2]Welsh Wound Innovation Centre, Ynysmaerdy, Wales

The impact of pressure ulcers

Pressure ulcers (PUs) are increasingly becoming recognized as an important healthcare challenge given both their impact on patient morbidity and the high incremental costs associated with PU prevention and treatment. The development and validation of patient-reported outcome tools well reflects the multiple ways that having a PU can cause problems for patients [1,2] with pain, wound exudate, odor, sleep disturbance, energy, reduced movement, changes to activities of daily living, impaired emotional well-being, self-awareness of having a wound, and participation in social activities, the main domains addressed within self-report tools. Recent analysis of data initially gathered from UK General Practitioner surgeries and then linked with other health records has helped to clarify the economic costs of PUs within the United Kingdom. Guest and colleagues [3] reported the annual cost of a range of wound types to the UK National Health Service (NHS), and while PUs comprised only 5% of all wounds, their treatment costs (£531.14) consumed 10% of the total NHS spent on wounds (£5.3 billion each year). In a similar review of linked databases, high expenditure on PUs was also reported by Phillips et al. [4] where PU treatment was one of the six key drivers of wound expenditure within NHS Wales.

Many authors have also associated PUs with increased patient mortality. Grey and Harding [5] noted a fivefold increase in the mortality of elderly patients who developed PUs with an in-hospital mortality for these patients of 25%–33%; however, it is unclear how these statistics were derived. Similar broad statistics suggest that annually 60,000 people die in the United States as a consequence of PUs [6]; again the origin of this estimate is unclear. Allman [7] reported a prospective observational study of 286 patients over 55 years old, who were expected to stay in hospital for at least 5 days and confined to bed or a chair or had a hip fracture. The overall mortality rate within 1 year of admission to the study was 40.9%, while the 1-year mortality for those with a PU was 59.4%. However, PU development was not found to be an independent risk factor for death after adjustment for comorbidities, disease severity, and nutritional state, suggesting that the development of a PU reflected underlying disease rather than a primary cause of death. In a similar observational study, while elderly patients with PUs survived for shorter periods of time within a skilled nursing facility, it was not possible to associated death with PU development given the presence of other comorbidities [8].

Changing terms and definitions of pressure ulcers?

While the link between PUs and patient mortality deserves greater consideration, it is evident that PUs may reduce quality of life and impose high financial burdens upon healthcare systems. One challenge to developing greater insights into the impact of PUs has been the lack of agreement among both clinicians and the research community over the terms and definitions of pressure damage! There may be few clinical issues that have changed name so often over a relatively short period of time; in the 1970s "bedsore" was in common use and generally remains so when discussing pressure damage with the public; by the 1980s "pressure sore" was introduced to communicate that pressure damage did not only arise from lying in bed, a decade later, and "pressure ulcer" largely replaced "pressure sore" to break the inherent assumption that pressure damage was always painful. In the last decade "pressure injury" has grown in popularity to reflect that not all manifestations of pressure damage involve physical breaks in the skin. Underlying these changes in overall terminology have been debates around the appropriate classification of different appearances of pressure damage. Detailed discussion of these amendments to PU classification lies beyond the scope of this chapter; however, one controversy will form the main focus of this contribution.

Is the wound a pressure ulcer or a diabetic foot ulcer? Does this matter?

The classification of pressure damage becomes increasingly controversial when considering wounds at the heel of patients with any type of diabetes. The heels are the second most common anatomical site where PUs develop [9] while wounds upon the feet of patients with diabetes are often managed as diabetic foot ulcers (DFUs). So how should a wound on the heel of a patient with diabetes be classified? Are these wounds PUs or DFUs? It may appear an academic argument given broad similarities in the etiology of PUs and DFU [10,11]; however, guidance on the management of the two wound types differs. For example, if the heel wound is classified as a PU then few of these wounds would be assessed by a podiatrist or specialist nurse [10] while few "PUs" at the heel would receive a vascular assessment. So superficially the definition of a heel wound as a PU or a DFU may have unintended consequences for patients. In reality the classification of a heel ulcer as a DFU or PU may not create significant impact for patients given the "inefficient and inadequate management" of diabetic foot wounds, at least in the United Kingdom [12]. Guest et al. [12] reported upon 130 patients with newly diagnosed DFU drawn from a database holding records of patients who attended a sample of General Practitioner practices across the United Kingdom. Patients were eligible for inclusion if they were over 18 years old, had a diagnosis of diabetes, and had a DFU with at least 12 months continuous medical history from its first identification. Among the cohort only 5% of the patients saw a podiatrist while the same proportion received a pressure off-loading device despite national guidance recommending these interventions. *Perhaps the choice of classification of a heel wound as either a PU or DFU may have few implications in imperfect healthcare systems?*

Is diabetes a risk factor for heel pressure ulcers?

Delmore and colleagues [13] reported a small retrospective case—control single-center investigation of risk factors for heel PUs among patients in hospital. The model was developed based on data from 37 patients with hospital-acquired heel PUs and 300 patients with no heel PUs; the initial model was then validated against a second cohort that included 12 patients with nosocomial heel PUs and 68 with no heel PUs. The model identified four independent risk factors for heel PU development: diagnosis of diabetes, immobility, vascular disease, and a Braden score under 18. In a second study, Delmore et al. [14] expanded their retrospective investigation to include data from multiple hospital sites with a model development population of 323 patients with nosocomial heel PUs and 1374 patients with no heel PUs. This larger sample suggested seven independent risk factors for heel PU development; diabetes and vascular disease were retained from the earlier model with five new risk factors—impaired nutrition, increasing age, mechanical ventilation, surgery, and poor perfusion. In both studies [13,14] diabetes and vascular disease were independent risk factors for heel PU development. Gaubert-Dahan et al. [15] reported upon 210 elderly hospital patients, of whom 26 had heel PUs at admission, of which 6 were full-thickness wounds. The patients were divided into three groups based on their ability to perceive tactile stimulation on the foot (light, moderate, and severe neuropathy) and their neuropathy symptom score; only two patients with light neuropathy had heel PUs (category I) while all six patients with severe PUs had either moderate or severe neuropathy. In this study there was no association between diabetes and the neuropathy groupings with, for example, 103 nondiabetic patients having moderate neuropathy. The cross-sectional design of this study may have obscured the association between diabetes and heel PUs but does highlight the important role of the sensory nerves in adjusting cutaneous microcirculation in response to pressure [16] and the risk of heel PUs where such adjustment is absent. A small case—control study [17] matched 15 patients with category II to IV heel PUs and 15 age, gender, and ethnicity-matched controls with no heel ulcers. Each subject had their ankle brachial pressure index (ABPI) measured with indices below 0.9 and over 1.3 considered abnormal; 12/15 subjects with heel PUs had abnormal ABPI while only 5/15 controls also had abnormal ABPI. The presence of diabetes was not controlled across the two groups. While this small study supports the contention that heel PUs may be associated with vascular disease the small numbers precluded any clear relationship between low ABPI (<0.8) and the presence of heel ulcers with five and three patients having low ABPI in the case and control groups, respectively. *It would appear that diabetes may be a risk factor for heel PU development while predisposition to heel ulcers may be complemented using measures of sensory neuropathy and potentially a vascular assessment.*

Do heel ulcers in diabetes heal?

Örneholm and colleagues [17] noted a general consensus that a patient with diabetes and a heel ulcer generally experiences amputation above the ankle. Seven-hundred and sixty-eight diabetic patients with a heel ulcer were followed from entry into a multidisciplinary DFU clinic between 1983 and 2013; of these 504 healed without amputation and only 72 required major amputation before their wound healed. The remaining 192 patients died before their ulcers had healed. The heel ulcers treated within the center were typically superficial and of short duration; 85% were superficial at first assessment with a median wound duration of 4 weeks at inclusion. *The results highlight that early referral to a specialist DFU clinic may help achieve positive outcomes for diabetics with heel ulcers.*

Do heel ulcers in diabetics heal slower than other wound types?

There are few studies that have directly compared the healing of DFU and other wound etiologies. Brem et al. [18] followed a nonrandomized cohort of 23 consecutive patients with treatment conducted within a single teaching hospital and a tertiary care center. Of the cohort, 10 had DFU and 13 had PU. All patients were treated with a living skin construct and those with a PU were provided with an active support surface while in bed. Subjects with a DFU were provided with crutches but were stated to be commonly nonconcordant with their use. Of the 10 subjects with DFU, 7 showed complete healing of their wound. In the PU group, 7/13 healed with none of the PU present on the foot. The wounds were typically small (PU surface area ranged from 0.8 to 22.56 cm^2 with 6/13 of the PU being no larger than 6 cm^2). The duration of time subjects experienced their wounds, the time to healing, and the severity of the PU were unreported. This case series suggests that DFU will heal similar to generally small PU; however, gaps in the reporting of data within this study reduce confidence in its conclusions. In a later study, Brem and colleagues [19] reported upon 40 consecutive elderly patients (age 65—102 years) with a mix of DFU, venous leg ulcers, and PU all treated following the same protocol that included off-loading for PU and DFU, compression for venous leg ulcers. All nonviable tissue was debrided, and all wounds were covered with a living skin construct. Of the 40 patients, 29 were reported to have healed in 6 months; however, the number of patients entering the study with DFU, PU, and leg ulcers was not reported nor was healing presented by wound etiology. In this cohort the 11 nonhealing patients had wounds that extended to bone or had deep soft tissue infections or had a significant delay before presentation for treatment. In an earlier study, Karanfilian and colleagues [20] followed the outcome of treatment among 48 male patients (with 59 limbs) affected by ischemic ulcerations or forefoot or digit gangrene. Of the cohort, 37 and 22 limbs were in diabetic and nondiabetic patients, respectively. The presentation of wound outcome data is challenging given that 20 of the limbs (23.9%) experienced metatarsal or digital amputations with the distribution unreported between diabetic and nondiabetic patients. The time to complete healing was unreported but stated to be 4 months in unspecified cases. Wound management within the cohort was briefly described and consisted of debridement and then healed by secondary intention or covered with split-thickness skin grafts; the number of grafted wounds was unreported. Karanfilian et al. reported that a higher proportion of the wounds healed in nondiabetic patients; 15/22 (68.2%) healed in the nondiabetic group while 47.0% (16/34) healed in the diabetics. *These cohort studies indicate that foot wounds in diabetic patients can heal outside specialist foot clinics although weaknesses in the reporting of outcome data weaken the conclusions that can be drawn from these small studies.*

What can be inferred about the healing of heel ulcers in diabetics and nondiabetics from wound registry data?

Several authors have reported the use of wound registry data to gain insights into the prevention and healing of common wounds (for example, Clark et al. [21,22], Essex et al. [23], Wilcox et al. [24], Jung et al. [25], and Jokinen-Gordon et al. [26], among others). Proponents of wound registry data note the advantages of gathering data upon "real-world" patients and their treatment with no artificial inclusion and exclusion criteria leading to key patients being excluded from formal study designs such as randomized controlled trials. However, the collection of wound registry data is not without significant challenges (Clark et al. [27]); developing a wound registry can be expensive where dedicated data collectors are employed to gather data (Clark et al. [21,22]) and consume considerable time to gather sufficient data to facilitate analysis of specific interventions or processes. Where clinicians complete data collection there may be issues around data quality and missing data; for example, specialist clinics may see patients infrequently with significant gaps between patient and wound assessments; there will also be loss to follow-up and incomplete data upon healing where patients transfer from specialist clinics to general care as their wounds improve. Finally, the complexity of gathering data upon routine wound care may prevent comparisons between interventions given that they may be used only infrequently.

Two wound registries were investigated to see whether their data could help address the healing of heel ulcers in patients with and without diabetes. The first registry (Clark et al. [21,22]) collected data upon PU prevention and treatment across four acute care hospitals in England. All hospital in-patients were eligible to participate with limited exclusion criteria (aged under 16 years, unable to provide informed consent, no assent to participate available from family members, and admitted to any of the following specialities: psychiatry, ophthalmology, gynecology, pediatrics, obstetrics, and mental illness and expected to stay in hospital for at least 2 days). Ten full-time data collectors gathered data across the four sites between July 1996 and May 1998. To avoid overemphasis upon those patients at highest risk of developing PUs, the number of admissions to hospital within discrete bands of PU risk assessment scores (minimal, low, high, and very high risk as measured using the Waterlow tool [28]) was calculated over the 6 months preceding the collection of data. The data collectors recruited a convenience sample of patients that reflected the proportion of patients at minimal, low, high, and very high risk per speciality to ensure the registry population mirrored the vulnerability to pressure ulceration of the general hospital patient population. All data collectors met at 3-month intervals to resolve any challenges around PU classification or risk assessment while formal estimates of their reliability when measuring PUs were assessed through measurement of three-dimensional wound models. PUs were classified according to Clark and Cullum [29] where a category I ulcer marked nonblanchable erythema, where tissue damage was observed a category II PU was a superficial break in the skin; full-thickness wounds were considered a category III ulcer where no wound cavity was observed while a category IV PU presented as a cavity wound. These rather crude wound descriptions prompted description of category I and II PUs as *superficial* wounds and category III and IV ulcers as *severe* wounds. All data were entered into a relational database and exported to SPSS Version 8.0 for analysis; no data that could directly identify any specific individual was entered into the database. Data were collected daily for all patients until they were discharged from hospital, died, or withdrew for any reason.

In total 2507 patients were entered into the registry; the mean age was 65.3 years (range 16—103 years) and 55% (n = 1380) were females. The median length of stay in hospital was 8 days (range 3—243 days) and patients were followed for a total of 29,611 patient-days in hospital. Of the registry population, 727 (28.9%) were at minimal risk of developing PUs with 22.9% (n = 576) and 18.7% (n = 470) at "high" or "very high" risk of PU development, respectively. One hundred patients (3.9%) were admitted to hospital with PUs; of these 24 had severe PUs with 4 admitted as a direct consequence of their wound (in-patient stay of 66 days). One hundred and seventeen patients (4.7%) developed PUs while in hospital. Table 1.1 details the characteristics of the PUs either present on admission or that developed during the hospital stay. Most of the PUs were superficial (82.5%; n = 179) and typically located at the sacrum (n = 141, 64.9%).

It could be argued that the 1996—98 registry data may bear no resemblance of the numbers and characteristics of patients with PU in acute care in the United Kingdom almost 25 years later. Table 1.1 compares the characteristics of patients and PUs from the 119—98 registry data with similar data collected in 2015 across all acute and community hospitals in Wales (Clark et al. [30]). The 2015 survey was conducted following the methodology developed by the European Pressure Ulcer Advisory Panel [9]. While the similarities between the two sets of PU data may be purely circumstantial, and noting the two sets of data used different classifications of PUs, both data sets indicate broadly similar prevalence proportions, comparable proportions of patients that developed their PU in hospital, and similar proportions with full-thickness wounds. Where

Table 1.1 Characteristics of patients and pressure ulcers from the 1996–98 registry data [21,22] and 2015 cross-sectional study [30].

	1996–98 [21,22]	2015 [30]
Age	65.3 (17.8); mean (SD)[a]	80–89 mode age range band
Prevalence	217/2507 (8.6%)	748/8365 (8.9%)
% Developed in hospital	117/217 (53.9%)	302/601 (50.2%); 65 origin unknown
% At high to very high risk of developing pressure ulcers[b]	1046/2507 (41.7%)	3597/6957 (51.7%)
Category		
I	88 (40.6%)	219 (32.9%)
II	91 (41.9%)	330 (49.5%)
III	25 (11.5%)	81 (12.2%)
IV	13 (6.0%)	38 (5.7%)
Location		
Sacrum	141 (65.0%)	268 (40.4%)
Heel/Foot	35 (16.1%)	183 (27.6%)
Ischial	17 (7.8%)	123 (18.5%)
Trochanter	14 (6.4%)	7 (1.0%)
Other	10 (4.6%)	82 (12.4%)

[a]SD, Standard Deviation.
[b]Risk measured using the Waterlow scale.

the two data sets diverge potential reasons for the differences can be proposed; the 2015 data set suggests an older population with elevated risk of developing PUs compared with the earlier registry data reflecting the changes in the UK population since 1990 [31]. The proportion of PUs at the sacrum appeared to decrease between the two data sets perhaps indicating successful prevention of wounds at this anatomical location? The proportion of PU at the trochanters in 2015 reduced from the 1996–98 data perhaps reflecting the trend toward 30° positioning rather than hip-to-hip turning. The proportion of heel and ischial ulcers increased between the two data sets and increased attention to PU prevention as these locations would appear to be required. *While over 20 years old the 1996–98 registry data appear to mirror more recent PU investigations and discrepancies can be explained through changes in the patient population and in clinical practices. There remains the potential for the broad similarities between UK PU data collected in the late 1990s and the current decade to be purely coincidental.*

While the registry data appear substantial with over 2500 patients, specific analysis of questions, such as do foot and heel wounds respond to treatment differently among diabetic and nondiabetic patients, reveals one the challenges of registry data use. From the 2507 patients entered into the registry, there were only 59 who had diabetes and PUs.

Of these 59 patients with diabetes and PU, only 8 (13.5%) had their wounds on the foot or heel. A reasonably large registry population only held small numbers of people with wounds at specific anatomical locations and with a specific comorbidity such as diabetes.

Table 1.2 presents the proportion of PU of each category that presented at the heel or foot, the length of stay in hospital along with the proportion that healed, and the number of diabetic patients. Only 10%–12% of the superficial PU presented at the heel or foot with almost 50% of category III PU found on the foot or heel. As the severity of the PU at the heel or foot increased, the length of follow-up of patients decreased. This finding was not attributable to increased mortality among the group with severe PU on the foot or heel. Across the registry only 59/2507 (2.3%) patients died in hospital and of these 19 had PU with only 4 (2 category III that developed in hospital and 2 category IV wounds present on admission) of these being severe PU. Few foot and heel PU healed; with 4/35 (11.4%) reported to have healed during the patients' stay in hospital. No diabetic patient had a foot or heel PU that healed during their stay in hospital. The low healing rates probably mark the early discharge of patients from hospital regardless of whether they had PU; of the registry population 2108/2507 (84.1%) returned to their own homes after discharge including 120 patients with PU. One-third (n = 5) of patients with severe foot or heel wounds was diabetic. While numbers are small there was no apparent association between being diabetic and having an unhealed foot or heel PU (Fisher's Exact T = 0.55). *The 1996–98 registry data did not include a large number of diabetic patients with heel or foot PUs and provide limited information upon the outcomes of treatment of PU in diabetic patients.*

The second wound registry investigated to see whether heel ulcers in diabetic patients healed slower than in nondiabetics covered the work of a single team of clinicians working in a cluster of out-patient wound clinics in South Wales [27]. Each clinic visit prompted the completion of a 110-item questionnaire using electronic pen and paper that transmitted the data to a central database with data analysis performed using Microsoft Access 2019 (Microsoft Corporation, USA). Fig. 1.1 illustrates a section of the questionnaire used to populate the South Wales wound registry. Data were collected on 1782

Table 1.2 Characteristics of the patients with heel or foot wounds within the 1996–98 registry.

	Pressure ulcers on the foot and heel			
	Pressure ulcer category			
	I	II	III	IV
Proportion of all PU that presented at the foot or heel	9/88 (10.2%)	11/91 (12.0%)	12/25 (48.0%)	3/13 (23.1%)
Follow-up (days)	38.6 (3–227)	24.3 (10–65)	21.0 (6–67)	15.3 (8–21)
Healed: Unhealed	3:6	1:10	0:12	0:3
Diabetes	0/9 (0%)	3/11 (27.3%)	4/12 (33.3%)	1/3 (33.3%)

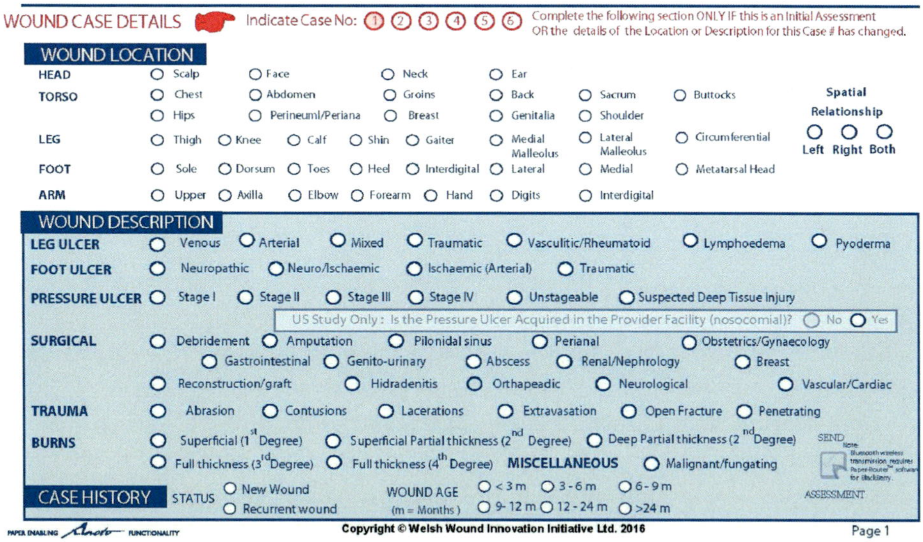
WOUND CASE DETAILS Indicate Case No: 1 2 3 4 5 6 Complete the following section ONLY IF this is an Initial Assessment OR the details of the Location or Description for this Case # has changed.

WOUND LOCATION

HEAD O Scalp O Face O Neck O Ear

TORSO O Chest O Abdomen O Groins O Back O Sacrum O Buttocks

O Hips O Perineuml/Periana O Breast O Genitalia O Shoulder

LEG O Thigh O Knee O Calf O Shin O Gaiter O Medial Malleolus O Lateral Malleolus O Circumferential

FOOT O Sole O Dorsum O Toes O Heel O Interdigital O Lateral O Medial O Metatarsal Head

ARM O Upper O Axilla O Elbow O Forearm O Hand O Digits O Interdigital

Spatial Relationship O Left O Right O Both

WOUND DESCRIPTION

LEG ULCER O Venous O Arterial O Mixed O Traumatic O Vasculitic/Rheumatoid O Lymphoedema O Pyoderma

FOOT ULCER O Neuropathic O Neuro/Ischaemic O Ischaemic (Arterial) O Traumatic

PRESSURE ULCER O Stage I O Stage II O Stage III O Stage IV O Unstageable O Suspected Deep Tissue Injury

US Study Only: Is the Pressure Ulcer Acquired in the Provider Facility (nosocomial)? O No O Yes

SURGICAL O Debridement O Amputation O Pilonidal sinus O Perianal O Obstetrics/Gynaecology O Gastrointestinal O Genito-urinary O Abscess O Renal/Nephrology O Breast O Reconstruction/graft O Hidradenitis O Orthapeadic O Neurological O Vascular/Cardiac

TRAUMA O Abrasion O Contusions O Lacerations O Extravasation O Open Fracture O Penetrating

BURNS O Superficial (1st Degree) O Superficial Partial thickness (2nd Degree) O Deep Partial thickness (2nd Degree) O Full thickness (3rd Degree) O Full thickness (4th Degree)

MISCELLANEOUS O Malignant/fungating

CASE HISTORY STATUS O New Wound O Recurrent wound

WOUND AGE (m = Months) O <3 m O 3-6 m O 6-9 m O 9-12 m O 12-24 m O >24 m

SEND ASSESSMENT

PAPER ENABLING Anoto FUNCTIONALITY

Copyright © Welsh Wound Innovation Initiative Ltd. 2016

Page 1

Figure 1.1 Section of the questionnaire used to collect wound registry data in South Wales wound clinics and within clinical studies managed by the Welsh Wound Innovation Initiative.

patients (950 females and 832 males) treated from April 26, 2012, to February 24, 2016, after which time the registry was relocated to a new database. Wound dimensions were recorded at first and final clinic visits for 1014 (56.9%) patients. Given that the wound clinics served a patient population with challenging wounds, few patients were seen in the clinics until complete healing occurred with the majority returning to their usual clinic settings as their wounds improved. An arbitrary wound size of 0.5 cm^3 was selected to reflect "healing" within the specialist wound clinics and on this basis 613/1014 (60.6%) patients' wounds healed. Eighty-nine patients presented in the specialist clinics with PU (4.99%; 95% Confidence Interval 4.07–6.10) and of these 61 made more than one clinic visit. Unlike the general Welsh hospital population [30], the majority of the PU patients with multiple clinic visits were male (n = 41; 67.2%), relatively young (mean 61.8 years, standard deviation 17.5) and presented mainly with full-thickness PUs (Category III n = 20; Category IV n = 35). The 61 patients with PU were typically nondiabetic (n = 54) with only 7 diabetic patients. None of the diabetic patients had PUs on their feet or heels. Few of the PU treated within the wound clinics "healed" with 13 (21.3%) reduced to below 0.5 cm^3. The scarcity of patients with diabetes within the South Wales wound registry may well reflect the successful implementation of local care pathways where patients with diabetes and wounds are referred to specialist foot clinics that did not participate in the registry data collection. It is unsurprising that few of the severe PUs seen in the wound clinics healed given that only 21% of category IV PU was reported to have healed following a year's treatment in UK community settings [32].

Conclusions

The debate regarding whether a heel ulcer in a diabetic patient should be classified as a PU or a DFU is likely to continue for some time; early referral to podiatrists and dedicated foot clinics may suggest that in diabetic patients a heel wound should be considered as a DFU, not a PU. However, these benefits may not accrue in inefficient healthcare systems, such as the United Kingdom, where no more than 5% of those patients with a DFU actually access specialist foot care.

Early referral to dedicated foot clinics does appear to offer improved outcomes for diabetics with foot wounds. However, there is scant evidence that heel and foot wounds of diabetics heal slower than similar wounds in nondiabetics where treated in general acute care. The use of wound registries to access evidence of the routine effectiveness of wound treatments is growing but the two registries described in this chapter clearly show that successful use of registry data will require extensive data to be available to allow robust interrogation around interventions and procedures to be undertaken. Registry data may be a promising route toward understanding the effects of everyday wound care but reaching the volume of data required to sustain meaningful analysis is likely to be time-consuming, expensive and may ultimately be restricted to a small number of centers. The recent use of linked databases based on community-delivered care [3,4,12,32] perhaps offers a solution to the slow development of specialist registries but challenges remain regarding the granularity of the data held in the linked databases along with the initial requirement that the patient's wound be entered into the database by their General Practitioner. There may be benefit in bringing together the use of linked databases with focused registry development to provide both breadth and granularity to questions such as the impact of viewing heel ulcers as PUs or as diabetic foot wounds.

References

[1] Gorecki C, Brown JM, Cano S, et al. Development and validation of a new patient-reported outcome measure for patients with pressure ulcers: the PU-QOL instrument. Health Qual Life Outcome 2013;11:95.
[2] Rutherford C, Brown JM, Smith I, et al. A patient-reported pressure ulcer health-related quality of life instrument for use in prevention trials (PU-QOL-P): psychometric evaluation. Health Qual Life Outcome 2018;16:227.
[3] Guest JF, Ayoub N, McIlwraith T, et al. Health economic burden that different wound types impose on the UK's National Health Service. Int Wound J 2017;14:322—30.
[4] Phillips CJ, Humphreys I, Fletcher J, et al. Estimating the costs associated with the management of patients with chronic wounds using linked routine data. Int Wound J 2015;13:1193—7.
[5] Grey JE, Harding KG. Pressure ulcers. BMJ 2006;332(7539):472—5.
[6] Bauer K, Rock K, Nazzal M, et al. Pressure ulcers in the United State's inpatient population from 2008 to 2012: results from a retrospective nationwide study. Ostomy Wound Manag 2016;62(11):30—8.
[7] Allman R. The impact of pressure ulcers on health care costs and mortality. Adv Wound Care 1998;11:2.
[8] Jaul E, Calderon-Margalit R. Systemic factors and mortality in elderly patients with pressure ulcers. Int Wound J 2015;12:254—9.

[9] Vanderwee K, Clark M, Dealey C, Gunningberg L, Defloor T. Pressure ulcer prevalence in Europe: a pilot study. J Eval Clin Pract 2007;13(2):227—35.

[10] Vowden P, Vowden K. Diabetic foot ulcer or pressure ulcer? This is the question. Diabet Foot J 2015; 18(2):62—6.

[11] Ousey K, Chadwick P, Cook L. Diabetic foot or pressure ulcer on the foot? Wounds U K 2011;7(3): 105—8.

[12] Guest JF, Fuller GW, Vowden P. Diabetic foot ulcer management in clinical practice in the UK: costs and outcomes. Int Wound J 2018;15:43—52.

[13] Delmore B, Lebovits S, Suggs B, et al. Risk factors associated with heel pressure ulcers in hospitalized patients. J Wound Ostomy Cont Nurs 2015;42(3):242—8.

[14] Delmore B, Ayello EA, Smith D, et al. Refining heel pressure injury risk factors in the hospitalized patient. Adv Skin Wound Care 2019;32(11):512—9.

[15] Gaubert-Dahan M-L, Castro-Lionard K, Blanchon MA, et al. Severe sensory neuropathy increases risk of heel pressure ulcer in older adults. J Am Geriatr Soc 2013;61(11):2050—2.

[16] Fromy B, Sigaudo-Roussel D, Gaubert-Dahan ML, et al. Aging-associated sensory neuropathy alters pressure -induced vasodilation in humans. J Invest Dermatol 2010;130:849—55.

[17] Örneholm H, Apelqvist J, Larsson J, et al. Heel ulcers do heal in patients with diabetes. Int Wound J 2017;14:629—35.

[18] Brem H, Balledux J, Bloom T, et al. Healing of diabetic foot ulcers and pressure ulcers with human skin equivalent. Arch Surg 2000;135:627—34.

[19] Brem H, Tomic-Canic M, Tarnovskaya A, et al. Healing of elderly patients with diabetic foot ulcers, venous stasis ulcers, and pressure ulcers. Surg Technol Int 2003;11:161—7.

[20] Karanfilian RG, Lynch TG, Zirul VT, et al. The value of laser Doppler velocimetry and transcutaneous oxygen tension determination in predicting healing of ischemic forefoot ulcerations and amputations in diabetic and nondiabetic patients. J Vasc Surg 1986;4:511—6.

[21] Clark M, Benbow M, Butcher M, et al. Collecting pressure ulcer prevention and management outcomes: 1. Br J Nurs 2002;11(4). 230,232,234,236,238.

[22] Clark M, Benbow M, Butcher M, et al. Collecting pressure ulcer prevention and management outcomes: 2. Br J Nurs 2002;11(5):310—4.

[23] Essex HN, Clark M, Sims J, et al. Health-related quality of life in hospital inpatients with pressure ulceration: assessment using generic health-related quality of life measures. Wound Repair Regen 2009; 17(6):797—805.

[24] Wilcox JR, Carter MJ, Covington S. Frequency of debridements and time to heal: a retrospective cohort study of 312 744 wounds. JAMA Dermatol 2013;149(9):1050—8.

[25] Jung K, Covington S, Sen CK, et al. Rapid identification of slow healing wounds. Wound Repair Regen 2016;24(1):181—8.

[26] Jokinen-Gordon H, Barry RC, Watson B, et al. A retrospective analysis of adverse events in hyperbaric oxygen therapy (2012-2015): lessons learned from 1.5 million treatments. Adv Skin Wound Care 2017;30(3):125—9.

[27] Clark M, Walkley N, Harding KG. Outcomes of wound treatment in specialist wound clinics. Amsterdam: Paper presented at European Wound Management Association; 2017.

[28] Waterlow J. Pressure sores: a risk assessment card. Nurs Times 1985;81(48):49—55.

[29] Clark M, Cullum N. Matching patient need for pressure ulcer prevention with the supply of pressure-redistributing mattresses. J Adv Nurs 1992;17:310—6.

[30] Clark M, Semple MJ, Ivins N, Mahoney K, Harding K. National audit of pressure ulcers and incontinence-associated dermatitis in hospitals across Wales: a cross-sectional study. BMJ Open 2017. Aug 21;7(8):e015616. https://doi.org/10.1136/bmjopen-2016-015616.

[31] Office for National Statistics. The changing UK population; n.d. Accessed at: www.ons.gov.uk/peoplepopulationandcommunity/populationandmigration/migrationwithintheuk/articles/thechangingukpopulation/2015-01-15 on June 4th 2020.

[32] Guest JF, Fuller GW, Vowden P, et al. Cohort study evaluating pressure ulcer management in clinical practice in the UK following initial presentation in the community: costs and outcomes. BMJ Open 2018;8(7):e021769. https://doi.org/10.1136/bmjopen-2018-021769. Published 2018 July 25.

CHAPTER 2

Diabetic foot ulcers and their wound management

Norihiko Ohura[1], Katsuya Hisamichi[2]
[1]Kyorin University School of Medicine, Tokyo, Japan; [2]Shimokitazawa Hospital Chairman, Tokyo, Japan

Introduction

The incidence of diabetic foot ulcer (DFU) has increased due to the worldwide prevalence of diabetes mellitus and the prolonged life expectancy of diabetic patients [1]. DFU is a severe diabetic complication that reduces capability in activities of daily living (ADL) and significantly impairs quality of life. DFU is a complex diverse pathology associated with ischemia that consists of microangiopathy and peripheral arterial disease (PAD), infection in the deep tissues, and neuropathic foot deformity [2].

DFU is the leading cause of infection and amputation in diabetic patients, with a lower limb amputated due to diabetes every 30 s [3]. In patients with DFU, ulceration is the greatest cause of nontraumatic minor and major lower limb amputation, with patients with DFU having a 25-fold higher risk of lower limb amputation in comparison to the general population [4]. In the United States, 15%−25% of patients with diabetes are at risk of developing DFU [5].

The global prevalence of DFU is 6.3% (95%CI: 5.4%−7.3%). The prevalence was higher in males (4.5%, 95%CI: 3.7%−5.2%) than in females (3.5%, 95%CI: 2.8%−4.2%), and higher in type 2 diabetic patients (6.4%, 95%CI: 4.6%−8.1%) than in type 1 diabetics (5.5%, 95%CI: 3.2%−7.7%). Zhang reported that patients with DFU were older, had a lower body mass index, longer diabetic duration, and had higher rates of hypertension, diabetic retinopathy, and smoking history in comparison to patients without DFU [1].

There are two types of management for DFU: preventive foot care, aimed at preventing ulceration, and foot care aimed at preventing the deterioration of wounds and promoting wound healing (WH). Many patients who visit a clinic for wound treatment have callus, ulceration, and gangrene (infection and necrotic wound) due to severe foot deformity, ischemia, and infection. Neuropathy is the cause of foot deformity. According to a review by Lipsky [6], although the presence of neuropathy predisposes a patient to the development of wounds that become infected and which may delay the time at which the patient becomes aware of the infection, it probably does not affect either the severity of the infection or the approach to treatment.

However, the presence of PAD, also found in most patients with diabetic foot infection (DFI), is almost certainly associated with worse outcomes in patients with infection [7,8].

Surgical treatment (debridement and amputation) or endovascular treatment (EVT) is required if infection or ischemia is present. These two therapeutic approaches are often the focus of attention in DFU treatment; however, in reality, there are eight stages of treatment: 1. assessment of the wound, 2. wound bed preparation (WBP)/debridement and amputation [9—11], 3. revascularization, 4. treatment to promote WH, 5. reconstruction of the soft tissue, 6. offloading/prosthetic device [12], 7. rehabilitation, and 8. prevention of recurrence/preventive foot care [13].

According to the cohort study of the investigation of the characteristics of diabetic patients with a foot ulcer in 14 European hospitals in 10 countries, PAD was diagnosed in 49% of the subjects, infection in 58%. One-third had both PAD and infection [14].

Two hundred forty-seven patients with DFUs and without previous major amputation consecutively presenting to a single diabetes center were included in another cohort study. Mean patient age was 68.8 years, 58.7% were males, and 55.5% had PAD [15].

In recent years, the Global Vascular Guidelines (GVGs) recommended that chronic limb-threatening ischemia (CLTI) be evaluated using the WIfI (wound, ischemia, foot infection) classification [16—19].

Thus, it is important to evaluate the pathological condition, such as the degree of ischemia and infection, the range, depth, and size of the wound as soon as possible before starting wound treatment for diabetic ulcer and gangrene. The WHS guideline recommends the use of the WIfI (wound/ischemia/foot infection) classification system as a means of stratifying the risk of amputation and benefit of revascularization in patients with DFU and [5].

When the pathological assessment is delayed, it will be difficult to treat, and in addition to becoming impossible to salvage the limb, it often becomes difficult to save the patient's life. After the assessment of the wound, treatment is performed according to the DFU practice pathway (Fig. 2.1).

Assessment of the wound

The DFU practice pathway starts from a wound assessment that evaluates ischemia and infection (Fig. 2.1) because ischemia and infection are the most important assessments in the management of DFU [7,8]. Since the worsening of DFU occurs rapidly, it is necessary to perform these two evaluations quickly and put their results on the pathway as soon as possible. The speed of the diagnosis is also important.

(1) Assessment of ischemia

In particular, it is important not to overlook ischemic lesions because the treatment policy changes greatly depending on the presence or absence of ischemia. In other words, the key to the assessment of DFU is to extract PAD and critical limb ischemia (CLI) from diabetic foot lesions.

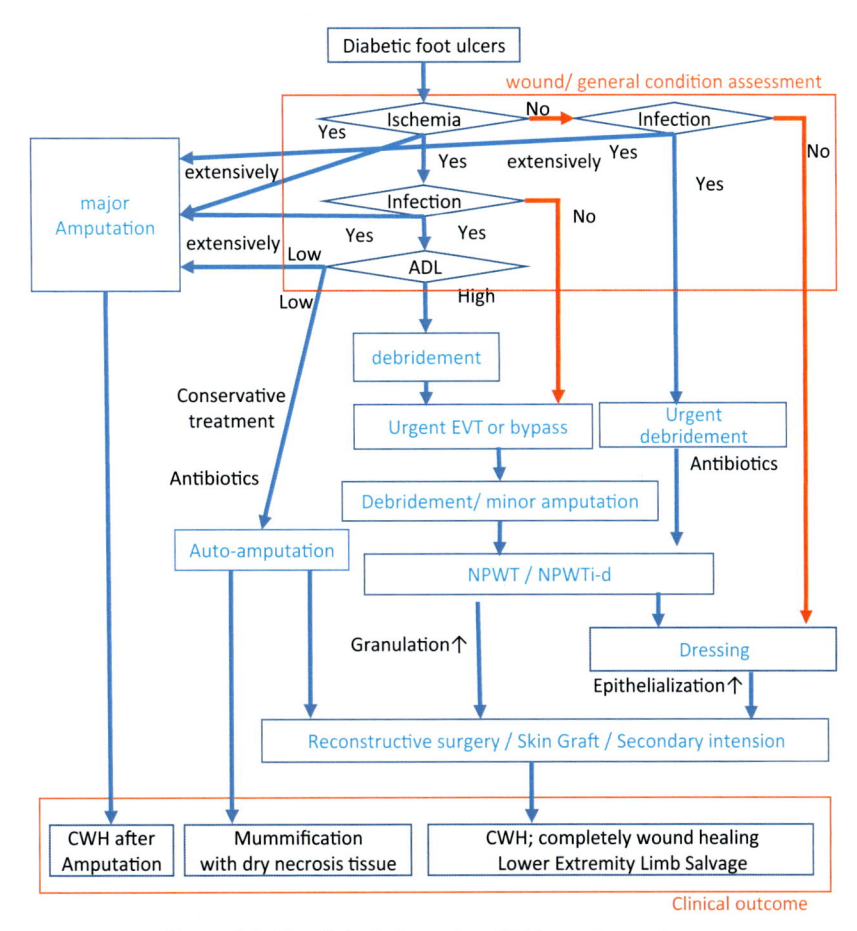

Figure 2.1 *The diabetic foot ulcer (DFU) practice pathway.*

If CLTI is to be treated aggressively, revascularization should be performed urgently, before topical treatment. The WH process will not be initiated without revascularization (e.g., EVT). In addition, ischemic necrosis progresses when surgical debridement is performed in an ischemic state without revascularization.

The ankle brachial index (ABI), skin perfusion pressure (SPP), and transcutaneous oxygen pressure (TcPO$_2$) are used as indicators of objective ischemia. An ABI of >0.9 is considered normal and rules out a diagnosis of PAD. An ABI of >1.3 suggests noncompressible (calcified) arteries [5].

TcPO$_2$ or SPP is used to determine targets for revascularization because WH does not progress if the TcPO$_2$ or SPP is ≤ 40 mmHg or less, and revascularization is performed with the goal of improving the TcPO$_2$ or SPP to >40 mmHg [18].

(2) Assessment of infection

Intervention for ischemia is important; however, early intervention for extensive and severe infection is also important. In cases where the blood flow is preserved, severe infection is often observed, and debridement or major amputation is urgently performed to prevent progression to gas gangrene or sepsis [20].

During the initial surgery, damage control surgery is performed in which only the tissue with an obvious infection is excised, and the wound is not closed without reconstruction. Thereafter, reconstructive surgery is performed after the general condition has stabled.

In addition to surgical debridement, intravenous antibiotics are important. On the other hand, when a patient suffers from both infection and ischemia, especially if the infection is significant, prioritizing revascularization for ischemia may result in the rapid deterioration of the patient's general condition due to sepsis or ischemia—reperfusion injury. It is necessary to comprehensively decide whether or not to perform revascularization before debridement (Fig. 2.2A and B).

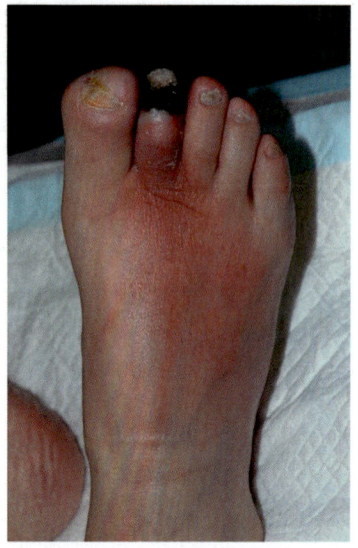

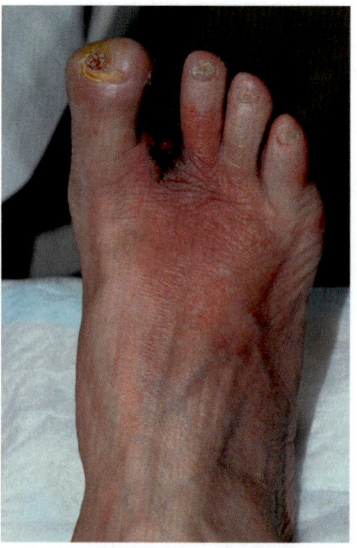

Figure 2.2 *Management of patients with both infection and ischemia.* The foot of a 75-year-old diabetic male with second toe dry gangrene and severe infection due to critical limb ischemia. (A) Erythema extending >2 cm from the edge of a wound was observed on the dorsum of the foot, and erythema was also observed which was consistent with the subcutaneous vein. The entire foot was swollen. It was classified as a Class 2B category according to the new skin and soft tissue infections system (SSTI). The second toe was amputated before revascularization. Revascularization in the presence with a Moderate Class 2B infection can spread the bacterial infection throughout the body and thus result in the onset of sepsis. Therefore, this time, we planned to revascularize the vessels of below-knee lesions after controlling the infection. He was hospitalized and bedridden and was given intravenous antibiotics, based on the new classification system, SSTIs. (B) After a week, the swelling of the entire foot improved and the area of erythema shrank, and therefore the infection could be successfully controlled. However, necrosis of the wound margin progressed, indicating that ischemia was progressing. Endovascular treatment was performed 1 week after amputation.

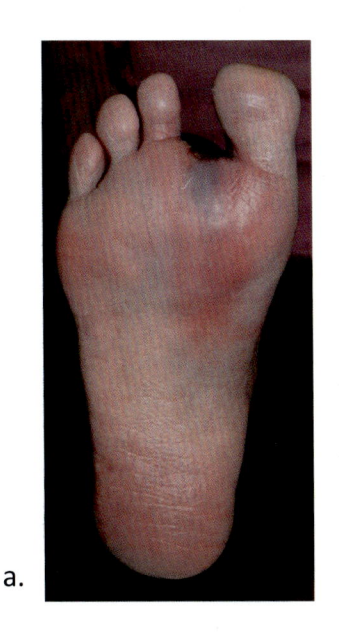

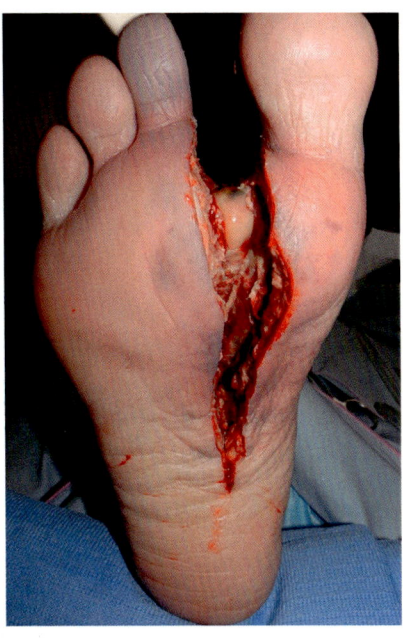

Figure 2.3 *Management of patients with both infection and ischemia.* The foot of a 75-year-old diabetic male with second toe dry gangrene and severe infection due to critical limb ischemia (CLI) after second toe amputation and endovascular treatment. (A) Erythema was observed on the middle plantar of the foot. This means an infection of deep tissue, flexor tendon, and muscle in the plantar central compartment. The bacterial infection progressed proximally through the tendon sheath from the amputation stump. The plantar of the foot should be incised immediately to drain the pus. It is necessary to understand that CLI often leads to a deterioration of infection after revascularization. (B) An incision was made in the plantar aponeurosis of the plantar, and the compartment was released and pus was drained while the necrotic tendon was debrided.

It is important to determine the timing of debridement and revascularization in a medium-term to long-term treatment plan (Figs. 2.4–2.9).

Lipsky et al. proposed new classification for skin and soft tissue infection (SSTI) modeled on the subset of DFI [6]. This new classification system is clinically more useful and valid than the classification advocated by the IDSA. The main problem with the current system of the IDSA is that the "moderate" infection category is quite broad and heterogeneous. To address this issue, and to make an improved system for the classification of SSTI, a new classification system divided the moderate designation into (1) Class A (local infections that are more "horizontally" extensive with erythema extending >2 cm from the rim of a wound) and (2) Class B (infections that are more "vertically" extensive, extending below the subcutaneous tissue).

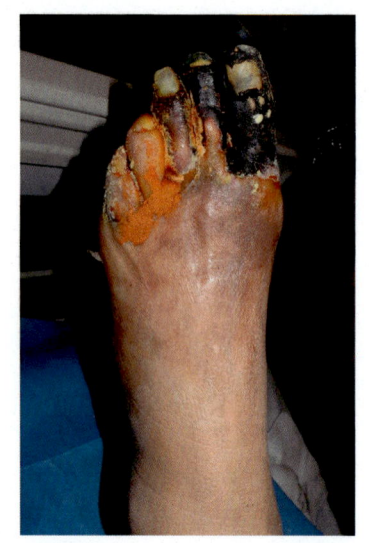

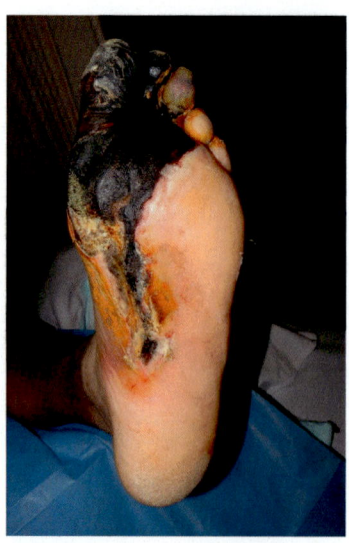

Figure 2.4 *Treatment course for patients with severe ischemia and extensive infection: severe gas-forming necrotizing fasciitis with critical limb ischemia (CLI).* A 61-year-old male DM CKD on HD. (A and B) The patient had dry gangrene; there was pus beneath the area of necrosis. The patient suffered from gas gangrene with CLI, namely, extensive infection and ischemia. His foot was not swollen due to severe ischemia with an SPP value of 20 mmHg on the dorsum of the foot. Therefore, transmetatarsal amputation was urgently performed with ankle block anesthesia.

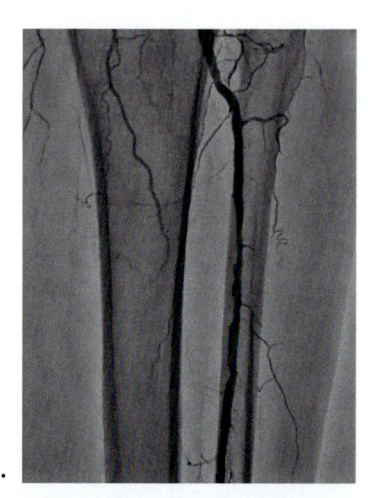

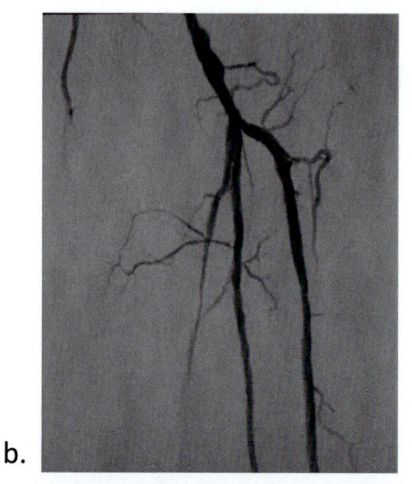

Figure 2.5 *Severe gas-forming necrotizing fasciitis with critical limb ischemia.* Endovascular treatment (EVT) was performed after carrying out emergency debridement. (A) Angiogram at the first visit. Complete total occlusion of the posterior tibial artery and the peroneal artery was observed. (B) Final angiogram after EVT. The posterior tibial artery demonstrated 25% patency and the peroneal artery showed 99% patency.

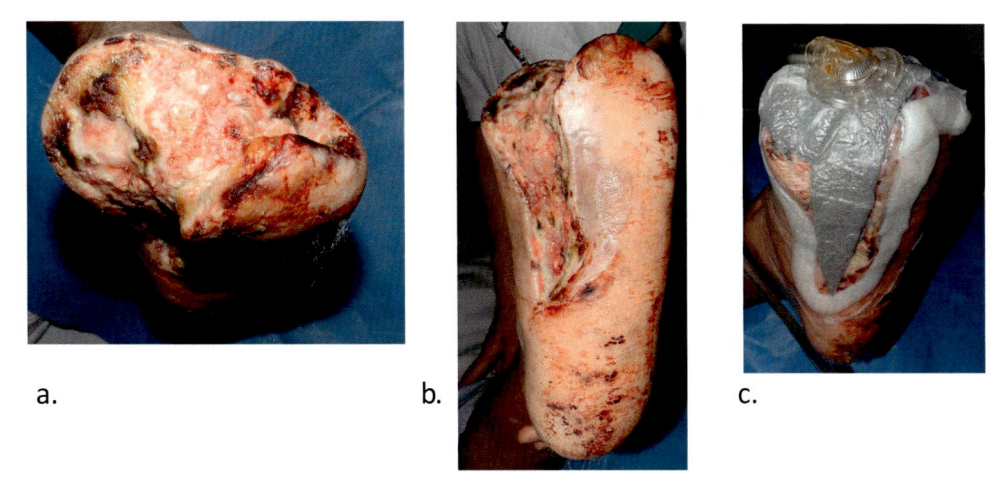

Diabetic foot ulcers and their wound management

Figure 2.6 *Severe gas-forming necrotizing fasciitis with critical limb ischemia.* The state of the patient at 2 weeks after debridement and endovascular treatment. (A and B) Eighty percent of the necrotic tissue had been removed and granulation tissue formation had begun. Local infection was controlled. (C) NPWTi-d with ROCF-CC dressing was started.

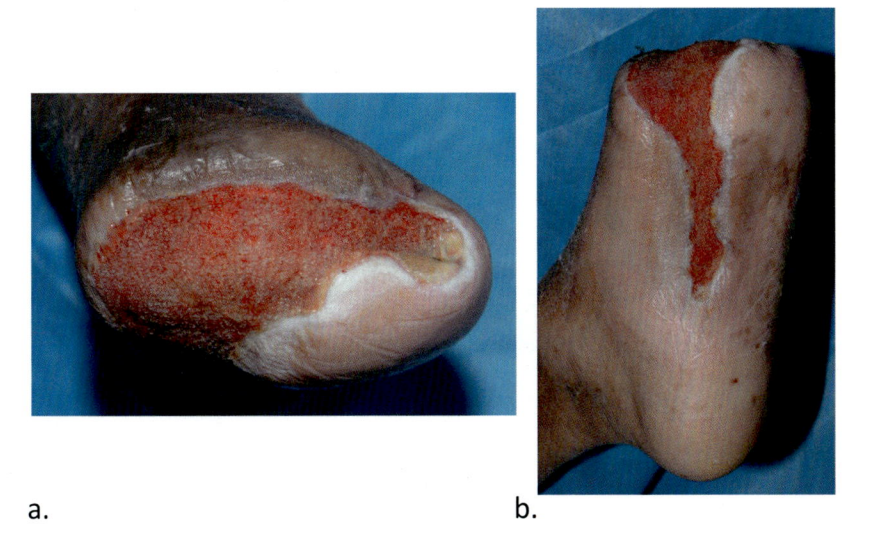

Figure 2.7 *Severe gas-forming necrotizing fasciitis with critical limb ischemia.* Three weeks after the start of NPWTi-d with ROCF-CC dressing, the treatment was changed to conventional NPWT. 10 weeks after debridement and endovascular treatment. (A and B) Granulation tissue had formed which had a sufficient blood flow, the infection had been controlled, and the patient was thus indicated to undergo skin grafting.

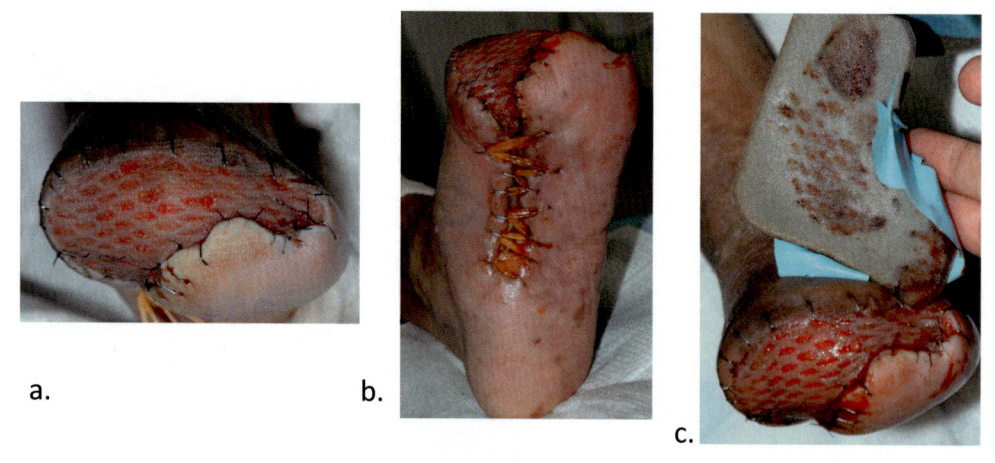

Figure 2.8 *Severe gas-forming necrotizing fasciitis with critical limb ischemia.* One week after split-thickness skin grafting. (A) The skin grafts had completely taken. (B) The wound on the plantar was narrow, so we used a rubber band technique to mechanically shrink the wound area. (C) A silver-containing foam dressing was applied for small meshed wounds.

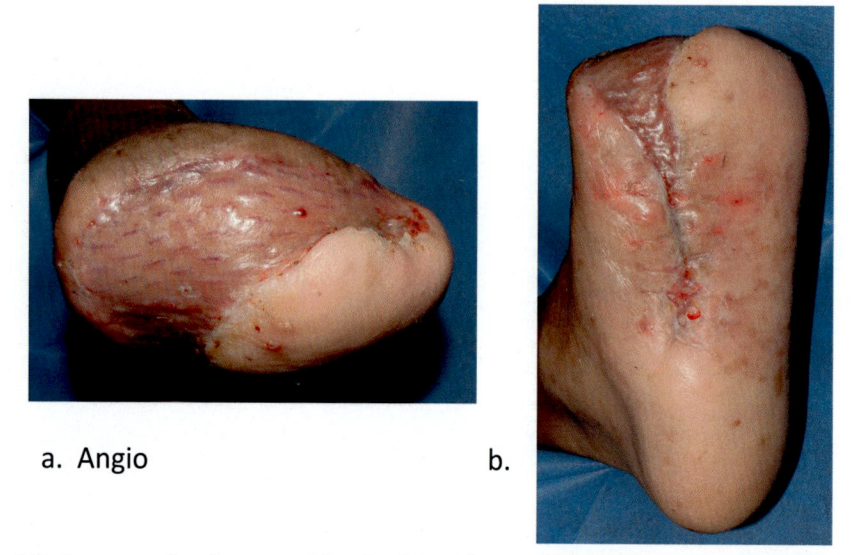

Figure 2.9 *Severe gas-forming necrotizing fasciitis with critical limb ischemia.* (A and B) At 3 months after debridement and endovascular treatment, the wound had almost completely healed.

(3) DFU treatment plan based on the evaluation of the ADL capability

CLI is divided into cases of active treatment, nonadaptation of aggressive treatment, and cases of difficult revascularization. When deciding whether or not to actively treat CLI, an evaluation of ADL, such as "whether patients were able to walk or stand up," before the development of ulceration is also used as a reference. Patients with postcerebral infarction paralysis or those in a long-term bedridden state are not able to walk or stand, even with aggressive lower limb salvage treatment. Aggressive lower limb salvage treatment is not indicated for such patients. If revascularization is not indicated, or if revascularization fails, it is impossible to perform aggressive limb salvage treatment, and major amputation (e.g., lower leg amputation or thigh amputation) will be selected if the patient's general condition is stable. In cases in which major amputation is difficult due to a poor general condition or in cases involving a patient in a long-term bedridden state, the necrotic area is dried and mummified, and the infection is controlled by applying an antibacterial ointment agent, aiming for autoamputation of the necrotic tissue (Fig. 2.10). In other words, the goal of treatment is not healing, but coexistence with dry gangrene (mummification).

Wound bed preparation/debridement and amputation

The concept of WBP is used globally for the management of chronic wounds such as DFU and gangrene with sufficient blood flow, after revascularization [9–11]. WBP is defined as the management of a wound in order to promote natural healing or facilitate alternative methods to achieve healing, using treatments such as skin grafting or the application of dermal matrices or other skin coverage products [10]. The TIME (Tissue, Inflammation/infection, Moisture imbalance, Epithelial edge advancement) acronym describes four aspects of WBP that need to be systematically addressed in order for WH to take place [10].

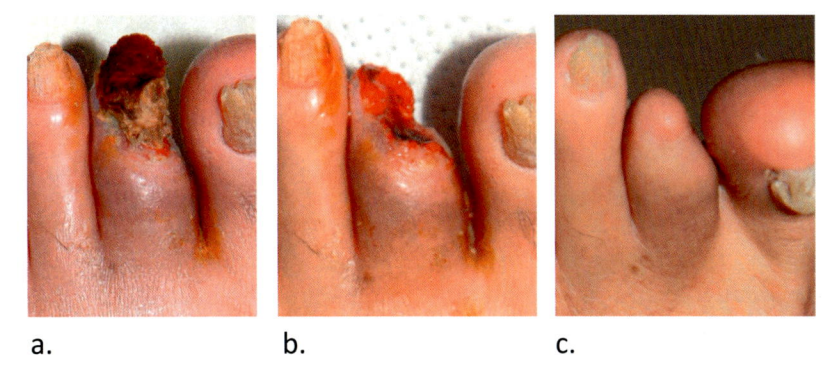

Figure 2.10 *The second toe autoamputation.* The foot of a 77-year-old diabetic female with dry gangrene without infection due to critical limb ischemia. After suffering a cerebral infarction, she became bedridden and had a low ADL activity due to paralysis. There is no indication for aggressive revascularization, and the management plan is to perform autoamputation with conservative treatment. (A) Dry necrosis of the second toe demonstrated a clear boundary between healthy tissue and necrotic tissue. (B) The dry necrotic tissue was spontaneously, automatically amputated. (C) Four months later, the wound had completely epithelialized and healed.

In the WBP concept, debridement and infection control are important items. First, amputation is performed and the necrotic area is debrided. The wound is then cleaned and antibacterial agents are used to control infection. Devitalized tissue provides a safe haven for bacterial proliferation, a barrier that prevents antibiotics from reaching bacterial pathogens. In addition, it limits the ability of the body's cellular defenses to fight infection. The removal of devitalized tissue reduces the bacterial bioburden [5].

If infection is suspected in a debrided ulcer, or if epithelialization from the margin does not progress within 2 weeks of debridement and the initiation of offloading therapy, the type and level of infection in a debrided diabetic ulcer should be determined by tissue biopsy or by a validated quantitative swab technique [5].

(1) Minor amputation/debridement

In the case of the minor amputation, the amputation level is determined from the viewpoint of WH, the blood flow status, and the walking function. Surgical intervention such as either minor amputation or debridement is performed under conditions of a sufficient blood flow, namely, the objective index, $TcPO_2$ value, and SPP value of >30−40 mmHg are maintained [18]. In wounds where blood flow is maintained and infection is controlled, debridement triggers WH.

For forefoot lesions, we are usually careful not to destroy the lateral arch of the metatarsal head. In addition, there are communication branches, arterial arches that connect the lateral plantar artery to the dorsal pedal artery between the first metatarsal bone and the second metatarsal bone; thus, this network is preserved as much as possible. By preserving the communication branches of this arch, it is possible to perform a distal puncture and insert a wire from the forefoot at the time of re-EVT in cases of restenosis. When treating a wound, it is necessary to consider what anatomical tissue is most important in revascularization and what should not be removed.

It is therefore essential that the revascularization team and the wound team work closely together. The plantar tissue has an anatomical and biomechanical structure that can endure the loading of weight, while other soft tissues do not. Therefore, it is reasonable to make an effort to preserve the plantar tissue as much as possible in the case of debridement.

In plantar lesions where the blood flow is preserved, the accumulation of pus may be observed under the plantar aponeurosis. There are three compartments in the plantar, and infection cannot be controlled unless the compartment is incised and opened for the drainage of pus. It is important to make sure that the plantar aponeurosis and fascia are incised until they are removed (Fig. 2.3A and B).

For minor amputations, the presence or absence of osteomyelitis is evaluated by plain X-ray or MRI, and if osteomyelitis is found, the sequestrum is completely excised using a Luer forceps. Minor amputations include toe amputation, toe fissure amputation, metatarsal amputation, Lisfranc joint amputation, and Chopart amputation.

Revascularization

According to the IWGDF guideline, all patients with diabetes and foot ulceration should be clinically examined (by relevant history and palpation of foot pulses) for the presence of [18].

Revascularization should always be considered for a patient with a DFU and PAD, irrespective of the results of bedside tests, when the ulcer does not heal within 4–6 weeks despite optimal management [5]. Wound care providers should always consider arterial insufficiency.

Urgently assess and treat patients with signs or symptoms of PAD and DFI, as they are at particularly high risk for major limb amputation (Fig. 2.1).

(1) Strategy of revascularization

When significant arterial disease is present after performing an objective blood flow evaluation, such as ABI, successful treatment requires that arterial insufficiency be addressed with revascularization. Revascularization includes EVT or surgical distal bypass. In practice, a CT angiogram or MR angiogram is taken, and after confirming the location and target area of the obstructive lesion, the decision to perform EVT or surgical distal bypass is made.

When performing revascularization in a patient with a DFU, aim to restore direct blood flow to at least one of the foot arteries, preferably the artery that supplies the anatomic region of the ulcer. After the procedure, the effectiveness is evaluated with an objective measurement of perfusion [5].

(2) Angiosome theory

Historically, the aim of revascularization in patients with PAD has been to achieve inline pulsatile flow to the foot, usually by targeting the best vessel available. However, more recently, the angiosome-directed approach has been advocated; however, this remains a subject of much debate [21,22].

According to this theory, the foot can be divided into three-dimensional blocks of tissue, each with its own feeding artery [23].

Direct revascularization would result in a restoration of pulsatile blood flow through the feeding artery to the area where the ulcer is located, while with indirect revascularization flow is restored through collateral vessels deriving from the neighboring angiosomes.

By targeting revascularization at the vessel directly supplying the anatomical area (angiosome) of tissue loss, the theory is that this will be a more effective method of revascularization than simply targeting the best vessel, which may not supply the area of tissue loss. A recent retrospective study of endovascular limb salvage attempts in patients with DFU suggested that indirect angiosome revascularization was associated with poorer outcomes than direct revascularization [24].

However, due to lack of clear definitions and factors such as selection bias, the effectiveness of the angiosome concept in patients with diabetes is unknown [21,25—27]. Based on previous reports, angiosome-guided EVT tends to improve WH rather than amputation-free survival and overall survival [28].

Particularly in patients with diabetes, who usually have poor collaterals, the restoration of flow to an artery directly supplying the affected area seems to be the best approach during an endovascular procedure [22].

Successfully opening one or more occluded vessels is not the same as a clinically successful procedure and before the procedure is terminated, blood flow to the ulcerated area should therefore be assessed. If feasible, the opening of multiple arteries may be useful, provided at least one of the arteries directly supplies the ischemic area directly [21].

Treatments for the promotion of wound healing

(1) Negative pressure wound therapy

Once infection is controlled, topical negative pressure wound therapy (NPWT) is used to promote granulation [29,30]. NPWT is a global standard therapy for chronic wounds, such as DFU [30]. It is a treatment that loads mechanical stress on the cells involved in WH such as fibroblasts, to increase cell division and proliferation [31,32]. Clinically, it physically contracts the wound margin, promotes granulation tissue formation, drains exudate, and reduces edema around the wound [31].

Recently, NPWT with instillation and dwelling time (NPWTi-d [V.A.C. VERAFLO Therapy; KCI, now part of 3M Company, San Antonio, Texas]) using a reticulated open-cell foam dressing with through holes (ROCF-CC; V.A.C. VERAFLO CLEANSE CHOICE Dressing; KCI, now part of 3M Company) has become available in the clinical setting. NPWTi-d is a revolutionary technological advancement of NPWT that includes the periodic instillation of a topical solution onto the wound bed, thereby utilizing both negative pressure and the instillation of topical wound solutions for wound cleansing and the removal of infectious materials [33,34], Furthermore, NPWTi-d using an ROCF-CC dressing incorporates an NPWT device that makes it easier to remove necrotic tissue in comparison to NPWTi-d (Fig. 2.6C) [34]. NPWTi-d has a definite therapeutic effect on all four aspects (TIME) of WBP.

Granulation is an intermediate goal in the WH process. When most of the wound is covered with good granulation tissue, it usually heals within a few weeks after simple reconstructive surgery (e.g., skin grafting) (Figs. 2.4—2.9).

(2) Dressing

In order to promote WH, it is important to maintain a moist environment by dressing the wound [35,36].

Following this report, efforts were made to develop dressings that maintained a moist environment. However, DFUs, which are susceptible to infection, often become infected when dressing is used to make the wound occlusive. To avoid this, it is necessary to select a dressing according to the amount of exudate in the wound and the choice of dressing should provide an appropriate moisture balance, avoid maceration of the skin edges, prevent leakage, and be easy to apply and remove. However, the volume of exudate is not the only factor to consider, as there is evidence that the chronic wound fluid composition is as important as the volume of exudate [10]. Chronic wound fluid contains proteases, such as MMPs, that inhibit WH. Protease-modulating dressings may be appropriate to control wound proteases and their development has focused on reducing the levels of MMPs by absorbing wound exudate, holding proteases within the dressing structure, and inactivating the excess MMPs [10]. There is evidence that collagen/oxidized protease—modulating dressings may increase healing rates in DFUs [37].

NPWT is the standard of care for DFU wound treatment, which has efficacy in promoting the formation of granulation tissue. However, the ability to epithelialize is not weaker than the ability to form granulation tissue; thus, once the entire wound is covered with granulation tissue, NPWT is finished and a dressing is usually applied to promote epithelialization (Fig. 2.11). The conditions under which a dressing can be used for DFU are that infection is controlled and that sufficient blood flow is maintained in the foot. Although the conditions are limited in this way, when almost the entire wound is covered with granulation tissue after NPWT, the condition is considered optimal for the initiation of dressing.

Reconstruction of the soft tissue

The ultimate goal of all wound physicians and surgeons is to eliminate exudate and close the wound. Eventually, the wound will heal with one of the following: secondary

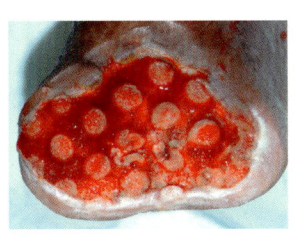

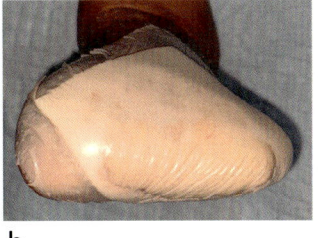

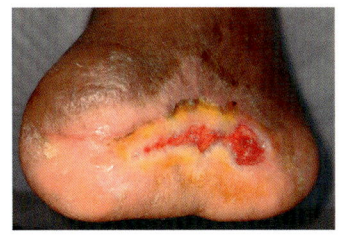

Figure 2.11 *Conservative treatment using foam dressing.* The foot of a 67-year-old diabetic male with critical limb ischemia after being treated by revascularization, transmetatarsal amputation, and NPWT. (A) NPWTi-d with ROCF-CC dressing. The stump wound of the forefoot was almost 95% covered with granulation tissue after using NPWTi-d with ROCF-CC dressing. (B) In order to prioritize gait rehabilitation more than performing a skin graft, conservative treatment with foam dressing was started in the homecare setting. (C) After 3 months, the epithelialization became accelerated owing to the foam dressing.

intention, primary or secondary wound closure, local flap transfer, distant flap transfer, free flap transfer, and amputation. These have a hierarchy, which is referred to as the "reconstruction ladder" in the field of plastic surgery (Fig. 2.12) [38,39].

DFUs are slow to heal and often cause infection. The longer a wound is present, the higher the risk of infection. Thus, the wound should be closed as soon as possible.

Regarding the treatment strategy for CLI, EVT of the native ischemic foot is the first option [19]. However, among patients with isolated infrapopliteal lesions who undergo EVT alone, the 3-month angiographic restenosis rate is 73%, which is extremely high [40]. CLI patients also need to achieve complete WH as soon as possible in terms of revascularization. Thus, the surgeon chooses reconstructive surgery.

(1) Skin graft

Skin grafts that incorporate all of the dermis are termed full-thickness skin grafts (FTSGs). If only a portion of the dermis is included, the term split-thickness skin graft (STSG) is used. Skin grafts offer the surgeon a quick and easy method for wound closure. STSGs are the most common method of wound closure (Figs. 2.4–2.9). Skin grafting is associated with a higher WH rate and shorter WH period in comparison to standard care [41–46].

A meta-analysis of the outcomes of STSGs reported that the healing rate of DFUs following STSG was as high as 85.5%, with a relatively short WH period of 5.3 weeks and a recurrence rate of 4.2%. Such values are considered to be far better than those reported after standard care with a mean follow-up period of more than 2 years [46].

| Free flap |
| Tissue expansion |
| Distant flap |
| Local flap |
| Dermal matrix |
| Skin graft |
| NPWT |
| Secondary intention |
| Primary closure |

Figure 2.12 A reconstructive ladder.

Generally, a thicker graft means better mechanical, functional, and aesthetic properties; however, the failure rate of thicker grafts is a higher than that of thinner grafts [4]. Since skin grafts lack their own blood supply, this process is entirely dependent on the wound bed blood supply. Skin grafts do not survive on infected wounds. Thus, it is important to perform WBP and revascularization before applying a skin graft (Figs. 2.4−2.9).

(2) Free flap transfer reconstruction

    (2-1) Free flap transfer for patients with DFU

Skin graft sites and secondary intention are difficult to heal in cases with factors such as exposed bone (Figs. 2.13−2.15), a corrupt midfoot joint (rupture of the joint capsule) (Figs. 3.16−3.18), soft tissue defects beneath the plantar and calcaneal lesion of the loading weight, or an exposed Achilles tendon. A pedicled flap and free flap transfer are usually used for tissue reconstruction in such cases. Especially, the reconstruction of the heel, which is heavily weight loaded, is the best indication of using free flap transfer (Fig. 2.19).

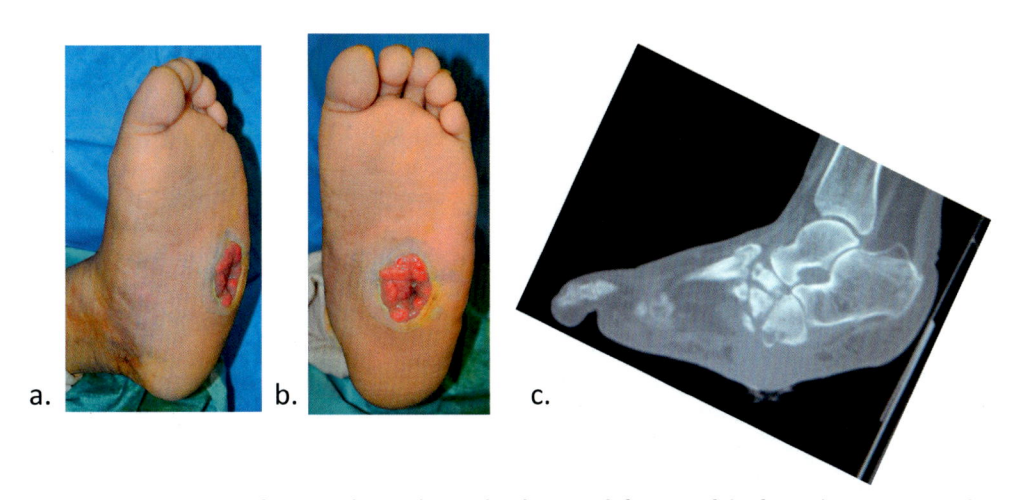

Figure 2.13 *Recurrent ulceration beneath a rocker-bottom deformity of the foot.* Charcot neuroarthropathy foot without ischemia. The foot of a 65-year-old diabetic female with recurrent ulceration beneath a rocker-bottom deformity of the foot due to neuroarthropathy. (A and B) The recurrent ulcer was demonstrated on the plantar midfoot without infection. She was using an adhesive felt padding which cut to the shape of the planter surface excluding the site of the ulcer due to the offloading. However, she experienced repeated ulcer recurrences. (C) On the computed tomography image, a destructive cuneiform bone and dislocation of the metatarsal bone were demonstrated.

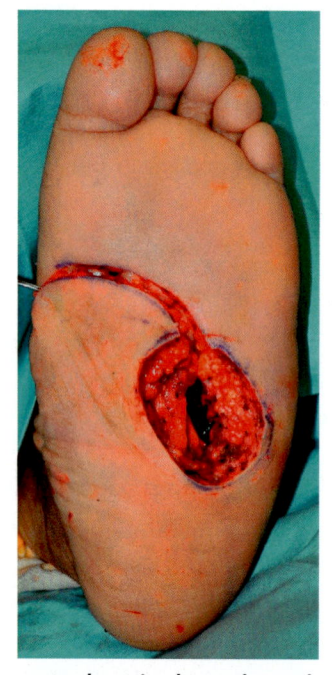

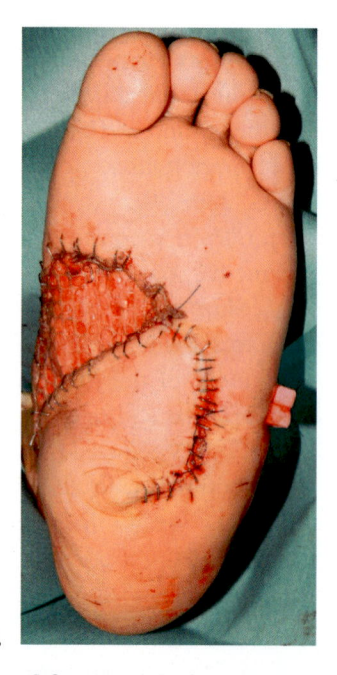

Figure 2.14 *Recurrent ulceration beneath a rocker-bottom deformity of the foot.* (A and B) A pedicled medial plantar flap. Without ischemia, the pedicled medial plantar flap showed a recurrent ulcer beneath the Charcot foot because plantar tissue defects need to be reconstructed using the same plantar soft tissue. The ulcer and abscess were treated by bursectomy and the bony prominence was resected using a bone chisel. Thereafter, the pedicled medial plantar flap was elevated and a mesh skin graft was performed on the donor from whom the flap had been harvested.

Free flap transfer is the top step of the reconstruction ladder. Diabetic foot reconstruction using a free flap is associated with a high chance of success and significantly increases the 5-year survival rate. Risk factors, such as PAD, a history of angioplasty in the extremity, and the use of immunosuppressive agents after transplantation may increase the risk of flap loss [47].

(2-2) Free flap transfer for patients with CLI

In CLI, free flap transfer is selected after revascularization because local pedicled flaps can cause new ischemic necrosis in the foot. Briggs et al. first reported a method of salvage therapy for patients with CLI using a combination of distal bypass and free tissue transfer [48]. When selecting a recipient vessel for anastomosis to the free flap, it is reasonable to choose a vessel that will provide long-term patency.

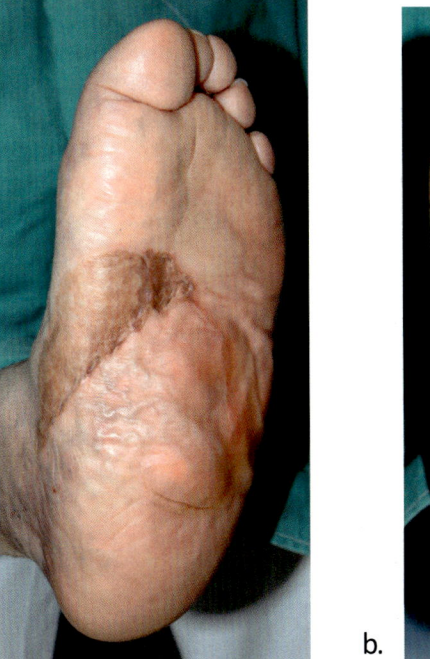

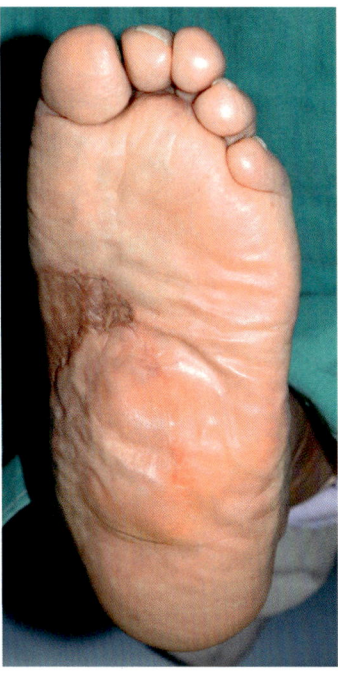

Figure 2.15 *Recurrent ulceration beneath a rocker-bottom deformity of the foot.* (A and B) The plantar ulcer that had repeatedly recurred has now not recurred for over 10 years.

If an artery with a high probability of restenosis that does not result in long-term patency is selected as the recipient vessel for flee flap, the restenosis of the recipient vessel blood flow to the flap and thereby cause the flap to become necrotic.

Mimoun et al. advocated that free tissue transfer plays a new role as "a nutrient flap" in cases of [49,50]. They reported that the performance of free tissue transfer provided the nutrient effect of the flap (due to the supplementary blood supply), effective venous return, and the progressive development of a new distal capillary bed.

Tukiainen et al. reported the long-term outcomes of cases involving the treatment of CLI by a combination of distal bypass surgery and free tissue transfer. In their study, the 1- and 5-year leg salvage rates were 73% and 66%, the 1- and 5-year survival rates were 91% and 63%, and the 1- and 5-year amputation-free survival rates were 70% and 41%, respectively [51].

Using free flap transfer in this way, it has become possible to salvage the feet of patients with DFU or CLI with wide-area defects for which major amputation would have traditionally been indicated.

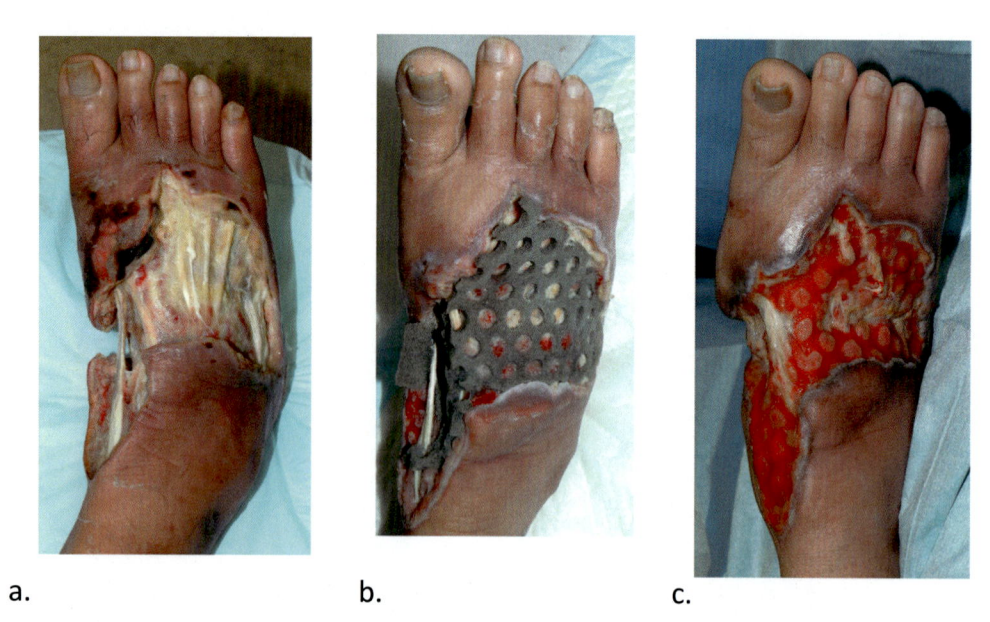

Figure 2.16 *Severe gas-forming necrotizing fasciitis without ischemia.* The foot of a 53-year-old diabetic male with severe gas-forming necrotizing fasciitis without ischemia. Fever and an elevated white blood cells count were observed, and the patient was considered to have a Class 3 infection. The SPP value was 60 mmHg, and the blood flow to the foot was maintained. (A) Urgent debridement was performed. Wound irrigation and maintenance debridement were performed daily. Systemic infection was controlled by intravenous antibiotics and hyperbaric oxygen therapy. (B) NPWTi-d with cleans choice foam was applied. (C) 4 weeks later, granulation tissue had covered 80% of the wound. However, part of the Lisfranc joint was released and opened by debridement. Therefore, we planned to carry out reconstruction using a free anterolateral thigh flap.

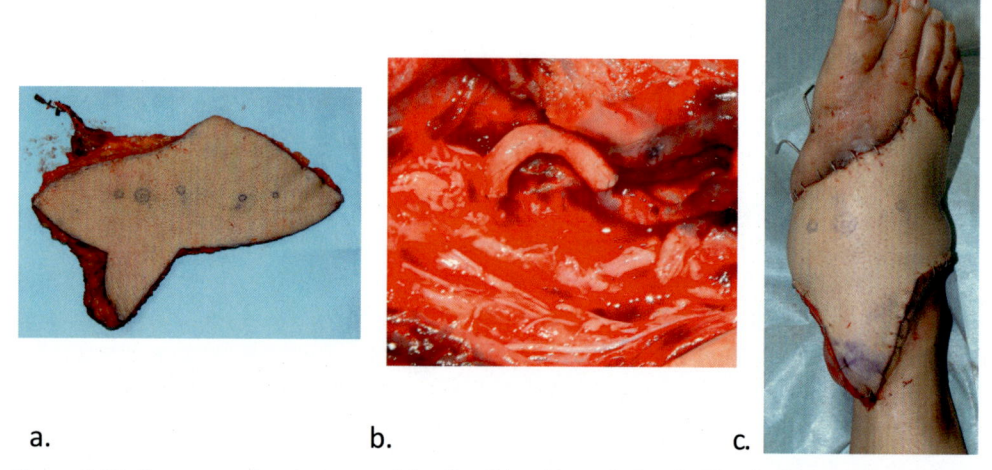

Figure 2.17 *Severe gas-forming necrotizing fasciitis without ischemia.* The findings during surgery. (A) An anterolateral thigh flap (11.5 × 25.5 cm) was harvested and separated from the left thigh. (B) The descending branch of the lateral circumflex femoral artery was anastomosed to the anterior tibial artery using end-to-side anastomosis under microscopy. (C) The Lisfranc joint, which had been released, was fixed with wire.

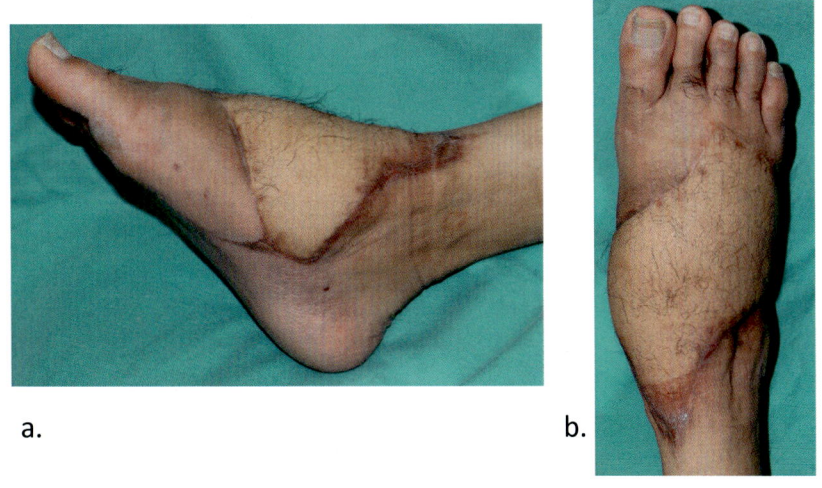

Diabetic foot ulcers and their wound management 31

a. b.

Figure 2.18 *Severe gas-forming necrotizing fasciitis without ischemia.* (A and B) Two months later, the free ALT flap transfer was found to be successful and the wound has completely healed including the Lisfranc joint. The patient's gait function could be maintained.

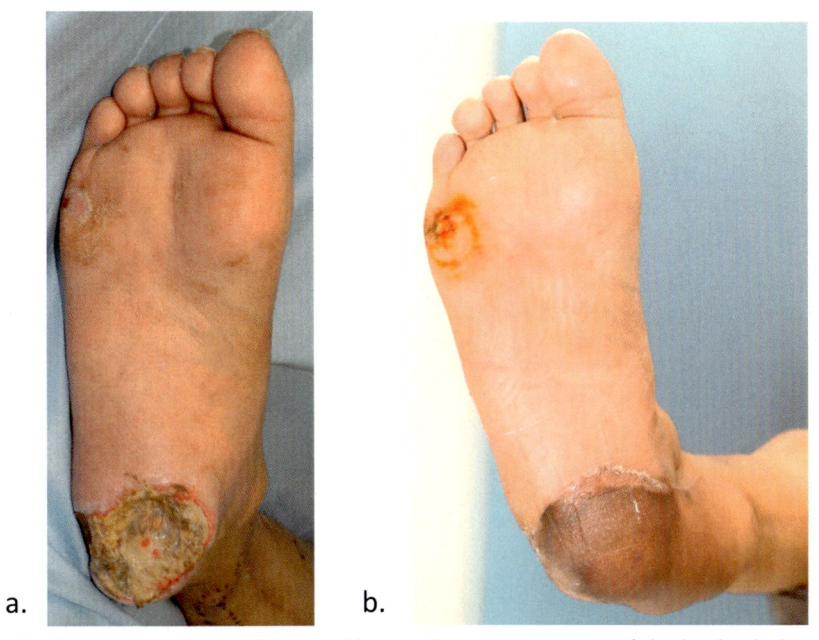

a. b.

Figure 2.19 *The reconstruction of the heel lesion.* The reconstruction of the heel, which is heavily weight loaded, is the best indication of using free flap transfer. (A) An ulceration with necrotic tissue was demonstrated on the heel. (B). The heel was reconstructed using the latissimus dorsi muscle flap.

Summary

Lower extremity limb salvage is important; however, when making efforts to treat the foot, the physician must not overlook a worsening general condition or disuse syndrome. It does not make sense to save a foot if the patient will be bedridden. It is necessary to constantly recognize the general condition and mobility of the patient and to set goals fluidly.

References

[1] Zhang P, Lu J, Jing Y, Tang S, Zhu D, Bi Y. Global epidemiology of diabetic foot ulceration: a systematic review and meta-analysis. Ann Med 2017;49(2):106–16.
[2] Apelqvist J. Diagnostics and treatment of the diabetic foot. Endocrine 2012;4141(3):384–97.
[3] Boulton AJ, Vileikyte L, Ragnarson-Tennvall G, Apelqvist J. The global burden of diabetic foot disease. Lancet 2005;366(9498):1719–24.
[4] Smuđ-Orehovec S, Mance M, Halužan D, Vrbanović-Mijatović V, Mijatović D. Defect reconstruction of an infected diabetic foot using split- and full-thickness skin grafts with adjuvant negative pressure wound therapy: a case report and review of the literature. Wounds 2018;30(11):E108–15.
[5] Lavery LA, Davis KE, Berriman SJ, Braun L, Nichols A, Kim PJ, Margolis D, Peters EJ, Attinger C. WHS guidelines update: diabetic foot ulcer treatment guidelines. Wound Repair Regen 2016;24(1):112–26.
[6] Lipsky BA, Silverman MH, Joseph WS. A proposed new classification of skin and soft tissue infections modeled on the subset of diabetic foot infection. Open Forum Infect Dis 2016;4(1):ofw255. 7.
[7] Prompers L, Schaper N, Apelqvist J, Edmonds M, Jude E, Mauricio D, Uccioli L, Urbancic V, Bakker K, Holstein P, Jirkovska A, Piaggesi A, Ragnarson-Tennvall G, Reike H, Spraul M, Acker KV, Baal JV, Merode FV, Ferreira I, Huijberts M. Prediction of outcome in individuals with diabetic foot ulcers: focus on the differences between individuals with and without peripheral arterial disease. The EURODIALE study. Diabetologia 2008;51:747–55.
[8] Edmonds M. Double trouble: infection and ischemia in the diabetic foot. Int J Low Extrem Wounds 2009;8:62–3.
[9] Schultz GS, Sibbald RG, Falanga V, Ayello EA, Dowsett C, Harding K, Romanelli M, Stacey MC, Teot L, Vanscheidt W. Wound bed preparation: a systematic approach to wound management. Wound Repair Regen 2003;11(Suppl. 1):S1–28.
[10] Harries RL, Bosanquet DC, Harding KG. Wound bed preparation: TIME for an update. Int Wound J 2016;13(Suppl. 3):8–14.
[11] Leaper DJ, Schultz G, Carville K, Fletcher J, Swanson T, Drake R. Extending the TIME concept: what have we learned in the past 10 years?(*). Int Wound J 2012;9(Suppl. 2):1–19.
[12] Bus SA, van Deursen RW, Armstrong DG, Lewis JE, Caravaggi CF, Cavanagh PR, International Working Group on the Diabetic Foot. Footwear and offloading interventions to prevent and heal foot ulcers and reduce plantar pressure in patients with diabetes: a systematic review. Diabetes Metab Res Rev 2016;32(Suppl. 1):99–118.
[13] Armstrong DG, Boulton AJM, Bus SA. Diabetic foot ulcers and their recurrence. N Engl J Med 2017;376(24):2367–75. 15.
[14] Prompers L, Huijberts M, Apelqvist J, Jude E, Piaggesi A, Bakker K, Edmonds M, Holstein P, Jirkovska A, Mauricio D, Tennvall GR, Reike H, Spraul M, Uccioli L, Urbancic V, Acker KV, Baal JV, Merode FV, Schaper N. High prevalence of ischaemia, infection and serious comorbidity in patients with diabetic foot disease in Europe. Baseline results from the Eurodiale study. Diabetologia 2007;50(1):18–25.

[15] Morbach S, Furchert H, Gröblinghoff U, Hoffmeier H, Kersten K, Klauke GT, Klemp U, Roden T, Icks A, Haastert B, Rümenapf G, Abbas ZG, Bharara M, Armstrong DG. Long-term prognosis of diabetic foot patients and their limbs: amputation and death over the course of a decade. Diabetes Care 2012;35(10):2021−7.

[16] Conte MS, Bradbury AW, Kolh P, White JV, Dick F, Fitridge R, Mills JL, Ricco JB, Suresh KR, Murad MH, GVG Writing Group. Global vascular guidelines on the management of chronic limb-threatening ischemia. J Vasc Surg 2019;69(6S):3S−125S.

[17] Mills Sr JL, Conte MS, Armstrong DG, Pomposelli FB, Schanzer A, Sidawy AN, Andros G, The Society for Vascular Surgery Lower Extremity Guidelines Committee. The society for vascular surgery lower extremity threatened limb classification system: risk stratification based on wound, ischemia, and foot infection (WIfI). J Vasc Surg 2014;59(1):220−34.

[18] Hinchliffe RJ, Forsythe RO, Apelqvist J, et al. International Working Group on the Diabetic Foot (IWGDF). Guidelines on diagnosis, prognosis, and management of peripheral artery disease in patients with foot ulcers and diabetes (IWGDF 2019 update). Diabetes Metab Res Rev 2020;36(Suppl. 1): e3276.

[19] Conte MS, GVG writing group for the joint guidelines of the society for vascular surgery (SVS), European society for vascular surgery (ESVS), and eorld federation of vascular societies (WFVS), et al. Global vascular guidelines on the management of chronic limb-threatening ischemia. Eur J Vasc Endovasc Surg 2019;58. S1-S109.e33.

[20] Stevens DL, Bisno AL, Chambers HF, Dellinger EP, Goldstein EJ, Gorbach SL, Hirschmann JV, Kaplan SL, Montoya JG, Wade JC. Practice guidelines for the diagnosis and management of skin and soft tissue infections: 2014 update by the Infectious Diseases Society of America. Clin Infect Dis 2014;59(2):147−59. 15.

[21] Stimpson AL, Dilaver N, Bosanquet DC, Ambler GK, Twine CP. Angiosome specific revascularisation: does the evidence support it? Eur J Vasc Endovasc Surg 2019;57(2):311−7.

[22] Jongsma H, Bekken JA, Akkersdijk GP, Hoeks SE, Verhagen HJ, Fioole B. Angiosome-directed revascularization in patients with critical limb ischemia. J Vasc Surg 2017;65(4):1208−19.

[23] Attinger CE, Evans KK, Bulan E, Blume P, Cooper P. Angiosomes of the foot and ankle and clinical implications for limb salvage: reconstruction, incisions, and revascularization. Plast Reconstr Surg 2006;117(7 Suppl. l):261S−93S.

[24] Lo ZJ, Lin Z, Pua U, Quek LHH, Tan BP, Punamiya S, Tan GWL, Narayanan S, Chandrasekar S. Diabetic foot limb salvage-a series of 809 attempts and predictors for endovascular limb salvage failure. Ann Vasc Surg 2018;49:9−16.

[25] Khor BYC, Price P. The comparative efficacy of angiosome-directed and indirect revascularisation strategies to aidhealing of chronic foot wounds in patients with co-morbid diabetes mellitus and critical limb ischaemia: a literature review. J Foot Ankle Res 2017;10:26.

[26] Alexandrescu V, Hubermont G. The challenging topic of diabetic foot revascularization: does the angiosome-guided angioplasty may improve outcome. J Cardiovasc Surg 2012;53(1):3−12.

[27] Lejay A, Georg Y, Tartaglia E, Gaertner S, Geny B, Thaveau F, Chakfe N. Long-term outcomes of direct and indirect below-the-knee open revascularization based on the angiosome concept in diabetic patients with critical limb ischemia. Ann Vasc Surg 2014;28(4):983−9.

[28] Hata Y, Iida O, Mano T. Is angiosome-guided endovascular therapy worthwhile? Ann Vasc Dis 2019; 12(3):315−8.

[29] Argenta LC, Morykwas MJ. Vacuum-assisted closure: a new method for wound control and treatment. Clin Exp 1997;38(6):563−76. Discussion 577.

[30] Armstrong DG, Lavery LA, Diabetic Foot Study Consortium. Negative pressure wound therapy after partial diabetic foot amputation: a multicentre, randomised controlled trial. Lancet 2005;366(9498): 1704−10.

[31] Orgill DP, Manders EK, Sumpio BE, Lee RC, Attinger CE, Gurtner GC, Ehrlich HP. The mechanisms of action of vacuum assisted closure: more to learn. Surgery 2009;146(1):40−51.

[32] Scherer SS, Pietramaggiori G, Mathews JC, Prsa MJ, Huang S, Orgill DP. The mechanism of action of the vacuum-assisted closure device. Plast Reconstr Surg 2008;122(3):786−97.

[33] Aycart MA, Eble DJ, Ross KM, Orgill DP. Mechanisms of action of instillation and dwell negative pressure wound therapy with case reports of clinical applications. Cureus 2018;10(9):e3377. https://doi.org/10.7759/cureus.3377.

[34] Kim PJ, Attinger CE, Constantine T, Crist BD, Faust E, Hirche CR, Lavery LA, Messina VJ, Ohura N, Punch LJ, Wirth GA, Younis I, Téot L. Negative pressure wound therapy with instillation: international consensus guidelines update. Int Wound J 2020;17(1):174—86.

[35] Winter GD. Effect of air exposure and occlusion on experimental human skin wounds. Nature 1963; 200:378—9.

[36] Winter GD, Scales JT. Effect of air drying and dressings on the surface of a wound. Nature 1963;197: 91—2.

[37] Veves A, Sheehan P, Pham HT. A randomized, controlled trial of promogran (a collagen/oxidized regenerated cellulose dressing) vs standard treatment in the management of diabetic foot ulcers. Arch Surg 2002;137:822—7.

[38] Janis JE, Kwon RK, Attinger CE. The new reconstructive ladder: modifications to the traditional model. Plast Reconstr Surg 2011;127(Suppl. 1):205S—12S.

[39] Gottlieb LJ, Krieger LM. From the reconstructive ladder to the reconstructive elevator. Plast Reconstr Surg 1994;93(7):1503—4.

[40] Iida O, Soga Y, Kawasaki D, Hirano K, Yamaoka T, Suzuki K, Miyashita Y, Yokoi H, Takahara M, Uematsu M. Angiographic restenosis and its clinical impact after infrapopliteal angioplasty. Eur J Vasc Endovasc Surg 2012;44(4):425—31.

[41] Ramanujam CL, Stapleton JJ, Kilpadi KL, Rodriguez RH, Jeffries LC, Zgonis T. Split-thickness skin grafts for closure of diabetic foot and ankle wounds: a retrospective review of 83 patients. Foot Ankle Spec 2010;3(5):231—40.

[42] Ramanujam CL, Han D, Fowler S, Kilpadi K, Zgonis T. Impact of diabetes and comorbidities on split-thickness skin grafts for foot wounds. J Am Podiatr Med Assoc 2013;103(3):223—32.

[43] Rose JF, Giovinco N, Mills JL, Najafi B, Pappalardo J, Armstrong DG. Split-thickness skin grafting the high-risk diabetic foot. J Vasc Surg 2014;59(6):1657—63.

[44] Sanniec K, Nguyen T, van Asten S, Fontaine J, Lavery LA. Split-thickness skin grafts to the foot and ankle of diabetic patients. J Am Podiatr Med Assoc 2017;107(5):365—8.

[45] Vilcu M, Catrina E, Pătraæcu T. Split-thickness skin grafting in the treatment of surgically operated diabetic foot. A retrospective 2-year study. Mod Med 2014;21:158—64.

[46] Yammine K, Assi C. A meta-analysis of the outcomes of split-thickness skin graft on diabetic leg and foot ulcers. Int J Low Extrem Wounds 2019;18(1):23—30.

[47] Oh TS, Lee HS, Hong JP. Diabetic foot reconstruction using free flaps increases 5-year-survival rate. J Plast Reconstr Aesthetic Surg 2013;66(2):243—50.

[48] Briggs SE, Banis Jr JC, Kaebnick H, Silverberg B, Acland RD. Distal revascularization and microvascular free tissue transfer: an alternative to amputation in ischemic lesions of the lower extremity. J Vasc Surg 1985;2(6):806—11.

[49] Mimoun M, Hilligot P, Baux S. The nutrient flap: a new concept of the role of the flap and application to the salvage of arteriosclerotic lower limbs. Plast Reconstr Surg 1989;84(3):458—67.

[50] Horch RE, Lang W, Arkudas A, Taeger C, Kneser U, Schmitz M, Beier JP. Nutrient free flaps with vascular bypasses for extremity salvage in patients with chronic limb ischemia. J Cardiovasc Surg 2014; 55(2 Suppl. 1):265—72.

[51] Tukiainen E, Kallio M, Lepäntalo M. Advanced leg salvage of the critically ischemic leg with major tissue loss by vascular and plastic surgeon teamwork: long-term outcome. Ann Surg 2006;244(6): 949—57.

CHAPTER 3

Computational modeling of the plantar tissue stresses induced by the clinical practice of off-loading of the diabetic foot

Hadar Shaulian[1], Amit Gefen[2], Alon Wolf[1]
[1]Faculty of Mechanical Engineering, Technion-Israel Institute of Technology, Haifa, Israel; [2]Department of Biomedical Engineering, Faculty of Engineering, Tel Aviv University, Tel Aviv, Israel

Introduction

Diabetic foot ulceration is among the most common, serious, and destructive complications of diabetes worldwide. The foot ulcers are mainly caused by peripheral neuropathy, lead to difficulties in ambulation, and are detrimental to the quality of life of the affected individuals [1]. Over the course of their disease, 25% of the people with diabetes will develop a foot ulcer, the ultimately leading cause of lower extremity amputation in up to 80% of the cases, especially when wound infection or osteomyelitis is involved [2]. Lower extremity amputations are the costliest and most feared consequence of a foot ulcer, comprising up to one-third of the direct cost of diabetes care [3].

Although heel ulcerations are less frequent than metatarsal ulcerations, they are more challenging to treat, with limb salvage success rate of two to three times less likely than that seen with forefoot metatarsal ulceration [4]. Retrospective and prospective studies have shown that focal pressure and repetitive normal and shear stresses at a given location are the causative factors in the development of foot ulcerations [5,6]. During gait, these forces cause elevated local stresses of 50%–60% of body weight on the heel pad, leading to an increased risk of foot ulcer development [7]. Therefore, optimal off-loading of the foot ulcer locations, while minimizing loads redistributed to peripheral tissues, is an essential component in preventing and treating foot ulcers [8].

Pressure reduction is often aided by the use of standard therapeutic shoes or custom-made insoles to off-load high plantar pressures and to accommodate foot deformities [8,9]. One specific common medical off-loading device is a custom-made insole designed with a hole under the active wound site, to reduce normal and shear stresses on the ulcer and redistribute them among other more peripheral foot regions [10]. The size, location, and shape of the hole are subjectively based on visual indication and experience of the therapist, and no scientific guidelines for optimal objective design of the off-loading hole have been established. The most common geometry of the hole is "step-shaped," with relatively sharp edges (Fig. 3.2). However, there are valid concerns that an aperture applied around the wound base increases shear and vertical forces at the peripheries of the wound (the periwound tissues), which escalates the risk of developing secondary ulcers at stress concentration areas, a phenomenon also known as the "edge-effect" [11].

Foot ulcers often start internally and progress outward [12], thus examining the foot internal layer stresses should be an essential part of any foot ulcer related study. Therefore, we used computational, finite element (FE) analysis which is a powerful tool that allows evaluation of internal tissue loads and has been used extensively by research groups for this purpose [12,13]. Most studies compare between models using peak foot plantar pressures or peak strains. However, using the peak as a quantitative value might lead to inaccurate results and biased conclusions, as the point loads are influenced by individual elements in the mesh. Moreover, the volumetric exposure of the tissue to the loading (e.g., the size of a stress concentration) is not considered. Here, we present a patient-specific method, based on FE analysis and available patient information, to investigate the internal tissue loads induced by this off-loading method.

Methods
Geometry
To examine foot internal loads during walking on "step-shaped" off-loading support, we developed a generic FE, three-dimensional (3D), and anatomically realistic model of a human heel, including the heel bones (calcaneus bone, lower aspect of the tibia and fibula, and posterior aspect of the talus and cuboid), cartilage, Achilles tendon (AT), soft tissues of the heel, and the skin (Fig. 3.1B). To develop the above geometry of a patient's heel, we used an existing left foot computerized tomography (CT) scan (iCT 256, Philips). The 486 CT scan slices were taken every 0.8 mm with the subject in a supine position in the scanner. The ScanIP module of Simpleware [14] was used to segment the different tissues from the CT data set and then truncate the model volume leaving the contact area of the foot with the ground during heel strike (Fig. 3.1A). A baseline flat support and off-loading hole-support models were both designed using Solidworks (Dassault Systèmes, CS, France) and applied to the heel model at the preprocessing stage in PreView of FEBio (Ver.1.19.0) [15]. The friction coefficient between the foot and the support was taken as 0.6 [16].

Mechanical properties
The mechanical behavior and properties of all tissues were adopted from the literature (Table 3.1). Specifically, the bones and AT were assumed to be isotropic linear-elastic materials with elastic moduli of 7300 MPa and 0.173 kPa, and Poisson's ratios of 0.3 and 0.49, respectively [17,18]. Skin and soft tissues were assumed to behave as viscoelastic solids. The hyperelastic component of this viscoelastic behavior was considered to be the Neo-Hookean with a strain energy density function W Eq. (3.1):

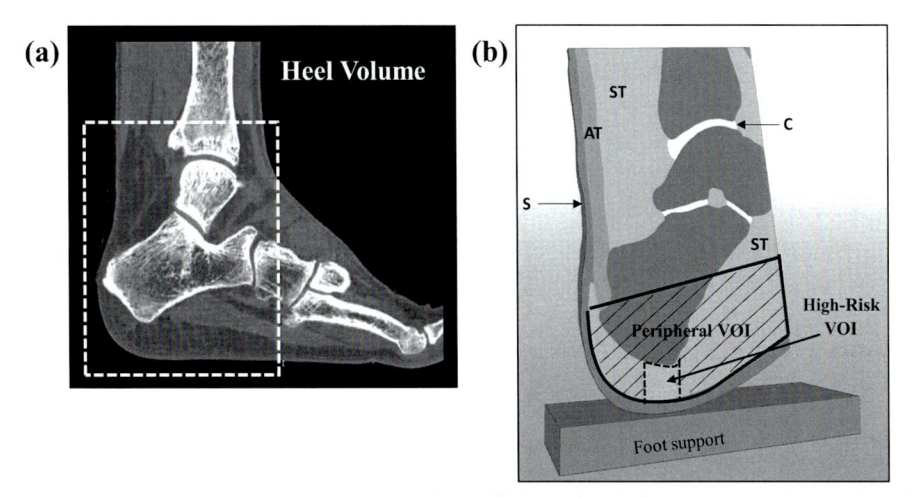

Figure 3.1 Finite element computational modeling of left heel with foot support: (A) A CT slice from the 3D scan set which was used for constructing the model geometry. The modeled volume is marked by a dashed perimeter. (B) A sagittal section of the 3D model of the heel. AT, Achilles tendon; C, cartilage; S, skin; ST, soft tissue. The peripheral VOI is indicated by diagonal lines, and the high-risk VOI is indicated by a square with a dashed perimeter, surrounded by the peripheral VOI.

$$W = \frac{\mu}{2}(I_1 - 3) - \mu ln(J) + \frac{1}{2}\lambda(\ln J)^2 \qquad (3.1)$$

where I_1 is the first invariant of the right Cauchy-Green deformation tensor, J is the determinant of the deformation gradient tensor, and μ and λ are Lame parameters [19]. The viscous component was simulated using the Prony series of stress relaxation functions G [15] (Eq. 3.2):

$$G(t) = 1 + \sum_{i=1}^{2} \gamma_i \exp\left(-\frac{t}{\tau_i}\right) \qquad (3.2)$$

where γ_i and τ_i are the tissue-specific material constants, specified in Table 3.1, and t is time.

Table 3.1 Mechanical properties of the model components and characteristics of the finite element mesh.

Model component	Elastic modulus [MPa]	Poisson's ratio	γ_1	τ_1	γ_2	τ_2
Skin	0.0954	0.495	0.086	0.212	0.214	4.680
Soft tissue	0.306	0.495	0.399	2.040	0.124	76.960
Achilles tendon	0.173	0.49	—	—	—	—
Bones	7300	0.3	—	—	—	—
Cartilage	1.01	0.4	—	—	—	—
Foot supports	10	0.3	—	—	—	—

Ethylene-vinyl acetate (EVA) material was used for both foot supports with nonlinear isotropic materials. The large deformation behavior of EVA was described using a compressible Neo-Hookean material model [15,20].

Boundary conditions and numerical method

Boundary conditions were chosen to simulate the foot during heel strike on the off-loading support. The initial angle between the heel and the support was approximated by 10 degrees to simulate the foot position at heel-strike [17]. The superior surfaces of the tibia, fibula, and soft tissue were fully fixed to simulate the effects of the constraints from superior-lying tissues. To simulate a ground reaction force (GRF) applied to the heel during heel-strike [7], displacements in the range of 7−10 mm of the flat support toward the heel were used.

All meshing was performed using the ScanIP + FE module of Simpleware [14]. All elements were of the tetrahedral type. The FE simulations were all set up using PreView of FEBio (Ver. 1.19), analyzed using the Pardiso linear solver of FEBio (Ver. 2.8.3) and post-processed using Matlab (Mathworks, Natick, MA, US) and PostView of FEBio (Ver. 2.3) [21]. The runtime of each simulation was approximately 40 min using a 64-bit Windows 10 based workstation with an Intel Core i7-6700 3.4 GHz CPU and 24 GB of RAM.

Outcome measures

For each analysis we calculated the volumetric exposures of the heel soft tissues to von Mises (effective) (σ_{vM}) and shear (σ_s) Cauchy stresses, with the von Mises stress and shear stress defined by Eqs. (3.3) and (3.4), respectively:

$$\sigma_{vM} = \sqrt{\sigma_{xx}^2 + \sigma_{yy}^2 + \sigma_{zz}^2 - \sigma_{xx}\sigma_{yy} - \sigma_{yy}\sigma_{zz} - \sigma_{zz}\sigma_{xx} + 3\left(\sigma_{xy}^2 + \sigma_{yz}^2 + \sigma_{zx}^2\right)} \quad (3.3)$$

$$\sigma_s = \max\left(\frac{|\sigma_1 - \sigma_2|}{2}, \frac{|\sigma_2 - \sigma_3|}{2}, \frac{|\sigma_3 - \sigma_1|}{2}\right) \quad (3.4)$$

where σ_{xx}, σ_{yy}, σ_{zz} are the normal stresses in the x, y, or z directions, respectively, σ_{xy}, σ_{yz}, σ_{zx} are the shear stresses, and $\sigma_1, \sigma_2, \sigma_3$ are the principal values of the Cauchy stress tensor σ.

We evaluated the biomechanical efficacy of the hole-support configurations as the reduction in the "total stress concentration exposure" (TSCE) value, defined as the area bounded between the soft tissue von Mises and shear stress volumetric exposure curve and the horizontal (stress) axis. We calculated the TSCE for the highest half (i.e., above the median value) of the calculated stress range, from two volumes of interest (VOIs): high-risk VOI and peripheral VOI. The high-risk VOI represented the soft tissues at a high risk of ulceration and defined as a cylindrical area with skin projection area of approximately $2cm^2$ [22], located at the soft tissue layer on the sole of the heel under the sharpest point at the surface of the calcaneus bone (where the highest interface pressures are usually measured on a flat support) [23,24]. The peripheral VOI was defined as the lower part of the heel soft tissues surrounding the high-risk area (Fig. 3.1B).

Results

Plantar contact pressure distributions on the soft tissues with flat support and "step-shaped" off-loading support are presented from an inferior view in Fig. 3.2A. Distributions of von Mises stresses are shown in cross-sectional views in Fig. 3.2B. Skin contact pressures were in the range of 0–500 kPa and the von Mises soft tissue stresses were in the range of 0–200 kPa. Contact pressure values agreed well with those obtained from previous computational and in vivo experimental studies that measured interface foot pressures during gait [24,25], which thereby validates the present modeling. Comparison of the volumetric tissue exposures to von Mises and shear stress and the TSCE criterion in both high-risk VOI and peripheral VOI (Figs. 3.3 and 3.4) demonstrates that with "step-shaped" configuration, which is often used by clinicians, the pressure on the high-risk VOI relieves; however, the TSCE in the soft tissue surrounding the high-risk VOI increases. Evidently, this off-loading design causes the mechanical loads in tissues to transfer through the hole edges in a relatively narrow "ring," causing stress concentrations in this area, hence increasing the risk of developing secondary ulcers in surrounding tissues.

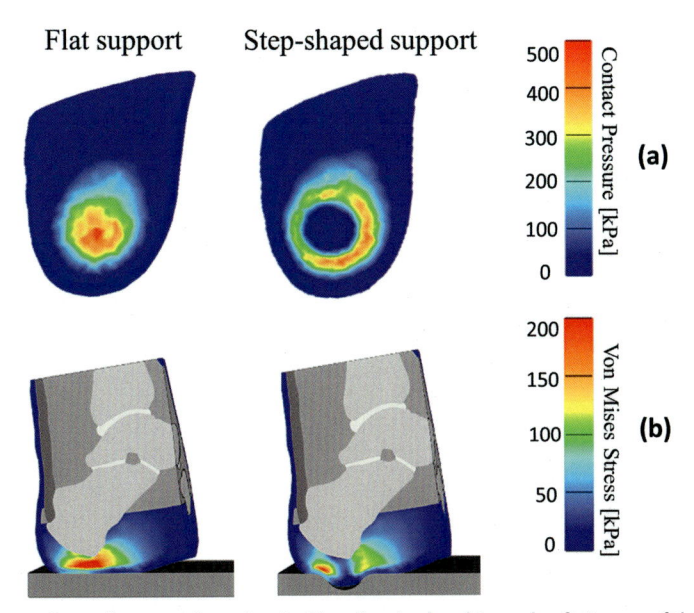

Figure 3.2 Comparison of states of mechanical loading in the skin and soft tissues of the heel with flat support (right column) and "step-shaped" support (left column). (A) Plantar contact pressure distributions in a bottom view. (B) Von Mises stress distributions in sagittal cross-sectional view.

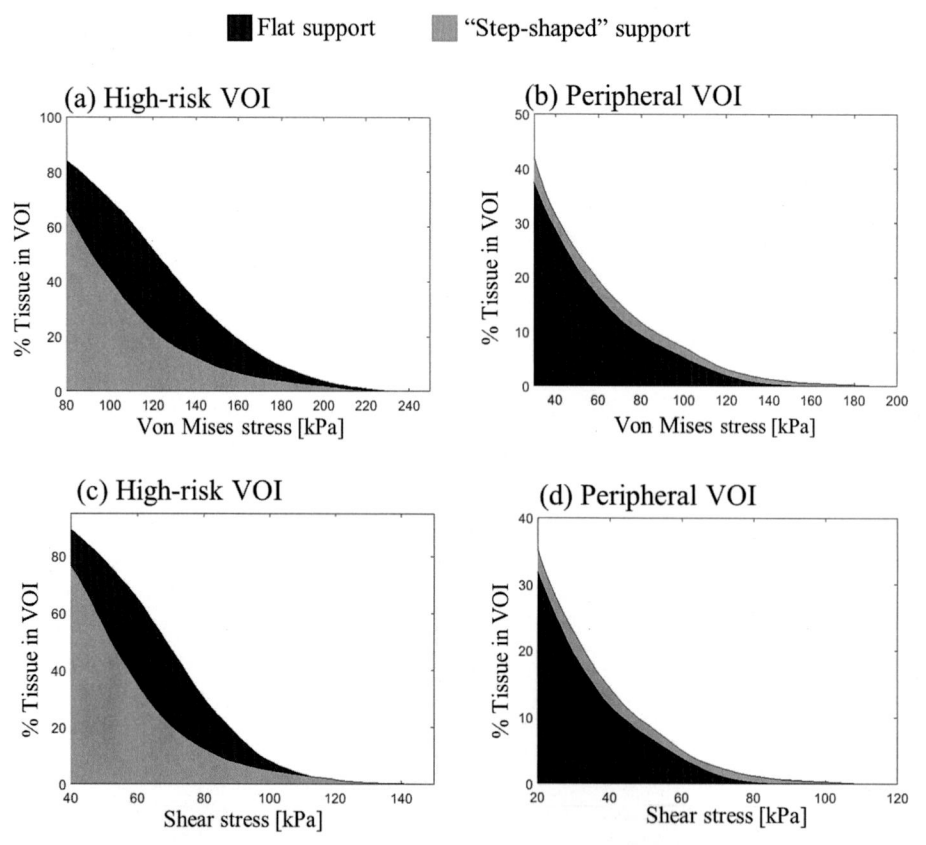

Figure 3.3 Cumulative percentage of soft tissue exposure to von Mises stress (A,B) and shear stress (C,D) with high-risk VOI (left column) and peripheral VOI (right column) when the heel was loaded with a flat support and a "step-shaped" off-loading hole support.

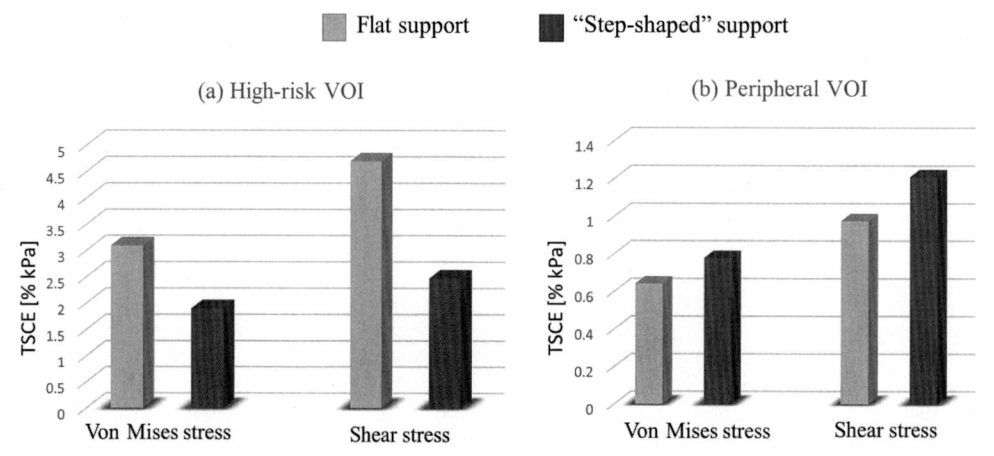

Figure 3.4 Comparisons of total tissue stress concentration exposures (TSCEs) derived from the histogram curves in Fig. 3.3, for (A) high-risk VOI and (B) peripheral VOI, and for von Mises and shear stress. The TSCE for a given VOI is defined as the area bounded between the corresponding stress curve in Fig. 3.3 and the horizontal (stress) axis for the highest half of the calculated stress range (to focus on exposures to elevated or focal tissue stresses).

Discussion

In the present work, we report a method to compare the biomechanical efficacy of off-loading supports on redistributing soft tissue stresses in the plantar foot using the TSCE criterion to identify heel stress concentration areas. This method enables a standardized, methodological, and quantitative comparison of many potential foot-support configurations considering the volumetric exposure of the soft tissues to elevated stresses at multiple desired areas. Comparing the two specific volume indices of high-risk VOI and peripheral VOI of the heel soft tissues shows that these tissue volumes serve as good indicators for evaluating the loading state in plantar heel tissues and subsequently, preventing foot ulcers and treating existing ulcers. Our novel practical scientific analysis method has the potential for streamlining the examination of biomechanical efficiency of novel off-loading solutions to optimally prevent and treat diabetic ulcers at any foot location. Finally, our generic FE method and modeling framework can be customized to multiple patient-specific parameters obtained using noninvasive measurements such as MRI, ultrasound, and GRFs. Therefore, our present method enables patient-specific examination of biomechanical efficiency of off-loading solutions to prevent and treat diabetic foot ulcers.

References

[1] Boulton AJM, Vileikyte L, Ragnarson-Tennvall G, Apelqvist J. The global burden of diabetic foot disease. Lancet 2005;366:1719−24.
[2] Burgess EM, Pecoraro RE, Reiber GE. Pathways to diabetic limb amputation. Basis for prevention. Diabetes Care 1990;13(5):513−21.
[3] Balducci G, Sacchetti M, Haxhi J, Orlando G, D'Errico V, Fallucca S, et al. Physical exercise as therapy for type 2 diabetes mellitus. Diabetes Metab Res Rev 2014;30(1):13−23.
[4] Cevera JJ, Bolton LL, Kerstein MD. Options for diabetic patients with chronic heel ulcer. J Diabet Complicat 1997;11:358−66.
[5] Veves A, Murray HJ, Young MJ, Boulton AJM. The risk of foot ulceration in diabetic patients with high foot pressure: a prospective study. Diabetologia 1992;35(7):660−3.
[6] Pham H, Armstrong DG, Harvey C, Harkless LB, Giurini JM, Veves A. Screening techniques to identify people at high risk for diabetic foot ulceration: a prospective multicenter trial. Diabetes Care 2000;23(5):606−11.
[7] Shaulian H, Solomonow-Avnon D, Herman A, Rozen N, Haim A, Wolf A. The effect of center of pressure alteration on the ground reaction force during gait: a statistical model. Gait Posture July 2018;66:107−13.
[8] Cavanagh PR, Bus SA. Off-loading the diabetic foot for ulcer prevention and healing. J Vasc Surg 2010;52(3 Suppl.):37S−43S.
[9] Singh N, Armstrong D, Lipsky B. Preventing foot ulcers in patients with diabetes. J Am Med Assoc 2005;293(2):217−28.
[10] Wu SC, Jensen JL, Weber AK, Robinson DE, Armstrong DG. Use of pressure offloading devices in diabetic foot ulcers do we practice what we preach? Diabetes Care 2008;31(11):2118−9.
[11] Athanasiou KA, Armstrong DG. The edge effect: how and why wounds grow in size and depth. Clin Podiatr Med Surg 1998;15(1):105−8.
[12] Levy A, Frank MBO, Gefen A. The biomechanical efficacy of dressings in preventing heel ulcers. J Tissue Viability 2015;24(1):1−11.
[13] Chen WP, Tang FT, Ju CW. Stress distribution of the foot during midstance to push off in barefoot gait: a 3D finite element analysis. Clin Biomech 2001;16:614−20.

[14] Simpleware Ltd. ScanIP, +FE, +NURBS and +CAD refer- ence guide ver. 5.1. 2012. http://www.simpleware.com/software.
[15] FEBio: finite element for biomechanics, theory manual ver. 2.5. 2016. http://mrl.sci.utah.edu/software/febio.
[16] Zhang M, Mak AFT, Mak AFT. In vivo friction properties of human skin. Prosthet Orthot Int 1999;23:135—41.
[17] Gefen A, Megido-Ravid M, Itzchak Y, Arcan M. Biomechanical analysis of the three-dimensional foot structure during gait: a basic tool for clinical applications. J Biomech Eng December 2000;122(6):630.
[18] Kuo P-L, Li P-C, Li M-L. Elastic properties of tendon measured by two different approaches. Ultrasound Med Biol September 2001;27(9):1275—84.
[19] Bonet J, Wood RD. Nonlinear continuum mechanics for finite element analysis. Cambridge Univ. Press; 1997.
[20] Even-Tzur N, Weisz E, Hirsch-Falk Y, Gefen A. Role of EVA viscoelastic properties in the protective performance of a sport shoe: computational studies. Bio Med Mater Eng 2006;16(5):289—99.
[21] Maas SA, Ellis BJ, Ateshian GA, Weiss JA. FEBio: finite elements for biomechanics. J Biomech Eng 2012;134(1):011005.
[22] Oyibo So, et al. The effects of ulcer size and site, patient's age, sex and type and duration of diabetes on the outcome of diabetic foot ulcers. Diabet Med 2001;18. 1336—138.
[23] Zammit GV, Menz HB, Munteanu SE. Reliability of the TekScan MatScan®system for the measurement of plantar forces and pressures during barefoot level walking in healthy adults. J Foot Ankle Res 2010;3(1):1—9.
[24] Akrami M, Qian Z, Zou Z, Howard D, Nester CJ, Ren L. Subject-specific finite element modelling of the human foot complex during walking: sensitivity analysis of material properties, boundary and loading conditions. Biomech Model Mechanobiol 2018;17(2):559—76.
[25] Soames RW. Foot pressure patterns during gait. J Biomed Eng 1985;7(2):120—6.

CHAPTER 4

Modeling effects of sustained bodyweight forces on adipose tissue microstructures and adipocytes in diabesity

Maayan Lustig, Golan Amrani[a], Adi Lustig[a], Liran Azaria[a], Raz Margi[a], Yoni Koren[a], Avraham Kolel[a], Nurit Bar-Shai[a], Avior Exsol[a], Maya Atias[a], Amit Gefen

Department of Biomedical Engineering, Faculty of Engineering, Tel Aviv University, Tel Aviv, Israel

Introduction

Obesity is an epidemic [1], associated with several serious chronic diseases, including hyperlipidemia, hypertension, and coronary atherosclerotic heart disease. In particular, obesity can be associated with type 2 diabetes, which has been referred to as diabesity, characterized by critical increased fibrosis in adipose tissue [2]. Fibrosis is the result of excessive deposition of extracellular matrix (ECM) proteins, namely, the deposition of collagen fibers. In obesity, the development of fibrosis increases the rigidity of adipose tissue and could result in obesity-related health problems [3], such as chronic inflammation [4].

The adipose tissue is composed primarily from lipid-filled adipocytes and their progenitor cells (preadipocytes), embedded in an extensive three-dimensional network of collagenous ECM, laced with a network of blood vessels [3,5].

The interlobular septa are a bundle of collagen type I structures which provides structural support in the subtissue level (Fig. 4.1). These interlobular septa, one to several millimeter in length with diameters ranging from 10 nm (single fiber) up to 30 μm (bundle of fibers), envelop groups of adipocytes within the tissue structure, dividing it to mechanical reinforced units called lobules [6,7].

Collagen type IV fibers encapsulate each adipocyte cell's basement membrane surface, creating pericellular collagen—elastin matrix with width of 1—6 μm in diameter, while providing cushioning effect in absorbing most of the mechanical impacts applied on the cells while also acting as thermal insulation [3,5]. However, among the various collagen types, collagens IV and VI are highly expressed in differentiated adipose tissues [8,9], taking on most of the tension during mechanical loading [3]. Along with collagen type V, collagen type VI is a thin, fine, mesh-like network of microfibrils, interwoven in a lacework pattern, ranging between 0.2 and 1.5 μm in diameter [10]. These branching

---

[a] Equal contribution

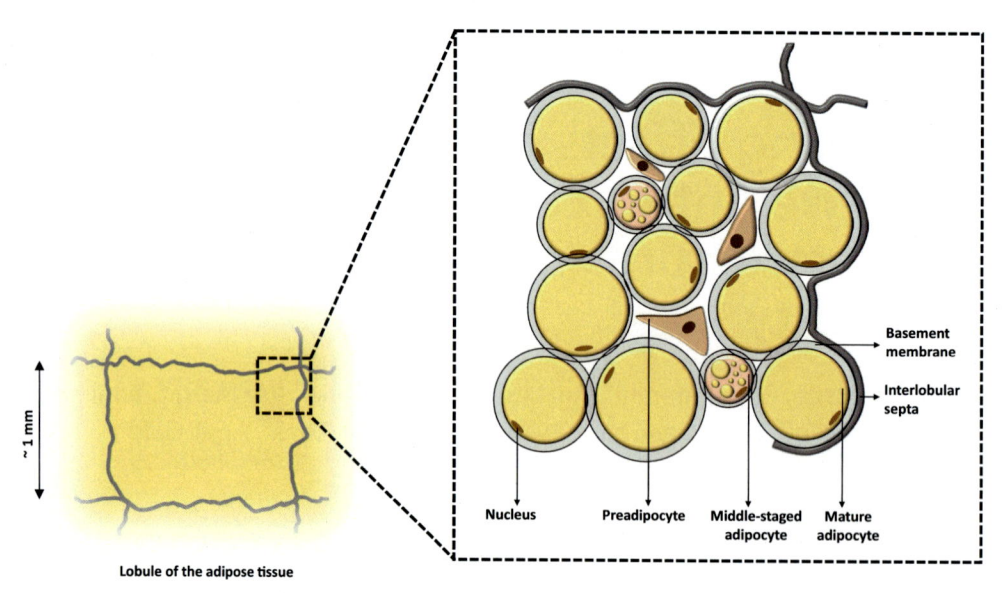

Figure 4.1 Schematic representation of a lobule of the adipose tissue which contains adipocytes in different stages of differentiation, basement membrane which surrounds the cells and interlobular septa, surrounding each lobule.

microfibrils cross-link the thick triple-helical type I collagen fiber bundles, while also interfacing those bundles with the type IV fibers of the basement membrane cells [5,10]. Type VI fibers are often associated with elastin microfibrils, which together act as a continuous reticular fiber network that distributes stress forces uniformly in and across tissues [10].

Previous results published by our group have shown that collagen proteins reorganize throughout the differentiation process. In particular, when preadipocytes begin differentiating into mature adipocytes, collagen types I, III, and V, which are associated with cell adhesion, are recapitulated into types IV (as basement membrane) and VI. This process alters the ECM microstructure and stiffness and in turn causes changes in cell cytoskeleton. These changes in the niche microenvironment are associated with changes in cell morphology and cell fate, which lead to a feedback loop of increasing stiffness of the cells [9]. Therefore, obese disease states may involve elevated amounts of type VI collagen in the adipose tissue, which is suggested to be one of the fibrotic components of tissue that restricts tissue expandability, increases stiffness, and regulates adipogenesis [5,11].

The development of obesity triggers two conditions: hypertrophy, which increases cell size, and hyperplasia, which increases the number of cells [12]. When cell volume of mature adipocyte increases there is a correlated increase in ECM volume [13,14]. It was further proposed that there is a physical limit to adipocyte growth and that this limit is determined by the ability of the cells to maintain its surrounding ECM, which protects the cell against damage [13]. These studies have shown hypertrophy of adipocytes prevent proper oxygen supply to the cell contents leading to hypoxia, apoptosis, and eventually generating an inflammatory response. This inflammation is now thought to be part of obesity-related metabolic dysfunctions like type 2 diabetes [13,15]. Hyperplasia results from recruitment of new adipocytes from preadipocytes, which have proliferated and differentiated [12].

It has been shown previously that a distinct difference between lean healthy and obese unhealthy patients can be observed in the collagen distribution within the patient's white adipose tissue. Obese patients are more prone to accumulation of collagen type I, as well as to high levels of fibrosis, which contributes to this hyper storage process [16]. As an outcome of the above microlevel structures, the interlobular septa thickens, fiber structures become denser, and fiber diameter increases [4]. At the macroscopic level, the effect of enlarged interlobular septa on adipose tissue elasticity is negligible. However, this local change in the cell niche may influence the cell microenvironment and therefore cell phenotype, since the cell senses different mechanical signals [9,17,18].

Therefore, in order to study these pathophysiological processes, we developed mesoscale level models of the adipose tissue that simulates lean and obese states, to better understand the effect that forces applying on the tissue have on the cells in lean and obese states. We used finite element (FE) modeling, which is a powerful bioengineering research method that facilitates evaluation of internal tissue loads and further allows the isolation of biomechanically driven processes and their characteristics.

Methods
Interlobular septa thickening

In order to examine the biomechanical effects of interlobular septa thickening on the mechanical states of adipocyte cells within an adipose tissue, three FE model variants of adipose tissue were developed: (*i*) a lean, healthy adipose tissue, represented by a thin (regular) interlobular septa, (*ii*) an obese, unhealthy adipose tissue, represented by a thick interlobular septa, and (*iii*) a no-interlobular septa condition used as a reference (Fig. 4.2). These mechanical states in the adipose tissue were simulated in pure compression loading of the model, as detailed below.

Geometry

The geometry represents a 200×200 μm^2 section of fat tissue with a thickness of 4 μm, containing a mature adipocyte, an adipocyte during differentiation (middle-staged adipocyte) and a preadipocyte with a fibroblast-like shape, embedded in the ECM (Fig. 4.2). The cells are bounded by a plasma membrane (PM) containing cytoplasm and a nucleus. Mature and middle-staged adipocytes include 1 and 15 intracellular lipid droplets (LDs),

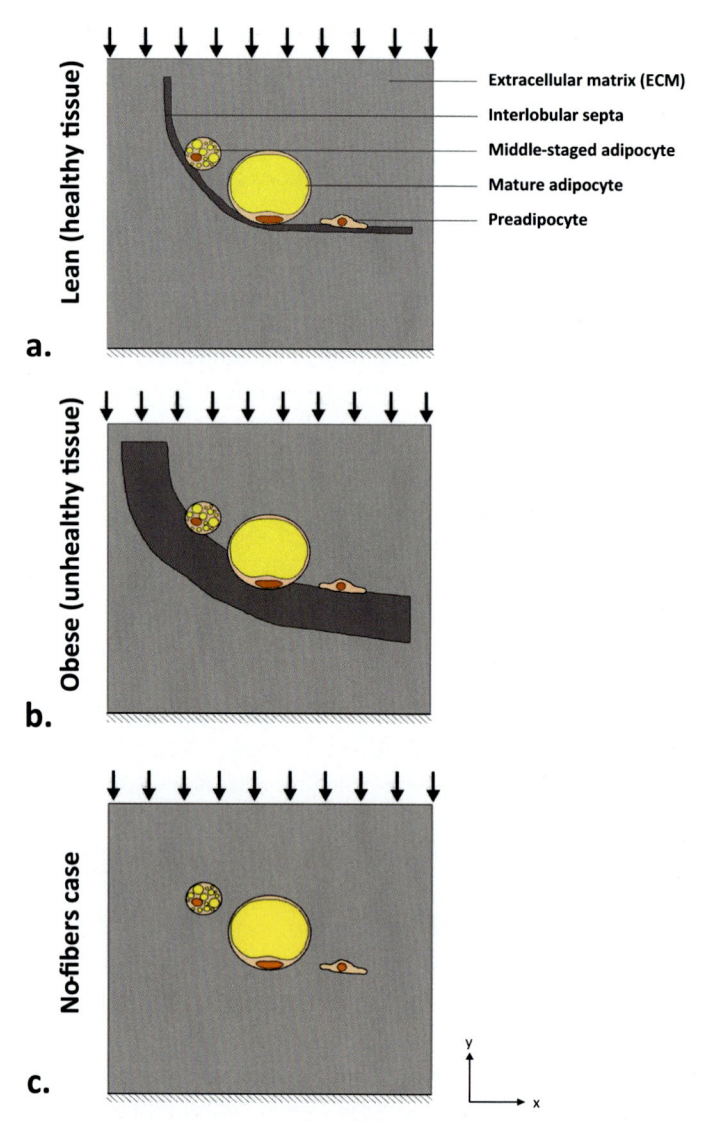

Figure 4.2 Geometry and boundary conditions applied to the computational model simulating the interlobular septa thickening effect. Each variant of the adipose tissue model included a mature adipocyte, a middle-staged adipocyte, and a preadipocyte all surrounded by extracellular matrix (ECM). The three model variants were: (A) lean, healthy, adipose tissue, represented by a thin interlobular septum, (B) obese, unhealthy, adipose tissue, represented by a thick interlobular septum, and (C) a reference case with no interlobular septa. For all model variants, the bottom is fixed for translations and rotations while downward displacement is applied on top.

respectively, with cell diameters of 50 and 20 μm, respectively. Preadipocyte length is 25 μm. The geometry is based on a previous study and measurements were taken from microscopic images of adipocyte cell cultures [19]. The adipose tissues include interlobular septa of 5 and 30 μm thick, to simulate lean (healthy) and obese (unhealthy) states, respectively. All models were created and meshed using Synopsys' Simpleware software package (Synopsys Inc., Mountain View, CA).

Boundary conditions

To examine the effects of interlobular septa thickening on cell deformations and loads, a uniform downward displacement of 20 μm was applied on the upper part of the model, while the bottom of the model was fixed for all translations and rotations, consequently inducing compressive load on the cells. Tied interfaces were defined between all model components. The displacements of 20 μm represents a load of ~10% which is in the physiological range for a weight-bearing adipose tissue [20].

Mechanical properties

Constitutive laws and mechanical properties of all model components included in the adipose tissue, i.e., the cytoplasm, LDs, nucleus, PM, interlobular septa, and ECM, were adopted from the literature and are listed in Table 4.1. All model components were assumed to be nonlinear isotropic materials, with their large deformation behavior described using an uncoupled Neo-Hookean constitutive model with the following strain energy density function W (Eq. 4.1):

$$W = \frac{G_{ins}}{2}\left(\lambda_1^2 + \lambda_2^2 + \lambda_3^2 - 3\right) + \frac{1}{2}K(\ln J)^2 \qquad (4.1)$$

where G_{ins} is the instantaneous shear modulus, λ_i ($i = 1, 2, 3$) are the principal stretch ratios, K is the bulk modulus, and $J = \det(F)$ where F is the deformation gradient tensor.

Table 4.1 Mechanical properties of model components and characteristics of FE mesh.

Model component	Elastic modulus [kPa]	Poisson's ratio	Number of elements
Cytoplasm[a]	1.8	0.45	30,969–31,115
Nucleus[a]	12	0.45	89,671–91,273
LDs[a]	6	0.45	11,396–14,705
PM[a]	6	0.46	33,598–33,787
ECM[a]	0.5	0.38	16,654–18,057
Interlobular septa[b]	1×10^6	0.46	22,654–27,717

[a]Ben-Or Frank et al. [19].
[b]Comley et al. [6].

Numerical method
Four-node linear tetrahedral elements were used in all model components which were meshed using the Scan-IP module of Simpleware (Synopsys Inc., Mountain View, CA). Numbers of elements in each of the model components are specified in Table 4.1. The FE simulations were set up using PreView of FEBio (Ver.2.1.3, University of Utah, Salt Lake City, UT), analyzed using the Pardiso linear solver of FEBio (Ver. 2.8.5), and postprocessed using the PostView module of FEBio (Ver. 2.3.2) [21]. The runtime of each model variant was up to 30 min, using a 64-bit Windows 10—based workstation with an Intel Core i7-8700 3.20 GHz CPU and 32 GB of RAM.

Biomechanical outcome measures
The applied load induced stress gradients throughout the adjacent cells and ECM in all cell entities. In each model variant, the effective stresses (von Mises) were calculated and compared, separately for each healthy and pathological state using stress exposure histogram charts, where the distribution of stress magnitudes in the elements representing the PM and LD organelles is presented.

Mechanical loads in the mesoscale of lean and obese tissues
In order to examine the biomechanical effects of adipose tissue compression due to an external compressive load, in lean (healthy) and obese (unhealthy) states on adipocyte cells, four FE model variants of the adipose tissue were developed: (*i*) early lean, healthy state; (*ii*) advanced lean, healthy state; (*iii*) early obese, unhealthy state; and (*iv*) advanced obese, unhealthy state. These mechanical states in the adipose tissue were simulated in pure compression loading of the model, as detailed below.

Geometry
The geometry represents a 340×265 μm^2 section of the fat tissue, containing ECM framework and round adipocytes (Fig. 4.3). The two healthy (lean) configurations (Fig. 4.3A and B) represent prefibrotic ECM organization, while the two obese configurations (Fig. 4.3C and D) represent fibrotic/disease states. For each state, one configuration represents an early stage differentiation (Fig. 4.3A and C), which includes adipocytes in various differentiation levels (n = 10), while the other configuration represents advanced differentiation (Fig. 4.3B and D), which includes only mature adipocytes (n = 10); specifically, this geometry included five mature adipocytes: one middle-stage and four initial-stage adipocytes. This was specified by the ratio of the number of adipocytes in three differentiation stages (initial, early, and middle stages), as was observed in adipose tissue by Shoham et al. [22]. Cell diameters were set to 20, 30, and 50 μm for initial, middle, and advanced differentiation stage adipocytes, respectively [23].

Boundary conditions
Nodes that constituted the upper edge of the network were subjected to a static and uniform external compression force, so the global strain of the network would be ∼3%,

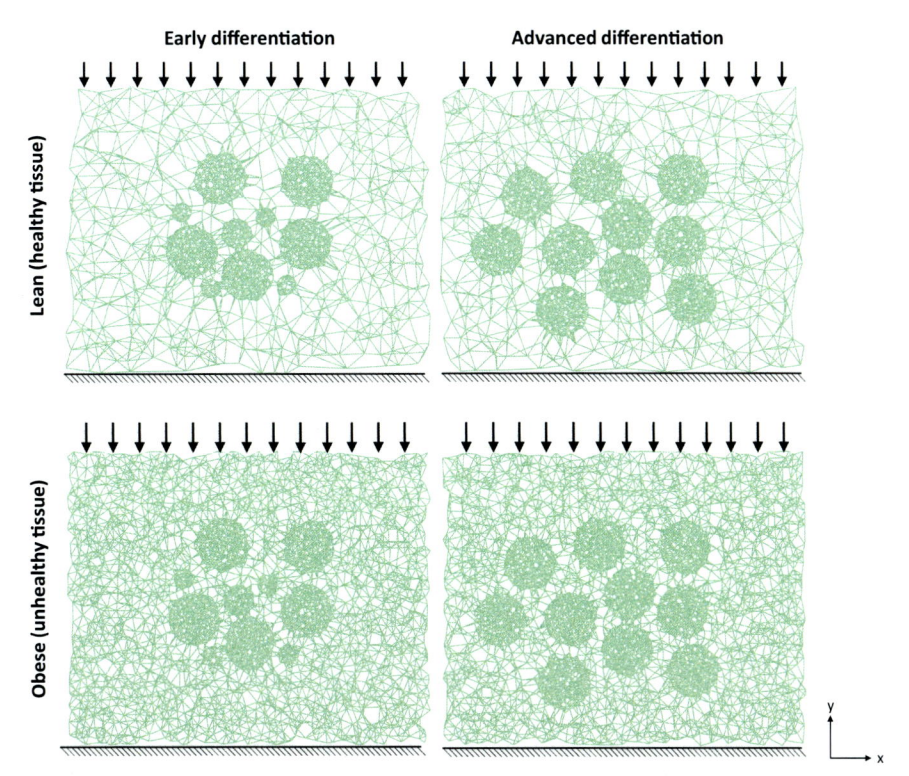

Figure 4.3 Geometry and boundary conditions applied to the computational model simulating the effects of mechanical loads in the mesoscale of lean (healthy) and obese (unhealthy) tissues. Each variant of the adipose tissue model included adipocytes in different differentiation stages surrounded by extracellular matrix (ECM) fibers. The lean variants (A, B) comprise a structure of fibers and cells, while the obese variants (C, D) comprise a denser structure of fibers. The early differentiation variants (A, C) include cells in various differentiation stages (initial, early, and middle stages), while the advanced differentiation variants (B, D) include only mature adipocytes. For all model variants, the bottom is fixed for translations and rotations while downward displacement is applied on top.

while nodes on the bottom edge were fixed for all translations and rotations, consequently inducing compressive load on the fibers and cells. Tied interfaces were defined between all model components (Fig. 4.3).

Mechanical properties

The ECM elements were assumed to behave as hyperelastic materials with Marlow strain energy density functions, exhibiting strain-stiffening under tension. The elastic modulus assigned to each one of the fibers is given by Eq. (4.2):

$$E = \begin{cases} E^*, & \varepsilon < \varepsilon_s \\ E^* e^{\left[\frac{\varepsilon - \varepsilon_s}{\varepsilon_0}\right]}, & \varepsilon > \varepsilon_s \end{cases} \quad (4.2)$$

while the values used in the model are: $E = 115$ kPa as the given elastic modulus, $\varepsilon_s = 2\%$ is the tensile strain above which strain stiffening occurs, and $\varepsilon_o = 4\%$ is the strain-stiffening coefficient. The mechanical behavior presented in Eq. (15.2) is largely based on computational models described in previous studies that convey discrete fibrous networks [24,25]. The values for fiber elastic modulus, fiber-stiffening threshold strain, and strain-stiffening coefficient were based on fitting of bulk stiffness between network simulations and in vitro experiments of collagen 1.2 mg/mL gels [26].

In order to simulate the mechanical properties of the adipocytes, the elements constituting the cells were modeled as an isotropic linear elastic material with a Poisson's ratio of 0.45 and elastic modulus 3.84, 4.92, and 5.76 kPa for initial, middle, and advanced differentiation stage adipocytes, respectively.

The fiber thickness for lean tissue was set to 0.5 μm and for obese tissue as 0.8 μm [27]. Additionally, the fiber densities used in our model were calculated as follows (Eq. 4.3):

$$\text{fiber density} = \frac{\left(\sum_{i}^{n} l_i\right) \cdot d}{A} \quad (4.3)$$

where l_i is the length of the fiber i located within the region of interest (ROI), n is the total number of fibers within the ROI, d is the fiber diameter, and A is the area of the region. Using Eq. (4.3), two different ROIs from each state were tested iteratively to define the fiber densities of each network. The fiber densities for lean and obese states were determined by previous experimental measurements on collagen VI in adipose tissue (5% and 25%, respectively) [11]. The mean fiber length between every two nodes was modeled based on these fiber densities and resulted in approximately 25 μm in the healthy state and 7.5 μm in the obese state.

In order to validate our model, we performed uniaxial stretch simulations of the overall structure (ECM and cells) and obtained elastic moduli of 1.8–1.9 kPa for the two lean states and 4.9–5.0 kPa for the two obese states (Fig. 4.4). We are able to see that the lean elastic moduli are within the range of experimentally measured human omental adipose tissue from a previous study (2.9 ± 1.5 kPa) [3].

Numerical method and construction of an isotropic and homogeneous fibrous network

The overall framework was designed to be isotropic (in fiber orientation) and homogenous (in fiber density) at the scale of a single cell, in order to replicate typical collagen networks, which, as mentioned, are a main component in adipose tissues [9,28]. The nodes were modeled as freely rotating hinges. Each individual fiber was modeled as a two-node linear truss element with a circular cross-section, undergoing uniaxial tension or compression.

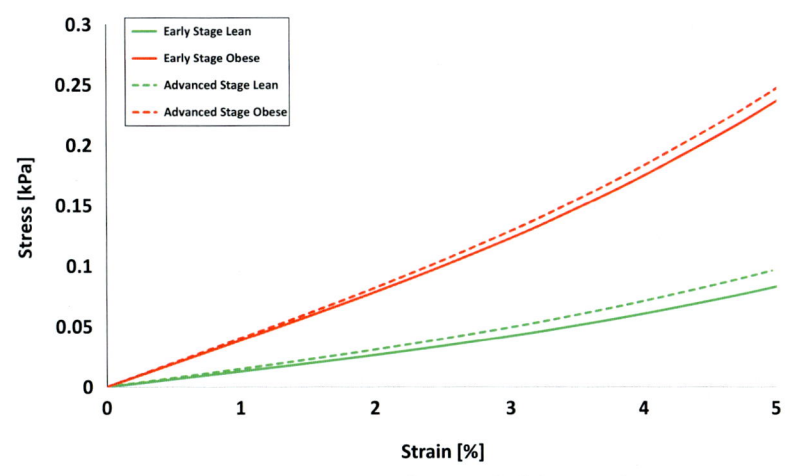

Figure 4.4 Model validation using stress–strain relations of all four configurations. *Green lines* indicate lean state and *red lines* indicate obese state. *Solid lines* represent early-stage simulations while dotted lines represent advanced-stage simulations. The lean elastic moduli (resulting from slope calculation) are within the range of experimentally measured human omental adipose tissue.

In order to construct this framework, nodes are inserted randomly in two dimensions throughout a circular domain of radius R_b by following a uniform distribution. Pairs of nodes are then connected by single fibers generated following a minimum cost algorithm as follows. The probability that a potential fiber connecting two neighboring nodes would be generated is determined by a cost function, P. In every iteration, each node looks at the 30 nearest neighbors in its vicinity as potential fibers to be generated. The cost function associated with each potential fiber is given by Eq. (4.4):

$$P = N_{ij} + aC_{ij} + b(A_i + A_j) \qquad (4.4)$$

where two constants a and b determine the relative importance of each term. A potential fiber can be generated only if its cost function, P, is negative. The three terms in Eq. (4.4) are explained as follows.

(1) N_{ij} is the degree of nearness of each node i to the other node j. For example, if node i is the second nearest neighbor of node j, and node j is the third nearest neighbor of node i, then $N_{ij} = 2 + 3 = 5$.

(2) $C_{ij} = c_i + c_j - 2c_{opt}$ takes into account of the current connectivity numbers c_i and c_j of nodes i and j, respectively, with c_{opt} being a chosen optimal connectivity (connectivity number is the number of fibers connected to a single node). We can set the average connectivity of the network by tuning c_{opt}. The higher c_{opt} is, the smaller C_{ij} becomes and so the probability for creating the fiber gets higher.

(3) The last term relates to the maximum free angle of each node, representing the local geometry around the nodes. Here A_i is the difference in maximum free angle after and before adding the new potential fiber to node i. A negative A_i increases the probability that a fiber would be generated.

Table 4.2 Number of fibers constituting the cells and ECM networks in the four configurations.

	Early stage lean	Early stage obese	Advanced stage lean	Advanced stage obese
Cell1	3,683	3,818	3,686	3,778
Cell2	3,681	3,778	3,667	3,791
Cell3	3,695	3,794	3,678	3,777
Cell4	3,699	3,788	3,699	3,791
Cell5	3,682	3,787	3,696	3,772
Cell6	586	596	3,687	3,791
Cell7	574	589	3,701	3,812
Cell8	583	594	3,681	3,766
Cell9	583	600	3,677	3,773
Cell10	1,321	1,344	3,694	3,747
Total number—cells	22,087	22,688	36,866	37,798
ECM	1,688	6,615	1,761	6,139

In summary, in every iteration, each node is connected by a new fiber to the node with the minimal negative cost (in comparison to the other 29 nearest neighbors). The iteration process finishes when no new fiber can be added. Thus, the generated networks are therefore always isotropic. Nodes tend to be connected by fibers that are oriented in a wide range of directions, in comparison to other methods where many fibers are aligned, mostly in some particular direction. This increases the isotropy and homogeneity of the networks.

The cells were meshed by high-density linear elastic fibers (highly isotropic and homogeneous), as an approximation for the adipocyte's cytoskeleton. The other organelles of the adipocytes were not considered since the cells' functionality in this work was to resist compression and to sustain mechanical loads, thus having an effective stiffness, which is determined by this mesh. The number of ECM fibers and cell fibers in each model configuration is specified in Table 4.2.

Matlab R2019a was used to create the network geometry and architecture. The FE software Abaqus/CAE 2017 (*Dassault Systèmes Simulia*) was used to model the mechanical properties. The uniaxial stretching and compression simulations of the ECM, cells, and the whole structures were performed with the software's implicit static solver.

Biomechanical outcome measures

The applied load induced stress gradients throughout the adjacent cells and ECM in all cell entities. In each model variant, the effective stresses (von Mises) were calculated and compared, separately for each healthy and pathological state.

Results
Interlobular septa thickening

Effective stress distribution in each individual cell following compression of the tissue reveals that higher stress values were more abundant in the obese state (thick septa) for all cell types (average stresses of 71 and 77 Pa for the PM and LDs, respectively), in comparison to the same cells simulated in the lean state (thin septa, average stresses of 61 and 75 Pa for the PM and LDs, respectively). In addition, stress distribution was less uniform in the obese state, yielding stress concentrations in proximity to the thicker septum (Fig. 4.5).

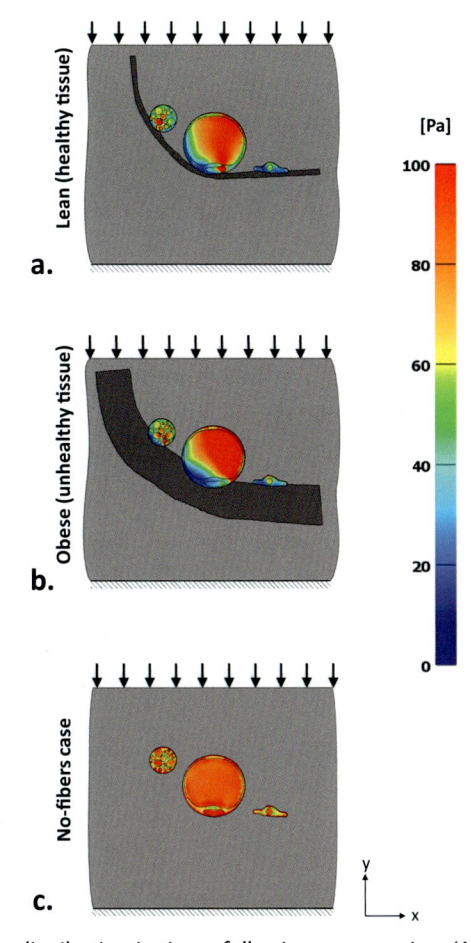

Figure 4.5 Effective stress distribution in tissue following compression. (A) Lean, healthy adipose tissue (thin septa); (B). Obese, unhealthy adipose tissue (thick septa); (C). a reference case (no septa). The stress distributions indicate that the highest stresses within the cells were observed in the no-septa case, followed by the unhealthy (thick-septa) case, and the lowest stresses were observed in the healthy (thin-septa) case.

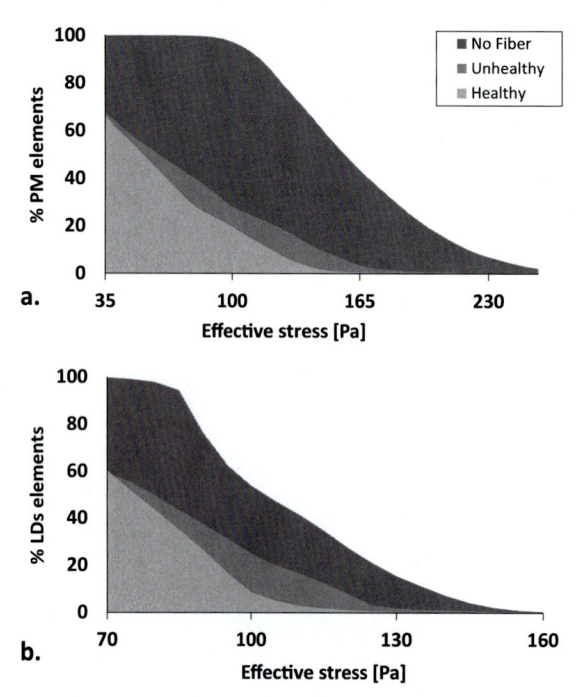

Figure 4.6 Stress exposure histogram charts of stress distribution magnitudes in the plasma membrane (PM) and lipid droplets (LDs) demonstrating that for both, more elements of the model were exposed to higher stresses in the reference case, with respect to the obese (unhealthy) and lean (healthy) cases. In other words, the presence of the septa reduces effective stress levels, but a thicker septum leads to elevated stresses when compared to the healthy case.

The no-septa model variant, which was used as a reference case, yielded substantially higher stress levels in the cells with respect to the lean and obese variants (average of 163 and 107 Pa for the PM and LDs, respectively). Stress distributions demonstrate that presence of interlobular septa helps in shielding stress in the cells, compared to the no-septa case. However, thickening of the septa induces higher stress values. Correspondingly, histograms demonstrating the effective stress distributions specifically in the LDs and PM reveal that for both, more elements of the model were exposed to higher stresses in the reference case, with respect to the obese and lean cases (Fig. 4.6), meaning that the presence of the septa reduces effective stress levels, but a thicker septa will cause elevation of stresses.

Mechanical loads in the mesoscale of lean and obese tissues

In all four configurations, the ECM fibers sustain higher stresses in comparison to the cells (Fig. 4.7). The ECM average stresses were 1249 Pa for the lean states and 1247 Pa for the obese states, while the cell average stresses were 262 Pa for the lean states and 376 Pa for

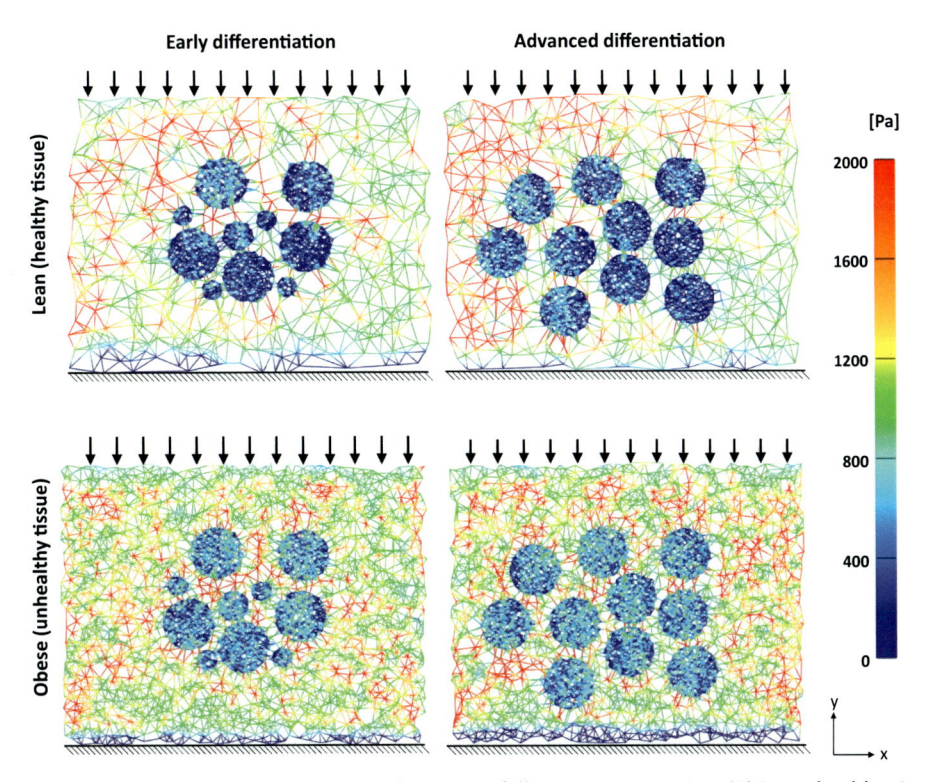

Figure 4.7 Effective stress distributions in the tissues following compression. (A) Lean, healthy tissue, early differentiation; (B) lean, healthy tissue, advanced differentiation; (C) obese, unhealthy tissue, early differentiation; (D) obese, unhealthy tissue, advanced differentiation. In all four configurations, the ECM fibers sustain higher stresses in comparison to the cells and the difference between the average cell stresses of lean (healthy) and obese (unhealthy) states (for both early and advanced stages) was about 6.5 times higher than the difference between advanced and early stages (for both obese and lean states).

the obese states. The stresses carried by the ECM were about four times higher than the stresses carried by the cells in both lean and obese tissues. This behavior demonstrates that since the ECM fibers are stiffer than the cells, they therefore carry most of the loads in the tissue, a potential indication that the ECM functions as a mechanical buffer to the cells.

The comparison between the advanced- (Fig. 4.7B and D) and early-stage models (Fig. 4.7A and C), versus between obese (Fig. 4.7C and D) and lean tissue (Fig. 4.7A and B), shows that modeling various differentiation stages of cells has little effect on the results of stress comparisons between obese and lean states. The average effective stresses for the ECM fibers were 1249 and 1247 Pa, respectively. This confirms the ECM senses relatively small changes in mechanical stresses between lean and obese states; however, we do observe a shift in effective stresses sensed by the individual ECM fibers between lean early (1198 Pa) and advanced (1300 Pa) differentiation stages. The average effective stresses for lean state cells in early stage and advanced stage were 248 and 278 Pa, respectively. The average effective stresses for obese state cells in early stage and advanced stage were 368 and 383 Pa, respectively.

The difference between the average cell stresses of lean and obese states (for both early and advanced stages) was about 6.5 times higher than the difference between advanced and early stages (for both obese and lean states). By comparing effective stresses experienced by the cells in the four configurations, we can see that the mean stresses of obese state cells (373 Pa) were higher than those of the lean state cells (256 Pa).

Discussion

Interlobular septa thickening

In this study, we used FE modeling to simulate a schematic adipose tissue containing three cells in different differentiation stages in proximity to interlobular septum in order to analyze reaction of the cells to external loads and to evaluate if there is a protective effect of the septa on the cells, both in healthy (lean) and pathological conditions (Figs. 4.2 and 4.5). Mechanical interactions in subtissue environments are key factors to intracellular processes [29]. In this context, pathological changes may disrupt the mechanical balance and affect other processes downstream. The results presented here demonstrate a clear protective effect of the interlobular septum on adjacent cells in healthy conditions [6].

In the obese condition, simulated with a thick septum, we can see an elevation in stress values in the cells. Instead of supporting the cells externally, the thick septum bounds the cells in a rigid envelope, lowering stress rates in those areas to almost zero, while creating weak points in the interface regions between free and bound areas. These weak points are subjected to high values of stress and high deforming rates due to the nonuniform load distribution (Fig. 4.5).

Our results indicate that excessive interlobular septa contents within the adipose tissue may lead to greater distortion of PMs in adipocytes when the tissue is under compression (Figs. 4.5 and 4.6). Such irregular, chronic deformation of the PM may activate PM receptors in adipocytes that influence signaling pathways under static loading, e.g., MEK/MAPK [23,30—33]. Furthermore, it was previously shown that increases in external static stresses applied on differentiating adipocytes accelerate the adipogenesis process [9,20,34—38]. This acceleration is triggered due to the positive feedback loop contributing to the onset and development of sarcopenic obesity: the vicious cycle of adipocytes [23,39]. Shoham et al. showed that distortion of the PM in adipocytes, resulting from chronic loading of the tissue, is very likely to stimulate an increase in LD content (due to progressing adipogenesis) [23,37]. These changes are expected to even further encourage adipogenesis in a positive feedback loop, since the LDs themselves contribute to creating a stiffer environment, which increases the stress levels sensed by the cells, hence promoting more differentiation [19,22]. This is consistent with the results presented here, where in the case of obesity (i.e., interlobular septa thickening) adipocytes were exposed to higher stresses. Therefore, it may trigger and promote additional adipogenesis.

Mechanical loads in the mesoscale of lean and obese tissues

We used FE modeling to study two different states of adipose tissue (lean and obese) under compression, while focusing on the ECM structure and the density of the fibers in different stages of differentiation. Compression loads were applied, and mechanical stresses were observed and analyzed. The results presented here revealed that total average stresses for overall tissue structures were found to be greater in the obese models (538 Pa) when compared to lean models (323 Pa) for both advanced and early stage. In addition, when examining the cells individually, they sustained significantly lower average stresses in lean tissue 262 ± 37 Pa than in obese tissue 375 ± 23 Pa. These mechanical stresses were found to be greater in the ECM than in the cells; thus we conclude that the ECM sustains most of the mechanical loading. The comparison of the effects to cells and the ECM allows us to conclude that the ECM mechanically buffers the cells from external compression. However, when the fibers of the ECM of obese state tissue become even denser and stiffer through fibrosis processes, as shown in previous studies [9], the ECM mechanical buffering becomes less effective. Moreover, in lean tissue, prior to fibrosis responses, the transition from early to advanced differentiation stages shows a shift in stresses (1198 and 1300 Pa, respectively). This potentially represents the beginning stages of the disease cycle described above and in previous studies [23,37,39].

During adipogenesis, fibril ECM collagens I, III, and V reorganize into ECM basement collagen IV and collagen VI. As these structural alterations take place, cell shape changes gradually, from elongated fibroblasts to spherical adipocytes [9,34–36,40]. The cells accumulate more LDs, which over time also enlarge and merge together, eventually forming one large LD (lipogenesis). This leads to changes in the cell mechanosignaling, affecting the rigidity of the ECM environment through fibrosis, thus altering the forces on the cells. This process has been shown to lead to modifications in content, organization, stiffness, and protein secretion, which also influences cell mechanotransduction, metabolism, fate, and function in an ongoing vicious cycle [9,23,39].

Physiologically, the ECM stiffening leads to even further adipogenesis (as seen in the comparison between lean differentiation stages), lipid accumulation (cell size increase), and eventually more damage to the overall tissue structure, as shown in previous studies [38]. This continues the inflammatory response, leading to chronic tissue dysfunction (obese state). This cell—ECM interdependent interaction could further explain the inter/extracellular stress distributions as well as tissue mechanotransduction in metabolic disorders.

Obesity and diabetes

The present study examines the effect of mechanical loads on adipose tissue and adipocytes in obesity using FE modeling where the state of obesity was simulated as subtissues with stiffer and denser ECM components. Still, it should also be noted that obesity is strongly correlated with diabetes [41–44]. It was previously shown that obese adults demonstrate reduced glucose disposal as well as impairment in insulin action on oxidation of fatty acids, leading to insulin resistance and abnormal lipolysis [45]. The relationship between obesity, diabetes, and mechanical loads was previously demonstrated in a living cell-scale model system [34,35], in which adipogenesis was monitored in 3T3-L1 cells that were exposed to high (450 mg/dL, hyperglycemia) and low (100 mg/dL, physiological) glucose concentrations; the cells were simultaneously exposed to static, chronic substrate tensile deformations. It was revealed that high glucose concentrations and substrate tensile strains delivered to adipocytes accelerated lipid production. This again indicates the presence of the vicious cycle process in adipose tissue also in the context of diabesity and mechanical loads on the tissue.

FE modeling limitations

We used the FE methodology to simulate different conditions of fat tissue; it should be noted that FE modeling inevitably involves assumptions and limitations. First, the simulated adipose tissues are simplified models which represent a small segment of tissue with 3–10 cells with ideal structure of round cells and ECM. The simulations involving interlobular septa thickening did not take into account the fibrous structure of the ECM, while the simulations of mechanical loads in the mesoscale of lean and obese tissue did not consider the inner structure of the cells nor the various biologically relevant geometries of real cells. Additional simplifications used in our models pertain to the physical locations of cells within the ECM and their concentration in the center of the bulk. Despite these limitations, FE modeling allows for an approximation of complex and irregular geometrical shapes in order to conduct comparisons of different physiological states while isolating various conditions for analysis.

Conclusion

In conclusion, the FE modeling results obtained in this study revealed that adipose tissue's mechanical state is greatly affected by its pathophysiological condition. We revealed that the ECM and the interlobular septa have a protective effect on the cells in healthy state. However, an unhealthy, fibrous tissue provides a stiffer microenvironment for the cells which, in turn, may affect the cell fate and phenotype. Specifically, in fat tissue, a stiffer microenvironment enhances obesity through the vicious cycle in which static cell deformations on the cells promote more adipogenesis and fat accumulation which may alter the functionality of the adipose tissue, development of inflammation, impaired metabolism, and altered responsiveness to signals. This implies the importance of prevention of obesity which has severe consequences on health and quality of life.

Acknowledgments

This research work was supported by the Israel Science Foundation (Grant no. 1266/16). The authors would like to thank Dr. Ayelet Lesman (School of Mechanical Engineering, Faculty of Engineering, Tel Aviv University) for providing the computational resources for the modeling of mechanical loads in the mesoscale of tissues.

References

[1] Caballero B. The global epidemic of obesity: an overview. Epidemiol Rev 2007;29(1):1—5.
[2] Divoux A, Tordjman J, Lacasa D, Veyrie N, Hugol D, Aissat A, Basdevant A, Guerre-Millo M, Poitou C, Zucker JD, Bedossa P, Clément K. Fibrosis in human adipose tissue: composition, distribution, and link with lipid metabolism and fat mass loss. Diabetes 2010;59(11):2817—25.
[3] Alkhouli N, Mansfield J, Green E, Bel J, Knight B, Liversedge N, Tham JC, Welbourn R, Shore AC, Kos K, Winlove CP. The mechanical properties of human adipose tissues and their relationships to the structure and composition of the extracellular matrix. Am J Physiol Endocrinol Metab 2013;305(12):E1427—35.
[4] Sun K, Tordjman J, Clément K, Scherer PE. Fibrosis and adipose tissue dysfunction. Cell Metabol 2013;18(4):470—7.
[5] Chun TH. Peri-adipocyte ECM remodeling in obesity and adipose tissue fibrosis. Adipocyte 2012;1(2):89—95.
[6] Comley K, Fleck NA. A micromechanical model for the Young's modulus of adipose tissue. Int J Solid Struct 2010;47(21):2982—90.
[7] Abrahamson DR. Recent studies on the structure and pathology of basement membranes. J Pathol 1986;149(4):257—78.
[8] Pasarica M, Gowronska-Kozak B, Burk D, Remedios I, Hymel D, Gimble J, Ravussin E, Bray GA, Smith SR. Adipose tissue collagen VI in obesity. J Clin Endocrinol Metab 2009;94(12):5155—62.
[9] Mor-Yossef Moldovan L, Lustig M, Naftaly A, Mardamshina M, Geiger T, Gefen A, Benayahu D. Cell shape alteration during adipogenesis is associated with coordinated matrix cues. J Cell Physiol 2019;234(4):3850—63.
[10] Ushiki T. Collagen fibers, reticular fibers and elastic fibers. A comprehensive understanding from a morphological viewpoint. Arch Histol Cytol 2002;65(2):109—26.
[11] Spencer M, Yao-Borengasser A, Unal R, Rasouli N, Gurley CM, Zhu B, Peterson CA, Kern PA. Adipose tissue macrophages in insulin-resistant subjects are associated with collagen VI and fibrosis and demonstrate alternative activation. Am J Physiol Endocrinol Metab 2010;299(6):E1016—27.
[12] Hausman DB, DiGirolamo M, Bartness TJ, Hausman GJ, Martin RJ. The biology of white adipocyte proliferation. Obes Rev 2001;2(4):239—54.
[13] Mariman EC, Wang P. Adipocyte extracellular matrix composition, dynamics and role in obesity. Cell Mol Life Sci 2010;67(8):1277—92.
[14] Halberg N, Wernstedt-Asterholm I, Scherer PE. The adipocyte as an endocrine cell. Endocrinol Metab Clin N Am 2008;37(3):753—68.
[15] Gozal D, Gileles-Hillel A, Cortese R, Li Y, Almendros I, Qiao Z, Khalyfa AA, Andrade J, Khalyfa A. Visceral white adipose tissue after chronic intermittent and sustained hypoxia in mice. Am J Respir Cell Mol Biol 2017;56(4):477—87.
[16] Katzengold R, Shoham N, Benayahu D, Gefen A. Simulating single cell experiments in mechanical testing of adipocytes. Biomech Model Mechanobiol 2015;14(3):537—47.
[17] Dupont S, Morsut L, Aragona M, Enzo E, Giulitti S, Cordenonsi M, Zanconato F, Le Digabel J, Forcato M, Bicciato S, Elvassore N, Piccolo S. Role of YAP/TAZ in mechanotransduction. Nature 2011;474(7350):179—83.
[18] Shih YR, Tseng KF, Lai HY, Lin CH, Lee OK. Matrix stiffness regulation of integrin-mediated mechanotransduction during osteogenic differentiation of human mesenchymal stem cells. J Bone Miner Res 2011;26(4):730—8.

[19] Ben-Or Frank M, Shoham N, Benayahu D, Gefen A. Effects of accumulation of lipid droplets on load transfer between and within adipocytes. Biomech Model Mechanobiol 2015;14(1):15−28.
[20] Levy A, Enzer S, Shoham N, Zaretsky U, Gefen A. Large, but not small sustained tensile strains stimulate adipogenesis in culture. Ann Biomed Eng 2012;40(5):1052−60.
[21] Maas SA, Ellis BJ, Ateshian GA, Weiss JA. FEBio: finite elements for biomechanics. J Biomech Eng 2012;134(1):011005.
[22] Shoham N, Levy A, Shabshin N, Benayahu D, Gefen A. A multiscale modeling framework for studying the mechanobiology of sarcopenic obesity. Biomech Model Mechanobiol 2017;16(1):275−95.
[23] Shoham N, Mor-Yossef Moldovan L, Benayahu D, Gefen A. Multiscale modeling of tissue-engineered fat: is there a deformation-driven positive feedback loop in adipogenesis? Tissue Eng 2015;21(7−8):1354−63.
[24] Liang L, Jones C, Chen S, Sun B, Jiao Y. Heterogeneous force network in 3D cellularized collagen networks. Phys Biol 2016;13(6):066001.
[25] Sopher RS, Tokash H, Natan S, Sharabi M, Shelah O, Tchaicheeyan O, Lesman A. Nonlinear elasticity of the ECM fibers facilitates efficient intercellular communication. Biophys J 2018;115(7):1357−70.
[26] Steinwachs J, Metzner C, Skodzek K, Lang N, Thievessen I, Mark C, Münster S, Aifantis KE, Fabry B. Three-dimensional force microscopy of cells in biopolymer networks. Nat Methods 2016;13(2):171−6.
[27] Seo BR, Bhardwaj P, Choi S, Gonzalez J, Eguiluz RC, Wang K, Mohanan S, Morris PG, Du B, Zhou XK, Vahdat LT, Verma A, Elemento O, Hudis CA, Williams RM, Gourdon D, Dannenberg AJ, Fischbach C. Obesity-dependent changes in interstitial ECM mechanics promote breast tumorigenesis. Sci Transl Med 2015;7(301). 301ra130.
[28] Vader D, Kabla A, Weitz D, Mahadevan L. Strain-induced alignment in collagen gels. PLoS One 2009;4(6):e5902.
[29] Janmey PA, Miller RT. Mechanisms of mechanical signaling in development and disease. J Cell Sci 2011;124(Pt 1):9−18.
[30] Shoham N, Gefen A. Deformations, mechanical strains and stresses across the different hierarchical scales in weight-bearing soft tissues. J Tissue Viability 2012;21(2):39−46.
[31] Shoham N, Gefen A. Mechanotransduction in adipocytes. J Biomech 2012;45(1):1−8.
[32] Tanabe Y, Koga M, Saito M, Matsunaga Y, Nakayama K. Inhibition of adipocyte differentiation by mechanical stretching through ERK-mediated downregulation of PPARγ2. J Cell Sci 2004;117(Pt 16):3605−14.
[33] Elsner JJ, Gefen A. Is obesity a risk factor for deep tissue injury in patients with spinal cord injury? J Biomech 2008;41(16):3322−31.
[34] Lustig M, Gefen A, Benayahu D. Adipogenesis and lipid production in adipocytes subjected to sustained tensile deformations and elevated glucose concentration: a living cell-scale model system of diabesity. Biomech Model Mechanobiol 2018;17(3):903−13.
[35] Lustig M, Moldovan Mor Yossef L, Gefen A, Benayahu D. Adipogenesis of 3T3L1 cells subjected to tensile deformations under various glucose concentrations. In: Gefen A, Weihs D, editors. Computer methods in biomechanics and biomedical engineering. Cham: Springer; 2018. p. 171−4.
[36] Lustig M, Feng Q, Payan Y, Gefen A, Benayahu D. Noninvasive continuous monitoring of adipocyte differentiation: from macro to micro scales. Microsc Microanal 2019;25(1):119−28.
[37] Shoham N, Girshovitz P, Katzengold R, Shaked NT, Benayahu D, Gefen A. Adipocyte stiffness increases with accumulation of lipid droplets. Biophys J 2014;106(6):1421−31.
[38] Shoham N, Gottlieb R, Sharabani-Yosef O, Zaretsky U, Benayahu D, Gefen A. Static mechanical stretching accelerates lipid production in 3T3-L1 adipocytes by activating the MEK signaling pathway. Am J Physiol Cell Physiol 2012;302(2):C429−41.
[39] Hara Y, Wakino S, Tanabe Y, Saito M, Tokuyama H, Washida N, Tatematsu S, Yoshioka K, Homma K, Hasegawa K, Minakuchi H, Fujimura K, Hosoya K, Hayashi K, Nakayama K, Itoh H. Rho and Rho-kinase activity in adipocytes contributes to a vicious cycle in obesity that may involve mechanical stretch. Sci Signal 2011;4(157). ra3−ra3.

[40] Lustig M, Zadka Y, Levitsky I, Gefen A, Benayahu D. Adipocytes migration is altered through differentiation. Microsc Microanal 2019;25(05):1195—200.
[41] Steppan CM, Bailey ST, Bhat S, Brown EJ, Banerjee RR, Wright CM, Patel HR, Ahima RS, Lazar MA. The hormone resistin links obesity to diabetes. Nature 2001;409(6818):307—12.
[42] Hossain P, Kawar B, El Nahas M. Obesity and diabetes in the developing world - a growing challenge. N Engl J Med 2007;356(3):213—5.
[43] Dandona P, Aljada A, Bandyopadhyay A. Inflammation: the link between insulin resistance, obesity and diabetes. Trends Immunol 2004;25(1):4—7.
[44] Verma S, Hussain ME. Obesity and diabetes: an update. Diabetes Metab Syndr 2017;11(1):73—9.
[45] Saltiel AR, Kahn CR. Insulin signalling and the regulation of glucose and lipid metabolism. Nature 2001;414(6865):799—806.

CHAPTER 5

Mechanisms underlying vascular stiffening in obesity, insulin resistance, and type 2 diabetes

Michael A. Hill[1,2], Yan Yang[1], Zhe Sun[1,2], Liping Zhang[1,2], James R. Sowers[1,2,3,4]

[1]Dalton Cardiovascular Research Center, University of Missouri, Columbia, MO, United States; [2]Department of Medical Pharmacology and Physiology, University of Missouri School of Medicine, Columbia, MO, United States; [3]Diabetes and Cardiovascular Center, University of Missouri School of Medicine, Columbia, MO, United States; [4]Department of Medicine, University of Missouri School of Medicine, Columbia, MO, United States

Abbreviations

AGEs advanced glycation end products
Ang II angiotensin II
BMI body mass index
CVD cardiovascular disease
ECM extracellular matrix
ENaC epithelial sodium channel
EnNaC endothelial sodium channel
eNOS endothelial nitric oxide synthase
HIF$_{1\alpha}$ hypoxia-inducible factor-1 alpha
HOMA-IR homeostatic model assessment of insulin resistance
IL interleukin
INS insulin
INSR insulin receptor
MR mineralocorticoid receptor
mTOR mammalian target of rapamycin
NADPH oxidase nicotinamide adenine dinucleotide phosphate oxidase
NO nitric oxide
NOS nitric oxide synthase
PVAT perivascular adipose tissue
PWV pulse wave velocity
RAGE receptor for advanced glycation end products
ROS reactive oxygen species
SGK1 serum- and glucocorticoid-activated kinase 1
T2DM type 2 diabetes mellitus
TG2 transglutaminase 2
TNFα tumor necrosis factor alpha

Introduction

Increased age-related arterial stiffening, typically measured as increased pulse wave velocity (PWV), is a predictor of future cardiovascular disease (CVD) events and all-cause mortality [1–3]. Further, arterial stiffening is commonly associated with isolated systolic hypertension, a powerful CVD risk factor, particularly after the fifth decade of life. In addition to being an aging phenomenon, arterial stiffening is accelerated in those with type 2 diabetes mellitus (T2DM) and the cardiometabolic syndrome, both disorders manifesting insulin (INS) resistance [4–7]. Importantly, INS resistance is associated with increased arterial stiffness independent of systolic blood pressure and glycemic status [8,9]. Of further note, increased arterial stiffening is especially apparent in obese and diabetic females who tend to lose the normal protection against CVD that is afforded in lean and nondiabetic women prior to the onset of menopause [5,10,11].

As coexisting morbidities such as hypertension and hyperglycemia are often associated with vascular stiffening it has been difficult to definitively establish direct causative relationships between obesity or INS resistance and arterial stiffening. Nevertheless, accumulating information points to an important early role for obesity and INS resistance in the genesis of arterial stiffening. To this point, fibrotic changes have been identified in the walls of subcutaneous arteries of obese individuals exhibiting both normal and elevated blood pressure levels [12]. Similarly, prepubescent children with obesity, but classified as having an absence or only one indicator of the cardiometabolic syndrome, exhibited increased arterial stiffness [13]. Importantly, the increase in stiffness correlated with both increased body mass index (BMI) and insulin resistance (HOMA-IR model) [13]. While similar results have been reported in other studies a recent study did not find a relationship between childhood obesity and increased PWV [14]. Nevertheless, extant observations indicate that increases in vascular stiffness in obese and INS-resistant persons occur prior to development of hypertension or clinical T2DM. Consistent with these observations in humans, diet-induced obesity and INS resistance in rodents are associated with increased vascular stiffening before the development of hypertension or hyperglycemia [15]. On the basis of this evidence obesity-associated and INS resistance—associated vascular stiffness appears to be an early risk factor for the development of CVD and signals a potential point of intervention. The importance of establishing causative links between obesity, INS resistance, and vascular stiffening is further heightened by the increasing prevalence of obesity in Western and developing societies [16–18].

From a mechanistic standpoint arterial stiffening is a complex process involving changes to the cellular and/or the extracellular matrix (ECM) components of the vascular wall [19,20], as well as interactions between these compartments. Contributions also emanate from specialized structures associated with the vascular wall including the perivascular adipose tissue (PVAT)/adventitia and the endothelial glycocalyx. In addition, the degree of stiffening differs throughout the arterial tree due to variation in local hemodynamics, differences in vascular structure, and mechanical factors. While vascular stiffness is typically thought of in terms of large arteries changes to the mechanical properties of small arteries/arterioles also impact vessel structure and function. In the aorta, stiffening results in an earlier propagation of the reflected wave from the periphery to the ascending aorta during systole, as opposed to diastole, with a resultant augmentation of aortic systolic pressure and pulse pressure and decreased diastolic coronary perfusion pressure [21]. Aortic stiffening further allows propagation of higher pulse pressures into the microcirculation which leads to remodeling of arterioles and impairment of blood flow autoregulation in organs such as the brain and kidneys [22–24]. Stiffening of small arteries alters the mechanosensing properties of small arteries negatively impacting local blood flow regulatory mechanisms such as the myogenic response [25,26]. This brief review will consider the factors implicated in enhanced vascular stiffening in obesity, INS resistance, and T2D (Fig. 5.1). Common to many

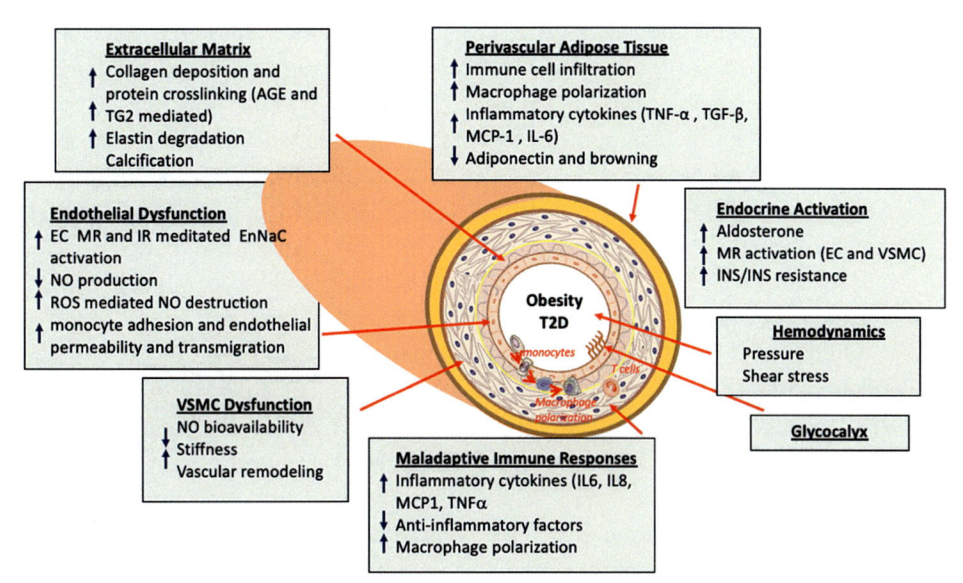

Figure 5.1 Multiple mechanisms are implicated in the pathogenesis of vascular stiffening in cardiometabolic syndrome and type 2 diabetes. These mechanisms involve dysregulation of the extracellular matrix, cellular components or the vessel wall (endothelial, smooth muscle, immune, and periadventitial cells), endocrine signaling, and hemodynamics. Of importance a number of these elements interact to alter the mechanical properties of arteries, including increasing vascular stiffness.

of the stiffening mechanisms are decreased nitric oxide (NO) bioavailability, immune cell activation, increases in reactive oxygen species (ROS), and inflammatory mediators which promote fibrosis of the vascular wall. Further, some mechanisms may be causative while others may be secondary to changes in variables such as hemodynamics (e.g., increased blood pressure and pressure overload) indicating a need to consider temporal relationships for these parameters.

Extracellular matrix and vascular stiffening in obesity and diabetes

i. Vascular Wall Extracellular Matrix

The ECM component of the vascular wall provides structural support, and it also contributes to vascular function via bidirectional signaling through cell surface binding molecules such as integrins, integrating force/contractile transmission, as well as being a reservoir for growth factors important in tissue repair/remodeling [27–30]. The ECM while consisting predominantly of fibrillar collagens (types I and III) and elastin also contains type IV collagen (basement membrane), laminins, fibronectin, and glycosaminoglycans. Considerable heterogeneity in the arrangement of ECM proteins occurs across the vessel wall and along the vasculature tree [31]. Further, the structure of the ECM shows considerable plasticity, with remodeling occurring during development and in response to injury and pathophysiological states.

An altered matrix phenotype occurs in response to a wide range of factors including ROS, advanced glycation end products (AGEs), proinflammatory factors, neurohumoral factors, and mechanical stimuli. The normal ECM, while consisting predominantly of fibrillar collagens (types I and III) and elastin, also contains Type IV collagen (basement membrane), laminins, fibronectin and glycosaminoglycans. [32].

ii. Changes in Protein Content

Obesity and T2DM are associated with increased deposition of ECM including collagen and fibronectin. Increased levels of collagen, especially type I collagen, are associated with increased passive stiffening of the arterial wall. As the most abundant proteins in the vessel wall, fibrillar collagen and elastin determine the curvilinear passive stress–strain relationship observed in arteries with elastin fibers contributing at low distending pressures and collagen fibers being recruited at higher pressures.

Levels of ECM proteins are partially dependent on a balance of synthetic and degradative pathways. Increased synthesis of collagen and fibronectin occurs in obesity and diabetes through a number of mechanisms involving inflammation, increased ROS production Rho kinase and protein kinase C activation [33].

ECM protein levels are also dependent on matrix-modifying enzymes, in particular matrix metalloproteinases (MMPs) and tissue inhibitors of matrix metalloproteinases (TIMPs). MMP-2 and MMP-9 and elastase activity are increased in obesity and T2DM leading to fragmentation of mature elastin and thus increases in vessel stiffness. Interestingly, TIMP 2 knockout mice consuming a high fat diet exhibit an obese phenotype, with heightened INS resistance (males) and hyperinsulinemia (males and females) [34]. These data support a relationship between INS resistance and alterations in the MMP/TIMP ratio in the pathogenesis of arterial stiffening.

Deposition of mature ECM proteins involves considerable extracellular processing, assembly, and involvement of other accessory proteins. This can result in a mismatch of observations at the mRNA, protein monomer, and mature cross-linked states. As an example, mRNA for the tropoelastin monomer of elastin can increase in response to injury/inflammation without necessarily leading to increased deposition of functional elastin within the artery wall [35]. Similarly, mature elastin can be modified by pathological cross-linking and, as mentioned previously, partial degradation contributing to elastin breaks.

iii. Protein Modification and Cross-linking

In addition to overt changes in ECM protein content the complex structure of mature ECM protein fibers can be altered, potentially impacting the passive stiffness properties of the vessel wall. In this regard, fragmentation of elastin fibers has been suggested to decrease the elastic properties of the vessel wall causing an increase in material stiffness. Further, pathological cross-linking of ECM proteins can increase the stiffness properties of the vascular wall.

a. Advanced Glycation End Product Formation

Of relevance to vascular stiffening is the ability of glucose to nonenzymatically modify macromolecules, in particular, proteins to exert effects on tissue structure and function. Specifically, the carbonyl groups of reducing sugars including fructose, ribose, galactose, glucose, and mannose are known to interact with amino groups of ECM proteins to form AGEs. The epsilon amino group of free lysine and the delta guanido group of free arginine residues in proteins of the ECM can be modified under hyperglycemic conditions to ultimately form AGEs through the "Maillard reaction." The Maillard reaction is a multistep process yielding a substantial number of different by-products and end products including irreversible AGEs.

AGEs may impact the mechanical properties of the vessel wall through a number of mechanisms. These include cross-linking and decreasing turnover of existing ECM proteins impacting their physical characteristics; altering normal ECM–cell interactions; and through signaling mechanisms arising from their binding with the receptor for AGEs (RAGE). For example, as ECM proteins bind integrins on the cell surface of smooth muscle cells, normal cell adhesion mechanisms may be altered. In this regard, fibronectin that is modified by AGEs shows increased binding to smooth muscle cells [36]. RAGE activation by AGEs leads to activation of NADPH and PKC, increased ROS production, and decreased bioavailable NO contributing to an inflammatory environment. Subsequent activation of the transcription factor NF-κB leads to increased expression of a number of profibrotic and inflammatory factors and further generation of ROS [37].

As a result of these adverse AGE-mediated effects considerable interest has been raised in approaches for preventing AGE production, breaking of already-formed AGE cross-links, and inhibiting of RAGE-mediated signaling. Prevention of AGE formation with agents such as aminoguanidine has been shown to decrease stiffness and improve distensibility of small arteries in experimental models of diabetes [25]. However, more extensive human studies need to be conducted with agents demonstrating suitable safety and efficacy profiles.

b. Transglutaminase 2

Tissue transglutaminase 2 (TG2) is a member of the transglutaminase superfamily of enzymes that is widely expressed in the vasculature [38]. TG2 has intracellular effects as well as being secreted to exert extracellular actions. TG2 is multifunctional in nature acting as a connective tissue cross-linking enzyme as well as exhibiting GTPase, kinase, and scaffolding actions [39,40]. ECM protein substrates for TG2 include collagen, fibronectin, and laminin [41].

TG2 activity is decreased by NO and is activated and secreted in states of decreased bioavailability of NO including diet-induced obesity and INS-resistant models [42]. Further, increased tissue TG2 activity contributes to central vascular stiffness in eNOS knockout mice [43]. Hyperglycemia-induced transcriptional regulation of TGM2 expression has been shown to be involved in small artery remodeling in obese diabetic minipigs [44]. Of further relevance, increased TG2 activity in diet-induced obesity not only increases large artery stiffness but is implicated in remodeling of resistance arteries leading to decreased distensibility and impaired vasomotor function [42].

c. Lysyl oxidases

Lysyl oxidases (LOXs) are enzymes involved in the normal cross-linking of fibrous collagen and elastin molecules, and thus modulate the tensile and elastic properties exhibited by blood vessels. An additional consequence of increased LOX activity is the generation of ROS including hydrogen peroxide (H_2O_2) and superoxide (O_2^-) [45,45]. Interestingly, the increased stiffness and abnormal elastin structure of mesenteric arteries from mice transgenic for LOX were prevented by several ROS inhibition strategies consistent with increased stiffness being a function of increased ROS generation [45]. It was further observed that LOX activity was elevated in experimental hypertension (Ang II infusion in mice and in spontaneously hypertensive rats) and that inhibition of LOX in these models ameliorated changes in vascular stiffening and elastin structure [45].

Expression of LOX is upregulated through hypoxia-inducible factor 1 alpha ($HIF_{1\alpha}$) and inflammation in adipose tissue [46]. Increased $HIF_{1\alpha}$ production in adipose tissue in obese states results from a relative hypoxia secondary to adipocyte hypertrophy. Increased $HIF_{1\alpha}$ production emanates from adipocytes, vascular cells, and macrophages. Importantly, increased $HIF_{1\alpha}$ production is linked to obesity-induced fibrosis providing a potential mechanism for increased vascular and cardiac stiffening, the latter contributing to heart failure with preserved ejection fraction [47].

iv. Calcification

Also implicated in vascular stiffening is calcification within the artery wall. Studies show that medial calcification is not necessarily associated with atherosclerosis while intimal calcification is typically associated with atheroma development [48,49]. Calcification occurs through both cellular (osteogenic) differentiation and ECM mineralization [50]. With respect to an osteogenic process, VSMCs dedifferentiate to an osteogenic phenotype promoting calcification and increased vascular stiffness [51]. Aortic calcification is associated with increased vascular stiffness and isolated systolic hypertension in normal subjects [50,52]. Medial calcification is further exacerbated in obesity, T2DM, and chronic kidney disease [50,52]. In obesity and T2DM increased calcification is associated with INS resistance [53] and maybe mechanistically be linked to increased AGEs, ROS, and inflammation which promote osteogenic differentiation [54].

Elastic fiber degradation has been linked to ectopic calcium phosphate mineral deposition (calcification) in the arterial walls, particularly in the medial layer [49,55,56]. Susceptibility of elastin fibers to calcification is increased by MMP-2, MMP-9, and elastase activation and such proteolytic actions have been shown to be associated with elevated systolic blood pressure and increased vascular stiffness in human subjects [57]. Interestingly, MMP-9 and serum elastase correlated with vascular stiffness in both subjects with isolated systolic hypertension and apparently normal controls [57]. Further, elastin breakdown products stimulate osteogenic differentiation and inflammation and arterial stiffening [58].

Cellular contributions to vascular stiffening in obesity and diabetes

While emphasis has historically been placed on the role of the ECM in vascular stiffening it has become apparent that the cellular components of the vessel wall contribute significantly to the stiffening process. Indeed, both ECs and VSMCs provide active mechanisms for affecting changes in the mechanical properties of the vessel wall. Such mechanisms involve autocrine and paracrine signaling events as well as dynamic modulation of binding interactions between ECM proteins and VSMCs. Both cell types also modulate their intrinsic stiffness/mechanical properties through polymerization of the cortical actin cytoskeleton. The relatively recent availability of biophysical techniques such as atomic force microscopy (AFM) has facilitated studies of these cellular properties [59—61]. In addition to the roles played by these cell types it is likely that other cells associated with the vessel wall (including immune cells, perivascular fat, adventitial fibroblasts, and perivascular nerves) modulate vascular stiffness.

i. Endothelial Cells

There is growing interest in the role of ECs in regulating stiffness of the arterial wall, with evidence emerging that ECs can impact vascular stiffness by a variety of mechanisms including through reduced NO-mediated increases of vascular tone, vascular permeability, secretion of vasoactive and inflammatory factors, and the recruitment of monocytes and macrophages [6]. Indeed, vascular stiffness in humans is increased by systemic and local administration of eNOS inhibitors [62—64] while NO donors have been shown to decrease stiffness [63,65,66].

As described in subsequent sections stiffening of ECs appears to be associated with cortical actin polymerization which leads to impaired ability to release NO which then contributes to a number of events leading to stiffening of the artery wall [6]. These altered mechanical properties of the EC and associated reductions in bioavailable NO increase vascular stiffness independent of a direct physical contribution of endothelial stiffening to the overall material stiffness of the wall.

ii. Vascular Smooth Muscle

VSMCs impact stiffness of the vessel wall through both changes in contractile tone and through their inherent cellular stiffness. Inherent cell stiffness is determined, in part, by the dynamic properties of the cortical cytoskeleton and through modulation of adhesion to ECM proteins [67]. Inherent changes in intrinsic smooth muscle cell stiffness have been shown in aging and hypertension [67,68]. These properties are also modulated by vasoactive agents including Ang II and NO with the former increasing cellular stiffness and adhesion while NO decreases these parameters [59]. The decrease in bioavailable NO in obesity has been hypothesized to contribute to increased stiffness of the vessel wall.

As mentioned earlier obesity, INS resistance, and T2DM are associated with changes in the ECM, for example, through both protein deposition and posttranslational modification. Increased vascular fibronectin has been consistently demonstrated in obesity, hypertension, and T2DM [47,69,70]. Further, this is associated with increased expression of the fibronectin-binding protein smooth muscle α5 integrin, providing an ECM—VSMC link to increases in vessel stiffness. Conversely, angiotensin-converting enzyme inhibition decreased fibronectin and $_\alpha$5 integrin expression and carotid artery stiffness in hypertensive rats.

Phenotypic changes in VSMCs that result in increased proliferation, migration, and senescence have also been implicated in vascular stiffening and have been extensively reviewed elsewhere [19,20]. Importantly, the VSMC phenotype is partly dependent on its local mechanical environment interactions with the ECM and cell stretch/tension changes. Stiffening of the ECM, itself, leads to stiffening and changes in cellular phenotype of VSMCs [71,72], illustrating the complex interaction between components of the artery wall.

iii. Immune Cells

The roles played by immune cells in obesity and cardiometabolic disorders, including their role in increased arterial stiffening, are receiving increased attention [73,74]. Particular emphasis is being placed on immune function and hypertension interactions in increasing arterial stiffness [75]. The inflammatory environment in obesity, INS resistance, and T2D results in increased expression of adhesion molecules and the chemoattractant MCP-1. This, in turn, enhances uptake of monocytes and macrophages into the vessel wall and PVAT. Activation of inflammatory M1 macrophages leads to increased production of ROS and reactive nitrogen species together with increased expression of inflammatory cytokines (including TNFα, IL-6, and MCP-1) [76]. Production of inflammatory cytokines, particularly within adipose tissue, further contributes to the systemic level of INS resistance impairing vascular INS metabolic signaling. This pattern of inflammation and INS resistance is further exacerbated by the T helper cells (Th1) [77] and possibly the presence of excess dendritic cells in visceral and perivascular adipose tissue [78].

The exact role of immune cells in associated inflammation and tissue remodeling/fibrosis in cardiometabolic diseases is very complex with additional cell types adding to the inflammatory environment (e.g., CD4+ Th17 cells secreting IL-17) and others playing a more antiinflammatory role (T reg cells through secretion of IL-10 TGF-β) [79]. β-cells may also contribute to hypertension, and thus arterial stiffening, through the production of autoantibodies [79], including antibodies targeted at receptors (including angiotensin and adrenergic receptors) thereby modulating vasoactivity [80].

A further immune cell-related component is the inflammasome, an intracellular multiprotein complex of the innate immune system that promotes the secretion of the inflammatory cytokines such as interleukin 1β (IL-1β) and interleukin

18 (IL-18). The NLR family pyrin domain-containing protein 3 (NLRP3) inflammasome is the most widely characterized of these protein complexes contributing to IL-1β and IL-18 secretion. IL-1β and IL-18 have been implicated in several facets of the cardiometabolic syndrome, including obesity, hypertension, and INS resistance. Of specific relevance the NLRP3 inflammasome is reported to have a major role in endothelial dysfunction [81]. T2D subjects treated with metformin showed a decreased production of proinflammatory factors in leukocytes and visceral fat consistent with downregulation of the NLRP3 inflammasome. Further, uric acid has been reported to activate the inflammasome in mouse models of obesity and increase vascular stiffness. Allopurinol reduced hyperuricemia and decreased indices of inflammation and vascular stiffening [82], consistent with a causative role for inflammasome activation in this model.

iv. Adventitia and Perivascular Adipose Tissue

While serving as a structural support for arteries, PVAT secretes a variety of paracrine-acting molecules that impact vascular function and structure. Such factors can be both antiinflammatory and inflammatory in nature while also exerting vasodilator and vasoconstrictive effects. In obesity and T2D the balance shifts toward an increased inflammatory and vasoconstrictor actions. In this regard, obesity, INS resistance, and T2D are associated with decreased levels of adiponectin and increased production of proinflammatory molecules such as TNFα and IL-6 [73,78].

PVAT is also a source for immune cells which may contribute to vascular stiffening. Importantly, immune cell infiltration and chronic low-grade inflammation in PVAT is now recognized as a hallmark of obesity and T2D. In addition to increased numbers of immune cells, changes in cell phenotype are also evident with macrophages shifting their distribution from the more antiinflammatory (M2) to the inflammatory (M1) subtype. In db/db diabetic mice an increased number of dendritic cells in the stromal vascular fraction of adipose tissue leads to the release of inflammatory factors which impair normal mechanisms of vascular endothelial—dependent vasodilation (Fig. 5.2) [78].

Of additional relevance PVAT can undergo transformation from brown to white fat with associated changes in the production of inflammatory cytokines. White fat is more associated with the production of factors which have been implicated in vascular stiffening. As increases in perivascular white fat predominate in obesity, INS resistance, and T2D this provides another possible mechanism underlying increases in vascular stiffness.

The adventitial layer of the vascular wall contains a number of cell types including adventitial fibroblasts and nerves. Paracrine signaling through the release/secretion of ECM proteins, vasoactive mediators, and growth factors have the potential to impact

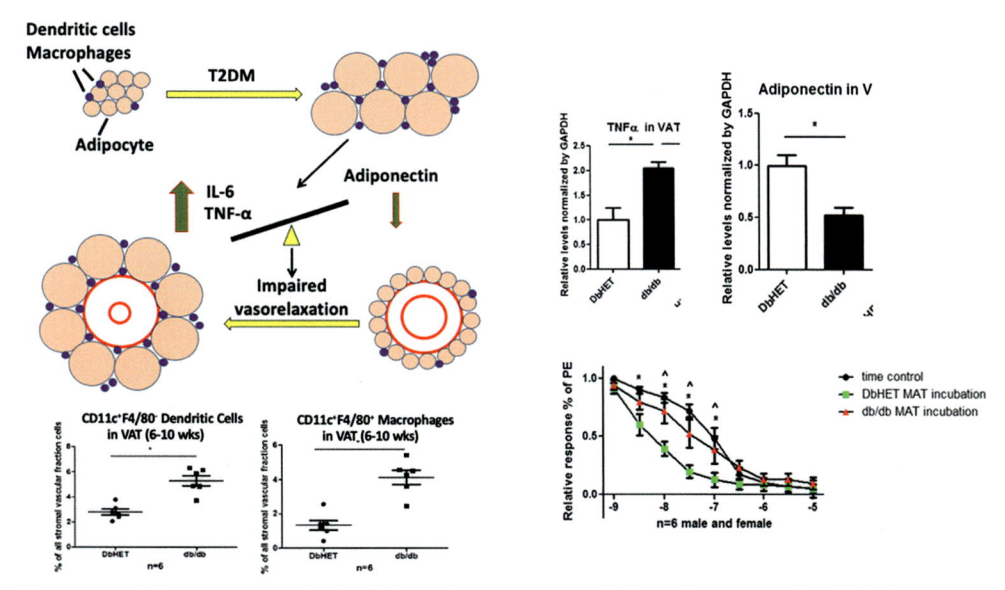

Figure 5.2 Adipocyte expansion in diabetes leads to accumulation of immune cells within the stromal vascular fraction. Increased production of inflammatory factors (including IL-6 and TNFα) and decreased production of antiinflammatory factors (e.g., adiponectin) impair the normal anticontractile activity conferred by perivascular adipose tissue.

structure and function of the vascular wall and thus arterial stiffness. Based on the above discussion PVAT contributes to vascular inflammation and stiffening—processes typically described as being accelerated and enhanced in obesity, INS resistance, and T2D.

While little direct experimental information is available to date, adventitial nerves may impact stiffening of the vascular wall through effects on VSMC vasomotor tone and also through the trophic actions of released transmitters and cotransmitters.

Metabolic and endocrine activation of vascular stiffening in obesity, INS resistance, and diabetes

Studies have shown that both strategies aimed at improving INS sensitivity as well as mineralocorticoid receptor (MR) antagonism and improve CVD outcomes. Randomized controlled clinical trials such as RALES, EPHESUS, and EMPHASIS showed decreased morbidity and mortality in heart failure patients administered MRAs [83–86]. Thus, considerable interest has developed in the adverse CV role(s) played by heightened activation of vascular MRs and hyperinsulinemia in states of INS resistance including obesity and T2D. This has been further fueled by the demonstration that vascular ECs, in particular, possess functional MRs and INS receptors (IRs). Further, by analogy with mechanisms described in renal tubule epithelial cells (collecting duct) it is apparent that MR and IRs may interact through shared signaling pathways (Fig. 5.3). Specifically, MR and INR activation leads to mTOR signaling, with downstream

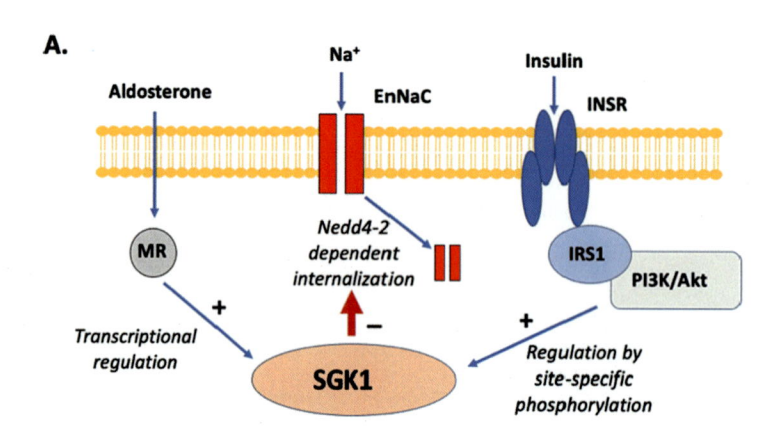

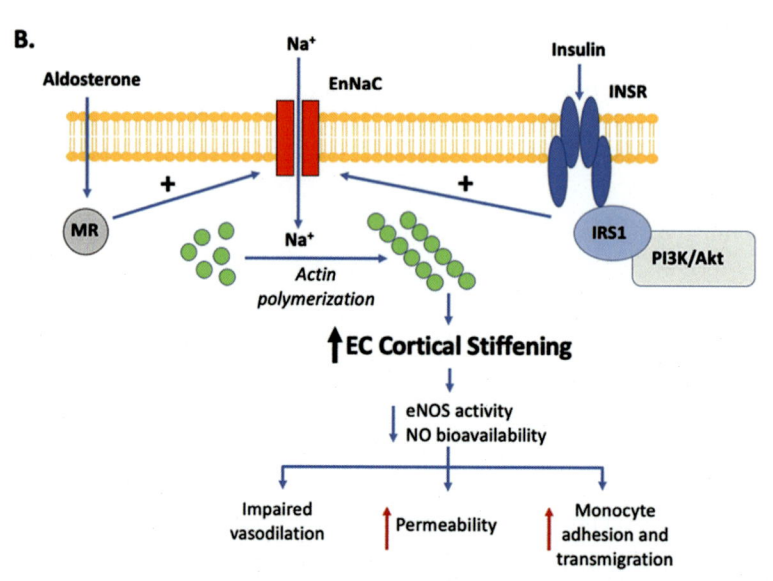

Figure 5.3 (A) Insulin- and aldosterone-mediated signaling in endothelial cells converge to activate SGK1, inhibiting EnNaC Nedd4-2-mediated internalization. This results in increased channel in the plasma membrane and an increase in its activity. (B) Increased EnNaC current leads to actin polymerization and cell stiffening which is associated with decreased eNOS activity and NO bioavailability. Decreased NO bioavailability impairs endothelial-dependent mechanisms of vasodilation, increases vascular permeability, and facilitates the transmigration of immune cells.

stimulation of SGK1 and resultant activation of the endothelial Na^+ channel (EnNaC) which promotes EC and vascular stiffness (Fig. 5.3). The following sections briefly consider the synergy that occurs between MR and INS signaling to activate EnNaC as a precursor to arterial stiffening.

i. Mineralocorticoid Receptor Activation

Although MRs have been well characterized in terms of their role in the renal aldosterone (ALDO)-sensitive distal tubular epithelium [87], more recent studies have emphasized their existence and importance in cells of the vascular wall, including ECs [88,89], VSMCs [90,91], and macrophages [92,93]. MR are also evident in adipocytes [94] and fibroblasts [95] which associate with the vascular wall. Increased vascular MR activation is now accepted as playing a significant role in the etiology of CV and renal diseases [96–98]. For example, enhanced VSMC MR activation has been implicated in age-dependent increases in arterial pressure and vascular fibrosis, in part resulting from altered expression of voltage-gated Ca^{2+} channels [90,99] Further, ECMR activation has been shown to decrease NO bioavailability while concurrently potentiating inflammatory responses and oxidative stress, particularly in situations at risk for CVD such as diet-induced obesity and associated INS resistance [90,100]. Activation of the MR in ECs and VSMCs typically occurs following response to ALDO and not glucocorticoids as the enzyme 11-beta-hydroxysteroid dehydrogenase 2 (11β-HSD2) is expressed in both vascular cell types converting glucocorticoids into inactive metabolites [101–103].

ii. Insulin Signaling

Persons with INS resistance and hyperinsulinemia show increased arterial stiffness independently of confounding factors such as hypertension and hyperglycemia, and hyperinsulinemia precedes the development of CVD [104–106]. Increased arterial stiffness relates to INS resistance even in the absence of the development of overt clinically defined T2D. Increased stiffness with INS resistance, as demonstrated by fasting hyperinsulinemia, is evident prior to the development of persistent hypertension (ref). Collectively these observations suggest that insulin resistance and hyperinsulinemia occur early in the time course for development of arterial stiffness.

iii. Endothelial Na+ channel

The amiloride-sensitive epithelial Na^+ channel (ENaC) is a known target for MR activation in the renal distal nephron tubules of the kidney [107–111]. Following activation, the MR binds the ENaC promoter element in epithelial cells increasing expression of the Na^+ ion channel and its insertion into the plasma membrane. In addition to ALDO-mediated MR activation, angiotensin II exerts an effect on ENaC open probability (Po) through NADPH-mediated production of reactive oxygen species [112]. Further, INS also increases ENaC activities in renal epithelial cells. Functionally, ENaC in the distal nephron then acts to fine-tune the reabsorption of filtered Na^+ and regulate extracellular fluid volume. ENaC has also been extensively studied in epithelial cells of the respiratory and gastrointestinal tracts initially and various glands, where the channel similarly regulates transepithelial cell Na^+ movement [113–115]. More recently the existence of the ENaC has been demonstrated in ECs and has been specifically termed EnNaC [116–118].

ENaC is member of the ENaC/degenerin family of cation-selective channels with the functional trimeric channel generally being composed of three subunits termed α, β, and γ [119–121]. In some species and tissues an additional δ subunit has been cloned and characterized [122]. Po of the channel is determined by subunit composition with the α-subunit being typically considered as the functional core of the channel [123]. Additional conformations of the channel have been suggested to include δ,β,γ and β,γ complexes in association with ASIC proteins (additional members of the degenerin family) [117,124,125]. The existence and functional significance of these alternate conformations in the vasculature are currently uncertain. The heteromeric channel is characterized by each subunit having two transmembrane domains and a large extracellular loop that regulates channel gating to a number of stimuli including proteases, extracellular Na^+, and, as mentioned above, shear stress [125–131]. In contrast to the large extracellular domain ENaC has a relatively small intracellular domain with the C- and N-termini located intracellularly [107]. The intracellular domain contains regulatory sites including for palmitoylation and phospholipids [107,132,133]. The intracellular domain also binds the cortical actin cytoskeleton, or actin-binding proteins such as filamin, likely contributing to the channel's mechanosensory properties [134–136]. Mechanosensory properties of vascular ENaC may also be affected through colocalization with integrins and interactions with extracellular matrix proteins [137]. Conceivably, the flow-sensitive properties of EnNaC are also affected by interactions between Na^+ channel activation and glycocalyx disruption and shedding [130,138,139], a process implicated in the regulation and effects of other ion channels [140]. In addition to these mechanosensory regulatory modes, as previously noted ENaC activity is modulated through a number of hormonal mechanisms (including ALDO, angiotensin II, and INS) and paracrine factors (including NO and ATP) [141–143].

Increased EnNaC activity leads to EC cell stiffening, decreased NO bioavailability, and impaired vasodilator function.

In studies of cultured human umbilical vein ECs it has been shown that ALDO promotes increased cell surface expression of the αEnNaC subunit as measured by quantum dot immunofluorescence and western blotting [144]. The ALDO-mediated effect was prevented by spironolactone consistent with a role for MR activation [144]. AFM measurements further showed that ALDO increased cell surface area, an assumed consequence of Na^+ entry via ENaC as the effect was blocked by amiloride [144]. This group also demonstrated that ALDO-mediated Na^+ entry led to stiffening of the EC cortical actin cytoskeleton as measured by indentation AFM which, in turn, impaired NO production through inhibition of eNOS activity. In addition, it was argued that under the physiological conditions the negative charge-rich cell surface proteoglycan layer, the glycocalyx, acts to limit Na^+ access to EnNaC while under certain pathological conditions this protective barrier is disrupted favoring enhanced EnNaC activity [145]. Collectively, under conditions of enhanced EnNaC activity this series of events was proposed to favor development of increased smooth muscle tone and a vasoconstricted state in an intact artery.

These cell-based studies provided the rationale for in vivo mouse studies of MR activation, specifically via chronic infusion of ALDO or feeding of an obesogenic Western diet (high in refined carbohydrates and saturated fat) which is associated with increased INS resistance, elevated plasma ALDO and INS levels, and increased CV MR expression [100]. Mice fed the Western diet showed increased endothelial and aortic stiffness, impaired endothelial-dependent vasorelaxation, aortic fibrosis, aortic oxidative stress, and increased vascular expression of EnNaC [146]. Treatment of the mice with a low dose of amiloride (1 mg/kg/day, administered in the drinking water) prevented or significantly attenuated these diet-induced abnormalities [146]. Further, in vivo evidence for a role for EnNaC was provided by studies of the global ENaC gain-of-function Liddle mice which showed increased stiffening of the cortical actin cytoskeleton in aortic ECs [147].

To gain further insight into the vascular role played by EnNaC we have characterized a mouse model with endothelial cell—specific deletion of the α pore-forming subunit of ENaC [6,100,116]. As the deletion strategy involved the use of Cre recombinase expression by the Tie2 promoter studies in macrophages was performed to confirm that αEnNaC was not affected in hematopoietic cells. Further, blood pressure and renal Na^+ handling were unaffected. Consistent with a role for EnNaC in sensing shear stress these mice were shown to have impaired flow-dependent dilation in small arteries of the intestinal mesentery [116]. The lack of an effect of the channel protein deletion on systemic blood pressure under physiological conditions may reflect compensatory mechanisms due to the multiple systems capable of altering blood pressure. In contrast, we have observed that the ENaC deletion does reduce the increase in blood pressure that occurs in response to MR activation by subcutaneous deoxycorticosterone acetate (DOCA) administration and consumption of 1% NaCl (unpublished data). Interestingly, this has parallels with the ECMR model where a marked phenotype is not evident under basal conditions but can be unmasked by a stressor such as diet-induced obesity or aldosterone infusion [6,100]. Further demonstrating a role in pathophysiological states, it has been shown that in renal ischemia/perfusion injury, a situation characterized by decreased NO bioavailability, deletion of αENaC prevents renal tubular injury and renal dysfunction [148].

As mentioned earlier obesity and INS resistance in both human subjects and experimental dietary obesity mouse models lead to increased ALDO and INS levels and increased vascular MR expression [149,150], resulting in increased EnNaC protein expression and its translocation to the cell membrane, an effect inhibited by MR antagonists in cultured umbilical vein ECs [151,152]. To more directly demonstrate that enhanced EnNaC activation underlies increases in EC stiffness and decreased NO-mediated vascular relaxation Western diet feeding studies were conducted in αEnNaC−/− mouse models and their littermate controls [6,153]. Sixteen weeks of the Western diet resulted in increased EnNaC activation as shown by increased Na^+ currents measured in isolated ECs [6,153]. Increased endothelial stiffness was demonstrated in segments of aorta by an AFM indentation protocol and this increase in endothelial stiffness was associated with reduced bioavailable NO and impaired aortic ring endothelium-dependent relaxation [6,153]. These Western diet–induced abnormalities, along with vascular remodeling and fibrosis, were all significantly attenuated in the $αEnNaC^{-/-}$ mouse. From a mechanistic standpoint these studies showed that Western diet feeding resulted in a heightened inflammatory response that was associated with reduced eNOS phosphorylation/activation and diminished NO production and bioavailability [7]. These latter events likely emanated from increased EnNaC activity leading to polymerization of cortical actin fibers, subsequently reducing eNOS activity, and decreasing NO production leading to increased vascular stiffness [6] (Fig. 5.3).

iv. ALDO and INS Interact to Regulate EnNaC via mTORC2/SGK1 Signaling

Studies in renal epithelial cells have shown that ENaC is regulated both by post-translational modification, including through protease activation, and by trafficking of subunits to the cell membrane. Inactivation occurs, in part, by ENaC internalization and ubiquitination/destruction pathways [107]. As component of this pathway, interest has centered on the role on mTOR signaling (Fig. 5.3). mTORC1 and mTORC2 are multiprotein complexes which regulate multiple biological processes including metabolism, ion transport, and cellular growth/proliferation. mTORC1 is increased by MR activation and ROS to phosphorylate S6K1 which then phosphorylates IRS-1 on ser307 and other serine residues to inhibit IRS-1 tyrosine phosphorylation and downstream INS metabolic signaling (INS resistance) and NO production [5,34]. This, in turn, leads to hyperinsulinemia which likely promotes arterial wall fibrosis and increased stiffness [154] (Figs. 5.3 and 5.4).

Evidence also exists to support the notion that mTORC2 and downstream serum- and glucocorticoid-regulated kinase 1 (SGK1) signaling constitute a convergence point that links obesity, metabolic disturbances, vascular stiffening, and hypertension. Studies of renal epithelial cells indicate that activation of mTORC2 causes phosphorylation of SGK1 decreasing activity of ubiquitin ligase, Nedd4-2, mediated ubiquitination and increasing plasma membrane localization and activation of ENaC [155–157]. Additional studies have also suggested that SGK1 activation mediates direct stimulatory effects of ALDO on ENaC production and membrane localization, although the exact mechanisms are uncertain [158,159]. Recent studies have shown that ECMR activation following consumption of a Western diet stimulates mTORC2 and SGK1 signaling pathways in ECs known to contribute to membrane localization and activation of EnNaC [100]. Further, in preliminary unpublished studies to confirm the relevance of these pathways in vascular cells, we have found that mTORC2 and SGK1 inhibition attenuates amiloride-sensitive Na^+ currents in isolated ECs. Also, in vitro treatment of ECs with ALDO and INS leads to mTORC2 activation and increased

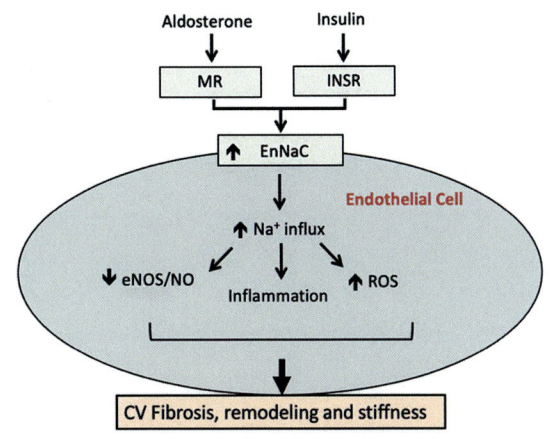

Figure 5.4 In cardiometabolic syndrome and T2D, insulin and mineralocorticoid signaling interact to increase the activity of the Na^+ ion channel, EnNaC. Increased Na^+ current is associated with decreased NO bioavailability, inflammation, and increased ROS generation. Collectively these pathological factors contribute to maladaptive responses including cardiovascular (CV) fibrosis, vascular remodeling, and increased arterial stiffness.

expression of the alpha subunit of EnNaC. These observations are consistent with studies conducted with renal tubule cells showing that INS also activates ENaC via mTOR and SGK1 signaling [160–162]. Additionally, INS and ALDO have additive effects on activation of ENaC in these renal derived cells. Consistent with this, we have recently found that INS exerts EnNaC stimulatory effects in isolated ECs that are related to activation of the mTOR2/SGK1 signaling pathway. In addition, ECs cultured in the presence of ALDO (10 nM) for 48 h show enhanced Na^+ current in response to acutely applied INS (100 nM) suggesting an additive action of these signaling pathways on ENaC activity (unpublished observations). While further studies are required to establish concentration–response and temporal relationships between these hormones, these preliminary observations in vascular ECs provide insight into integrative mechanisms by which INS resistance/hyperinsulinemia, elevated ALDO, and associated increases in ECMR and EC IR signaling activate EnNaC as an instigator for increased CV stiffness in INS-resistant states such as obesity and T2DM.

Conclusion

Arterial stiffening is an early indicator of vascular dysfunction and impending CVD in persons with the cardiometabolic syndrome and T2DM. A common denominator in these disorders is INS resistance which leads to hyperinsulinemia. This hyperinsulinemia/increased INS receptor activation further synergizes with enhanced MR activation through stimulation of SGK1 and enhanced activity of EnNaC. Activation of EnNaC leads to decreased NO bioavailability, inflammation, excessive generation of ROS, and immune cell activation (Fig. 5.4). Together with additional mechanisms that contribute to ROS production and impaired NOS activity, a self-perpetuating cycle of low-grade inflammation is established leading to fibrosis, increased vascular stiffness, and associated CVD.

References

[1] Laurent S, Boutouyrie P, Asmar R, et al. Aortic stiffness is an independent predictor of all-cause and cardiovascular mortality in hypertensive patients. Hypertension 2001;37(5):1236–41.
[2] Vlachopoulos C, Aznaouridis K, Stefanadis C. Prediction of cardiovascular events and all-cause mortality with arterial stiffness: a systematic review and meta-analysis. J Am Coll Cardiol 2010;55(13):1318–27.
[3] Mitchell GF, Hwang SJ, Vasan RS, et al. Arterial stiffness and cardiovascular events: the framingham heart study. Circulation 2010;121(4):505–11.
[4] Jia G, Hill MA, Sowers JR. Diabetic cardiomyopathy: an update of mechanisms contributing to this clinical entity. Circ Res 2018;122(4):624–38.

[5] Kim HL, Lee JM, Seo JB, et al. The effects of metabolic syndrome and its components on arterial stiffness in relation to gender. J Cardiol 2015;65(3):243—9.
[6] Sowers JR, Habibi J, Aroor AR, et al. Epithelial sodium channels in endothelial cells mediate diet-induced endothelium stiffness and impaired vascular relaxation in obese female mice. Metabolism 2019;99:57—66.
[7] Castro JP, El-Atat FA, McFarlane SI, Aneja A, Sowers JR. Cardiometabolic syndrome: pathophysiology and treatment. Curr Hypertens Rep 2003;5(5):393—401.
[8] Seo HS, Kang TS, Park S, et al. Insulin resistance is associated with arterial stiffness in nondiabetic hypertensives independent of metabolic status. Hypertens Res 2005;28(12):945—51.
[9] Sengstock DM, Vaitkevicius PV, Supiano MA. Arterial stiffness is related to insulin resistance in nondiabetic hypertensive older adults. J Clin Endocrinol Metab 2005;90(5):2823—7.
[10] Masding MG, Stears AJ, Burdge GC, Wootton SA, Sandeman DD. Premenopausal advantages in postprandial lipid metabolism are lost in women with type 2 diabetes. Diabetes Care 2003;26(12):3243—9.
[11] Madonna R, Balistreri CR, De Rosa S, et al. Impact of sex differences and diabetes on coronary atherosclerosis and ischemic heart disease. J Clin Med 2019;8(1).
[12] Celik T, Yuksel UC, Fici F, et al. Vascular inflammation and aortic stiffness relate to early left ventricular diastolic dysfunction in prehypertension. Blood Pres 2013;22(2):94—100.
[13] Ruiz-Moreno MI, Vilches-Perez A, Gallardo-Escribano C, et al. Metabolically healthy obesity: presence of arterial stiffness in the prepubescent population. Int J Environ Res Publ Health 2020;17(19).
[14] Jakab AE, Hidvegi EV, Illyes M, et al. Childhood obesity: does it have any effect on young arteries? Front Pediatr 2020;8:389.
[15] Weisbrod RM, Shiang T, Al Sayah L, et al. Arterial stiffening precedes systolic hypertension in diet-induced obesity. Hypertension 2013;62(6):1105—10.
[16] Ward ZJ, Bleich SN, Cradock AL, et al. Projected U.S. State-level prevalence of adult obesity and severe obesity. N Engl J Med 2019;381(25):2440—50.
[17] Gregg EW, Shaw JE. Global health effects of overweight and obesity. N Engl J Med 2017;377(1):80—1.
[18] Collaborators GBDO, Afshin A, Forouzanfar MH, et al. Health effects of overweight and obesity in 195 countries over 25 years. N Engl J Med 2017;377(1):13—27.
[19] Lacolley P, Regnault V, Avolio AP. Smooth muscle cell and arterial aging: basic and clinical aspects. Cardiovasc Res 2018;114(4):513—28.
[20] Lacolley P, Regnault V, Segers P, Laurent S. Vascular smooth muscle cells and arterial stiffening: relevance in development, aging, and disease. Physiol Rev 2017;97(4):1555—617.
[21] O'Rourke MF, Hashimoto J. Mechanical factors in arterial aging: a clinical perspective. J Am Coll Cardiol 2007;50(1):1—13.
[22] Levy BI, Schiffrin EL, Mourad JJ, et al. Impaired tissue perfusion: a pathology common to hypertension, obesity, and diabetes mellitus. Circulation 2008;118(9):968—76.
[23] Vianna LC, Deo SH, Jensen AK, Holwerda SW, Zimmerman MC, Fadel PJ. Impaired dynamic cerebral autoregulation at rest and during isometric exercise in type 2 diabetes patients. Am J Physiol Heart Circ Physiol 2015;308(7):H681—7.
[24] Khavandi K, Greenstein AS, Sonoyama K, et al. Myogenic tone and small artery remodelling: insight into diabetic nephropathy. Nephrol Dial Transplant 2009;24(2):361—9.
[25] Hill MA, Ege EA. Active and passive mechanical properties of isolated arterioles from STZ-induced diabetic rats. Effect of aminoguanidine treatment. Diabetes 1994;43(12):1450—6.
[26] Hill MA, Meininger GA. Impaired arteriolar myogenic reactivity in early experimental diabetes. Diabetes 1993;42(9):1226—32.
[27] Wagenseil JE, Mecham RP. Vascular extracellular matrix and arterial mechanics. Physiol Rev 2009;89(3):957—89.

[28] Jana S, Hu M, Shen M, Kassiri Z. Extracellular matrix, regional heterogeneity of the aorta, and aortic aneurysm. Exp Mol Med 2019;51(12):1—15.

[29] Hill MA, Meininger GA. Arteriolar vascular smooth muscle cells: mechanotransducers in a complex environment. Int J Biochem Cell Biol 2012;44(9):1505—10.

[30] Hill MA, Meininger GA. Small artery mechanobiology: roles of cellular and non-cellular elements. Microcirculation 2016;23(8):611—3.

[31] Hill MA, Nourian Z, Ho IL, Clifford PS, Martinez-Lemus L, Meininger GA. Small artery elastin distribution and architecture-focus on three dimensional organization. Microcirculation 2016;23(8):614—20.

[32] Kohn JC, Lampi MC, Reinhart-King CA. Age-related vascular stiffening: causes and consequences. Front Genet 2015;6:112.

[33] Giacco F, Brownlee M. Oxidative stress and diabetic complications. Circ Res 2010;107(9):1058—70.

[34] Jaworski DM, Sideleva O, Stradecki HM, et al. Sexually dimorphic diet-induced insulin resistance in obese tissue inhibitor of metalloproteinase-2 (TIMP-2)-deficient mice. Endocrinology 2011;152(4):1300—13.

[35] Wagenseil JE, Mecham RP. Elastin in large artery stiffness and hypertension. J Cardiovasc Transl Res 2012;5(3):264—73.

[36] Dhar S, Sun Z, Meininger GA, Hill MA. Nonenzymatic glycation interferes with fibronectin-integrin interactions in vascular smooth muscle cells. Microcirculation 2017;24(3).

[37] Barlovic DP, Soro-Paavonen A, Jandeleit-Dahm KA. RAGE biology, atherosclerosis and diabetes. Clin Sci (Lond) 2011;121(2):43—55.

[38] Steppan J, Bergman Y, Viegas K, et al. Tissue transglutaminase modulates vascular stiffness and function through crosslinking-dependent and crosslinking-independent functions. J Am Heart Assoc 2017;6(2).

[39] Tatsukawa H, Furutani Y, Hitomi K, Kojima S. Transglutaminase 2 has opposing roles in the regulation of cellular functions as well as cell growth and death. Cell Death Dis 2016;7(6):e2244.

[40] Lorand L, Graham RM. Transglutaminases: crosslinking enzymes with pleiotropic functions. Nat Rev Mol Cell Biol 2003;4(2):140—56.

[41] Wang Z, Griffin M. TG2, a novel extracellular protein with multiple functions. Amino Acids 2012;42(2—3):939—49.

[42] Aroor AR, Habibi J, Nistala R, et al. Diet-induced obesity promotes kidney endothelial stiffening and fibrosis dependent on the endothelial mineralocorticoid receptor. Hypertension 2019;73(4):849—58.

[43] Jung SM, Jandu S, Steppan J, et al. Increased tissue transglutaminase activity contributes to central vascular stiffness in eNOS knockout mice. Am J Physiol Heart Circ Physiol 2013;305(6):H803—10.

[44] Ludvigsen TP, Olsen LH, Pedersen HD, Christoffersen BO, Jensen LJ. Hyperglycemia-induced transcriptional regulation of ROCK1 and TGM2 expression is involved in small artery remodeling in obese diabetic Gottingen Minipigs. Clin Sci (Lond) 2019;133(24):2499—516.

[45] Martinez-Revelles S, Garcia-Redondo AB, Avendano MS, et al. Lysyl oxidase induces vascular oxidative stress and contributes to arterial stiffness and abnormal elastin structure in hypertension: role of p38MAPK. Antioxidants Redox Signal 2017;27(7):379—97.

[46] Pastel E, Price E, Sjoholm K, et al. Lysyl oxidase and adipose tissue dysfunction. Metabolism 2018;78:118—27.

[47] Leite S, Cerqueira RJ, Ibarrola J, et al. Arterial remodeling and dysfunction in the ZSF1 rat model of heart failure with preserved ejection fraction. Circ Heart Fail 2019;12(7). e005596.

[48] Cecelja M, Chowienczyk P. Molecular mechanisms of arterial stiffening. Pulse (Basel) 2016;4(1):43—8.

[49] Niederhoffer N, Lartaud-Idjouadiene I, Giummelly P, Duvivier C, Peslin R, Atkinson J. Calcification of medial elastic fibers and aortic elasticity. Hypertension 1997;29(4):999—1006.

[50] Chen Y, Zhao X, Wu H. Arterial stiffness: a focus on vascular calcification and its link to bone mineralization. Arterioscler Thromb Vasc Biol 2020;40(5):1078−93.
[51] Pikilidou M, Yavropoulou M, Antoniou M, Yovos J. The contribution of osteoprogenitor cells to arterial stiffness and hypertension. J Vasc Res 2015;52(1):32−40.
[52] Durham AL, Speer MY, Scatena M, Giachelli CM, Shanahan CM. Role of smooth muscle cells in vascular calcification: implications in atherosclerosis and arterial stiffness. Cardiovasc Res 2018;114(4):590−600.
[53] Ong KL, McClelland RL, Rye KA, et al. The relationship between insulin resistance and vascular calcification in coronary arteries, and the thoracic and abdominal aorta: the Multi-Ethnic Study of Atherosclerosis. Atherosclerosis 2014;236(2):257−62.
[54] Chistiakov DA, Sobenin IA, Orekhov AN, Bobryshev YV. Mechanisms of medial arterial calcification in diabetes. Curr Pharmaceut Des 2014;20(37):5870−83.
[55] Cocciolone AJ, Hawes JZ, Staiculescu MC, Johnson EO, Murshed M, Wagenseil JE. Elastin, arterial mechanics, and cardiovascular disease. Am J Physiol Heart Circ Physiol 2018;315(2):H189−205.
[56] Basalyga DM, Simionescu DT, Xiong W, Baxter BT, Starcher BC, Vyavahare NR. Elastin degradation and calcification in an abdominal aorta injury model: role of matrix metalloproteinases. Circulation 2004;110(22):3480−7.
[57] Yasmin, McEniery CM, Wallace S, et al. Matrix metalloproteinase-9 (MMP-9), MMP-2, and serum elastase activity are associated with systolic hypertension and arterial stiffness. Arterioscler Thromb Vasc Biol 2005;25(2):372.
[58] Antonicelli F, Bellon G, Debelle L, Hornebeck W. Elastin-elastases and inflamm-aging. Curr Top Dev Biol 2007;79:99−155.
[59] Hong Z, Sun Z, Li M, et al. Vasoactive agonists exert dynamic and coordinated effects on vascular smooth muscle cell elasticity, cytoskeletal remodelling and adhesion. J Physiol 2014;592(6):1249−66.
[60] Hong Z, Sun Z, Li Z, Mesquitta WT, Trzeciakowski JP, Meininger GA. Coordination of fibronectin adhesion with contraction and relaxation in microvascular smooth muscle. Cardiovasc Res 2012;96(1):73−80.
[61] Sun Z, Martinez-Lemus LA, Hill MA, Meininger GA. Extracellular matrix-specific focal adhesions in vascular smooth muscle produce mechanically active adhesion sites. Am J Physiol Cell Physiol 2008;295(1):C268−78.
[62] Bellien J, Favre J, Iacob M, et al. Arterial stiffness is regulated by nitric oxide and endothelium-derived hyperpolarizing factor during changes in blood flow in humans. Hypertension 2010;55(3):674−80.
[63] Noma K, Goto C, Nishioka K, et al. Roles of rho-associated kinase and oxidative stress in the pathogenesis of aortic stiffness. J Am Coll Cardiol 2007;49(6):698−705.
[64] Sugawara J, Komine H, Hayashi K, et al. Effect of systemic nitric oxide synthase inhibition on arterial stiffness in humans. Hypertens Res 2007;30(5):411−5.
[65] Shimizu K, Yamamoto T, Takahashi M, Sato S, Noike H, Shirai K. Effect of nitroglycerin administration on cardio-ankle vascular index. Vasc Health Risk Manag 2016;12:313−9.
[66] Liberts EA, Willoughby SR, Kennedy JA, Horowitz JD. Effects of perhexiline and nitroglycerin on vascular, neutrophil and platelet function in patients with stable angina pectoris. Eur J Pharmacol 2007;560(1):49−55.
[67] Sehgel NL, Vatner SF, Meininger GA. Smooth muscle cell stiffness syndrome"-revisiting the structural basis of arterial stiffness. Front Physiol 2015;6:335.
[68] Sehgel NL, Sun Z, Hong Z, et al. Augmented vascular smooth muscle cell stiffness and adhesion when hypertension is superimposed on aging. Hypertension 2015;65(2):370−7.
[69] Yu G, Zou H, Prewitt RL, Hill MA. Impaired arteriolar mechanotransduction in experimental diabetes mellitus. J Diabet Complicat 1999;13(5−6):235−42.
[70] Takasaki I, Chobanian AV, Sarzani R, Brecher P. Effect of hypertension on fibronectin expression in the rat aorta. J Biol Chem 1990;265(35):21935−9.

[71] McDaniel DP, Shaw GA, Elliott JT, et al. The stiffness of collagen fibrils influences vascular smooth muscle cell phenotype. Biophys J 2007;92(5):1759−69.
[72] Galbraith CG, Sheetz MP. Forces on adhesive contacts affect cell function. Curr Opin Cell Biol 1998;10(5):566−71.
[73] Aroor AR, McKarns S, Demarco VG, Jia G, Sowers JR. Maladaptive immune and inflammatory pathways lead to cardiovascular insulin resistance. Metabolism 2013;62(11):1543−52.
[74] Schiffrin EL. Immune mechanisms in hypertension and vascular injury. Clin Sci (Lond) 2014;126(4):267−74.
[75] Zanoli L, Briet M, Empana JP, et al. Vascular consequences of inflammation: a position statement from the ESH working group on vascular structure and function and the ARTERY society. J Hypertens 2020;38(9):1682−98.
[76] Aroor AR, Jia G, Sowers JR. Cellular mechanisms underlying obesity-induced arterial stiffness. Am J Physiol Regul Integr Comp Physiol 2018;314(3):R387−98.
[77] Kalupahana NS, Moustaid-Moussa N, Claycombe KJ. Immunity as a link between obesity and insulin resistance. Mol Aspect Med 2012;33(1):26−34.
[78] Qiu T, Li M, Tanner MA, et al. Depletion of dendritic cells in perivascular adipose tissue improves arterial relaxation responses in type 2 diabetic mice. Metabolism 2018;85:76−89.
[79] Mikolajczyk TP, Guzik TJ. Adaptive immunity in hypertension. Curr Hypertens Rep 2019;21(9):68.
[80] Kem DC, Li H, Velarde-Miranda C, et al. Autoimmune mechanisms activating the angiotensin AT1 receptor in 'primary' aldosteronism. J Clin Endocrinol Metab 2014;99(5):1790−7.
[81] Bai B, Yang Y, Wang Q, et al. NLRP3 inflammasome in endothelial dysfunction. Cell Death Dis 2020;11(9):776.
[82] Aroor AR, Jia G, Habibi J, et al. Uric acid promotes vascular stiffness, maladaptive inflammatory responses and proteinuria in western diet fed mice. Metabolism 2017;74:32−40.
[83] Effectiveness of spironolactone added to an angiotensin-converting enzyme inhibitor and a loop diuretic for severe chronic congestive heart failure (the Randomized Aldactone Evaluation Study [RALES]). Am J Cardiol 1996;78(8):902−7.
[84] Collier TJ, Pocock SJ, McMurray JJ, et al. The impact of eplerenone at different levels of risk in patients with systolic heart failure and mild symptoms: insight from a novel risk score for prognosis derived from the EMPHASIS-HF trial. Eur Heart J 2013;34(36):2823−9.
[85] Pitt B, Remme W, Zannad F, et al. Eplerenone, a selective aldosterone blocker, in patients with left ventricular dysfunction after myocardial infarction. N Engl J Med 2003;348(14):1309−21.
[86] Pitt B, Zannad F, Remme WJ, et al. The effect of spironolactone on morbidity and mortality in patients with severe heart failure. Randomized Aldactone Evaluation Study Investigators. N Engl J Med 1999;341(10):709−17.
[87] Funder JW. Aldosterone and mineralocorticoid receptors-physiology and pathophysiology. Int J Mol Sci 2017;18(5).
[88] Rickard AJ, Morgan J, Chrissobolis S, Miller AA, Sobey CG, Young MJ. Endothelial cell mineralocorticoid receptors regulate deoxycorticosterone/salt-mediated cardiac remodeling and vascular reactivity but not blood pressure. Hypertension 2014;63(5):1033−40.
[89] Schafer N, Lohmann C, Winnik S, et al. Endothelial mineralocorticoid receptor activation mediates endothelial dysfunction in diet-induced obesity. Eur Heart J 2013;34(45):3515−24.
[90] McCurley A, Pires PW, Bender SB, et al. Direct regulation of blood pressure by smooth muscle cell mineralocorticoid receptors. Nat Med 2012;18(9):1429−33.
[91] DuPont JJ, Hill MA, Bender SB, Jaisser F, Jaffe IZ. Aldosterone and vascular mineralocorticoid receptors: regulators of ion channels beyond the kidney. Hypertension 2014;63(4):632−7.
[92] Rickard AJ, Morgan J, Tesch G, Funder JW, Fuller PJ, Young MJ. Deletion of mineralocorticoid receptors from macrophages protects against deoxycorticosterone/salt-induced cardiac fibrosis and increased blood pressure. Hypertension 2009;54(3):537−43.

[93] Bienvenu LA, Morgan J, Rickard AJ, et al. Macrophage mineralocorticoid receptor signaling plays a key role in aldosterone-independent cardiac fibrosis. Endocrinology 2012;153(7):3416−25.

[94] Infante M, Armani A, Marzolla V, Fabbri A, Caprio M. Adipocyte mineralocorticoid receptor. Vitam Horm 2019;109:189−209.

[95] Brilla CG, Zhou G, Matsubara L, Weber KT. Collagen metabolism in cultured adult rat cardiac fibroblasts: response to angiotensin II and aldosterone. J Mol Cell Cardiol 1994;26(7):809−20.

[96] DuPont JJ, Jaffe IZ. 30 years of the mineralocorticoid receptor: the role of the mineralocorticoid receptor in the vasculature. J Endocrinol 2017;234(1):T67−82.

[97] Jia G, Jia Y, Sowers JR. Role of mineralocorticoid receptor activation in cardiac diastolic dysfunction. Biochim Biophys Acta (BBA) - Mol Basis Dis 2017;1863(8):2012−8.

[98] Feraco A, Marzolla V, Scuteri A, Armani A, Caprio M. Mineralocorticoid receptors in metabolic syndrome: from physiology to disease. Trends Endocrinol Metab 2020;31(3):205−17.

[99] DuPont JJ, McCurley A, Davel AP, et al. Vascular mineralocorticoid receptor regulates microRNA-155 to promote vasoconstriction and rising blood pressure with aging. JCI Insight 2016;1(14). e88942.

[100] Jia G, Habibi J, Aroor AR, et al. Endothelial mineralocorticoid receptor mediates diet-induced aortic stiffness in females. Circ Res 2016;118(6):935−43.

[101] Funder JW, Pearce PT, Smith R, Smith AI. Mineralocorticoid action: target tissue specificity is enzyme, not receptor, mediated. Science 1988;242(4878):583−5.

[102] Skott O, Uhrenholt TR, Schjerning J, Hansen PB, Rasmussen LE, Jensen BL. Rapid actions of aldosterone in vascular health and disease–friend or foe? Pharmacol Ther 2006;111(2):495−507.

[103] Jaffe IZ, Mendelsohn ME. Angiotensin II and aldosterone regulate gene transcription via functional mineralocortocoid receptors in human coronary artery smooth muscle cells. Circ Res 2005;96(6): 643−50.

[104] Bender SB, McGraw AP, Jaffe IZ, Sowers JR. Mineralocorticoid receptor-mediated vascular insulin resistance: an early contributor to diabetes-related vascular disease? Diabetes 2013;62(2):313−9.

[105] Muniyappa R, Sowers JR. Role of insulin resistance in endothelial dysfunction. Rev Endocr Metab Disord 2013;14(1):5−12.

[106] Seifalian AM, Filippatos TD, Joshi J, Mikhailidis DP. Obesity and arterial compliance alterations. Curr Vasc Pharmacol 2010;8(2):155−68.

[107] Kleyman TR, Kashlan OB, Hughey RP. Epithelial Na(+) channel regulation by extracellular and intracellular factors. Annu Rev Physiol 2018;80:263−81.

[108] Pearce D, Soundararajan R, Trimpert C, Kashlan OB, Deen PM, Kohan DE. Collecting duct principal cell transport processes and their regulation. Clin J Am Soc Nephrol 2015;10(1): 135−46.

[109] Mutchler SM, Kleyman TR. New insights regarding epithelial Na+ channel regulation and its role in the kidney, immune system and vasculature. Curr Opin Nephrol Hypertens 2019;28(2): 113−9.

[110] Waldmann R, Champigny G, Bassilana F, Voilley N, Lazdunski M. Molecular cloning and functional expression of a novel amiloride-sensitive Na+ channel. J Biol Chem 1995;270(46): 27411−4.

[111] Li Q, Fung E. Multifaceted functions of epithelial Na(+) channel in modulating blood pressure. Hypertension 2019;73(2):273−81.

[112] Mamenko M, Zaika O, Ilatovskaya DV, Staruschenko A, Pochynyuk O. Angiotensin II increases activity of the epithelial Na+ channel (ENaC) in distal nephron additively to aldosterone. J Biol Chem 2012;287(1):660−71.

[113] Matalon S, Bartoszewski R, Collawn JF. Role of epithelial sodium channels in the regulation of lung fluid homeostasis. Am J Physiol Lung Cell Mol Physiol 2015;309(11):L1229−38.

[114] Schild L, Kellenberger S. Structure function relationships of ENaC and its role in sodium handling. Adv Exp Med Biol 2001;502:305−14.

[115] Garty H, Palmer LG. Epithelial sodium channels: function, structure, and regulation. Physiol Rev 1997;77(2):359—96.
[116] Tarjus A, Maase M, Jeggle P, et al. The endothelial alphaENaC contributes to vascular endothelial function in vivo. PloS One 2017;12(9). e0185319.
[117] Drummond HA, Grifoni SC, Jernigan NL. A new trick for an old dogma: ENaC proteins as mechanotransducers in vascular smooth muscle. Physiology 2008;23:23—31.
[118] Golestaneh N, Klein C, Valamanesh F, Suarez G, Agarwal MK, Mirshahi M. Mineralocorticoid receptor-mediated signaling regulates the ion gated sodium channel in vascular endothelial cells and requires an intact cytoskeleton. Biochem Biophys Res Commun 2001;280(5):1300—6.
[119] Kellenberger S, Schild L. Epithelial sodium channel/degenerin family of ion channels: a variety of functions for a shared structure. Physiol Rev 2002;82(3):735—67.
[120] Boscardin E, Alijevic O, Hummler E, Frateschi S, Kellenberger S. The function and regulation of acid-sensing ion channels (ASICs) and the epithelial Na(+) channel (ENaC): IUPHAR Review 19. Br J Pharmacol 2016;173(18):2671—701.
[121] Kellenberger S, Schild L. International Union of Basic and Clinical Pharmacology. XCI. structure, function, and pharmacology of acid-sensing ion channels and the epithelial Na+ channel. Pharmacol Rev 2015;67(1):1—35.
[122] Ji HL, Zhao RZ, Chen ZX, Shetty S, Idell S, Matalon S. Delta ENaC: a novel divergent amiloride-inhibitable sodium channel. Am J Physiol Lung Cell Mol Physiol 2012;303(12):L1013—26.
[123] Fyfe GK, Canessa CM. Subunit composition determines the single channel kinetics of the epithelial sodium channel. J Gen Physiol 1998;112(4):423—32.
[124] Meltzer RH, Kapoor N, Qadri YJ, Anderson SJ, Fuller CM, Benos DJ. Heteromeric assembly of acid-sensitive ion channel and epithelial sodium channel subunits. J Biol Chem 2007;282(35):25548—59.
[125] Baldin JP, Barth D, Fronius M. Epithelial Na(+) channel (ENaC) formed by one or two subunits forms functional channels that respond to shear force. Front Physiol 2020;11:141.
[126] Kleyman TR, Carattino MD, Hughey RP. ENaC at the cutting edge: regulation of epithelial sodium channels by proteases. J Biol Chem 2009;284(31):20447—51.
[127] Carattino MD, Passero CJ, Steren CA, et al. Defining an inhibitory domain in the alpha-subunit of the epithelial sodium channel. Am J Physiol Ren Physiol 2008;294(1):F47—52.
[128] Abi-Antoun T, Shi S, Tolino LA, Kleyman TR, Carattino MD. Second transmembrane domain modulates epithelial sodium channel gating in response to shear stress. Am J Physiol Ren Physiol 2011;300(5):F1089—95.
[129] Althaus M, Bogdan R, Clauss WG, Fronius M. Mechano-sensitivity of epithelial sodium channels (ENaCs): laminar shear stress increases ion channel open probability. Faseb J 2007;21(10):2389—99.
[130] Knoepp F, Ashley Z, Barth D, et al. Shear force sensing of epithelial Na(+) channel (ENaC) relies on N-glycosylated asparagines in the palm and knuckle domains of alphaENaC. Proc Natl Acad Sci U S A 2020;117(1):717—26.
[131] Wang S, Meng F, Mohan S, Champaneri B, Gu Y. Functional ENaC channels expressed in endothelial cells: a new candidate for mediating shear force. Microcirculation 2009;16(3):276—87.
[132] Mukherjee A, Wang Z, Kinlough CL, et al. Specific palmitoyltransferases associate with and activate the epithelial sodium channel. J Biol Chem 2017;292(10):4152—63.
[133] Pochynyuk O, Bugaj V, Stockand JD. Physiologic regulation of the epithelial sodium channel by phosphatidylinositides. Curr Opin Nephrol Hypertens 2008;17(5):533—40.
[134] Martinac B. The ion channels to cytoskeleton connection as potential mechanism of mechanosensitivity. Biochim Biophys Acta 2014;1838(2):682—91.
[135] Mazzochi C, Bubien JK, Smith PR, Benos DJ. The carboxyl terminus of the alpha-subunit of the amiloride-sensitive epithelial sodium channel binds to F-actin. J Biol Chem 2006;281(10):6528—38.

[136] Wang Q, Dai XQ, Li Q, et al. Filamin interacts with epithelial sodium channel and inhibits its channel function. J Biol Chem 2013;288(1):264–73.
[137] Shakibaei M, Mobasheri A. Beta1-integrins co-localize with Na, K-ATPase, epithelial sodium channels (ENaC) and voltage activated calcium channels (VACC) in mechanoreceptor complexes of mouse limb-bud chondrocytes. Histol Histopathol 2003;18(2):343–51.
[138] Korte S, Wiesinger A, Straeter AS, Peters W, Oberleithner H, Kusche-Vihrog K. Firewall function of the endothelial glycocalyx in the regulation of sodium homeostasis. Pflügers Archiv 2012;463(2): 269–78.
[139] Ramiro-Diaz J, Barajas-Espinosa A, Chi-Ahumada E, et al. Luminal endothelial lectins with affinity for N-acetylglucosamine determine flow-induced cardiac and vascular paracrine-dependent responses. Am J Physiol Heart Circ Physiol 2010;299(3):H743–51.
[140] Fancher IS, Le Master E, Ahn SJ, et al. Impairment of flow-sensitive inwardly rectifying K(+) channels via disruption of glycocalyx mediates obesity-induced endothelial dysfunction. Arterioscler Thromb Vasc Biol 2020;40(9):e240–55.
[141] Mansley MK, Watt GB, Francis SL, et al. Dexamethasone and insulin activate serum and glucocorticoid-inducible kinase 1 (SGK1) via different molecular mechanisms in cortical collecting duct cells. Phys Rep 2016;4(10).
[142] Wang J, Barbry P, Maiyar AC, et al. SGK integrates insulin and mineralocorticoid regulation of epithelial sodium transport. Am J Physiol Ren Physiol 2001;280(2):F303–13.
[143] Blazer-Yost BL, Liu X, Helman SI. Hormonal regulation of ENaCs: insulin and aldosterone. Am J Physiol 1998;274(5):C1373–9.
[144] Kusche-Vihrog K, Sobczak K, Bangel N, et al. Aldosterone and amiloride alter ENaC abundance in vascular endothelium. Pflügers Archiv 2008;455(5):849–57.
[145] Kusche-Vihrog K, Jeggle P, Oberleithner H. The role of ENaC in vascular endothelium. Pflügers Archiv 2014;466(5):851–9.
[146] Martinez-Lemus LA, Aroor AR, Ramirez-Perez FI, et al. Amiloride improves endothelial function and reduces vascular stiffness in female mice fed a western diet. Front Physiol 2017;8:456.
[147] Jeggle P, Callies C, Tarjus A, et al. Epithelial sodium channel stiffens the vascular endothelium in vitro and in Liddle mice. Hypertension 2013;61(5):1053–9.
[148] Tarjus A, Gonzalez-Rivas C, Amador-Martinez I, et al. The absence of endothelial sodium channel alpha (alphaENaC) reduces renal ischemia/reperfusion injury. Int J Mol Sci 2019;20(13).
[149] Bentley-Lewis R, Adler GK, Perlstein T, et al. Body mass index predicts aldosterone production in normotensive adults on a high-salt diet. J Clin Endocrinol Metab 2007;92(11):4472–5.
[150] Huby AC, Antonova G, Groenendyk J, et al. Adipocyte-derived hormone leptin is a direct regulator of aldosterone secretion, which promotes endothelial dysfunction and cardiac fibrosis. Circulation 2015;132(22):2134–45.
[151] Druppel V, Kusche-Vihrog K, Grossmann C, et al. Long-term application of the aldosterone antagonist spironolactone prevents stiff endothelial cell syndrome. Faseb J 2013;27(9):3652–9.
[152] Korte S, Strater AS, Druppel V, et al. Feedforward activation of endothelial ENaC by high sodium. Faseb J 2014;28(9):4015–25.
[153] Jia G, Habibi J, Aroor AR, et al. Epithelial sodium channel in aldosterone-induced endothelium stiffness and aortic dysfunction. Hypertension 2018;72(3):731–8.
[154] Jia G, Aroor AR, Martinez-Lemus LA, Sowers JR. Overnutrition, mTOR signaling, and cardiovascular diseases. Am J Physiol Regul Integr Comp Physiol 2014;307(10):R1198–206.
[155] Kamynina E, Debonneville C, Bens M, Vandewalle A, Staub O. A novel mouse Nedd4 protein suppresses the activity of the epithelial Na+ channel. Faseb J 2001;15(1):204–14.
[156] Frindt G, Bertog M, Korbmacher C, Palmer LG. Ubiquitination of renal ENaC subunits in vivo. Am J Physiol Ren Physiol 2020;318(5):F1113–21.
[157] Gleason CE, Frindt G, Cheng CJ, et al. mTORC2 regulates renal tubule sodium uptake by promoting ENaC activity. J Clin Invest 2015;125(1):117–28.

[158] Diakov A, Korbmacher C. A novel pathway of epithelial sodium channel activation involves a serum- and glucocorticoid-inducible kinase consensus motif in the C terminus of the channel's alpha-subunit. J Biol Chem 2004;279(37):38134–42.
[159] Valinsky WC, Touyz RM, Shrier A. Aldosterone, SGK1, and ion channels in the kidney. Clin Sci (Lond) 2018;132(2):173–83.
[160] Blazer-Yost BL, Esterman MA, Vlahos CJ. Insulin-stimulated trafficking of ENaC in renal cells requires PI 3-kinase activity. Am J Physiol Cell Physiol 2003;284(6):C1645–53.
[161] Record RD, Froelich LL, Vlahos CJ, Blazer-Yost BL. Phosphatidylinositol 3-kinase activation is required for insulin-stimulated sodium transport in A6 cells. Am J Physiol 1998;274(4):E611–7.
[162] Record RD, Johnson M, Lee S, Blazer-Yost BL. Aldosterone and insulin stimulate amiloride-sensitive sodium transport in A6 cells by additive mechanisms. Am J Physiol 1996;271(4 Pt 1):C1079–84.

CHAPTER 6

Pathomechanics of diabetic foot ulceration: revisiting plantar shear and temperature

Metin Yavuz
Department of Applied Clinical Research, Prosthetics and Orthotics Program, University of Texas Southwestern Medical Center, Dallas, TX, United States

Biomechanics overview

Diabetic foot ulcers are known to have a biomechanical etiology. By biomechanics, we refer to the ground reaction forces that act on the foot sole by. Newton's third law of motion states that for every action (i.e., force) there is an opposite reaction (i.e., force) in the same magnitude. Since body weight is conveyed to the ground during standing and locomotion via the human foot, the ground counters that force in the same amount, however, in opposite reaction. We call this the vertical component of ground reaction forces. While standing still, the vertical ground reaction force is equal to the body weight which can be expressed as:

$$F_N = m \times a$$

where

F_N is normal force or simply vertical force experienced by the feet,

m is body mass, and

a is acceleration and is equal to gravitational acceleration or g during quiet standing.

When human beings ambulate, such as in walking, the ground reaction forces act on all three dimensions. This is because the foot experiences horizontal forces in the anteroposterior and mediolateral planes as well during walking. These horizontal forces are called shear forces. These forces can easily be measured by utilizing a force plate (Fig. 6.1).

Unfortunately, the use of ground reaction forces alone may not be clinically relevant in diabetic foot ulceration. An analogy is an inflated latex balloon. If we apply a force of 2 pounds on the balloon with our palms, the chances are that it will stay intact. If we apply the same amount of force via a pushpin, we all can expect to hear a loud "pop!". The difference between the two cases, while the amount of force is identical, is the surface area on which the force applies. Similar to our latex balloon, many materials, including tissue, are susceptible to failure under high force concentrations. The technical term for force concentration is "stress" and it is calculated by the following formula:

$$S = F/A$$

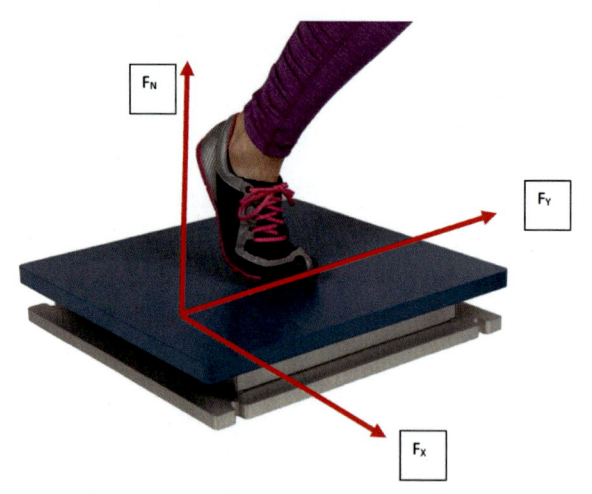

Figure 6.1 Ground reaction forces act on all three planes during ambulation. *FN*, Normal or vertical; *FX*, anteroposterior ground reaction forces; *FY*, mediolateral.

where
    F is the magnitude of force that is applied, and
    A is the surface area on which F acts.

In our balloon exercise, let us assume that the surface area of our palm is 5 in^2 and the surface area of the tip of the pushpin is 0.001 in^2. The stress that is applied on to the balloon via the palms will be 0.4 psi (2 lbs/5 in^2) compared to 2000 psi (2 lbs/ 0.001 in^2) in the pushpin case. Therefore, in situations where material failure is of concern, such as diabetic foot ulceration, it is important to quantify and assess stress magnitudes as opposed to force magnitudes only.

As a matter of fact, Hall and Brand stated that diabetic foot ulcers occur due to "repetitive moderate mechanical stresses in the presence of peripheral neuropathy" [1]. The word "repetitive" in this statement refers to ambulation, as ground reaction forces are exerted on the foot sole during each step that we take (Fig. 6.2). If this commonly accepted statement is true, it is indicated that diabetic foot ulcers occur due to loadbearing activity. In other words, if a diabetic patient at high risk for foot ulceration does not walk at all, then he or she will not develop a foot ulcer. This statement clearly emphasizes the significance of biomechanics in the formation of foot ulcers.

While a force plate can provide the magnitude and direction of ground reaction forces exerted on the foot sole, it cannot reveal how stresses are distributed. It is relatively easy to calculate the surface area of the human foot and then derive the stress magnitude by dividing the measured force amplitude by that surface area. However, it must be kept in mind that effective loadbearing area of the foot may change based on different variables. For example, when walking barefoot, the midfoot area usually does not experience

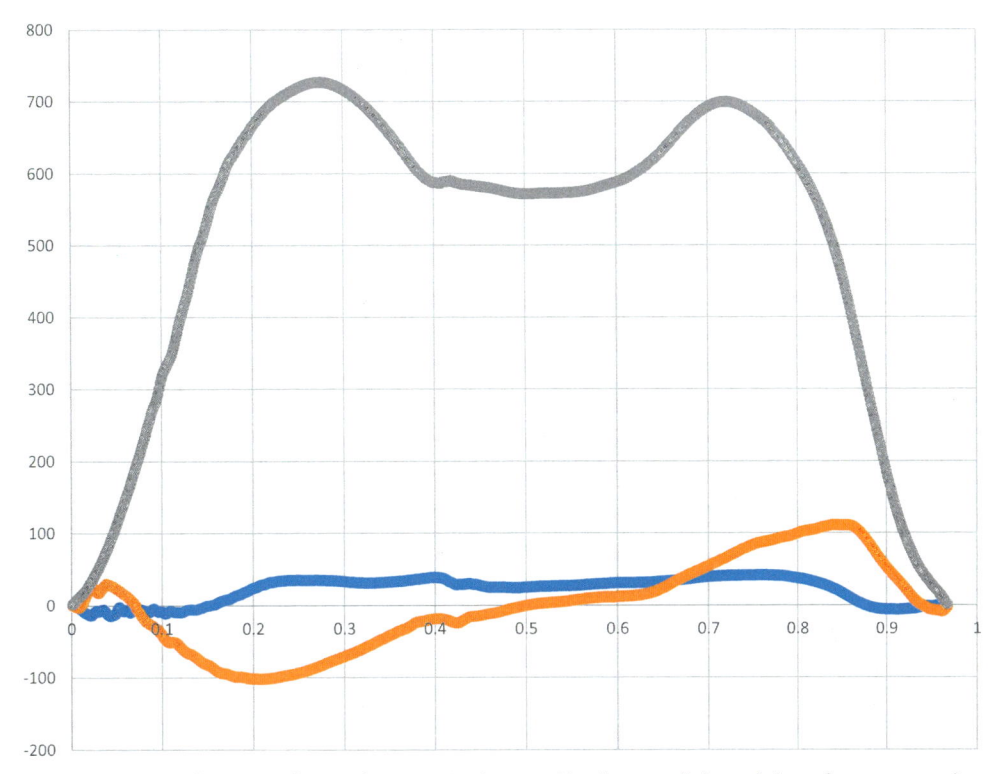

Figure 6.2 Ground reaction forces during a single step. *Blue line*: mediolateral shear force, *orange line*: anteroposterior shear force, *gray line*: vertical (normal) ground reaction force. Y axis unit is Newtons, X axis unit is seconds.

significant stress due to the presence of the foot arch. This is evident, for example, from footprints left on the sand when a healthy person walks on the beach while barefooted. The shape of the footprint is like a crescent moon, indicating that only a small area in the midfoot bears weight. Similarly, in an individual with a bony prominent area or a plantar callus, stresses localize in that area, rendering our initial formula-based stress calculation mostly useless. There may be other factors such as internal muscle activity, various foot deformities, etc., that may further alter the distribution of plantar forces and stresses and hence ultimately one needs to have a different kind of equipment in order to quantify plantar stresses.

It is worth mentioning here that the plantar stress associated with the vertical ground reaction force is termed "pressure." Since the term "pressure" is specific to the vertical or normal plantar stress, the use of "shear pressure" is absolutely incorrect. The author of this chapter has seen and heard such use many times, even from well-known scholars in diabetic foot research. The correct term for the horizontal plantar stress is "shear stress."

Plantar pressure

Modern plantar pressure measurement systems date back to as early as 1970s. Betts and Duckworth described a system which utilized a glass plate on which a special thin film was placed [2]. A monochrome camera captured light deflections coming from the film as a subject stood or walked across the glass plate. Measured light intensities were converted to force readings. A number of manuscripts utilized this system and reported barefoot plantar pressures as high as 1100 kPa in diabetic patients [3—6]. A number of other devices utilizing various pedobarographic approaches have been described and developed in those years as well [7—9]. By the year 1994, various companies had brought thinner capacitive and resistive sensor-based devices to the market [10]. Availability of commercial devices accelerated research that aimed at studying foot pressures as they related to diabetic foot ulcers. Initial studies indicated that pressures were elevated in diabetic patients with peripheral neuropathy compared to patients without neuropathy and/or healthy controls [11—14].

Pressure measurement systems can be categorized under two main groups: (i) pressure platforms which are plate-type devices that are placed on the ground and capture barefoot pressures, and (ii) in-shoe pressure systems which are usually insole-type devices. In-shoe devices, as the name suggests, are placed inside the shoes usually on top of the shoe insert and capture plantar pressures over a few steps. In this regard, the main difference between the two units is the foot condition: barefoot versus shod. Pressure platforms can capture data only from one single step. Many investigators traditionally collect data from multiple trials in order to average pressure values over a few steps or analyze the data in statistical approaches that utilize repeated measures. The advantage of the in-shoe systems is that data from multiple steps can be captured during a single walking trial. Investigators have traditionally reported pressure data from three or five midgait steps during normal gait. Traditionally, barefoot pressures were recorded for diagnostic purposes and to identify locations of peak plantar pressures. In-shoe systems are usually employed to assess various orthotic interventions, such as multidensity pressure-relieving insoles or metatarsal pads.

Regardless of the type of the pressure system, the acquired data are usually complex. Both barefoot and in-shoe systems employ an array of sensors in a discrete arrangement. Data are often collected at a minimum of 50 Hz. The resulting data set from a single step is a three-dimensional data array (Fig. 6.3), with third dimension being the time as the typical duration of a stance phase is about 0.5—1.0 s. This means that there are usually 25—50 frames of plantar pressure data collected every 20 ms over the stance duration. Since it is not practical to analyze and present dynamic pressure information, the data set is reduced to either a mountain pressure plot (Fig. 6.4) or simply a two-dimensional array of maximum pressure values. This array is usually called the peak pressure plot or profile.

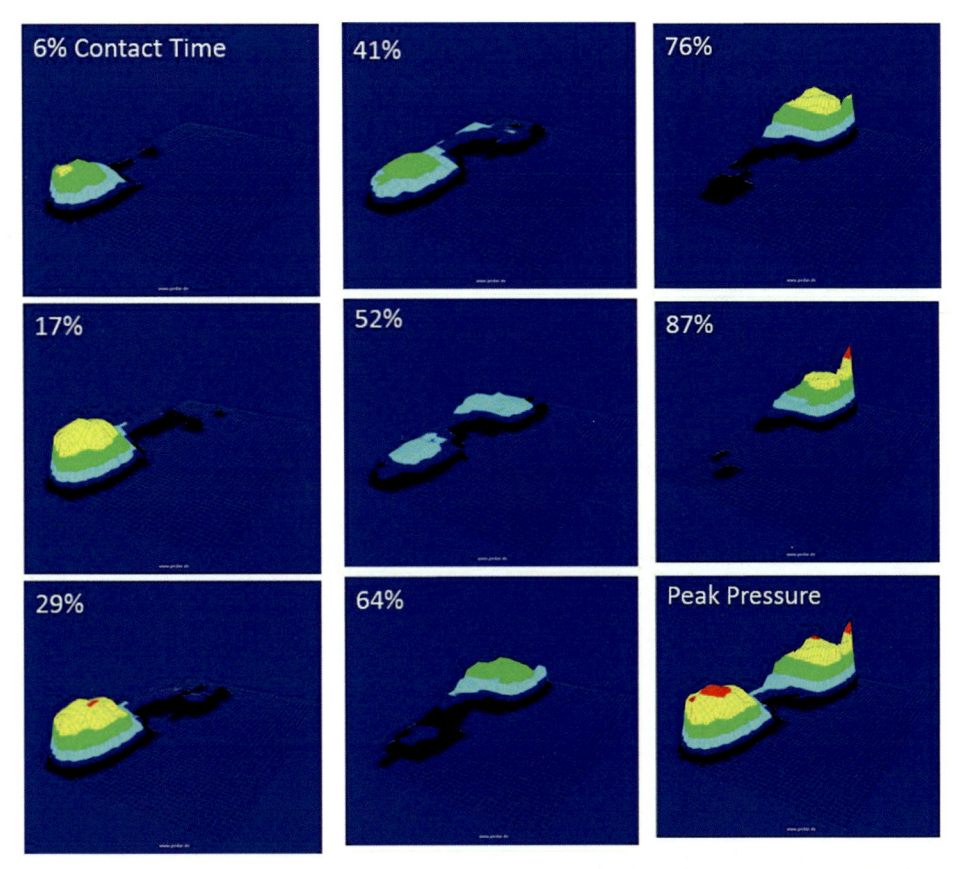

Figure 6.3 Progression of plantar pressures during the stance phase. Data were collected at 50 Hz, indicating that a two-dimensional matrix of data was collected 50 times every second. The last frame represents the maximum pressure values for each active sensor over the 50 time points.

A further reduction of the data is generally necessary for statistical analysis purposes. Moreover, the footprint can be masked into a few regions based on various masking techniques [15,16]. This approach enables the comparison of regional foot biomechanics across subject groups. The other alternative is choosing the maximum global pressure value from the peak pressure profile. For example, in Fig. 6.4, the global peak pressures for the right and left feet are 468 kPa (corresponding to the right hallux) and 365 kPa (corresponding to the left hallux), respectively. If a regional analysis is not provided in a manuscript, it can be assumed that the provided peak pressure values refer to the global peak values.

As we have discussed before, over the years, many studies have revealed that peak plantar pressures are higher in diabetic patients with neuropathy. This increase has been attributed to increased body mass [17–19], change in tissue and skin properties

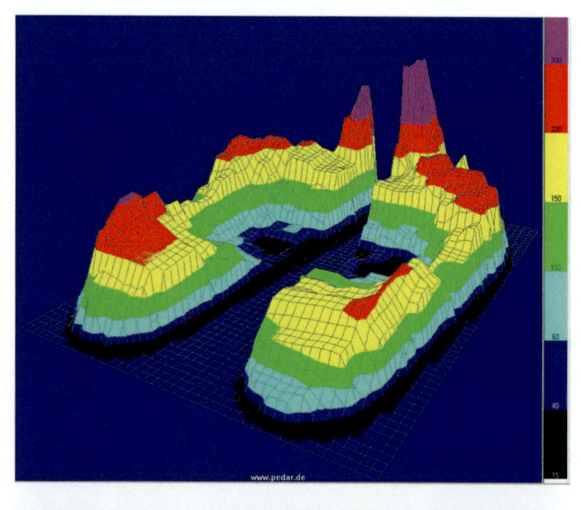

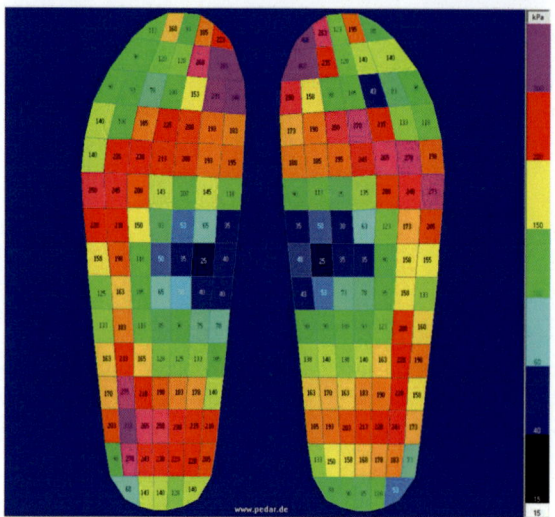

Figure 6.4 Mountain pressure plot of a subject as provided by Novel Pedar software (top). Peak pressure plots of both feet reduced from the mountain plot (bottom).

due to diabetes [20–23], limited joint mobility and deformities [24–27], plantar callosities [28,29], and a combination of these factors as well as others [30]. Therefore, high plantar pressure has been considered a risk factor for diabetic foot ulcers for a long time.

On the other hand, there are various issues associated with the clinical value of plantar pressures. Several studies tried to identify a critical pressure value that could serve as a prediction tool. Armstrong et al. quantified plantar pressures in a total of 219 diabetic subjects, of whom 76 had a history of ulceration [31]. They identified a significant increase in peak pressures in patients with ulcers compared to the control group. They calculated the diagnostic sensitivity and specificity values of three peak pressure values. For 350 kPa, they observed 98.6% sensitivity, whereas the specificity was only 6.7%. The peak value of 1100 kPa had high specificity at 98.0%, however poor sensitivity at only 15.7%. The optimum critical point was determined as 700 kPa which revealed a sensitivity of 70.0% and a specificity of 65.1%. The results of the study indicated that ulcers may develop at low pressure values, while existence of high pressures may not necessarily mean that a foot would ulcerate. In a similar, but larger, study with 1666 enrolled subjects, Lavery et al. investigated the same research question [32]. Their analyses indicated 875 kPa as the most optimum peak pressure value which carried a sensitivity of 63.5% and a specificity of 46.3%. The authors concluded that "foot pressure is a poor tool by itself" to predict diabetic foot ulcers.

If foot pressure were the most important causative factor in foot ulcer pathomechanics, one would expect to see that a significant portion of ulcers developed at peak pressure locations. To our knowledge, the only prospective study that reported the percent overlap between peak pressure locations and subsequent ulcer locations revealed that only 38% of plantar ulcers develop at foot sites that bear the maximum pressures [33].

Lastly, we would like to overview pressure-relieving insoles and footwear that have been on the market for a long time. Given the emphasis placed on foot pressures, many companies and research groups have developed and tested various pressure-relieving insoles in order to prevent foot ulcers. A review of such footwear has been published by Bus et al. initially in 2008 and then in 2016 [34,35], which unfortunately revealed that preventive value of such footwear in preventing ulcer recurrence was "meager" at best, while several randomized controlled trials indicated that if patients adhere to their prescribed footwear then prevention rates are higher.

While there is no clear cut-off value for peak pressures that may lead to tissue damage, Owings et al. recommended that until such a value can be established preventive footwear should target a maximum in-shoe pressure value of 200 kPa [36]. Several articles investigated the pressure-relieving characteristics of off-the-shelf athletic shoes, which have been shown to be used by more than 70% of diabetic patients and the most used type of shoe other than therapeutic shoes [37]. Lavery et al. reported that the athletic and comfort shoes that they studied reduced peak pressures down to the recommended level of peak pressures (i.e., <200 kPa) [38]. They concluded their study by stating that such shoes are as effective as commonly prescribed therapeutic shoes, and when used in conjunction with a viscoelastic insole, they were more effective in reducing foot pressures. Zwaferink et al. reported similar pressure reductions in subjects who walked with regular athletic shoes [39]. Given that a significant portion of diabetic patients have access to pressure-relieving insoles/footwear, whether in the form of customized devices or just regular athletic shoes, it could be anticipated that diabetic foot ulcer and related amputation rates must have been declining over the years. Unfortunately, recent data in the United States indicate just the opposite, as diabetic foot ulcer incidences and associated amputations have been on the rise [40]. This outcome indicates that designing preventive devices and/or footwear based only on pressure characteristics is not effective at all. Some investigators may argue here that adherence to prescribed footwear is a major factor playing into this outcome. However, we would argue back that if patients are not wearing their prescribed footwear, then they still would be wearing a comfort shoe or sneakers while outdoors which have been shown to provide good pressure relief. Of course, the situation may be different for habitually barefoot walking communities that are spread in the world.

The issues we have raised here clearly indicate that diabetic foot ulcers have a more complicated pathomechanics and taking only pressure into account in prevention of foot ulcers will not move the needle in ulceration and amputation rates.

Comparison with pressure injuries (ulcers)

We believe that many diabetic foot researchers have been influenced by the pressure ulcer research, which arguably have stronger foundations. Therefore, we believe it is worth comparing the pathomechanics of the two clinical problems here. According to the quick reference guide of the National Pressure Injury Advisory Panel (NPIAP) and European Pressure Ulcer Advisory Panel (EPUAP), a pressure injury (i.e., ulcer) is a local injury to the skin and/or underlying tissue as a result of pressure, or pressure in combination with shear [41]. Pressure ulcers usually occur over a bony prominent area and the underlying cause is usually cell autolysis due to ischemia which is caused by mechanical stresses. From a biomechanical perspective, what causes pressure ulcers is the continuous or unrelieved application of mechanical stress, in particular pressure. In this regard, the main difference between the pathomechanics of both types of ulcers is "continuous" or "unrelieved" versus "repetitive" application of stresses. Diabetic foot ulcers occur due to loadbearing activity of the body, which may include standing, walking, running, jumping, etc. If we take the most common type of activity, namely, walking, into account, while plantar stresses may restrict blood flow to the tissue during the stance phase, these loads are lifted during the swing phase which restore blood flow to the tissue. The emphasis on "repetitive moderate mechanical stress" should actually be a hint for describing "fatigue failure" of the tissue which we will discuss in the upcoming sections of this chapter. It should also be noted here that most ulcers that develop on the dorsum of the foot occur due to deformities and ischemia and may be treated as pressure ulcers. Dorsal ulcers account for about 15% of the ulcers that are observed in diabetic patients [42].

The microclimate of the interface environment while sitting or lying on a support surface has also been held responsible as a contributing factor in pressure ulcer etiology. While traditionally this has not been discussed widely in diabetic foot research, recent reports have pointed to this direction. We will also examine the effects of microclimate, in particular tissue temperature, in the following sections.

Shear stress

During human locomotion, shear stresses act tangentially in anteroposterior and mediolateral directions at the foot—ground interface, which transmit a sophisticated stress pattern to the sublayers of the tissue. The definition of shear is "to cut through something with or as if with a sharp instrument" according to the Merriam-Webster Dictionary. Early reports such as the one by Delbridge et al. recognized shear stress as a major causative factor behind ulceration even before shear stress could not be adequately measured; "It is probably this factor (i.e., shear) than vertical load that is responsible for the areas of breakdown that occur deep to the skin ..." [43]. Dr. Brand stated that ".shear stress was more damaging than normal stress" to conclude their animal study [44].

There have been a few early attempts in 1980s and 1990s to measure plantar shear stress distribution [45–48]. Authors of these investigations used an in-shoe system employing only three to five sensors that measured shear forces at the foot—shoe interface. However, these systems had some major drawbacks pertaining to the accuracy of the data. The systems developed by Tappin et al. and Laing et al. imposed attaching 2.7 mm thick transducers to the sole of the foot. Their results were almost certainly compromised, as these relatively thick sensors acted as "stress risers" under the foot. Another potential problem with these units was the choice for locations of shear transducers. Shear sensors were placed at peak pressure sites under the foot. However, it has been later shown that locations of peak pressure usually do not overlap with the maximum shear sites in the same foot. The device developed by Davis and his associates resembled the commercial pressure distribution platforms [49]. However, their device suffered from poor spatial resolution.

To our knowledge, the first report which provided a clear and validated representation of shear stress distribution under the foot [50] used an improved version of the device reported by Davis et al. [49]. We will refer to this device as the Cleveland Clinic Foundation (CCF) device in this chapter.

A number of intriguing results were reported with the help of this device. The first study revealed that peak pressure and shear locations do not overlap in about 80% diabetic neuropathic feet [51]. In 60% of the patients, the distance between the location of peak pressure and peak shear was reported to be > 2.5 cm (almost 1 in) (Fig. 6.5). A similar outcome was observed in healthy subjects. Authors claimed that their results had the potential to explain the discrepancy between the locations of peak pressure and ulcer occurrences. Another report utilizing the CCF device provided follow-up data in 2015 [52]. The study investigated plantar pressure and shear stresses in a small cohort of diabetic individuals with recently healed plantar ulcers. The results indicated that in three out of the eight subjects, the recently healed ulcers occurred at the peak shear site while peak pressure and peak shear occurred at different sites. In one subject, peak pressure, peak shear, and the ulcer all occurred under the hallux. Interestingly, in one other subject, peak

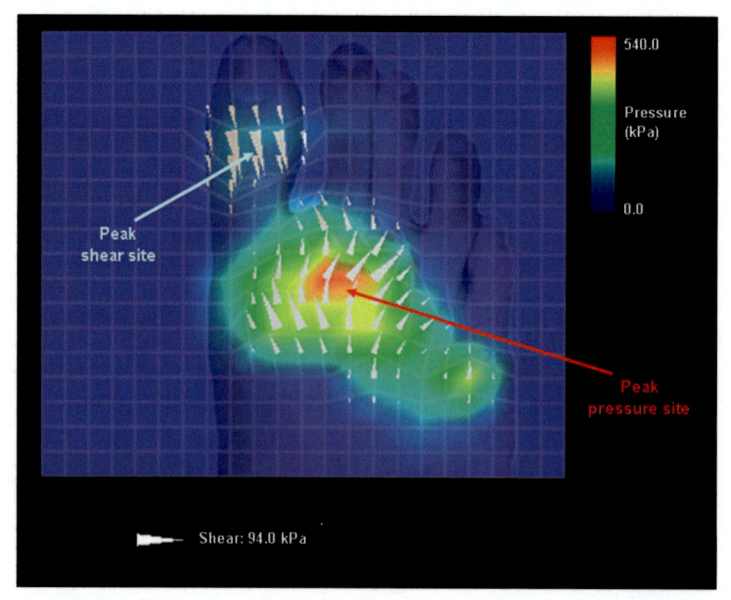

Figure 6.5 Locations of peak pressure and peak shear do not overlap in 80% of the diabetic patients.

pressure and peak shear were experienced underneath the fourth metatarsal head; however, the ulcer occurred under the first metatarsal head. Fig. 6.6 depicts a representative subject who had a recently healed ulcer underneath the first metatarsal head. This area experienced the peak shear stress, whereas peak pressure occurred underneath the third metatarsal head. The study had limitations such as the small sample size (N = 8) and adopting a retrospective approach; however, results indicated that shear stress may in fact be a significant clinical causative factor.

Three other studies reported the magnitudes of plantar shear stress experienced by the diabetic foot and compared peak shear values across various subject groups [53–55]. All three studies reported data collected with the CCF device. The 2008 article that appeared in the *Journal of Biomechanics* reported significantly higher ($P=.016$) peak shear in diabetic neuropathic subjects (92.1 kPa ± 31.6) compared to healthy individuals (70.0 kPa ± 19.8). The 2014 article confirmed these results as well as provided data from a second control group. The third article reported shear values from a cohort of individuals who had a history of ulceration. This group experienced shear stresses as high as 135.3 kPa (±60.6). A comparison with diabetic patients without an ulcer history (86.4 ± 30.3 kPa) revealed a statistically significant difference ($P=.0465$). Interestingly, the two groups did not differ in peak pressure values. Reulceration rate in patients with an ulcer history is quite high, around 40% [56], which may render the results of this study clinically meaningful.

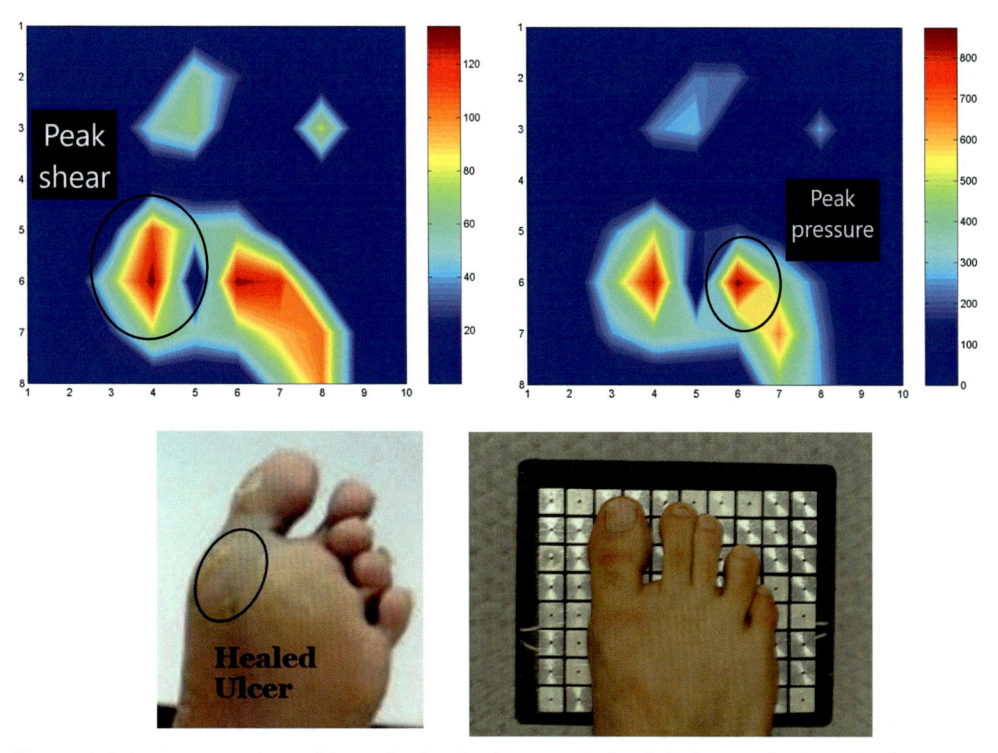

Figure 6.6 A representative subject who has had a recently healed ulcer under his right first metatarsal head (bottom left: a mirror image was provided for better visualization). Peak shear stress occurred right at this location (top left). Peak pressure was observed under the third metatarsal head (top right). Typical foot placement on the Cleveland Clinic Foundation (CCF) device (bottom right).

Another interesting observation regarding plantar shear was its biphasic character. Authors pointed out to the application of anteroposterior shear during the stance phase [53], which has a positive value in the initial stage of the stance phase due to braking forces and then a negative value during the push-off phase due to propulsion (Fig. 6.7). Their results revealed that the plantar surface experiences these cyclic forces in opposite directions with every step. The analogy of the breaking of a paper clip via fatigue failure was provided. Cyclic forces acting in reverse directions make it easy to break the paper clip into two pieces easily. Authors resembled this to the application of shear underneath and argued that plantar shear might have a similar damaging effect on the tissue. Authors also claimed that the unique biphasic characteristic of plantar shear is a better fit to the "repetitive moderate mechanical stresses" that are held responsible for ulceration, as shear was twice as repetitive as pressure.

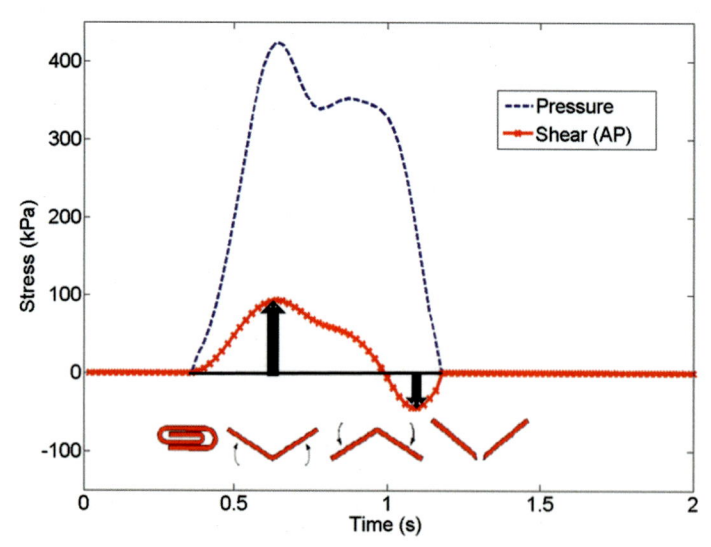

Figure 6.7 Biphasic character of plantar shear. Data were obtained with the Cleveland Clinic Foundation (CCF) device from a diabetic neuropathic subject. Data come from one single sensor corresponding to the first metatarsal head area. *Blue line* indicates the progression of pressure. *Red line* depicts anteroposterior shear stress. Note that shear stress has positive values until midstance due to braking forces. At midstance its magnitude is zero. In the push-off phase shear stress has a negative value, indicating that the force at the same location is now acting in opposite direction. Authors resembled the biphasic character of shear to braking of a paper clip by simply bending it back and forth. Mediolateral shear usually follows a similar pattern.

Authors also resembled plantar shear and its impact on diabetic foot lesions to the use of a chainsaw [55]. The application of shear via the running chain makes it a relatively easy job to cut a tree limb. However, if the engine is off, or when shear is removed, application of pressure only on the saw will most likely be useless in severing the tree limb. It was argued that shear is the main damaging mechanical force in the failure of the foot tissue.

Plantar callus and shear

Murray et al. have shown in a prospective study that the relative risk of developing an ulcer was 11.0 in patients with a callus, compared to 4.7 in patients with high plantar pressures [29]. A manuscript by Sage et al. reported that more than 80% of plantar ulcers were preceded by a callosity [57]. Another study found that removal of plantar calluses leads to an average of 29% reduction in foot pressures [28]. While the initial two studies indicated the significance of plantar callosities in foot ulceration, the results of the second study caused many investigators to believe that calluses formed because of high plantar pressures. Currently, removal of hyperkeratotic skin is a standard clinical care to prevent ulceration. However, the etiology of plantar calluses still needs to be solved.

Several early lichenification studies identified frictional shear as the primary causative factor behind hyperkeratosis [58,59]. If it is also true for foot skin, this fact would further increase the importance of shear stress in ulceration. It can be hypothesized that, in the presence of autonomic neuropathy and impaired sweating, high shear stresses, lead to calluses, which act as stress risers further increase pressure and shear at these locations. Therefore ulcers occur in these hyperkeratotic areas. The first part of this hypothesis has been partially confirmed by Amemiya et al. [60].

Temperature increase and shear

Another intriguing clinical implication of shear stress is that it causes the temperature of the tissue to rise [61,62]. Yavuz et al. revealed that plantar temperatures increase by about 5°C as a result of walking only for 10 min [61]. The authors correlated this temperature increase with shear stresses and reported a significant correlation. On the other hand, Bergtholdt and Brand, and Brand indicated that application of elevated plantar stresses would cause inflammation within the foot and hot spots may be a result of this inflammation [63,64]. In another study, Yavuz et al. partially confirmed this hypothesis [61]. Brand also indicated that "foot heats up before breaking down" [65]. Yavuz et al. identified two types of temperature increase in the diabetic foot: the acute and chronic increase [62]. Acute temperature increase is due to loadbearing activity and associated frictional shear. This is similar to the effect of kinetic friction on the palms when they are rubbed against each other. The chronic temperature increase occurs due to the repetitive and prolonged application of pressure and shear and associated inflammation [62].

The significance of high tissue temperatures has been discussed in detail in pressure ulcer literature. A computational model demonstrated the damaging effects of elevated tissue temperature in pressure ulcers [66]. The model showed that increases in ambient and tissue temperatures would substantially increase the risk for superficial pressure ulcers. A study by Sae-Sia et al. (2007) reported that skin temperature in 17 patients who developed pressure ulcers was approximately 1.2°C higher, which was statistically significant, than those who did not develop ulcers [67]. Perhaps, the most convincing study in this area was conducted by Kokate et al., who applied a constant amount of pressure on pigs utilizing instrumented discs that were maintained at different temperatures [68]. The discs were kept in place for 5 h every day. At the end of the 4-week experiment, the site that was maintained at 25°C exhibited no damage. At the 35°C site, moderate muscle damage was observed. Full epidermal necrosis, moderate dermal and subdermal, and severe muscle damage were observed at the warmest site that was maintained at 45°C. It is well known in the literature that metabolic rate doubles with every 10°C increase in temperature [69]. Yavuz et al. argued that in mechanically loaded tissue, such as the foot that experiences the whole body weight, increased metabolic rate would create an imbalance between oxygen demand and supply [70]. Several manuscripts have shown that application of only 100–120 mmHg (13–16 kPa) pressure is sufficient for blood vessels to get substantially occluded [71,72]. As a comparison, the diabetic foot may experience barefoot pressures as high as 1000 kPa and shod pressures as high as 300 kPa. Therefore, occlusion of blood flow in high temperatures might be quite problematic as the increased demand for oxygen may not be met. Exacerbating the problem, protective diabetic footwear is usually made of synthetic materials that have heat insulation characteristics, further impacting the microclimate within the shoe. In an ongoing study in the author's lab we have observed in-shoe plantar temperatures as high as 37°C, which indicates that the hazard threshold as indicated by Kokate et al. [68] may be easily reached after walking briefly.

While many studies recommended that foot temperatures be used as a predictive tool for ulceration, to our knowledge only a few indicated that temperature may be a major causative factor. Reports such as Bergtholdt and Brand, Lavery et al., Armstrong et al., and Frykberg et al. recommended that a bilateral asymmetry in foot temperatures can indicate ulceration [63,73–75].

The need to revisit the etiology

We believe that given the recent developments in assessment of plantar shear stresses and plantar temperatures, the etiology of diabetic foot ulcers should be revisited. Despite the widely available pressure-relieving footwear 130,000 amputations needed to be performed in 2020 in the United States due to diabetes [76]. Five-year mortality rates after a lower extremity amputation or even an ulceration are reported to be worse than some cancer types [77]. In order to prevent this deadly complication via more effective devices and methods, a better understanding of the ulceration pathology is crucial. We believe that shear stress and temperature should be treated as major causative factors and further researched.

To our knowledge while there have been attempts to estimate the distribution of shear stress as it relates to pressure ulcers, no article has reported actual shear stress measurements. Despite this, shear stress has been recognized as a significant risk factor in pressure ulcer formation in as early as 2009. Unfortunately, despite the availability of various manuscripts that reported shear stress magnitudes and how they might play a role in ulceration, shear is still not discussed adequately in the latest clinical guideline published by the International Working Group on Diabetic Foot [78]. We hereby ask the International Community here once again to stop oversimplifying the foot ulcer pathomechanics by focusing only on pressure and considering shear stress as a major causative factor.

Acknowledgments

The authors are grateful for the assistance provided by Ali Ersen, Ph.D., and Ophelie Puissegur, MS.

References

[1] Hall OC, Brand PW. The etiology of the neuropathic plantar ulcer: a review of the literature and a presentation of current concepts. JAPMA 1979;69(3):173−7.
[2] Betts RP, Duckworth T. A Device for measuring plantar pressures under the sole of the foot. Eng Med 1978;7(4):223−8.
[3] Betts RP, Franks CI, Duckworth T. Analysis of pressures and loads under the foot. I. Quantitation of the static distribution using the PET computer. Clin Phys Physiol Meas 1980;1(2):101.
[4] Betts RP, Franks CI, Duckworth T. Analysis of pressure and loads under the foot. II. Quantitation of the dynamic distribution. Clin Phys Physiol Meas 2001;1(2):113.
[5] Betts RP, Franks CI, Duckworth T, Burke J. Static and dynamic foot-pressure measurements in clinical orthopedics. Med Biol Eng Comput 1980;18(674).
[6] Boulton AJ, et al. Dynamic foot pressure and other studies as diagnostic and management aids in diabetic neuropathy. Diabetes Care 1983;6(1):26−33.
[7] Arcan M, Brull MA. A fundamental characteristics of the human body and foot, the foot-ground pressure pattern. J Biomech 1976;9(7):453−7.
[8] Nicol K. Measurement of pressure distribution on curved and soft surfaces. Biomech Basic & Appl Res 1986;3(1):543−50.
[9] Cavanagh PR, Ae M. A technique for the display of pressure distributions beneath the foot. J Biomech 1980;13(2):69−75.
[10] Cavanagh PR, Ulbrecht JS. Clinical plantar pressure measurement in diabetes: rationale and methodology. Foot 2004;4(3):123−5.
[11] Ctercteko GC, et al. Vertical forces acting on the feet of diabetic patients with neuropathic ulceration. Br J Surg 1981;68(9):608−14.
[12] Stess RM, Jensen SR, Mirmiran R. The role of dynamic plantar pressures in diabetic foot ulcers. Diabetes Care 1997;20(5):855−8.
[13] Stokes IA, Faris IB, Hutton WC. The neuropathic ulcer and loads on the foot in diabetic patients. Acta Orthop Scand 1975;46(5):839−47.
[14] Armstrong DG, Lavery LA. Elevated peak plantar pressures in patients who have Charcot arthropathy. J Bone Joint Surg Am 1998;80(3):365−9.
[15] Cavanagh PR, Rodgers MM. The arch index: a useful measure from footprints. J Biomech 1987;20(5):547−51.
[16] Forghany S, Bonanno DR, Menz MB, Landorf KB. An anatomically-based masking protocol for assessment of in-shoe plantar pressure measurement of the forefoot. J Foot Ankle Res 2018;11(31).
[17] Cavanagh PR, Sims DS, Sanders LJ. Body mass is a poor predictor of peak plantar pressure in diabetic men. Diabetes Care 1991;14(8):750−5.
[18] Ahroni JH, Boyko EJ, Forsberg RC. Clinical correlates of plantar pressure among diabetic veterans. Diabetes Care 1999;22(6):965−72.
[19] Vela SA, Lavery LA, Armstrong DG, Anaim AA. The effect of increased weight on peak pressures: implications for obesity and diabetic foot pathology. J Foot Ankle Surg 1998;37(5):448−9.
[20] Delbridge L, Ellis CS, Robertson K, Lequesne LP. Non-enzymatic glycosylation of keratin from the stratum coreneum of the diabetic foot. Br J Dermatol 1985;112(5):547−54.
[21] Hashmi F, Malone-Lee J, Hounsell E. Plantar skin type 2 diabetes: an investigation of protein glycation and biomechanical properties of plantar epidermis. Eur J Dermatol 2006;16(1):23−32.
[22] Jacob Thomas V, Mothiram Patil K, Radhakrishnan S, Narayanamurthy VB, Parivalavan R. The role of skin hardness, thickness and sensory loss on standing foor power in the development of plantar ulcers in patients with diabetes mellitus- a preliminary study. Int J Lower Extrem Wounds 2003;2(3):132−9.
[23] Klaesner JW, Hastings MK, Dequan Z, Lewis C, Mueller MJ. Plantar tissue stiffness in patients with diabetes mellitus and peripheral neuropathy. Arch Phys Med Rehabil 2002;83(12). 1976.801.
[24] Ledoux WR, Shofer JB, Ahroni JH, Smith DG, Sangeorzan BJ, Boyko EJ. Biomechanical differences among pes cavus, neutrally aligned, and pes planis feet in subjects with diabetes. Foot Ankle Int 2003;24(11):845−50.

[25] Masson EA, Hay EM, Stockley I, Veves A, Betts RP, Boulton AJ. Abnormal foot pressures alone may not cause ulceration. Diabet Med 1989;6(5):426—8.
[26] Ledoux WR, Shofer JB, Smith DG, Sullivan K, Hayes SG, Assal M, Reiber GE. Relationship between foot type, foot deformity and ulcer occurrence in the high-risk diabetic foot. J Rehabil Res Dev 2005;42(5):665—72.
[27] Birke JA, Franks BD, Foto JG. First ray joint limitation, pressure, and ulceration of the first metatarsal head in diabetes mellitus. Foot Ankle Int 1995;16(5):277—84.
[28] Young MJ, Cavanagh PR, Thomas G, Johnson MM, Murray H, Boutlon AJ. The effect of callus removal on dynamic plantar foot pressures in diabetic patients. Diabet Med 1992;9(1):55—7.
[29] Murray H, Boulton AJ, Young MJ, Hollis S. The association between callus formation, high pressures and neuropathy in diabetic foot ulceration. Diabet Med 1996;13(11):972—82.
[30] Cavanagh PR, Ulbrecht JS, Caputo GM. New developments in the biomechanics of the diabetic foot. Diabetes Metab Res Rev 2000;16(1):S6—10.
[31] Armstrong DG, Peters EJ, Athanasiou KA, Lavery LA. Is there a critical level of plantar foot pressure to identify patients at risk for neuropathic foot ulceration? J Foot Ankle Surg 1998;37(4):303—7.
[32] Lavery LA, Armstrong DG, Wunderlich RP, Tredwell J, Boulton AJ. Predictive value of foot pressure assessment as part of a population — based diabetes disease management program. Diabetes Care 2003;26(4):1069—73.
[33] Veves A, Murray HJ, Young MJ, Boulton AJ. The risk of foot ulceration in diabetic patients with high foot pressure: a prospective study. Diabetologia 1992;35(7):660—3.
[34] Bus SA, Valk GD, Van Deursen RW, Armstrong DG, Caravaggi C, Hlavacek P, Bakker K, Cavanagh PR. The effectiveness of footwear and offloading interventions to prevent and heal foot ulcers and reduce plantar pressure in diabetes: a systematic review. Diabetes Metab Res Rev 2008;24(1):S162—80.
[35] Bus SA, Valk GD, Van Deursen RW, Armstrong DG, Caravaggi C, Lewis JEA, International Working Group on Diabetic Foot. Footwear and offloading interventions to prevent and heal foot ulcers and reduce plantar pressure in patients with diabetes: a systematic review. Diabetes Metab Res Rev 2016;32(1):99—118.
[36] Owings TM, Apelqvist J, Stenstrom A, Becker M, Bus SA, Ulbrecht JS, Cavanagh PR. Plantar pressures in diabetic patients with foot ulcers which have remained healed. Diabet Med 2009;26(11):1141—6.
[37] Reiber GE, Smith DG, Wallace CM, Vath CA, Sullivan K, Hayes S, Yu O, Martin D, Maciejewski M. Footwear used by individuals with diabetes and history of foot ulcer. J Rehabil Res Dev 2002;39(5):615—22.
[38] Lavery LA, Vela SA, Fleischli JG, Armstrong DC, Lavery DC. Reducing plantar pressure in the neuropathic foot. A comparison of footwear. Diabetes Care 1997;20(11):1706—10.
[39] Zwaferink JBJ, Custers W, Paardekooper I, Berendsen HA, Bus SA. Optimizing footwear for the diabetic foot: data driven custom-made footwear concepts and their effect on pressure relief to prevent diabetic foot ulceration. PLoS One 2020;15(4).
[40] Geiss LS, Li Y, Hora I, Albright A, Rolka D, Gregg EW. Resurgence of diabetes — related non traumatic lower-extremity amputation in the young and middle-aged adult US population. Diabetes Care 2019;42(1):50—4.
[41] Prevention and Treatment of Pressure Ulcers: Quick Reference Guide. European pressure ulcer advisory panel and national pressure ulcer advisory Panel. 2016 [Accessed 15 September 20], https://www.epuap.org/wp-content/uploads/2016/10/quick-reference-guide-digital-npuap-epuap-pppia-jan2016.pdf.
[42] Edmonds ME, Blundell MP, Morris ME, Thomas EM, Cotton LT, Watkins PJ. Improved survival of diabetic foot: the role of a specialized foot clinic. Int J Med 1986;60(2):763—71.
[43] Delbridge L, Ctercteko G, Fowler C, et al. The aetiology of diabetic neuropathic ulceration of the foot. Br J Surg 1985;72(1):1—6.
[44] Brand PW. Tenderizing the foot. Foot Ankle Int 2003;24(6):457—61.
[45] Tappin JW, Pollard J, Beckett EA. Method of measuring shearing forces on the sole of the foot. Clin Phys Physiol Meas 2001;1(1):83.

[46] Laing P, Deogan H, Cogley D, Crerand S, Hammond P, Klenerman L. The development of the low profile liverpool shear transducer. Clin Phys Physiol Meas 1992;13(2):115—24.
[47] Akhlaghi F, Peper MG. In-shoe biaxial shear force measurement: the Kent shear system. Med Biol Eng Comput 1996;34(4):315—7.
[48] Hosein R, Lord M. A study of in-shoe plantar shear in normals. Clin Biomech 2000;15(1):46—53.
[49] Davis BL, Perry JE, Neth DC, Waters KC. A device for simultaneous measurement of pressure and shear force distribution on the plantar surface of the foot. Gait Posture 2002;15(1):93—104.
[50] Yavuz M, Botek G, Davis BL. Plantar shear stress distributions: comparing actual and predicted frictional forces at the foot-ground interface. J Biomech 2007;40(13):3045—9.
[51] Yavuz M, Erdemir A, Botek G, Hirschman GB, Bardsley L, Davis BL. Peak plantar pressure and shear locations: relevance to diabetic patients. Diabetes Care 2007;30(10):2643—5.
[52] Yavuz M, Master H, Garrett A, Lavery LA, Adams LS. Peak plantar shear and pressure and foot ulcer locations: a call to revisit ulceration pathomechanics. Diabetes Care 2015;38(11):e184—5.
[53] Yavuz M, Tajaddini A, Botek G, Davis BL. Temporal characteristics of plantar shear distribution: relevance to diabetic patients. J Biomech 2008;41(3):556—9.
[54] Yavuz M. American Society of Biomechanics Clinical Biomechanics Award 2012: plantar shear stress distributions in diabetic patients with and without neuropathy. Clin Biomech 2014;29(2):223—9.
[55] Yavuz M, Ersen A, Hartos J, Schwarz B, Garrett AG, Lavery LA, et al. Plantar shear stress in individuals with a history of diabetic foot ulcer: an emerging predictive marker for foot ulceration. Diabetes Care 2017;40(2):e14—5.
[56] Armstrong DG, Boulton AJ, Bus SA. Diabetic foot ulcers and their recurrence. N Engl J Med 2017; 376(24):2367—75.
[57] Sage RA, Webster JK, Fisher SG. Outpatient care and morbidity reduction in diabetic foot ulcers associated with chronic pressure callus. J Am Podiatr Med Assoc 2001;91(6):275—9.
[58] MacKenzie IC. The effects of frictional stimulation on mouse ear epidermis. J Invest Dermatol 1974; 63(2):194—8.
[59] Goldblum RW, Piper WN. Artificial lichenification produced by a scratching machine. J Invest Dermatol 1954;22(5):405—15.
[60] Amemiya A, Noguchi H, Oe H, Takehara K, Ohashi Y, Suzuki R, Yamauchi T, Kadowaki T, Sanada H, Mori T. Factors associated with callus formation in the plantar region through gait measurement in patients with diabetic neuropathy — an observation case — control study. Sensors 2020;20(17): E4863.
[61] Yavuz M, Brem RW, Glaros AG, Garrett A, Flyzik M, Lavery L, Davis BL, Hilario H, Adams LS. Association between plantar temperatures and triaxial stresses in individuals with diabetes. Diabetes Care 2015;38(11):e178—9.
[62] Yavuz M, Ersen A, Hartos J, Lavery LA, Wukich DK, Hirschman GB, Armstrong DG, Quiben MU, Adams LS. Temperature as a causative factor in diabetic foot ulcers: a call to revisit ulceration pathomechanics. J Am Podiatr Med Assoc 2019;109(5):345—50.
[63] Bergtholdt HT, Brand PW. Temperature assessment and plantar inflammation. Lepr Rev 1976;47(3): 211—9.
[64] Brand PW. Management of the insensitive limb. Phys Ther 1979;59(1):8—12.
[65] Boulton AJM. Diabetic foot — what can we learn from leprosy? Legacy of Dr. Paul W. Brand. Diab Met Res Rev 2012;28(1).
[66] Gefen A. How do microclimate factors affect the risk for superficial pressure ulcers: a mathematical modeling study? J Tissue Viability 2011;20(3):81—8.
[67] Sae-Sia W, Wipke-Tevis DD, Williams DA. The effect of clinically relevant pressure duration on sacral skin blood flow and temperature in patients after acute spinal cord injury. Arch Phys Med Rehabil 2007;88(12):p1673—80.
[68] Kokate JY, Leland KJ, Held AM, Hansen GL, Johnson BA, Wilke MS, Sparrow EM, Iaizzo P. Temperature-modulated pressure ulcers: a porcine model. Arch Phys Med Rehabil 1995;76(7): 666—73.
[69] Ruch RC, Patton HD. Energy metabolism. 19th ed. Philadelphia, PA: Saunders Press; 1965.

[70] Yavuz M, et al. Temperature- and pressure-regulating insoles for prevention of diabetic foot ulcers. J Foot Ankle Surg 2020;59(4):p685–8.

[71] Bennett L, Kavner D, Lee BK, Trainor FA. Shear vs pressure as causative factors in skin blood blow occlusion. Arch Phys Med Rehabil 1979;60(7):309–14.

[72] Manorama A, Meyer R, Wiseman R, Bush TR. Quantifying the effects of external shear loads on arterial and venous blood flow: implications for pressure ulcer development. Clin Biomech 2013;28(5):574–8.

[73] Lavery LA, Higgins KR, Lanctot DR, Constantinides GP, Zamorano RG, Armstrong DG, Athanasiou KA, Agrawal CM. Home monitoring of foot skin temperatures to prevent ulceration. Diabetes Care 2004;27(11):2642–7.

[74] Armstrong DG, Holtz-Neiderer K, Wendel C, Mohler MJ, Kimbriel HR, Lavery LA. Skin temperature monitoring reduces the risk for diabetic foot ulceration in high-risk patients. Am J Med 2007;120(12):1042–6.

[75] Frykberg RG, Gordon IL, Reyzelman AM, Cazzell SM, Fitzgerald RH, Rothernberg GM, Bloom JD, Petersen BJ, Linders DR, Nouvong A, Najafi B. Feasibility and efficacy of a smart mat technology to predict development of diabetec plantar ulcers. Diabetes Care 2017;40(7):973–80.

[76] National Diabetes Statistics Report. US Department of Health and Human Services. 2020. p. 10 [Accessed 15 September 20], https://www.cdc.gov/diabetes/pdfs/data/statistics/national-diabetes-statistics-report.pdf.

[77] Armstrong DG, Swerdlow MA, Armstrong AA, Conte MS, Padula WV, Bus SA. Five year mortality and direct costs of care for people with diabetic foot complications are comparable to cancer. J Foot Ankle Res 2020;13(1):16.

[78] Practical Guidelines. International working group on diabetic foot. 2019 [Accessed 15 September 20], https://iwgdfguidelines.org/guidelines/guidelines/.

CHAPTER 7

Novel technologies for detection and prevention of diabetic foot ulcers

Neil D. Reeves[1], Bill Cassidy[2], Caroline A. Abbott[1], Moi Hoon Yap[2]

[1]Research Centre for Musculoskeletal Science and Sports Medicine, Faculty of Science and Engineering, Manchester Metropolitan University, Manchester, United Kingdom; [2]Centre for Applied Computational Science, Faculty of Science and Engineering, Manchester Metropolitan University, Manchester, United Kingdom

Health and societal burden of diabetic foot ulcers

Diabetic foot ulcers present a major health and economic cost due to treatment/management of the condition and the associated amputations. Diabetic foot ulcers are a global issue and it is estimated that somewhere in the world, a limb is lost to diabetes every 20 s. Up to 34% of people with diabetes will develop an ulcer at some point in their lifetime [3], frequently leading to hospitalization, amputation, and negatively affecting a patient's quality of life. Recurrence rates are very high; when someone has developed a diabetic foot ulcer they have a 40% risk of it redeveloping within 1 year, rising to 60% risk within 3 years [3]. Over 8000 amputations occur in the United Kingdom every year due to the development of diabetic foot ulcers. In the United Kingdom, 80% of the National Health Service budget is spent on treating the complications of diabetes [43] and specifically diabetic foot ulcers and the associated amputations cost the National Health Service over £1.2 billion annually [44]. Prevention and early detection of diabetic foot ulcers are key to enabling an improved outlook for this common complication of diabetes.

Risk factors for the development of diabetic foot ulcers

Prospective studies consistently show that the major clinical risk factors for diabetic foot ulcer development include history of foot ulcer(s), peripheral neuropathy, foot deformities, and elevated plantar pressures [1,13,32]. Diabetic peripheral sensorimotor neuropathy is one of the major factors associated with the development of a diabetic foot ulcer. Diabetic neuropathy involves damage to the peripheral sensory and motor nerves and is length dependent, affecting the most peripheral nerves first. Diabetic neuropathy begins at the ends of the toes and then progresses proximally up the foot and ankle. One of the major issues for the development of diabetic foot ulcers is that diabetic peripheral neuropathy damages sensory nerves, in many cases causing a complete loss of peripheral sensation. Therefore, sensory nerves that would normally detect touch (haptic feedback) and pain are damaged and fail to provide this natural feedback to the person. Painful stimuli are in fact helpful in the sense that they provide feedback to avoid damage to the foot. Without this self-correcting feedback, the person with diabetic neuropathy may continue to cause damage to their foot over time.

Foot deformities occur in people with diabetic neuropathy due to selected atrophy and calcification of foot muscles. A number of different foot deformities may occur including prominent metatarsal heads and clawing of the toes, but the common result is the presence of bony prominences impairing the capacity to distribute loads more evenly across the foot, creating high-pressure areas at risk for ulceration. Other risk factors for diabetic foot ulceration include atrophy of the fat pad under the foot, which typically provides a cushioning of bony foot structures. Limited ankle dorsiflexion (reduced bending of the foot up toward the lower leg) occurs due to nonenzymatic glycation causing stiffening the Achilles tendon and ankle plantar flexor (calf) muscles on the back of the leg. This reduced ankle dorsiflexion means that higher pressures might be developed for longer on the forefoot area as the tibia does not rotate as freely over the standing foot during walking.

All of these risk factors discussed—diabetic peripheral neuropathy, foot deformities, plantar fat pad atrophy, and limited ankle dorsiflexion—combine to cause high pressures on the plantar surface (foot sole) of the foot. Cumulatively over time and without any natural feedback due to diabetic peripheral neuropathy to regulate these pressures, tissue breakdown occurs and a diabetic foot ulcer develops.

Foot pressures and diabetic foot ulcer development

Foot loading and pressures have been measured for a number of years in the laboratory setting using a range of different equipment (Fig. 7.1). Pressure plates are small platforms larger than length of the average foot that are placed on the floor, or embedded into a walkway. The person walks barefoot over the platform with a single foot striking the platform and plantar pressures are measured across the whole area of the foot. Force platforms are embedded into the floor and measure ground reaction forces and the position of the center of pressure under the foot. In the same manner as pressure plates, the person walks over the force platform with a single foot striking within the boundaries of the platform. Force platforms measure ground reaction forces from the foot striking the platform and the resultant ground reaction force arises from the center of pressure, which tracks along the foot from heel to toe during walking. Commercial pressure systems typically measure only vertical pressures (perpendicular to the foot), whereas force platforms measure vertical as well as shear forces. Although force platforms measure forces very accurately applied to the feet, they do not measure pressures. In-shoe sensors are thin sensors that fit inside people's shoes and are positioned under the feet to measure pressure developed across the whole plantar foot surface. In-shoe pressure systems allow for measurement from multiple steps during a data capture walking trial, whereas pressure platforms (and force platforms) capture a single foot strike during each walking trial. A simple but reliable measure for plantar pressure is based upon using carbon paper that forms a graded imprint of the foot as the person walks over it positioned on the floor

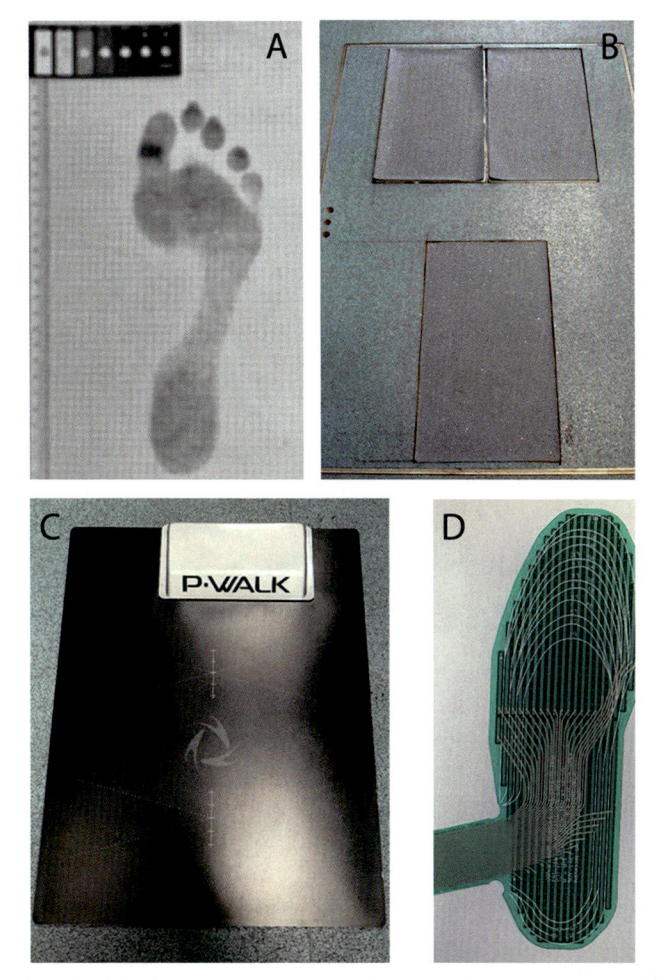

Figure 7.1 Examples of equipment used to measure forces and plantar pressure. (A) Graded carbon paper; (B) three ground-embedded force platforms; (C) pressure plate; (D) laboratory-based digital pressure sensors.

in a similar way to that described for force platforms and pressure plates [35]. This carbon paper method might be considered to be much more clinically relevant due to the low cost and non-technical requirements compared to other systems described here.

Foot pressure in people with diabetes is always higher than nondiabetic individuals, when measured barefoot, because of the exposure of bony prominences due to foot deformities and related factors causing high-pressure areas (for comparison of barefoot and in-shoe pressure studies, see review by Ref. [14]). In-shoe foot pressures are always lower than barefoot pressures and this fact underpins the rational for recommendations that everyone with diabetes (and especially those at high risk for diabetic foot ulceration) should wear footwear at all times (even in the home) to protect their feet and reduce their foot ulcer risk (*see subsequent section on footwear prevention strategies*).

As described earlier, a number of risk factors combine to mean that peak plantar pressures are much higher in the diabetic foot compared to someone without diabetes whether measured barefoot or in-shoe (for review, see Ref. [14]). Peak pressure is typically higher in the forefoot area, which also corresponds with the most common sites of foot ulcers, at the metatarsal heads and the big toe [7]. High peak plantar pressures and high pressure-time integrals (magnitude of pressure developed over a normalized time-period) have been shown to strongly correlate with diabetic foot ulcer development and those with a history of diabetic foot ulcers have higher pressures than people with diabetes but without a history of ulceration [14,31,34]. A few studies have investigated how well foot pressure measurements can predict those people who will go on to develop a foot ulcer in the future (for review, see Ref. [14]). These studies have shown some limited predictive ability, likely hindered by single time-points of measurement confined to laboratory-based methods, whereas real-world longitudinal measurements are really what is needed to improve predictive ability for diabetic foot ulcers.

A threshold level for in-shoe plantar pressure of 200 kPa has been suggested to identify patients at high risk for ulceration [27]. However, the wide spread of peak pressure values reported for people with diabetes both with and without plantar ulcers means that the range measured for each group overlaps substantially, limiting sensitivity and specificity of a defined threshold value. The difficulty of identifying a single pressure value is further compounded by the fact that this has been generated based upon cross-sectional, single time-point measurements confined to laboratory-based methods, which as highlighted above has its limitations. Furthermore, the 200 kPa threshold is not translatable for barefoot walking, as barefoot peak plantar pressures may be up to four times higher than in-shoe peak plantar pressures, largely due to the exposure of bony prominences caused by foot deformities. We could consider that future risk identification strategies should include site-specific in-shoe plantar thresholds, rather than a "whole in-shoe" threshold. This is because previous studies have shown that diabetic foot ulcers occur at only few, specific plantar sites, being most prevalent at the metatarsal heads and great toe [11,31,36], rather than occurring generally across the plantar surface. Site-specific thresholds may therefore provide a more targeted diabetic foot ulcer prevention approach.

Current standard prevention strategies

Preventing diabetic foot ulcers from occurring in the "at-risk" diabetic patient requires a holistic approach, involving regular contact between the patient, the caregiver, and the healthcare team, plus the provision of integrated foot care [3]. Current guidelines recommend that such patients should receive education about appropriate foot self-care, and that they should wear prescribed, properly fitting therapeutic footwear and insoles to accommodate foot deformities and offload plantar pressures sufficiently to prevent foot ulcers and their recurrence. In addition, orthotic interventions, such as toe silicone or (semi)rigid orthotic devices, may also help to reduce abundant callus, and therefore the patient's ulcer risk [9]. There is clinical evidence to support the use of custom-made footwear for providing plantar pressure relief and preventing diabetic foot ulcer recurrence [10,33]. However, there are some inconsistencies in the evidence for the effectiveness of therapeutic footwear on diabetic foot ulcer prevention, largely because of the absence of standardization of interventions and control conditions in randomized controlled trials [10,30]. One trial has shown that adequate adherence to daily wearing of such therapeutic footwear is required for effective ulcer prevention [10]; however, it appears that footwear interventions are often described as being associated with poor adherence, thus possibly explaining their limited effectiveness in preventing diabetic foot ulcer development [6].

The need for technology solutions for diabetic foot ulcer prevention

One of the primary consequences of diabetic peripheral neuropathy is the absence of any peripheral sensory perception at the foot and therefore the complete removal of any natural feedback mechanism. The natural feedback mechanism with intact peripheral sensation would typically take the form of someone walking, or standing in a particular way that placed undue pressure on a certain part of the foot, or they might detect the presence of a foreign body in their shoe. In all of these examples, with intact sensation, the person would feel the pain or uncomfortable nature of the situation and take action to modify the loading on the foot during walking or standing, or by removing their shoe to take out any foreign body. This action based on the natural sensory feedback would therefore protect the person's foot from any potential harm. In the case of someone with severe diabetic peripheral neuropathy, they have no sensory feedback and any high pressure developed in all of the above examples would continue without modification because of the lack of natural sensory feedback. Cumulatively over time, it is this exposure to high pressures on the foot (contributed to by all of the factors described in the Risk factors for the development of diabetic foot ulcers section) due to the lack of natural feedback that results in the development of diabetic foot ulcers. This situation presents the opportunity for novel technological interventions to help "substitute" the natural feedback and assist in providing this feedback to the person with diabetes using technology, rather than via the natural sensory nervous system. New technologies may also help in being vigilant screening for any signs of ulcer development and help in early detection.

Continuous "real-world" pressure feedback technology

The majority of studies using in-shoe or other laboratory techniques described above to measure foot pressures have provided only a "snapshot" of plantar pressure and foot loading in people with diabetes during a single laboratory visit. This highlights the limitation of most plantar pressure research in relation to understanding diabetic foot ulcer risk and approaches for prevention. While laboratory-based measurements provide valuable data, they may not be representative of daily foot loading patterns in people with diabetes and as such may miss vital information to help inform risk prediction and preventative strategies. All laboratory-based foot pressure studies have only assessed walking, whereas other daily activities such as standing could be important in diabetic foot ulcer development. A study using continuous daily activity monitoring showed that people with diabetic peripheral neuropathy spent relatively more time standing as they did walking [24] and therefore, standing time could be an important risk factor in cumulative foot loading, leading to diabetic foot ulcer risk. Real-world measurements of plantar pressure are needed to fully understand the nature of cumulative plantar loading, leading to diabetic foot ulceration and for informing preventative strategies.

Recent technology advancements have seen the development of new "smart/innovative" technologies that enable plantar pressure and other signals to be measured continuously throughout daily life [25]. These technologies use thin in-sole sensors of similar nature to those described above for laboratory-based analysis, but importantly are not tethered to cables limiting movement. In contrast, these wireless systems relay foot pressure information to a smartwatch or smartphone. The compromise that allows these innovative technologies to operate continuously in real-world settings is typically a lower sampling frequency compared to laboratory systems. While laboratory-based systems will sample at frequencies in excess of 100 Hz, real-world innovative technologies will typically have much lower sampling frequencies, but many will regard this as an appropriate compromise when such important cumulative loading data can be captured over a long period of time. Pressure data measured by in-shoe sensors is fed-back to smartwatches or phones to provide this information for the patient with diabetes to act upon.

A few small-scale, laboratory-based studies have examined the concept of using pressure feedback in people with diabetic peripheral neuropathy to modify foot loading and plantar pressures (see Ref. [14] for review). Most studies have used visual guides shown to participants to provide pressure feedback while walking with the aim of reducing high pressures. Most of these previous studies have been laboratory-based and focused on pressure data relating to one specific high-risk area of the foot only [15,28,29]. Provision of plantar pressure feedback from a single laboratory session has actually been shown to reduce high pressures on an at-risk area of the foot and furthermore, the effects of this single session remained over a follow-up lasting up to 10 days [15,28,29]. These studies, however, did not include the highest risk patients, since those with foot deformities were excluded. In contrast, when another study included patients with diabetic neuropathy and foot deformities, they did not find any persisting pressure reductions over a 1-week period based on a single laboratory pressure feedback session [42]. Although these studies indicate the potential concept of using plantar pressure feedback to reduce high pressures in people with diabetic peripheral neuropathy, they are confined to the laboratory and show relatively limited retention periods. These encouraging indications seem to justify the need for more continuous daily monitoring and feedback of foot pressures toward diabetic foot ulcer prevention.

Intelligent insole system: high-pressure feedback, offloading, and adherence

Preventing diabetic foot ulcers can only be effective if patients wear their prescribed footwear, yet several studies suggest that adherence is generally low, reporting that high-risk patients wear their prescribed therapeutic footwear during only one-third, or less, of their total daily activities [4,5]. A recent prospective study ($n = 12$) has examined "high-risk" patients' adherence to daily use of an intelligent insole system providing continuous daily monitoring and alert-based feedback of foot pressures when plantar pressure thresholds have been exceeded [26]. In brief, feedback of periods of high plantar pressure (detected by intelligent insoles worn in patients' footwear) during daily activities, plus encouragement to offload pressure from that region, was displayed to patients via a smartwatch displaying images of the feet and the affected foot regions. In terms of adherence to offloading, the question addressed was whether receiving a high number of pressure alerts per hour would cause user "fatigue" and lead to the person disengaging with the intelligent insole system, or conversely, whether a lower number of alerts per hour would lead to the perception the device was not working and subsequent user disengagement. Over a 3-month period, patients who had received at least one high-pressure alert every 2 hours were shown to be more adherent in terms of offloading high pressures compared to patients who received fewer pressure alerts. This supports the hypothesis that relatively frequent and timely pressure alerts help to encourage user engagement and improve patient action to offload these high pressures. Patients receiving more frequent alerts also wore their intelligent insole devices for longer (average 8 h per day) compared to patients receiving fewer alerts (3.6 h per day). Overall, this study shows that timely and frequent alerts via an intelligent insole system will optimize adherence to the technology and improve offloading of high pressures compared to patients receiving fewer alerts. This may be a useful educational tool to encourage adherence to technology and footwear, with implications for reducing the risk of developing diabetic foot ulcers.

Intelligent insole system: reduced diabetic foot ulcer incidence through continuous daily foot pressure feedback

A proof-of-concept randomized control trial has recently shown how continuous daily monitoring and feedback of foot pressures, using an intelligent insole system, can reduce incidence of diabetic foot ulcers over an 18-month follow-up in patients with diabetic peripheral neuropathy at high risk for ulceration [2]. In this study, 90 participants were recruited who all had diabetic neuropathy, a previous foot ulcer, and were therefore considered high risk for future diabetic foot ulcers. Participants were randomized to control and intervention groups with both groups receiving an intelligent insole system consisting of pressure-sensing insoles and a wrist-worn smartwatch (Fig. 7.2). The sensors measured pressure and relayed this information to a smartwatch where it was stored to enable information on adherence and pressure development during daily life. Only the intervention group received plantar pressure feedback to their smartwatch, whereas the control group did not receive any feedback to the watch and it acted as a sham device. The intervention group received auditory and vibratory alerts upon the development of high pressure. Patients were able to visualize the area of high pressure on the foot from

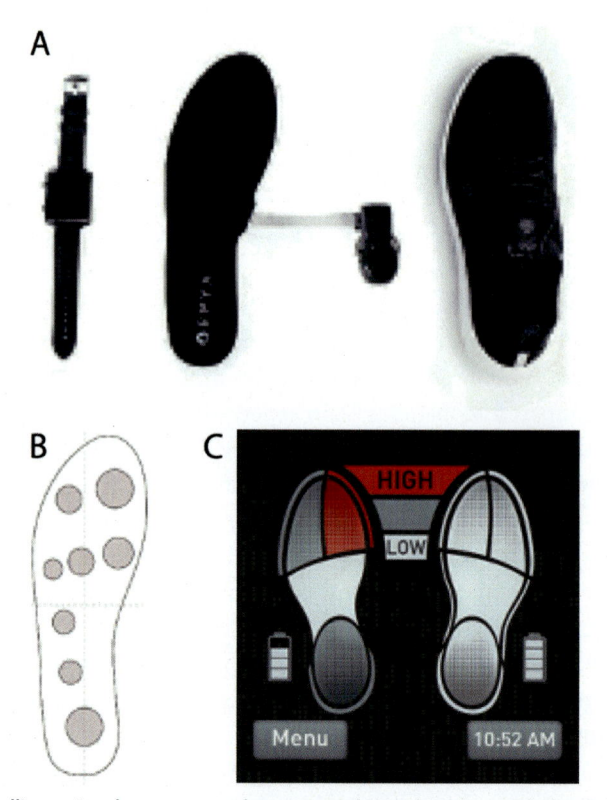

Figure 7.2 The intelligent insole system and smartwatch used in the study by Ref. [2]. (A) Intelligent insole system, comprising 0.6 mm flexible pressure-sensing inserts (placed underneath the patient's insoles) and smartwatch. (B) Schematic diagram showing the sites of the eight pressure sensors embedded within the insole. (C) Close-up of the smartwatch screen displaying user-friendly visual foot map, highlighting an area of high pressure on the front right side of the left foot. The device integrates pressure data over time and generates high-pressure alerts for the user so that they can adequately offload the affected area.

the foot map displayed on the smartwatch. Although the insole sensors contained eight pressure sensors per foot, each visual display of the foot was split into four regions for simplicity of display and understanding (Fig. 7.2). When a high pressure notification was received, patients had three simple steps to offload the foot and reduce the pressure on the affected foot region.

This intelligent insole system functions to effectively replace the natural feedback lost due to diabetic peripheral neuropathy to empower patients with the ability to now detect and respond to high pressures developed on their feet. Through this mechanism, the intervention was able to effectively reduce the incidence of plantar foot ulcers in the intervention group by 71% over the 18-month study period [2]. The intelligent insole system was able to measure and record pressure information in both intervention and control groups continuously over the 18-month follow-up period allowing for an understanding of how often patients wore the sensors and used this technology. In those patients who wore the devices more often and therefore showed better adherence to the technology intervention, ulcer incidence was reduced by 86%. It is well known from other studies in people at risk for diabetic foot ulceration that adherence to the intervention is a major determinant of ulcer prevention or healing.

This novel proof-of-concept trial shows the potential for intelligent insole systems to help "replace" the natural feedback lost due to diabetic peripheral neuropathy and as such enable the person to detect and respond to high plantar pressure loading that would otherwise lead to foot ulceration.

Using computer vision and artificial intelligence in diabetic foot ulcer detection

Another novel strategy in diabetic foot ulcer detection and prevention is through computer vision [37,39]. Wang et al. obtained diabetic foot ulcer photographs for serial analysis by using a mirror image capture box. However, this analysis requires physical contact between the capture box and the foot, which raises infection control issues. Unlike the image capture box, "FootSnap" [38] demonstrated the principle of using a mobile device in diabetic foot ulcer data capturing for serial analysis. To promote patient self-care, the "MyFootCare" mobile app [8] was developed with additional functions, such as personal goals, diaries, and notifications. The data capturing process involves placing the phone on the floor, the user then places their foot above the phone screen and the photo is automatically captured when the foot is at the correct position. Although this idea represents a highly inventive solution to the problem, the authors did not report the reliability of its functionality.

Existing computer vision research lacks automation and is not scalable. In recent years, researchers have explored the use of artificial intelligence (AI) and cloud computing to overcome such issues.

The role of artificial intelligence

AI is not a new or recent concept. As far back as the 1950s, ideas were formulated that continue to form the backbone of current AI research. The term AI was first coined in 1955 by John McCarthy, an American cognitive scientist, in a proposal for the 1956 Dartmouth Conference [23]. His aim was to investigate ways to make a machine capable of reason, abstract thought, problem-solving, and self-improvement. Even before McCarthy, in 1950 Alan Turing, an English mathematician and computer scientist, devised a method for determining if a computer system is capable of thinking like a human [17]. This test is still used today. As early as 1955 scientists working in the field developed a method to automate visual pattern recognition of alphabetical characters [22]. This system did not work very well, but it was an important milestone for AI research. Since that time, pattern recognition, and the recognition of objects, has become an important branch of AI. This particular field has become known as "machine learning," with "deep learning" being a subset of machine learning. Deep learning has become the primary tool for enabling computer systems to both classify and detect objects in images and video. Models are trained using ground truth data, which can then make predictions and calculate the probability of the prediction being true. The mechanisms for doing this have been inspired by the workings of networks of neurons within the human brain. This approach is concerned with probabilities applied to specific problem domains, as opposed to the deterministic approach that is traditionally used in computing. The problem can therefore be framed as "how likely is an instance of *a* to be an instance of *b*." Recently, AI has been used widely used in medical imaging, including diabetic foot ulcer detection and prevention.

For the diabetic foot, this can be realized in the automated analysis of foot images (or video) to determine the presence of a diabetic foot ulcer at different stages of development [21]. Deep learning systems are capable of extracting key features from images using layers known as convolutions [18]. Convolutions represent layers of an image that have been extracted using various filtering techniques. Deep learning models are trained to recognize classes of objects using large datasets of images that exhibit examples of the objects/features in question. At the end of the training process, the model is "frozen," and can then be used to infer the presence of the object/feature it was trained to recognize in images that it has not been exposed to previously. For diabetic foot ulcer detection, the output will be a set of coordinates representing either a simple or irregular polygon positioned on the area of the image where a diabetic foot ulcer has been detected, together with a confidence score (Fig. 7.3).

One of the main issues raised by the advent of machine learning is that large datasets are required in the training stage. Dataset sizes used for training deep learning models often run into the thousands, and in some cases tens of thousands of images. This poses numerous issues—the biggest being one of data acquisition. Training a machine learning

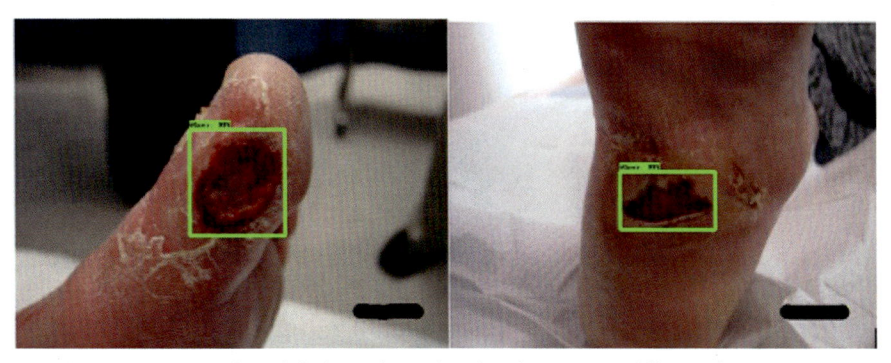

Figure 7.3 Examples of diabetic foot ulcer localization on different foot regions.

model to identify a diabetic foot ulcer requires thousands of images to act as the input for the model to learn from. To date, there are a limited publicly available dataset. The Manchester Metropolitan University and The Lancashire Teaching Hospitals [12,19,20] have shared digital diabetic foot ulcer image datasets with expert annotations. The aim is to encourage more researchers working in this domain to address this issue and conduct reproducible experiments. In addition to the creation of the dataset for diabetic foot ulcer detection [12,21], researchers are actively working on the dataset for diabetic foot ulcer classification; this includes ischemia and infection classifications [20]. Moving forward, further efforts in creating large-scale datasets with ground truth annotations [40,41] will be crucial to improve the reliability and understanding of the AI algorithms.

The role of cloud computing

Cloud computing is becoming an increasingly prominent component of modern computing. Cloud computing can be defined as the remote processing of data. This means that there is less emphasis on the computing performance of user hardware such as individual mobile devices, since the workload can be passed to a cloud service to undertake the bulk of the complex operations. Once the processing is complete, the results can be returned to the user's device, in some cases almost instantaneously. As far as the user is concerned, the processing has happened on their device. The cloud element is functionally transparent, with the only requirement being access to the internet. The benefits of such an architecture are enormous—users no longer have to be concerned about the limitations of their devices. Theoretically, a device from 10 years ago could be just as capable of consuming cloud services as a £1000 GBP smartphone manufactured in 2020. This uncoupling of computing requirements from user's devices has brought a paradigm shift in the way that systems are designed and used, and is having a dramatic effect on the ever-growing ubiquity of computing. On the one hand we have the unprecedented worldwide adoption of smartphones and mobile devices, and on the other we have a surge in available remote computing performance.

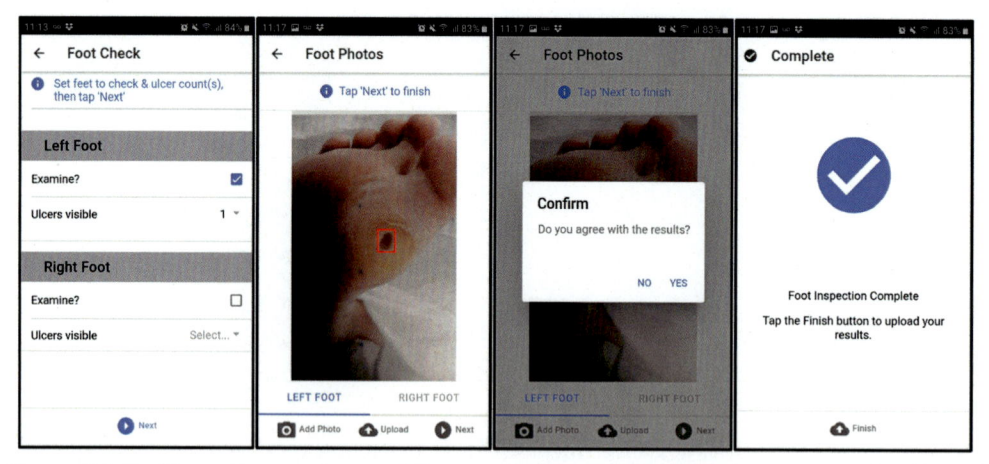

Figure 7.4 FootSnap AI—prototype mobile app used for sending foot images to a cloud-based service. The cloud service is hosted on Oracle Cloud Infrastructure, where an AI model is used to determine the presence of a diabetic foot ulcer. The results are returned to the app in a matter of seconds, highlighting the location of the detected diabetic foot ulcer by drawing a *red rectangle* around the wound. This prototype is currently being evaluated at Salford Royal Hospital and Lancashire Teaching Hospitals.

The pervasiveness of mobile and cloud computing provides the perfect opportunity for new technologies to fill the numerous gaps where healthcare shortfalls exist. Healthcare settings around the world are witnessing a significant growth in cases of diabetes. Many of these patients go on to develop further complications, which can include a diabetic foot ulcer. Early intervention is vital in the care of patients who suffer from this condition. In an attempt to address this issue, recent research has focused on the automated remote detection of diabetic foot ulcers using machine learning techniques. However, machine learning algorithms require significant computing power, well beyond the capability of most mobile devices. A natural solution to this problem is to offload the bulk of the processing to a cloud platform, with the results being returned to the patient's mobile device. Such proof-of-concept system known as "FootSnap AI" already exists and is being used to automatically detect diabetic foot ulcers [16] (see Fig. 7.4). FootSnap AI involves photographs being taken with any smartphone or tablet device, then uploaded to the cloud for automatic analysis, with results of the analysis returned to the patient's phone providing a diagnosis. This can take place in a matter of seconds, a process that might otherwise have taken days or weeks had the patient needed to visit clinical specialists.

The future of diabetic foot care

In the future, digital, AI, and cloud technologies will transform the way in which we prevent and manage diabetic foot ulcers. Digital and AI-based technologies will enable enhanced remote monitoring, optimized prevention and place greater opportunities for foot self-management in the hands of the patient. Digital, AI, and cloud technologies will empower patients with the tools to better manage their foot health and simultaneously relive the time and cost burden on health systems.

These technology advances will facilitate fewer patient clinics visits and reduced use of hospital services for outpatient appointments. This will provide major time and cost savings for patients, health professionals, and hospital services. Telemedicine systems are already being utilized, especially in light of the recent Covid-19 pandemic. Many patient—health professional interactions do not require the patient to be present at the clinic. Even in the absence of AI technologies, telemedicine systems can be utilized for routine check-ups and routine interactions with clinicians. This presents major time savings meaning that patients no longer need to spend time traveling to and from clinics, do not need to spend time in clinic waiting rooms, thus reducing the likelihood of transmitting infectious diseases. Intelligent insole systems and digital and AI technologies will need to demonstrate cost-effectiveness, but considering the direct healthcare costs and wider societal and quality of life costs, there is significant scope for these technologies to make an impact.

Combining cloud computing with AI will allow us to create new systems that people with diabetes can use themselves to help monitor their own foot health and to promote early interventions. This could have a dramatic impact on global healthcare systems. Deep learning models are not limited to inferring information from images and video. Other patient data, and data taken from other devices, such as sensors, could potentially be used to increase the accuracy of automated predictions. Such systems can be used to complement existing health programmes, and to help take the strain off overburdened healthcare resources.

Acknowledgments

We gratefully acknowledge a project grant from Diabetes UK, Reference number: 12/0004565, to fund years 1—3 of a clinical trial into intelligent insole systems for diabetic foot ulcer prevention, with year 4 funded by Orpyx Medical Technologies. We gratefully acknowledge the support of Oracle Corporation with an Innovation Accelerator Grant.

References

[1] Abbott CA, Carrington AL, Ashe H, Bath S, Every LC, Griffiths J, Hann AW, Hussein A, Jackson N, Johnson KE, Ryder CH, Torkington R, Van Ross ERE, Whalley AM, Widdows P, Williamson S, Boulton AJM, North-West Diabetes Foot Care Study. The North-West diabetes foot care study: incidence of, and risk factors for, new diabetic foot ulceration in a community-based patient cohort. Diabet Med 2002;19:377—84. https://doi.org/10.1046/j.1464-5491.2002.00698.x.

[2] Abbott CA, Chatwin KE, Foden P, Digital AHTL. Innovative intelligent insole system reduces diabetic foot ulcer recurrence at plantar sites: a prospective, randomised, proof-of-concept study. Lancet Digital Health 2019;1. https://doi.org/10.1016/S2589-7500(19)30128-1.

[3] Armstrong DG, Boulton AJM, Bus SA. Diabetic foot ulcers and their recurrence. N Engl J Med 2017; 376:2367—75. https://doi.org/10.1056/NEJMra1615439.

[4] Armstrong DG, Lavery LA, Kimbriel HR, Nixon BP, Boulton AJM. Activity patterns of patients with diabetic foot ulceration: patients with active ulceration may not adhere to a standard pressure off-loading regimen. Diabetes Care 2003;26:2595—7. https://doi.org/10.2337/diacare.26.9.2595.

[5] Armstrong DG, Nguyen HC, Lavery LA, van Schie CHM, Boulton AJM, Harkless LB. Off-loading the diabetic foot wound: a randomized clinical trial. Diabetes Care 2001;24:1019—22. https://doi.org/10.2337/diacare.24.6.1019.

[6] Binning J, Woodburn J, Bus SA, Barn R. Motivational interviewing to improve adherence behaviours for the prevention of diabetic foot ulceration. Diabetes Metab Res Rev 2019;35:e3105. https://doi.org/10.1002/dmrr.3105.

[7] Boulton AJM, Betts RP, Franks CI, Newrick PG, Ward JD, Duckworth T. Abnormalities of foot pressure in early diabetic neuropathy. Diabet Med 1987;4:225—8. https://doi.org/10.1111/j.1464-5491.1987.tb00867.x.

[8] Brown R, Ploderer B, Da Seng LS, Lazzarini P, van Netten J. MyFootCare: a mobile self-tracking tool to promote self-care amongst people with diabetic foot ulcers. In: Proceedings of the 29th Australian conference on computer-human interaction; November 2017. p. 462—6.

[9] Bus SA, Armstrong DG, Gooday C, Jarl G, Caravaggi C, Viswanathan V, Lazzarini PA. Guidelines on offloading foot ulcers in persons with diabetes (IWGDF 2019 update). Diabetes Metab Res Rev 2020; 36:353. https://doi.org/10.1002/dmrr.3274.

[10] Bus SA, van Deursen RW, Armstrong DG, Lewis JEA, Caravaggi CF, Cavanagh PR, International Working Group on the Diabetic Foot. Footwear and offloading interventions to prevent and heal foot ulcers and reduce plantar pressure in patients with diabetes: a systematic review. Diabetes Metab Res Rev 2016;32(Suppl. 1):99—118. https://doi.org/10.1002/dmrr.2702.

[11] Caselli A, Pham H, Giurini JM, Armstrong DG, Veves A. The forefoot-to-rearfoot plantar pressure ratio is increased in severe diabetic neuropathy and can predict foot ulceration. Diabetes Care 2002; 25:1066—71. https://doi.org/10.2337/diacare.25.6.1066.

[12] Cassidy B, Reeves ND, Joseph P, Gillespie D, O'Shea C, Rajbhandari S, Maiya AG, Frank E, Boulton A, Armstrong D, Wu J, Yap MH. DFUC2020: analysis towards diabetic foot ulcer detection. 2020. arXiv preprintarXiv:2004.11853.

[13] Crawford F, Cezard G, Chappell FM. The development and validation of a multivariable prognostic model to predict foot ulceration in diabetes using a systematic review and individual patient data meta-analyses. Diabet Med 2018;35:1480—93. https://doi.org/10.1111/dme.13797.

[14] Chatwin KE, Abbott CA, Boulton AJM, Bowling FL, Reeves ND. The role of foot pressure measurement in the prediction and prevention of diabetic foot ulceration-A comprehensive review. Diabetes Metab Res Rev 2020;36:e3258. https://doi.org/10.1002/dmrr.3258.

[15] de León Rodriguez D, Allet L, Golay A, Philippe J, Assal JP, Hauert CA, Pataky Z. Biofeedback can reduce foot pressure to a safe level and without causing new at-risk zones in patients with diabetes and peripheral neuropathy. Diabetes Metab Res Rev 2013;29:139—44. https://doi.org/10.1002/dmrr.2366.

[16] FootSnap-AI. Fully automated diabetic foot ulcer detection. 2020. https://dfu-challenge.github.io/footsnap/.

[17] French R. The turing test: the first 50 years. Trends Cogn Sci 2000;4:115—22. https://doi.org/10.1016/S1364-6613(00)01453-4.

[18] Goodfellow I, Bengio Y, Courville A, Bengio Y. Deep learning, vol. 1. Cambridge: MIT Press; 2016.

[19] Goyal M, Reeves ND, Davison AK, Rajbhandari S, Spragg J, Yap MH. DFUNet: convolutional neural networks for diabetic foot ulcer classification. IEEE Trans Emerg Top Comput Intell 2018: 1—12. https://doi.org/10.1109/TETCI.2018.2866254.

[20] Goyal M, Reeves ND, Rajbhandari S, Ahmad N, Wang C, Yap MH. Recognition of ischaemia and infection in diabetic foot ulcers: dataset and techniques. Comput Biol Med 2020:103616.

[21] Goyal M, Reeves ND, Rajbhandari S, Yap MH. Robust methodsfor real-time diabetic foot ulcer detection and localization on mobile devices. IEEE J Biomed Health Inf 2019;23:1730—41. https://doi.org/10.1109/JBHI.2018.2868656.

[22] McCorduck P, Cfe C. Machines who think: a personal inquiry into the history and prospects of artificial intelligence. CRC Press; 2004. p. 157—8.
[23] Moor J. The Dartmouth College artificial intelligence conference: the next fifty years. AI Mag 2006; 27:87—91.
[24] Najafi B, Crews RT, Wrobel JS. Importance of time spent standing for those at risk of diabetic foot ulceration. Diabetes Care 2010;33:2448—50. https://doi.org/10.2337/dc10-1224.
[25] Najafi B, Reeves ND, Armstrong DG. Leveraging smart technologies to improve the management of diabetic foot ulcers and extend ulcer-free days in remission. Diabetes Metab Res Rev 2020;36(Suppl. 1):e3239. https://doi.org/10.1002/dmrr.3239.
[26] Najafi B, Ron E, Enriquez A, Marin I, Razjouyan J, Armstrong DG. Smarter sole survival: will neuropathic patients at high risk for ulceration use a smart insole-based foot protection system? J Diabetes Sci Technol 2017:702—13. https://doi.org/10.1177/1932296816689105.
[27] Owings TM, Apelqvist J, Stenström A, Becker M, Bus SA, Kalpen A, Ulbrecht JS, Cavanagh PR. Plantar pressures in diabetic patients with foot ulcers which have remained healed. Diabet Med 2009;26:1141—6. https://doi.org/10.1111/j.1464-5491.2009.02835.x.
[28] Pataky Z, de León Rodriguez D, Allet L, Golay A, Assal M, Assal JP, Hauert CA. Biofeedback for foot offloading in diabetic patients with peripheral neuropathy. Diabet Med 2010;27:61—4. https://doi.org/10.1111/j.1464-5491.2009.02875.x.
[29] Pataky Z, Faravel L, Da Silva J, Assal J. A new ambulatory foot pressure device for patients with sensory impairment. A system for continuous measurement of plantar pressure and a feed-back alarm. J Biomech 2000;33:1135—8. https://doi.org/10.1016/s0021-9290(00)00082-8.
[30] Reiber GE, Smith DG, Wallace C, Sullivan K, Hayes S, Vath C, Maciejewski ML, Yu O, Heagerty PJ, LeMaster J. Effect of therapeutic footwear on foot reulceration in patients with diabetes: a randomized controlled trial. J Am Med Assoc 2002;287:2552—8. https://doi.org/10.1001/jama.287.19.2552.
[31] Stess RM, Jensen SR, Mirmiran R. The role of dynamic plantar pressures in diabetic foot ulcers. Diabetes Care 1997;20:855—8.
[32] Pham H, Armstrong DG, Harvey C, Harkless LB, Giurini JM, Veves A. Screening techniques to identify people at high risk for diabetic foot ulceration: a prospective multicenter trial. Diabetes Care 2000;23:606—11. https://doi.org/10.2337/diacare.23.5.606.
[33] Ulbrecht JS, Hurley T, Mauger DT, Cavanagh PR. Prevention of recurrent foot ulcers with plantar pressure-based in-shoe orthoses: the CareFUL prevention multicenter randomized controlled trial. Diabetes Care 2014;37:1982. https://doi.org/10.2337/dc13-2956.
[34] Veves A, Murray HJ, Young MJ, Boulton AJ. The risk of foot ulceration in diabetic patients with high foot pressure: a prospective study. Diabetologia 1992;35:660—3. https://doi.org/10.1007/s00125-015-3640-6.
[35] van Schie CH, Abbott CA, Vileikyte L, Shaw JE, Hollis S, Boulton AJ. A comparative study of the Podotrack, a simple semiquantitative plantar pressure measuring device, and the optical pedobarograph in the assessment of pressures under the diabetic foot. Diabet Med 1999;16:154—9. https://doi.org/10.1046/j.1464-5491.1999.00018.x.
[36] Waaijman R, de Haart M, Arts MLJ, Wever D, Verlouw AJWE, Nollet F, Bus SA. Risk factors for plantar foot ulcer recurrence in neuropathic diabetic patients. Diabetes Care 2014;37:1697—705. https://doi.org/10.2337/dc13-2470.
[37] Wang L, Pedersen PC, Agu E, Strong DM, Tulu B. Area determination of diabetic foot ulcer images using a cascaded two-stage SVM-based classification. IEEE Trans Biomed Eng 2016;64(9):2098—109.
[38] Yap MH, Chatwin KE, Ng CC, Abbott CA, Bowling FL, Rajbhandari S, Boulton AJ, Reeves ND. A new mobile application for standardizing diabetic foot images. J Diabetes Sci Technol 2018;12(1):169—73.
[39] Yap MH, Ng CC, Chatwin K, Abbott CA, Bowling FL, Boulton AJ, Reeves ND. Computer vision algorithms in the detection of diabetic foot ulceration: a new paradigm for diabetic foot care? J Diabetes Sci Technol 2016;10(2):612—3.
[40] Yap MH, Reeves N, Boulton A, Rajbhandari S, Armstrong D, Maiya AG, Najafi B, Frank E, Wu J. Diabetic foot ulcers grand challenge 2021. 2020. https://doi.org/10.5281/zenodo.3715020.

[41] Yap MH, Reeves ND, Boulton A, Rajbhandari S, Armstrong D, Maiya AG, Najafi B, Frank E, Wu J. Diabetic foot ulcers grand challenge 2020. 2020. https://doi.org/10.5281/zenodo.3715016.

[42] York RM, Perell-Gerson KL, Barr M, Durham J, Roper JM. Motor learning of a gait pattern to reduce forefoot plantar pressures in individuals with diabetic peripheral neuropathy. PMRJ 2009;1:434—41. https://doi.org/10.1016/j.pmrj.2009.03.001.

[43] Hex N, Bartlett C, Wright D, Taylor M, Varley D. Estimating the current and future costs of Type 1 and Type 2 diabetes in the UK, including direct health costs and indirect societal and productivity costs. Diabet Med 2012;29:855—62. https://doi.org/10.1111/j.1464-5491.2012.03698.x.

[44] Kerr M, Barron E, Chadwick P, Evans T, Kong WM, Rayman G, Sutton Smith M, Todd G, Young B, Jeffcoate WJ. The cost of diabetic foot ulcers and amputations to the National Health Service in England. Diabet Med 2019;36:995—1002. https://doi.org/10.1111/dme.13973.

CHAPTER 8

The role of tissue biomechanics in improving the clinical management of diabetic foot ulcers

Panagiotis Chatzistergos, Roozbeh Naemi, Nachiappan Chockalingam
Centre for Biomechanics and Rehabilitation Technologies, Staffordshire University, Stoke-on-Trent, United Kingdom

Introduction

Recent advances in medical imaging have opened the way for measuring tissue biomechanics as part of standard clinical practice. The unique capabilities of these new assessment modalities are already enhancing the clinical management of conditions known to affect tissue biomechanics, such as chronic liver disease [1]. Considering that diabetes [2–5] and diabetic foot ulceration (DFU) [6] affect the mechanical behavior of the soft tissues of the sole of the foot (plantar soft tissue), one could argue that measuring tissue biomechanics would also have a role to play in the clinical management of diabetic foot. This chapter will try to strike a balance between the potential benefits of measuring tissue biomechanics and the challenges of measuring tissue biomechanics in the clinic in light of the limitations of the available techniques.

When making the case for the benefits of using the assessment of soft tissue biomechanics as part of clinical practice, emphasis will be given on what can be learned, using tissue biomechanics, that cannot be learned using any other existing approach (The case for assessing tissue biomechanics section). The final section of this chapter will focus on specific techniques that have already been used in clinics to study the biomechanics of plantar soft tissue. Some of these techniques are still experimental and others are already developed for clinical practice. For each one of them the principles underpinning their function, their contribution to clinical research in the area of diabetic foot, and their limitations will be discussed (Clinically applicable methods to assess plantar soft tissue biomechanics section).

The topic of this chapter is clearly multidisciplinary in nature with a strong engineering component. However, prior knowledge of engineering theory should not be a prerequisite for reading it. The language and level of technical detail included in this chapter has been adapted to make its content accessible to diverse audiences. Moreover, the key engineering concepts that are needed to understand the content of this chapter are introduced and explained in Basic engineering concepts section.

Basic engineering concepts

All tissues within the human body, regardless of their function, are subjected to various mechanical loads. These loads can change the size or shape of the tissues (i.e., cause a deformation) and, under specific conditions, even damage them. The study of the mechanical characteristics of tissues enables us to understand and predict how a tissue will be affected by a specific load, which can be loosely described as mechanical behavior. That includes how the tissue will deform and whether it will get damaged/injured or not.

There are two specific aspects of mechanical behavior that are particularly relevant to tissue injury: stiffness and strength. Broadly speaking, stiffness offers an assessment of a tissue's ability to deform when it is loaded. In other words, stiffness tells us how easy or difficult it is to compress or stretch a tissue. On the other hand, strength is an assessment of the tissue's capacity to carry load without sustaining any damage. In other words, strength tells us how easy or difficult it is to injure a tissue.

In conventional engineering applications, stiffness and strength are measured with the help of standardized mechanical tests which involve cutting a sample of the material we want to study and then compressing or stretching it until the point of failure (Fig. 8.1). Measuring how much force is needed to deform the material to different degrees enables the calculation of a force—deformation graph. However, because the force that is needed to deform a sample by a given degree also depends on the dimensions and shape of the sample itself, the force—deformation graph, in essence, describes the mechanical behavior of the sample and not of the material the sample is made from. For example, it is easily understood that a thicker sample of Fig. 8.1 will need more force to be stretched compared to a thinner one that is made from the same material. To overcome this problem and to calculate the properties of the actual material we need to convert forces (F) into stresses (σ) and deformations (dL) into strains (ε).

Stress is a measure of how the externally applied load is experienced by each particle in the sample. In the case of the standardized sample of Fig. 8.1, the externally applied force is uniformly distributed between all particles of cross-section (A), so stress in this case can easily be calculated by dividing the value of the externally applied force over the area of the sample's cross-section (i.e., $\sigma = F/A$).

Strain, on the other hand, quantifies the change in dimensions of each particle in the sample relative to their original/undeformed size. Similar to stresses, strains are uniform in the case of the sample of Fig. 8.1, so strains can be calculated simply by dividing the overall change in length over the initial length of the sample (i.e., $\varepsilon = dL/L_0 = (L-L_0)/L_0$).

The stress—strain graph enables us to calculate parameters that define the material and are independent of the size and shape of the sample we used during testing. If the stress—strain graph is linear then the material can be described as being linearly elastic. The slope of the linear part of the stress—strain graph quantifies the stiffness of the material and is known as modulus of elasticity or Young's modulus (E).

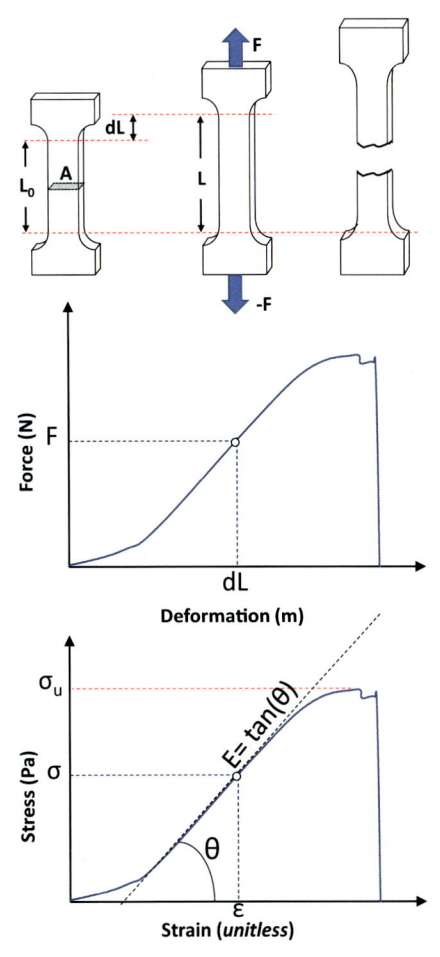

Figure 8.1 A typical standardized material sample (top) and typical force–deformation (middle) and stress–strain (bottom) graphs produced from mechanical testing. The sample is shown at different stages of testing, from left to right: (i) before testing (initial length: L_0, cross-sectional area: A), (ii) stretched by a force (F) that increased its length by dL, and (iii) after failure has occurred. The definition of the modulus of elasticity (E) and of ultimate stress (σ_u) is also shown on the stress–strain graph.

Depending on the application there are different ways to quantify strength: one of the simplest is to use the maximum value of stress in the stress–strain graph. This value is called ultimate stress (σ_u) and corresponds to the maximum stress that this material can carry before breaking.

Soft tissues are known to exhibit complex mechanical behavior. Among their key characteristics is that they can sustain very large deformations and that during loading these tissues are very soft at first but as deformation increases, they gradually get very stiff. The materials that exhibit this type of behavior are called hyperelastic and the phenomenon of increasing stiffness with increasing deformation is called strain stiffening. As a result of this complex behavior in many cases a simple measurement of the slope of the stress—strain graph might not be adequate to fully understand and quantify the stiffness of soft tissues.

For this reason the mechanical behavior of hyperelastic materials is usually described with the help of material models such as the Ogden hyperelastic model [7]. In its simplest form this model requires the calculation of two material coefficients (C and m) to fully define the tissue's stiffness and how it changes with strain. Coefficient C is related to the initial slope of the stress—strain graph while coefficient m is related to the tissue's strain-stiffening behavior [8,9].

Another characteristic of the mechanical behavior of soft tissues is that it is strongly dependent on time. Such materials are called viscoelastic and their stiffness is not affected only by their deformation status but also by the speed at which loading is applied. As a result, the stiffness of viscoelastic materials increases for more dynamic loading relative to static loading (strain rate stiffening). Another difference is that their resistance to static loading changes over time. When a purely elastic material is stretched, its resistance (i.e., reaction force) stays constant for the duration the stretch is kept constant. However, the resistance of a viscoelastic material to a constant deformation tends to gradually drop over time and its internal stresses tend to become reduced. If the deformation is kept constant for long enough then the tissue's resistance will stabilize. The time it takes for this stabilization to take place is called relaxation time. To describe these aspects of the tissue's behavior there is a need for a viscosity model. In one of its simplest forms this model is defined using one coefficient that is related to the tissue's strain rate—stiffening behavior (g) and one related to its relaxation time (t).

Readers who are interested in a more in-depth description of the complex viscohyperelastic mechanical behavior of the soft tissue of the foot and how this can be defined and modeled are encouraged to refer to a relevant paper by Behforootan et al. [8].

Despite their complexity, human tissues can be studied like any other material. However, their study is made significantly more difficult due to challenges that are unique to the area of tissue biomechanics. These include the availability of samples and the relevance of *in vitro* testing. In the vast majority of cases cutting a sample from the tissue we want to study is simply impossible. This is because of ethical but also practical reasons. At the same time *in vitro* testing can never truly replicate the *in vivo* conditions which, for many applications, renders the use of cadaveric or *ex vivo* samples irrelevant.

One of the methods to overcome this problem involves the use of computer modeling techniques such as finite element (FE) modeling. FE is a computational technique that enables the design of a mathematical model of the structure that we want to study. Once this mathematical model is completed it can then be used to perform virtual experiments to predict how a tissue would behave under different loading conditions and to explore aspects of tissue biomechanics that cannot be investigated experimentally. For more information on how FE modeling has been used and what has been learned from its use in the area of diabetic foot the readers are encouraged to refer to the review papers by Behforootan et al. [7] and Telfer et al. [10].

The case for assessing tissue biomechanics

Based on the current literature focusing on diabetic foot and also the broader research area of soft tissue injury, there appear to be two main topics which highlight the usefulness of tissue biomechanics to enhance the clinical management of DFU. The first one is to help improve our ability to predict ulcers and to assess ulceration risk for better patient stratification, and the second is to enhance DFU prevention through early diagnosis of internal tissue damage [11].

Risk of ulceration

The soft tissues of the sole of the foot (i.e., plantar soft tissue), which are one of the main tissues affected by ulceration, play a primarily mechanical role: They act as a shock absorber and promote a more even distribution of plantar load. Because of their purely mechanical role the concepts of stiffness and strength are of paramount important to understand the mechanisms of injury in this particular tissue.

DFU is clearly multifactorial, but the consensus is that it is triggered by mechanical trauma that goes unnoticed due to lack of sensation of pain in the foot caused by peripheral neuropathy [12]. Because of that, understanding the mechanisms that make mechanical trauma more or less likely should also enable us to understand the mechanisms that make DFU more or less likely.

If we see ulceration through the lens of tissue biomechanics, then we will find three possible changes that can make ulceration more or less likely: (a) change in tissue loading, (b) change in tissue strength, or (c) change in tissue stiffness. If loading in a tissue increases above the tissue's strength, then the tissue will get damaged/injured. In the absence of pain, overloading persists leading to tissue breakdown and ulceration.

People with diabetes and neuropathy are known to load their feet more heavily than their nondiabetic/nonneuropathic counterparts [13,14]. However, according to the remaining two mechanisms an overloading injury can also become more likely even without any increase in the magnitude of loading, but simply because something changed within the tissue itself. If the strength of a tissue is reduced, then the magnitude of loading that can be safely applied to the tissue is essentially lowered making injury more likely even without any increase in the magnitude of loading the tissue is experiencing. In extreme circumstances where the strength is substantially reduced, then we could potentially have an injury even with minimal external force. An example of changes in a tissue that could significantly affect its strength and therefore also the likelihood of tissue damage is the changes in the hydration of skin. Diabetes is known to lead to changes in the skin including diabetic xerosis or dry skin which has been associated with increased risk for skin damage and ulceration [15–17]. At the same time other studies in the area of pressure ulcers have also indicated that increased humidity in a load-bearing area of skin also has a detrimental effect on skin strength [18,19].

At the same time stiffness can also change in such a way that it could compromise the tissues' ability to uniformly distribute loading. In this case for the same net ground reaction force we have higher stresses developing inside the tissue, which lead again to increased risk for overloading injury and ulceration. This change in stiffness can be localized or it could be uniform; namely, softening or stiffening that either happens in a small localised area or occurs uniformly across the area of the foot. For example, the formation of callus can be described as localized stiffening of skin which magnifies the stresses developed in the neighboring tissues, thus increasing the likelihood of soft tissue injury and DFU.

There is also strong evidence in literature that plantar soft tissue tends to be stiffer [2–5] and harder [5] in people with diabetes, but the exact implications of these changes are not fully understood. In order to get a first assessment of the possible effect of altered tissue stiffness on the likelihood of ulceration we created an FE model of a heel and used it to simulate heel strike [8]. The material properties of the heel pad and loading conditions were taken from a young healthy individual and we used this model to assess how plantar pressure changes if the heel pad becomes softer or stiffer. For this analysis we assumed that a change in stiffness that reduced plantar pressure for the same external load would reduce the likelihood of overloading and therefore of ulceration, while a change that increased plantar pressure would also increase the likelihood of ulceration. This analysis indicated that when the heel pad becomes stiffer its ability to uniformly distribute plantar loading is also reduced making overloading more likely [8]. As it can be seen in Fig. 8.2 increasing the heel pad's stiffness makes the tissue less deformable (as expected) and at the same time it also increases the plantar pressure developed between the heel and the simulated ground. Overall it was found that lower deformability led to higher pressures (Fig. 8.2A).

At the same time we also found that it is not only how much the tissue deforms that's important, but how it deforms also has an effect [8]. Considering that the mechanical behavior of plantar soft tissue is nonlinear the effect of the shape of the stress–strain graph was also investigated. It was found that changes in the tissue that led to more intense strain stiffening (i.e., more nonlinear behavior) also reduced the tissue's ability to uniformly distribute plantar loading, thus increasing the risk for overloading (Fig. 8.2B). The conclusion of this study was that a change in tissue stiffness can indeed influence the risk for overloading and therefore the risk for tissue damage and ulceration.

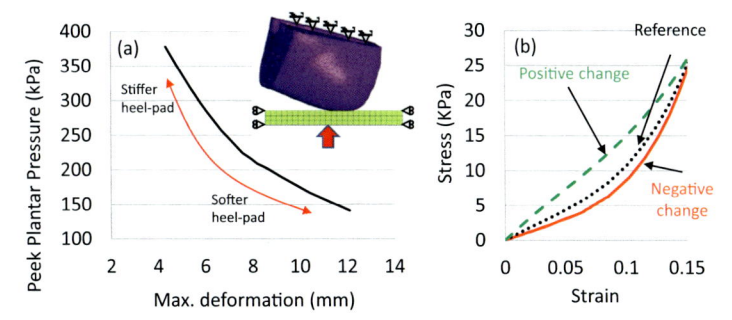

Figure 8.2 The FE model of heel strike and the effect of altered heel pad deformability on peak plantar pressure (A). In this case deformability is assessed with the help of the maximum deformation of the heel pad during heel strike. The changes in the shape of the stress–strain graph that led to increased (negative change) or decreased (positive change) plantar pressure for the same overall deformability are also shown (B). The reference heel pad stress–strain graph was measured for a young healthy individual. *(Adapted from Figs. 3 and 7. Behforootan S, Chatzistergos P, Chockalingam N, Naemi R. A clinically applicable non-invasive method to quantitatively assess the visco-hyperelastic properties of human heel pad, implications for assessing the risk of mechanical trauma. J Mech Behav Biomed Mater. 2017;68:287–95.)*

Detection of internal tissue damage

Any damage inside a tissue will also change the tissue's mechanical behavior, which in turn could be used to detect internal damage [20]. Moreover, changes in the tissue that are caused by the body's response to trauma (e.g., inflammation) could also have an impact on the tissue's mechanical characteristics. It can therefore be hypothesized that mapping changes in tissue mechanical properties could be used to detect internal trauma and to monitor healing. This hypothesis is supported by *in vivo* evidence which link overloading in the foot with altered soft tissue properties and altered tissue properties with pressure ulcer development [21–24]. A good example from the broader area of foot biomechanics is the heel pain syndrome, which is classically associated to an overuse injury. *In vivo* measurements have shown that the heel pads in heel pain syndrome exhibit different mechanical behavior compared to healthy heel pads [21]. Because people in this case can feel pain, studying tissue biomechanics is not needed to diagnose this specific condition. However, this example shows that an overuse injury of the plantar soft tissues can indeed be linked to changes in tissue mechanical properties, which are measurable and could potentially be used to detect the overuse injury in the absence of pain.

Similarly, in the case of pressure ulcers, deep tissue trauma is associated with localized stiffening of muscle tissue and palpation to detect localized stiffening is indicated by current guidelines as a useful method for early diagnosis [22]. Moreover, *in vivo* experiments using animal models have shown that measurements of soft tissue stiffness can be used to diagnose pressure ulcer development before the breakdown of the tissue becomes visible [24,25]. More specifically a study on the effect of deep tissue injury in the muscles of rats concluded that tissue damage leads to measurable localized increase in stiffness in the affected area (Fig. 8.3B and C) followed by a decrease in stiffness in 3—10 days postinjury [25]. In this particular study deep tissue injury was induced through prolonged compression (Fig. 8.3A) and changes in tissue stiffness were assessed with the help of magnetic resonance imaging (MRI). A second study that used ultrasound instead of MRI reached similar conclusions about the immediate effect of deep muscle injury on the tissue's stiffness [24].

Even though pressure ulcers are inherently different to DFU this evidence highlights the link between soft tissue trauma and altered mechanical properties and also the need for objective ways to measure them [23].

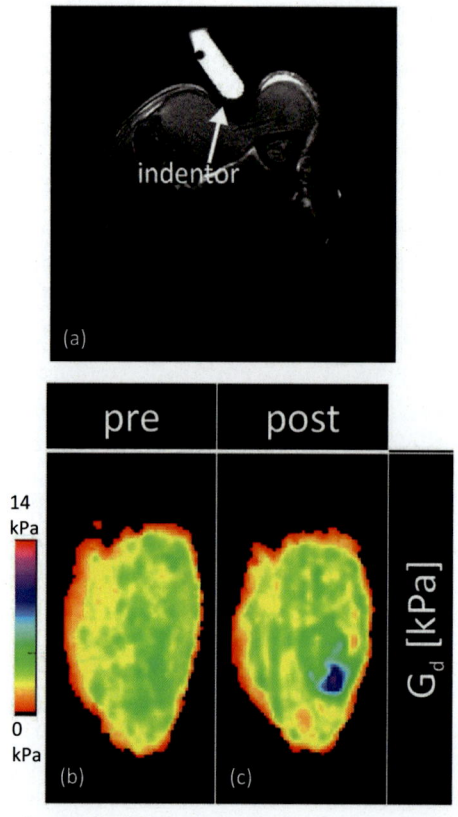

Figure 8.3 MR image of a rat's muscle during prolonged compression to induce deep tissue damage (A) and the maps of stiffness (G_d) before (B) and after injury (C) highlighting significant localized stiffening in the affected area. *(Adapted from Fig. 3. Nelissen JL, Oomens CWJ, Sinkus R, Strijkers GJ. Magnetic resonance elastography of skeletal muscle deep tissue injury. NMR Biomed. 2019;(April 2018):1—12.)*

The imaging techniques that were used in the abovementioned animal studies produced maps of stiffness inside the tissue that enabled the assessment of changes in tissue biomechanics over time as well as the detection of differences between regions (i.e., loaded/injured vs. unloaded/healthy) [24,25]. The capacity to image an aspect of a tissue's mechanical behaviour is called elastography. Two commonly used ultrasound-based elastography techniques will be discussed in the following section (Strain/ real time ultrasound elastography and Shear wave ultrasound elastography sections).

Clinically applicable methods to assess plantar soft tissue biomechanics

This section will focus on methods to assess clinically relevant aspects of the mechanical behavior of plantar soft tissue that have already been used in clinics in diabetic populations. Three different methods will be discussed in total: ultrasound indentation, which offers the most comprehensive assessment of tissue biomechanics but is still at an experimental stage of development, and two different ultrasound elastography techniques that are already part of clinical practice.

Ultrasound indentation

Ultrasound indentation is a noninvasive mechanical test where the soft tissues of the foot are compressed between an ultrasound probe and a bony surface [8,26–30]. An example of a device that has been used to perform this type of test is shown in Fig. 8.4 [8,28–30]. As it can be seen, the foot is constrained at one end leaving the plantar surface exposed. At the center of the system there is an ultrasound probe in series with a force sensor (load cell).

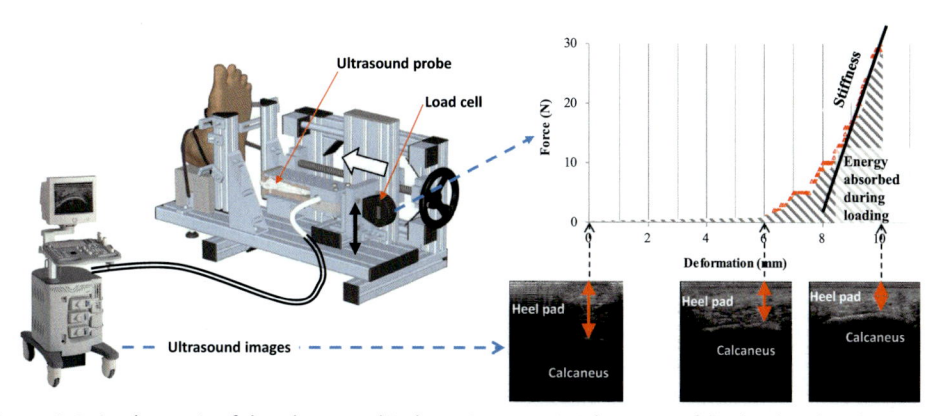

Figure 8.4 A schematic of the ultrasound indentation test. Axial images of the heel pad at the area of the apex of the calcaneus are used to measure the bulk deformation of heel pad. *(Adapted from Figs. 1 and 2. Chatzistergos PE, Naemi R, Sundar L, Ramachandran A, Chockalingam N. The relationship between the mechanical properties of heel-pad and common clinical measures associated with foot ulcers in patients with diabetes. J Diabetes Complications. 2014 Jan; 28(4):488–93.)*

When the instrumented ultrasound probe is pressed against the surface of the foot, the applied force is measured from the load cell and the resulted deformation in the plantar soft tissue is measured from the ultrasound images. Synchronizing the information taken from the force sensor and the ultrasound unit [8] allows drawing the force—deformation graph of the indentation test.

One of the key advantages of using ultrasound imaging is that it enables a direct measurement of tissue deformation that is not affected by any possible movement of the foot. Even though the foot is constrained, movement of the entire foot or of individual bone segments remains likely during testing. Such movement would cause errors in the measurement of tissue deformation if this was not based on medical imaging (e.g., assessed based on the displacement of the indenter).

In the case of the results shown in Fig. 8.4 the test is performed at the heel and deformation is measured based on the change in thickness of the entire heel pad (i.e., bulk deformation). However, this method can also be used to separately measure the deformation of individual layers in the plantar soft tissue [29].

This indentation device was used in a diabetic foot clinic to investigate the correlation between the mechanical characteristics of the heel pad of people with diabetes and clinical parameters used to monitor their health and also to assess ulceration risk [28]. The main outcome measures of these tests were the initial thickness of the heel pad, a measurement of stiffness, and a measurement of the energy that is absorbed by the tissue during loading. More specifically, stiffness was assessed by the slope of the linear part of the indentation force—deformation graph and the energy by the area below the graph (Fig. 8.4). This study showed that people with higher levels of blood sugar and triglycerides also tend to have stiffer heel pads [28]. This is the first direct demonstration of an association between the blood biochemical profile of people with diabetes and the mechanical properties of their plantar soft tissue.

Even though ultrasound indentation on its own can give valuable insight on plantar soft tissue biomechanics and how this is affected by diabetes, its relevance is somewhat restricted by the inherent limitations of the force—deformation graph, which is its main output. Everything that is measured on a force—deformation graph is representative only for the conditions for which this graph was created. These measurements can be used in comparative studies between people who were subjected to the same test, but they cannot, for example, be used to predict how the tissue would behave under different loading conditions.

As we saw in Basic engineering concepts section, in order to get a reliable assessment of the mechanical behavior of the heel pad as a material its stress—strain graph is needed. However, in this case the distribution of internal forces or deformations are not uniform, so we cannot calculate stresses or strains using the simplistic method one would use for the standardized sample of Fig. 8.1. Hence, a more advanced approach is needed.

As it was explained earlier in this section one of the key advantages of using an ultrasound probe as the indenter is that enhances the accuracy of the measured tissue deformation. However, an added benefit is that it also gives access to valuable information about the tissues' morphology. This information is not used for drawing the force—deformation graph but can open the way for advanced postprocessing to estimate the tissue's stress—strain graph. This process involves the reconstruction of the 3D geometry of the tested tissue and the design of a subject-specific FE model of the indentation test (Fig. 8.5). This model can then be used to calculate the tissue's stress—strain graph.

FE modeling is one of the most popular methods in biomechanics research. However, its use is still limited to the research domain and as a result has yet to make a direct impact on clinical practice [7,10]. In an attempt to design a method of FE modeling that could be used in clinics by people with no previous expertise in computer modeling, an automated method to design subject-specific FE models of heel pad indentation was created [8]. This method requires the force—deformation graph of the indentation test, the initial heel pad thickness, and two ultrasound images of the tested area: one in the frontal and one in the sagittal plane (Fig. 8.5A) as input. If the viscoelastic nature of the tissue is also of interest, then a stress relaxation test will have to be performed too. In this case a graph of the tissue's resistance to static deformation over time (stress relaxation graph) is imported along with the conventional indentation force—deformation graph. All needed input is collected using the system of Fig. 8.4.

As a next step a semiautomated process is used to outline the boundary between the calcaneus and heel pad on each one of the two images and to recreate the 3D bony surface by dragging one outline curve along the second (Fig. 8.5B). The 3D geometry of the

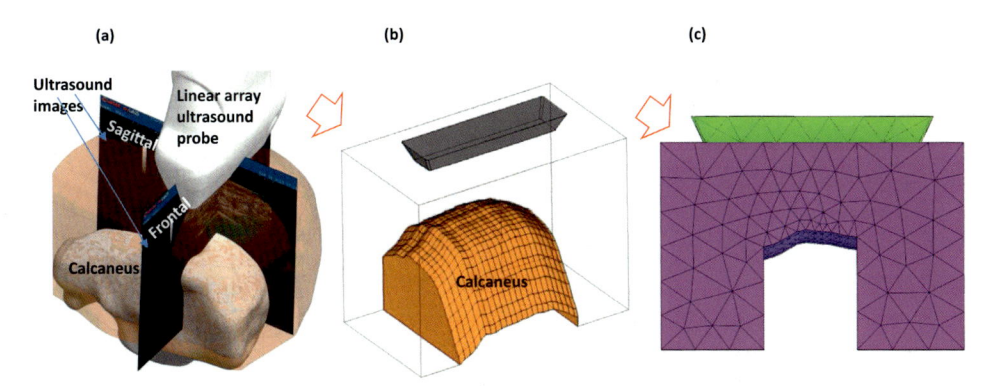

Figure 8.5 The process for creating subject-specific FE models of the indentation test. (A) Input ultrasound images; (B) reconstruction of the 3D shape of the calcaneus at the area of testing; (C) design of a subject-specific FE model of the indentation test. *(Adapted from Figs. 1 and 2. Behforootan S, Chatzistergos P, Chockalingam N, Naemi R. A clinically applicable non-invasive method to quantitatively assess the visco-hyperelastic properties of human heel pad, implications for assessing the risk of mechanical trauma. J Mech Behav Biomed Mater. 2017;68:287—95.)*

calcaneus is used together with the initial heel pad thickness by an FE algorithm to design the subject-specific model of the indentation test (Fig. 8.5C) and calculate (inverse engineer) its viscohyperelastic material coefficients.

In a direct application of FE modeling the material coefficients are already known and the model is used to predict how the tissue would deform during different loading scenarios (i.e., to predict the indentation force—deformation graph). In this case, however, the problem is reversed; namely, we know how the tissue deforms during the loading scenario that interests us but the material coefficients are not known. To solve this problem we start with a random set of values for the material coefficients and then use an optimization algorithm to find the coefficients that minimize the difference between the numerical and the experimental force—deformation graphs.

This method offers one of the most comprehensive assessments of the *in vivo* viscohyperelastic mechanical behavior of plantar soft tissue. Testing itself is noninvasive and can be fully automated, but it has been applied only for the heel and, in its current form, could also be cumbersome and time-consuming to use. This postprocessing procedure has been validated for healthy individuals and has the potential to be fully automated for use by people with no prior knowledge on computer modeling. However, further testing is needed to establish whether it is robust and reliable enough when used in people with diabetes in a clinical setting.

Strain/real-time ultrasound elastography

Ultrasound elastography is a generic term encompassing a number of different imaging modalities. The unifying feature for all these modalities is that they use ultrasound waves and that they provide some assessment of tissue stiffness. Strain elastography (which is also called real-time elastography) is one of the two most commonly used such modalities that have been integrated into clinical practice. The second modality is shear wave (SW) elastography which will be discussed in Shear wave ultrasound elastography section.

Strain elastography involves manually compressing and uncompressing the imaged tissue in cycles of repetitive low-amplitude loading. The system then measures the shift of different regions of the image during loading/unloading to produce a qualitative assessment of the relative deformability at different areas of the image. A typical output from strain elastography is a color map with different colors corresponding to different degrees of relative deformability. A typical example of the output of strain elastography can be seen in Fig. 8.6, where the areas of the image that appear to undergo the highest deformation (i.e., highest deformability or lowest stiffness) are painted green and the areas that appear to undergo the least deformation (i.e., lowest deformability or highest stiffness) are painted red.

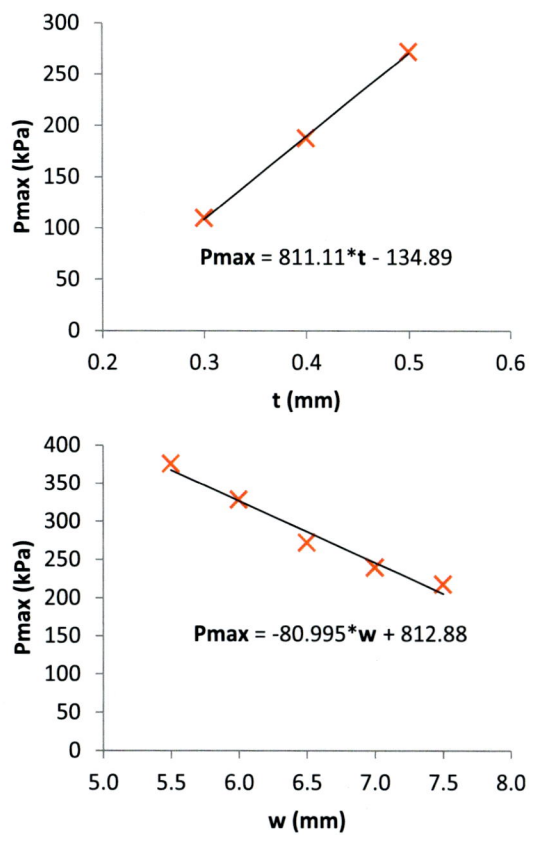

Figure 8.6 A typical output from ultrasound strain elastography at the heel. The heel pad is imaged at the frontal plane at the area of the apex of the calcaneus. The elliptical areas where relevant deformability is measured are shown on the image. Zone 2 corresponds to the standoff that is used as reference material and Zone 1 to the heel pad. The relative deformability of the heel pad over the reference is also indicated on the left of the ultrasound image. *(Adapted from Fig. 1, Naemi R, Chatzistergos P, Sundar L, Chockalingam N, Ramachandran A. Differences in the mechanical characteristics of plantar soft tissue between ulcerated and non-ulcerated foot. J Diabetes Complications. 2016;30(7):1293—9.)*

The strain elastography output map provides immediate feedback on how the different areas of the imaged tissue respond to the externally applied compression which can be very useful for detecting inhomogeneities (e.g. for the detection of tumors). At the same time because deformability is calculated relative to the most and least deformable tissues in the imaging window any comparison between different subjects or even between different imaging windows in the same subject becomes extremely challenging. A second major limitation of this technique, when used in its conventional form is that the calculation of relative deformability is based on the hypothesis that all areas in the imaging window are subjected to the same load and therefore any difference in their deformation is due to differences in stiffness. As a result, this technique cannot differentiate between the cases where a tissue does not deform because it is very stiff or simply because it is not loaded. When the effect of the applied force on the output of strain elastography was investigated it was found that the measured relevant deformability is actually significantly influenced by the magnitude of the applied force [6].

A possible relatively simple way to overcome the limitations that arise from the qualitative nature of strain elastography is the use of a standoff as reference material [6]. Standoffs are soft echolucent (i.e., transparent to ultrasound) materials that are used between the probe and the imaged tissue to improve coupling between these two surfaces. Using such a material during imaging enables the calculation of deformability relative to a known reference and can transform the qualitative output of strain elastography into a quantitative assessment of relevant stiffness which can then be used for comparison between people [6,11].

Moreover, as one can imagine when the tissue is compressed the soft standoff will get compressed too. This means that measuring changes in the thickness of the standoff during loading could also allow us to tell whether the compressive force that is applied during imaging deviates significantly from being uniform along the length of the probe and also to assess differences in applied loading between subjects [6]. In the absence of real-time clinical methods to control the applied force during imaging [31], the retrospective measurements of standoff deformation could be used to normalize the measured deformability and reduce the effect of possible differences in loading between people [6].

This assessment technique was used for 40 patients with diabetes and neuropathy to investigate whether there is a link between strain elastography's deformability and ulceration risk [11]. At baseline, normalized tissue thickness and normalized relative deformability were measured at the heel and first metatarsal mead (met-head) region along with a range of demographic and clinical parameters. Ulceration incidents were monitored and recorded for 12 months from baseline. After the end of the follow-up period a multivariate logistic regression analysis was performed to identify the parameters that contribute to predicting the risk of ulceration. During this analysis, covariates were added in consecutive blocks to find the model with the best sensitivity, specificity, and overall prediction accuracy [11]. The prognosis strength of the final model was assessed using the area under the receiver operating characteristics curve (confidence level: 95%). Finally the predicting ability of the produced model was compared against existing models from literature [32,33].

The ulceration incident rate in the tested population was 17.5% (7 people in total, M: 6/F:1). The results indicated that seven parameters contributed significantly to predicting the risk of ulceration and these included both the normalized thickness and deformability. The remaining six parameters were: history of ulcer, history of callus, insulin and oral hypoglycaemic agent use, duration of diabetes, and vibration perception threshold at the first metatarsal head. Adding a measurement of stiffness and thickness to the model improved its specificity by 3%, its sensitivity by 14%, and its prediction accuracy by

5% [11]. The prognosis strength of this model (89.7%), for the tested population, was higher than the models proposed by Boyko et al. [32] (74.8%) and Monteiro-Soares et al. [33] (63.2%).

Given the relatively small sample size in the study, these results cannot be used to infer which predictive model is best for assessing ulceration risk. However, the fact that strain elastography, with all its limitations, was capable to improve the predicting ability for DFU highlights the potential value of measurements of tissue biomechanics for better risk assessment and patient stratification [11].

Shear wave ultrasound elastography

SW elastography is a quantitative method that involves the generation of a high-intensity ultrasound pulse inside the tissue that displaces a column of the tissue and generates SWs. This high-intensity pulse in some respect plays the role of the manual compression that is needed in the case of strain elastography. So in this case there is no need for applying any loads to the tissue. Another significant difference is that stiffness is not assessed based on changes in the shift between frames but by tracking the propagation of the generated SWs. SWs are waves similar to the ripples created in water and their propagation can be tracked using ultrasound to measure their speed. At the end SW elastography provides a map of the speed with which these SWs propagate in different areas of the imaged tissue and based on this a map of elasticity modulus is also produced (Fig. 8.7).

Assuming that the imaged tissues are linearly elastic, homogenous and that they have the same properties in all directions (i.e., isotropic materials) allows the calculation of elastic modulus (E) from SW speed (C) using this very simple formula: $E = 3\rho C^2$, where ρ is the tissue's density (for soft tissues $\rho = \rho_{water} = 1000$ kg/m^3).

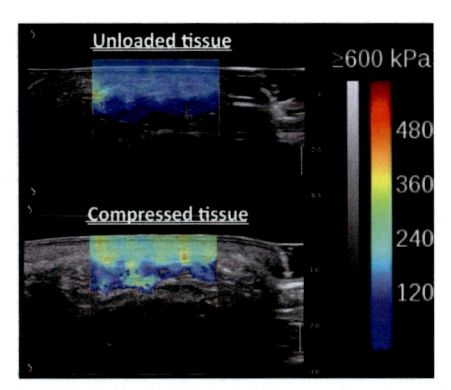

Figure 8.7 A typical output from SW elastography at the heel (frontal plane, area of the apex of the calcaneus). The maps of the estimated modulus of elasticity (E) are shown for the tissue fully unloaded (top) and when it is compressed (bottom) to highlight the effect of compression.

Even though the conditions of isotropy or linear elasticity appear to be relatively restrictive with regard to the materials where SW can be used, SW elastography has already been proven capable of enhancing diagnosis of specific conditions that are associated with altered tissue stiffness, such as chronic liver disease or breast cancer [1]. It has also been used to investigate the effect of diabetes on soft tissue properties [34,35]. However, the fact that no biological tissue is isotropic or linearly elastic also means that translating SW speed into a reliable measurement of stiffness can be very challenging. Therefore the physical meaning and clinical relevance of the outputs of SW elastography should be validated on a tissue- and application-specific basis.

In the case of plantar soft tissue, a first indication on the validity of the measured elasticity modulus from SW elastography was achieved by comparison against the previously presented comprehensive method that combined ultrasound indentation and FE modeling (Ultrasound indentation section). This comparison revealed a substantial and systematic underestimation of stiffness by SW elastography [29]. At the same time, however, an analysis of correlation also showed a significant positive correlation between the two methods. These two findings together indicate that in the case of plantar soft tissue, SW elastography should be able to reliably quantify differences and changes in stiffness, but the absolute values of the predicted mechanical properties should be used with caution [29].

Another characteristic of SW elastography is that it is affected by movement and by loading in the tissue. More specifically SW speed significantly increases with compression (Fig. 8.7). For this reason, SW elastography is always performed with minimum compression on the imaged tissue. Because of that it is clear that any assessment of stiffness is only relevant to the initial stiffness (i.e., the initial slope of its stress—strain graph) of the tissue and its nonlinear behavior, which as we saw earlier is also very important [8], is not assessed.

A method to overcome this limitation and to assess the nonlinear behavior of plantar soft tissue is with the help of the theory of acoustoelasticity [29,36,37]. Applying this theory enables us to use measurements of how SW speed changes with compression to calculate a parameter called nonlinear shear modulus which quantifies the tissue's strain-stiffening behavior. Together with the conventional measurement of SW speed for the unloaded tissue, which is related to the initial slope of the stress—strain graph, these two SW-based measurements can provide a complete assessment of the hyperelastic mechanical behavior of plantar soft tissue. For more information on how to apply this method the reader is encouraged to see the relevant paper by Chatzistergos et al. [29].

Concluding remarks

One of the key objectives of this chapter was to highlight that, altered tissue biomechanics can be a significant contributor to increased vulnerability to injury and ulceration that needs to be investigated further. Besides increased loading, changes in plantar soft tissue stiffness or strength can also make overloading injury more likely. Increased stiffness (regional or uniform), which we know that takes place in the diabetic foot, can compromise the ability of plantar soft tissue to uniformly distribute loading, leading to higher pressures and internal stresses for the same ground reaction forces [8]. At the same time reduced strength lowers the bar for what constitutes overloading, making soft tissue injury more likely.

This potential role of tissue biomechanics in ulceration has been hypothesized before but only recently the development of clinically applicable elastography techniques allowed testing this hypothesis in the clinic. A first clinical study on the predictive value of tissue biomechanics showed that measurement of tissue stiffness and thickness could indeed lead to better risk assessment [11]. Further research on this topic is needed to verify and establish the potential role tissue biomechanics could play in this direction.

Apart from the assessment of risk, being able to monitor changes in tissue biomechanics could also open the way for unique methods to detect deep tissue trauma in the plantar soft tissue, something which would significantly enhance our ability to study the phenomena that lead to ulceration and improve our understanding on its etiology.

Three specific methods to study plantar soft tissue biomechanics were discussed in the final part of this chapter. These methods have been used in clinics in people with diabetes but are characterized by varying levels of complexity, reliability, and clinical relevance.

Ultrasound indentation combined with computer modeling can offer the most comprehensive assessment of tissue biomechanics. However, its use can be time-consuming and labor-intensive. It has been demonstrated for the heel that this process can be automated but in its current form it appears still to be more suited for clinical research rather than clinical practice. Further research is needed to ensure that this method is robust enough when used in clinics as a "black box" by people with no previous knowledge on computer modeling or engineering theory.

Ultrasound elastography on the other hand is already part of clinical practice. However, not all elastography techniques are the same. Strain elastography offers a qualitative assessment of initial relative deformability in an imaging window. Because it is qualitative it cannot be used to monitor changes or compare between people. This substantial limitation can be overcome with the use of a standoff material as reference for stiffness [6,11] but as it stands the capacity of this technique for quantifying soft tissue biomechanics remains relatively limited.

SW elastography, on the other hand, offers a quantitative assessment of SW propagation speed which is linked to tissue stiffness. Even though it cannot directly measure stiffness it can be reliably used to study differences between people and monitor changes in the full nonlinear behavior of plantar soft tissue [29]. The assessment of nonlinear mechanical behaviour involves the use of acoustoelasticity theory which is not part of the standard use of SW elastography, but the capability is there. Although this technology, like any technology, has its technical limitations, it appears to be the most promising and has already introduced quantitative tissue biomechanics in everyday clinical practice.

As the reader might have already noticed, despite the importance of tissue strength in defining overloading and assessing the risk for injury, no mention was made about strength in the Clinically applicable methods section where existing techniques were discussed. That's because, currently there is no method to measure tissue strength in a noninvasive way. Our inability to study the strength of tissues *in vivo* is a major hurdle in our effort to understand the mechanics of injury. Even though this problem appears to have no apparent direct solution, changes in strength most likely will be coupled with changes in stiffness too. This means that studying the pattern of stiffness change might also provide information about changes in strength, which also highlights even more the need for reliable clinically applicable methods to detect, quantify, and monitor changes in plantar soft tissue stiffness.

References

[1] Cosgrove D, Piscaglia F, Bamber J, Bojunga J, Correas J-M, Gilja O, et al. EFSUMB guidelines and recommendations on the clinical use of ultrasound elastography. Part 2: clinical applications. Ultraschall in Med 2013;34(03):238−53.
[2] Pai S, Ledoux WR. The compressive mechanical properties of diabetic and non-diabetic plantar soft tissue. J Biomech 2010;43(9):1754−60.
[3] Chao CYL, Zheng Y-P, Cheing GLY. Epidermal thickness and biomechanical properties of plantar tissues in diabetic foot. Ultrasound Med Biol July 2011;37(7):1029−38.
[4] Klaesner JW, Hastings MK, Zou D, Lewis C, Mueller MJ. Plantar tissue stiffness in patients with diabetes mellitus and peripheral neuropathy. Arch Phys Med Rehabil 2002;83(12):1796−801.
[5] Piaggesi A, Romanelli M, Schipani E, Campi F, Magliaro A, Baccetti F, et al. Hardness of plantar skin in diabetic neuropathic feet. J Diabet Complicat 1999;13(3):129−34.
[6] Naemi R, Chatzistergos P, Sundar L, Chockalingam N, Ramachandran A. Differences in the mechanical characteristics of plantar soft tissue between ulcerated and non-ulcerated foot. J Diabet Complicat 2016;30(7):1293−9.
[7] Behforootan S, Chatzistergos P, Naemi R, Chockalingam N. Finite element modelling of the foot for clinical applications: a systematic review. Med Eng Phys 2017;39:1−11.
[8] Behforootan S, Chatzistergos P, Chockalingam N, Naemi R. A clinically applicable non-invasive method to quantitatively assess the visco-hyperelastic properties of human heel pad, implications for assessing the risk of mechanical trauma. J Mech Behav Biomed Mater 2017;68:287−95.
[9] Maas S, Rawlins D, Weiss J, Ateshian G. FEBio theory manual. Utah, US: University of Utah; 2015.
[10] Telfer S, Erdemir A, Woodburn J, Cavanagh PR. What has finite element analysis taught us about diabetic foot disease and its management? A systematic review. PLoS One January 2014;9(10):e109994.
[11] Naemi R, Chatzistergos P, Suresh S, Sundar L, Chockalingam N, Ramachandran A. Can plantar soft tissue mechanics enhance prognosis of diabetic foot ulcer? Diabet Res Clin Pract 2017;126:182−91.
[12] Cavanagh PR, Ulbrecht JS. The biomechancis of the foot in diabetes mellitus. In: Bowker JH, Pfeifer MA, editors. Levin and O'Neal's the diabetic foot. 7th ed. Philadelphia: Elsevier Health Sciences; 2008. p. 115−84.
[13] Sacco ICN, Hamamoto AN, Tonicelli LMG, Watari R, Ortega NRS, Sartor CD. Abnormalities of plantar pressure distribution in early, intermediate, and late stages of diabetic neuropathy. Gait Posture 2014;40(4):570−4.
[14] Boulton AJM, Betts RP, Franks CI, Newrick PG, Ward JD, Duckworth T. Abnormalities of foot pressure in early diabetic neuropathy. Diabet Med 1987;4(3):225−8.
[15] Piérard GE, Piérard-Franchimont C, Scheen A. Critical assessment of diabetic xerosis. Expert opinion on medical diagnostics, vol. 7. Informa Healthcare; 2013. p. 201−7.
[16] de Macedo GMC, Nunes S, Barreto T. Skin disorders in diabetes mellitus: an epidemiology and physiopathology review. Diabetol Metab Syndr 2016;8(1):63.

[17] Lechner A, Lahmann N, Neumann K, Blume-Peytavi U, Kottner J. Dry skin and pressure ulcer risk: a multi-center cross-sectional prevalence study in German hospitals and nursing homes. Int J Nurs Stud August 1, 2017;73:63—9.

[18] Gefen A. How do microclimate factors affect the risk for superficial pressure ulcers: a mathematical modeling study. J Tissue Viability 2011;20(3):81—8.

[19] Wildnauer RH, Bothwell JW, Douglass AB. Stratum corneum biomechanical properties. I. Influence of relative humidity on normal and extracted human stratum corneum. J Invest Dermatol 1971;56(1): 72—8.

[20] Gefen A, Gefen N, Linder-Ganz E, Margulies S. In Vivo muscle stiffening under bone compression promotes deep pressure sores. J Biomech Eng 2005;127(3):512—24.

[21] Rome K, Webb P, Unsworth A, Haslock I. Heel pad stiffness in runners with plantar heel pain. Clin Biomech 2001;16(10):901—5.

[22] Black JM, Brindle CT, Honaker JS. Differential diagnosis of suspected deep tissue injury. Int Wound J 2016;13(4):531—9.

[23] Nelissen JL, De Graaf L, Traa WA, Schreurs TJL, Moerman KM, Nederveen AJ, et al. A MRI-compatible combined mechanical loading and mr elastography setup to study deformation-induced skeletal muscle damage in rats. PLoS One 2017;12(1):e0169864.

[24] Deprez J, Brusseau E, Fromageau J, Cloutier G, Basset O. On the potential of ultrasound elastography for pressure ulcer early detection. Med Phys 2012;38(4):1943—50.

[25] Nelissen JL, Oomens CWJ, Sinkus R, Strijkers GJ. Magnetic resonance elastography of skeletal muscle deep tissue injury. NMR Biomed 2019;32(6):e4087.

[26] Erdemir A, Viveiros ML, Ulbrecht JS, Cavanagh PR. An inverse finite-element model of heel-pad indentation. J Biomech 2006;39(7):1279—86.

[27] Lin S-C, Chen CP-C, Tang SF-T, Chen C-W, Wang J-J, Hsu C-C, et al. Stress distribution within the plantar aponeurosis during walking - a dynamic finite element analysis. J Mech Med Biol 2014; 14(4):1450053.

[28] Chatzistergos PE, Naemi R, Sundar L, Ramachandran A, Chockalingam N. The relationship between the mechanical properties of heel-pad and common clinical measures associated with foot ulcers in patients with diabetes. J Diabet Complicat January 2014;28(4):488—93.

[29] Chatzistergos P, Behforootan S, Allan D, Naemi R, Chockalingam N. Shear wave elastography can assess the in-vivo nonlinear mechanical behavior of heel-pad. J Biomech 2018;28(80):114—50.

[30] Behforootan S, Chatzistergos PE, Chockalingam N, Naemi R. A simulation of the viscoelastic behaviour of heel pad during weight-bearing activities of daily living. Ann Biomed Eng 2017;45(12): 2750—61.

[31] Schimmoeller T, Colbrunn R, Nagle T, Lobosky M, Neumann EE, Owings TM, et al. Instrumentation of off-the-shelf ultrasound system for measurement of probe forces during freehand imaging. J Biomech 2019;23(83):117—24.

[32] Boyko EJ, Ahroni JH, Cohen V, Nelson KM, Heagerty PJ. Prediction of diabetic foot ulcer occurrence using commonly available clinical information: the Seattle Diabetic Foot Study. Diabet Care 2006;29(6):1202—7.

[33] Monteiro-Soares M, Dinis-Ribeiro M. External validation and optimisation of a model for predicting foot ulcers in patients with diabetes. Diabetologia 2010;53(7):1525—33.

[34] Püttmann S, Koch J, Steinacker JP, Schmidt SA, Seufferlein T, Kratzer W, et al. Ultrasound point shear wave elastography of the pancreas: comparison of patients with type 1 diabetes and healthy volunteers - results from a pilot study. BMC Med Imag 2018;13(18):52.

[35] Bob F, Grosu I, Sporea I, Bota S, Popescu A, Sima A, et al. Ultrasound-based shear wave elastography in the assessment of patients with diabetic kidney disease. Ultrasound Med Biol October 1, 2017; 43(10):2159—66.

[36] Latorre-Ossa H, Gennisson JL, De Brosses E, Tanter M. Quantitative imaging of nonlinear shear modulus by combining static elastography and shear wave elastography. IEEE Trans Ultrason Ferroelectrics Freq Contr 2012;59(4):833—9.

[37] Aristizabal S, Amador C, Nenadic IZ, Greenleaf JF, Urban MW. Application of acoustoelasticity to evaluate nonlinear modulus in ex vivo kidneys. IEEE Trans Ultrason Ferroelectrics Freq Contr 2017;65(2):188—200.

CHAPTER 9

The mechanobiology of adipocytes in the context of diabetes: much more than a fat depot

Shirley L. Yitzhak-David, Daphne Weihs
Faculty of Biomedical Engineering, Technion-Israel Institute of Technology, Haifa, Israel

Introduction

Diabetes mellitus is a group of metabolic disorders characterized by hyperglycemia, which is caused by defects in insulin action, its secretion, or both. Chronic hyperglycemia can lead to long-term damage and organ failure, and complications of diabetes can cause loss of vision, renal failure, foot ulcers, amputations, sexual dysfunction, cardiovascular symptoms, and more. The development of diabetes can be caused by destruction of pancreatic β-cells, leading to insulin deficiency, or by diminished tissue response to insulin; decreased tissue response to insulin results from problems with insulin secretion, e.g., due to damaged pancreatic β-cells, or response of the hormone pathway. Both processes may coexist, making the leading cause for hyperglycemia difficult to identify in patients.

The two main categories of diabetes are types 1 and 2, which have different functional origins. Type 1 is caused by a deficiency of insulin secretion due to loss of pancreatic β-cells. The loss of cells is caused by an autoimmune response and typically initiates in childhood or adolescence. Type 2 diabetes is caused by a combination of insulin resistance and compensatory insulin response and secretion. Type 2 diabetes is usually caused because of excess body weight and lack of physical activity [1,2], and is more typically found in adults. Both type 1 and type 2 diabetes are more common following excess weight gain and with increased body mass. The increased prevalence of obesity has increased the number of individuals diagnosed with type 2 diabetes. Obesity is defined by the accumulation of adipose tissue that causes health, physical, and psychosocial impairments for the individual [3]. Weight gain and fat accumulation are caused by calorie intake as well as many biological, biochemical, and genetic factors that influence the body energetics.

Adipose tissue and adipocytes

Adipose tissue or fat tissue is a main source of energy for the body and provides connectivity and cushioning between the organs and for the body in general. Normal fat levels change from men to women and, respectively, range between 9%—18% and 14%—28% of body weight. In athletes, 2%—3% of body weight is fat mass, compared to about

60%—70% in obese individuals [4]. Two types of adipose tissue with opposite functions have been characterized in humans: white and brown adipose tissue; those are defined by the type composition of mature adipocytes. While the white adipose tissue/cell stores energy in the form of triglycerides, brown adipose tissue specializes in heat production and energy dissipation and is usually found in newborns, helping them to adjust to cold temperatures [5]. Besides energy storage, the adipose tissue can secrete cytokines and hormones that are required for regulation of homeostasis and metabolism [4,6,7].

In this chapter, we will focus on white adipose tissue, which can markedly increase or decrease in volume after reaching adulthood. Adipose depots are usually located in three anatomical sites: subcutaneous, dermal, and intraperitoneal [4]. The adipose tissue is composed of about 50% mature adipocyte (fat) cells, with the rest being preadipocytes (undifferentiated yet committed cells), fibroblasts, vascular endothelial cells, and various immune cells [8]. Mature, differentiated white adipocytes contain a single lipid droplet (LD), or few large LDs that are mostly composed of triglycerides. The cells are spherical and vary in size (25—200 μm diameter), depending on the size of the LD [5,9], which also affects placement and size of other cellular organelles. The production of triglycerides in LDs occurs during the differentiation of preadipocytes to mature adipocytes. LDs form in the lumen of the endoplasmic reticulum [10] and then existing LDs fuse together into larger units [11].

Adipocytes are derived from mesenchymal stem cells (MSCs) in a multistep process of adipogenesis. MSCs are multipotent and can differentiate to various types of mesenchymal tissues depending on chemical, physical, and biological cues that they receive from their external environment. Those cues activate cascades that guide the MSCs to commit to a specific lineage such as bone or fat [12]. Commitment to the fat lineage induces formation of preadipocyte cells. While MSCs are elongated and spread on substrates, upon transition to preadipocytes, cells change their morphology to round [13]. In type 2 diabetes, as well as in obesity, two processes related to adipocytes occur successively: hyperplasia and hypertrophy; those processes affect, respectively, the number and size of adipocytes in tissue. In hyperplasia, adipogenesis or differentiation of adipocytes occurs from committed preadipocytes, which may also proliferate. Then in hypertrophy, already existing differentiated adipocytes expand the size of their LDs to store more triglycerides [14,15]. The hyperplasia and hypertrophy of the cells increase the mass of adipose tissue [4,6,7,16—18].

Mechanobiology of adipocytes

Recent studies propose that the mechanical environment of adipocytes affects the development of excess fat tissue, which may eventually lead to obesity [15,19]. Thus, conditions that control adipose tissue formation may be determined through changes in mechanostructure and function of cells induced in response to mechanical cues from the environment, their mechanobiology (Fig. 9.1) [97].

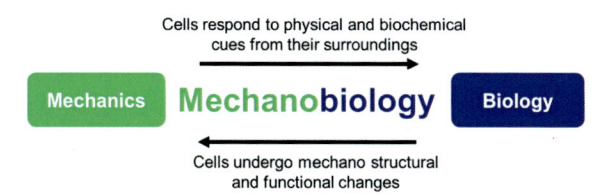

Figure 9.1 Mechanobiology is the interplay between the biological function and capacity of the cells and the mechanics of the cells and tissues and of their environment in micro- and macroscales. For examples, applying mechanical forces to the cells may change their mechanics and function, leading to cell damage/death or conversely to improved performance. *Reprinted with permission Weihs D. Bioengineering studies of cell migration in wound research. In: Gefen A, ed. Innovations and Emerging Technologies in Wound Care. Elsevier Academic Press; 2019:103-122. https://doi.org/10.1016/B978-0-12-815028-3.00006-7.*

The mechanobiological response of cells is induced by mechanotransduction processes that occur in (adipose) tissues, e.g., leading to remodeling of the cell cytoskeleton [20–24]. The differentiation process of adipocytes is affected by the magnitude of stresses and deformations [14,15,19]. Static or cyclic mechanical stress applied for extended time periods can accelerate or inhibit the adipogenesis process and the morphology of single adipocytes [15,25,26]. Furthermore, growing numbers or volume of adipocytes can influence the mechanical environment of their neighboring cells by application of mechanical forces, stresses, and deformation. The cell deformation is affected by structure and mechanics of single cells and the distribution of cell stiffness in a tissue [27,28].

Measurement of cell-level mechanics has been performed with various approaches Techniques such as atomic force microscopy (AFM) or micropipette aspiration have been used to evaluate single-cell mechanics. However, different protocols to determine the properties at the whole cell level affect the generalizability of those studies [29–31]. Concurrently, particle tracking studies provide a (currently inseparable) combined measure of the mechanics and dynamics of the intracellular environment [32–34]. Hence, different approaches are required to evaluate the mechanobiology of preadipocytes, differentiating and mature adipocytes, and the response of cells in tissue to externally applied mechanical loading.

The migration ability of the cells is directly related to their mechanobiology and is especially important for preadipocytes. The preadipocytes migrate when recruited to repair and regenerate damaged tissues and to replenish the capacity for fat storage. As the cells differentiate into mature adipocytes, they lose their migratory capacity, yet are still affected by the mechanobiology. Specifically, the forces and mechanical stresses applied by cells to their substrate during adhesion or migration may be evaluated. An approach that has been applied to measure cell–substrate forces during adhesion and migration is traction force microscopy [35,36].

Cell-applied forces—traction force microscopy

Traction force microscopy is a widely used method for measuring the forces applied by single cells [36–40] or groups of cells [41–44] on two-dimensional (2D) substrates with embedded fluorescent beads (Fig. 9.2). In traction force microscopy, cells are cultured and adhere to the substrates (Fig. 9.2A) and as they apply forces, the substrates are deformed, causing the beads to move. By calculating the deformations and bead displacements between different time points (Fig. 9.2C), the mechanical strains are calculated. Since the gels used as substrates are linear elastic, the mechanical stresses (Fig. 9.2D) and forces applied by the cells can be calculated [35].

By using the theory of elasticity [45] and assuming zero body forces and an infinite half-space elastic substrate, the Boussinesq solution determines through the Green's function, G, the connection between the applied traction stresses, T, and the resulting displacement (u) at each point (x):

$$u(x) = \int G(x, x') T(x') dx' \tag{9.1}$$

By solving the equation in the Fourier field, the convolution reduces to the multiplication [35]:

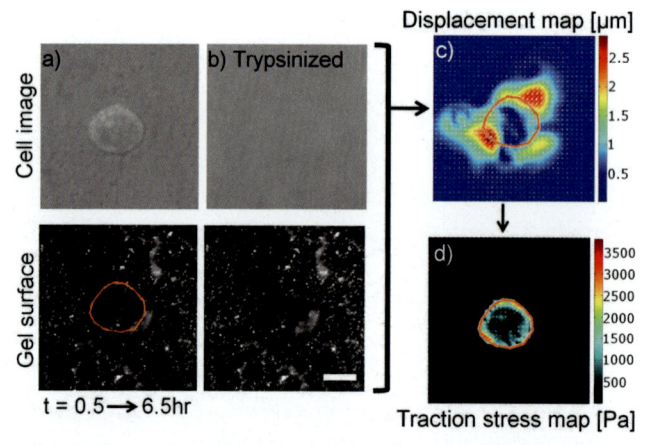

Figure 9.2 Traction force microscopy experiments. (A) Cells are seeded on a PAM gel (top) which has fluorescent beads embedded in its surface (bottom); the *orange line* is the cell boundary. Cells are imaged every 1 h up to 6.5 h after seeding, and gel deformations due to cell-applied forces are identified through bead displacement. (B) The cell is removed from the gel with trypsin, providing the undeformed, relaxed gel, reference state. (C) The displacement map of the beads between the cell-deformed and relaxed gel is calculated for each time point. (D) Traction stress map is calculated from the displacement map by using the gels' Young's modulus and Poisson's ratio. Scale bar is 20 μm. Reprinted with permission Massalha S, Weihs D. Metastatic breast cancer cells adhere strongly on varying stiffness substrates, initially without adjusting their morphology. Biomech Model Mechanobiol 2017; 16:961–70. https://doi.org/10.1007/s10237-016-0864-4.

$$\widetilde{u}(k) = \widetilde{G}(k)\widetilde{T}(k) \qquad (9.2)$$

Where the Green's function becomes:

$$\widetilde{G}(k) = \frac{2(1+\nu)}{Ek^3} \begin{pmatrix} (1-\nu)k^2 + \nu k_y^2 & -\nu k_x k_y \\ -\nu k_x k_y & (1-\nu)k^2 + \nu k_x^2 \end{pmatrix} \qquad (9.3)$$

with ν and E being, respectively, the Poisson's ratio and Young's modulus of the gel, and k is the radial wave vector. Since experiments provide measurements of the resulting displacements, and the underlying stresses are to be determined, it is required to solve the inverse problem:

$$\widetilde{T}(k) = FT_2^{-1}\left(\widetilde{G}(k)^{-1}\widetilde{u}(k)\right) \qquad (9.4)$$

The forces applied by a cell are only applied within its boundaries. The tractions are iteratively calculated as constrained inside the boundaries of the cell, while the forces outside are set to be zero [35]. To experimentally perform those calculations, images of the cells are collected alongside and images of fluorescent beads embedded in the gel surface (Fig. 9.2A—B). To obtain the stresses applied to the gel, cells are the removed and reference image of the relaxed gel obtained. The tractions are calculated from the difference in bead locations between the two sets of images (Fig. 9.2C—D).

From stem cells to mature adipocytes: mechanobiology of migration and differentiation

Signaling factors induce the conversion and lineage commitment of MSCs to preadipocytes which differentiate to adipocytes. Differentiating stem cells are highly affected by their biomechanical environment [46]; hence changes to substrate stiffness or application of mechanical stresses can determine the lineage fate. In the studies highlighted in this chapter [15,37,38,41,47] we demonstrate different mechanobiological aspects of the cells throughout the different stages of development. Traction force microscopy experiments were performed on gels in stiffness ranging from 2.4 to 4.2 kPa which is in the physiological range for soft tissues [48,49]. In vitro studies of adipose-related diseases and adipose engineering typically used the mouse embryonic 3T3-L1 preadipocyte cell line, as it mimics adipogenesis and other biological processes in adipose tissue [50—57].

Proliferation and expansion of embryonic stem cells in three dimensions

Formation of embryoid bodies is an important step in the development of embryonic stem cells (ESCs). Those bodies enable the formation of three-dimensional (3D) tissue-like structures, known as spheroids, that can later be differentiated [58]. Those 3D spheroids can be induced on top of 2D tunable substrates, but can also spontaneously form within 3D hydrogels [59]. Cell—cell interactions facilitated by E-cadherin are some of the biochemical events supporting embryoid body formation; E-cadherin is an essential component for the organization of the cytoskeleton during embryo development. These biochemical events regulate proliferation and differentiation of the pluripotent stem cells [60]. The uniformity of embryoid bodies and their differentiation may be controlled by tuning the properties of 2D substrates, e.g., stiffness, hydrophobicity [61,62].

The time evolution of cell—cell interactions during growth of a 3D colony was evaluated using murine ESCs seeded on a 2D gel substrate. ESCs were seeded on polyacrylamide gels with a Young's modulus of 4.2 kPa and the time-dependent size of the cell colonies and traction stresses they applied on the substrate were evaluated (Fig. 9.3). Over 1—3 days, the murine ESC colonies expanded on the gel horizontally and vertically, while forming 3D colonies of multiple cell layers; after 3 days colonies had merged on the gel and were no longer separable. While the colony size of the murine ESCs increased with time, the stresses applied on the gel decreased (Fig. 9.3A—B), most significantly on day 3 as compared to days 1—2.

The observed changes in cell—substrate interactions occurred due to the growth of the undifferentiated cells in the vertical direction. A control colony of human hepatoma cells (HepG2) was used, which are known to only expand laterally on a substrate. In the HepG2 cell colonies, the average traction stresses remained unchanged throughout 5 days of growth in 2D (Fig. 9.3C). To also ensure that the changes in applied mechanical stresses were not related to cell differentiation, the expression of pluripotency genes, Nanog and Oct-4, was evaluated. As the gene expression was unchanged throughout the 3-day evaluation time and differentiation did not occur, results were related to the mechanical response of the cells in the ESC colony.

The time-dependent reduction of surface-applied mechanical stresses was shown to result from the growth of the ESC colony in 3D. As the ESC colonies grew in all three directions, decrease in gene expression levels of proteins related to cell—substrate interactions was observed concurrently with increase in cell—cell interactions. Specifically, reduction of focal adhesion kinase that participates in cell—substrate junction formation [63] occurred together with increase in E-cadherin activities, while the colonies grew in all directions. E-cadherin is a transmembrane glycoprotein that is involved in cell—cell binding [60]. That is, the cells within the growing colonies relied more on interactions with each other than with the substrate for mechanical stability, potentially in the first stages of spheroid formation. The study demonstrated the ability of revealing in vitro development of 3D spheroids from 2D colonies of stem cells. As the spheroid grows, cell—cell interactions replace some of the cell—substrate interactions. The cells rely on each other to stabilize the structure of the colony.

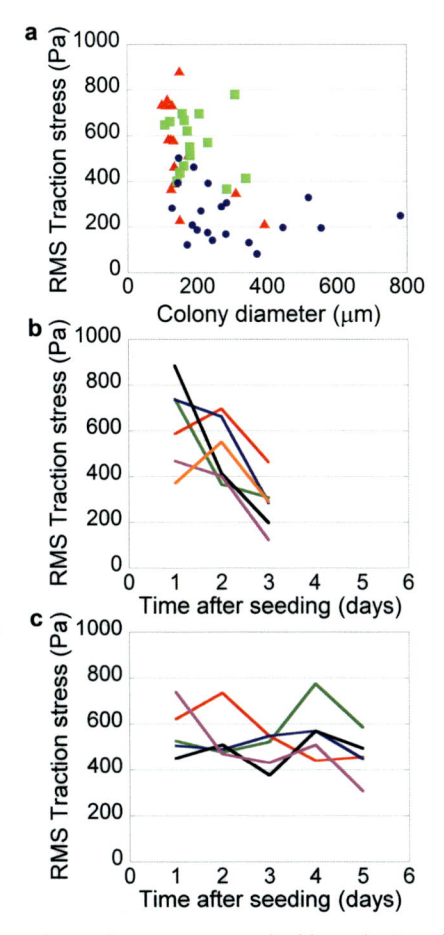

Figure 9.3 Evaluation of time-dependent stresses applied by colonies of mESCs and HepG2 cells. (A) Root mean square traction stress of mESCs measured on random fields of view in the colony as a function of overall colony diameter, as measured at three different times after seeding: days 1 (*triangle*), 2 (*square*), and 3 (*circle*). (B) The mESCs applied traction stresses normalized to the day 1 stress of each colony; each line is a single, monitored colony. The tractions decrease by statistically significant 47.6% on day 3 ($P < .01$). (C) The HepG2 applied traction stresses normalized to the day 1 stress of each colony. In HepG2 cells stresses remained at 523 ± 108 Pa throughout the 5 days of observation. *(Reprinted with permission Teo A, Lim M, Weihs D. Embryonic stem cells growing in 3-dimensions shift from reliance on the substrate to each other for mechanical support. J Biomech 2015;48:1777−1781. https://doi.org/10.1016/j.jbiomech.2015.05.009).*

Migration of single preadipocytes

Stem cells can remain migratory after commitment to specific lineage, prior to differentiation. Preadipocytes are undifferentiated stem cells, committed to the adipose lineage and remaining migratory. They can thus be recruited and mobilized by the body to specific sites where an increase in fat storage capacity is required, e.g., in tissue regeneration or wound healing. In wound healing, preadipocytes reach sites of damage where they differentiate into mature adipocytes to fill any gaps in tissues.

Migrating cells, including preadipocytes, apply adhesive traction forces to their surroundings to crawl on surfaces [64–68]. Specifically, cell migration includes several complex processes: extension of membrane protrusions, structural polarization of the cell, and retraction of the cell rear; those processes require detachment and reattachment and force application to the substrate. Thus, cell migration relies on mechanical interactions between the cells and their microenvironment [41]. Aptly, it was shown that the substrate stiffness can affect cell migration [69], and that externally applied, low-level, sustained mechanical strains can accelerate cell migration into a gap [70,71].

The migration of preadipocytes and their mechanical interaction with a 2D substrate have recently been evaluated [37]. Mouse preadipocytes (3T3-L1 cells) were seeded on polyacrylamide gels with Young's modulus of 2.44 kPa. Following cell adherence, individual cells were imaged at intervals of 20 min (Fig. 9.4A–B). Migration of single cells was evaluated through their speed and directionality (Fig. 9.4C) and by the magnitude of traction forces applied by the cells during migration (Fig. 9.4B).

A bounded asymmetry was observed in the directionality of the traction forces applied by migrating cells (Fig. 9.4E). The migratory cells exhibited an elongated morphology, as they extended on one end and then retracted the other one. To determine the asymmetry in force application, the elongated cell morphology was fit to an ellipse, which provided directionality of the cell. The minor axis of the ellipse was used to define the left/right sides of the migrating cell in each image (Fig. 9.4D); cells migrated toward the left or right sides. An asymmetry index was defined (Eq. 9.5) through the total traction forces (summed magnitudes) obtained for the left (F_{totL}) and right (F_{totR}) sides of the cell:

$$Asymmetry[\%] = \frac{F_{totL} - F_{totR}}{F_{totL} + F_{totR}} \qquad (9.5)$$

The difference in forces between the two sides divided by the total traction force applied by the cell provides the asymmetry index. While the total traction force applied by the cells varied widely (100–800 nN), the asymmetry between the sides was bounded to 33% for all total traction forces (Fig. 9.4E).

The effective speeds of the cells, migrating with no chemoattractant, exhibited a lognormal distribution (Fig. 9.4C); such distributions were previously observed for intracellular particle motion [32,33]. The migration speed of the cells was calculated by the difference between locations of the geometrical center of mass of the cell area in successive images (the time interval). The mean adipocyte migration speed that was measured between the evaluated time intervals (20 min) was 0.27 μm/min, which is in the range of previously measured migration speeds (0.19–0.31 μm/min) of 3T3 mouse fibroblasts on

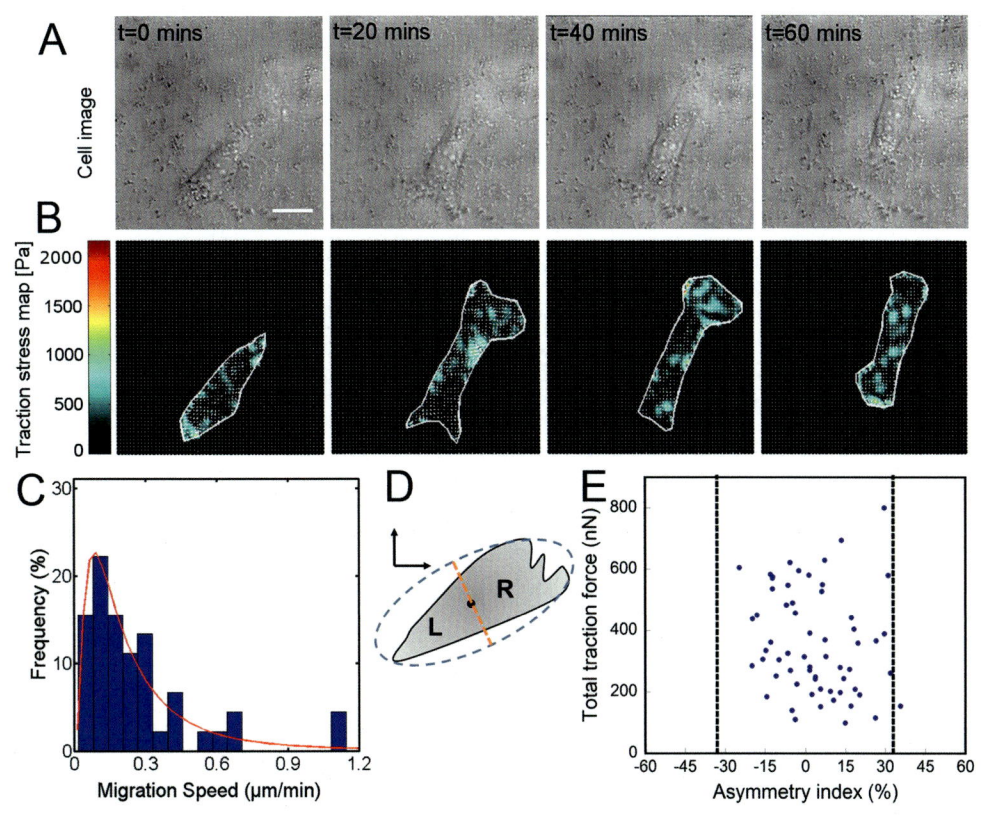

Figure 9.4 (A–B) Single preadipocytes migrating on 2.44 kPa stiffness polyacrylamide gel; cell is migrating from bottom left to top right corners. Top row shows the differential interference contrast images of the cells and bottom row provides the maps of the traction stresses applied by the cell at time point with 20 min intervals. Stresses are confined within the manually marked cell boundaries (*white line*). Scale bar is 20 μm and color bar indicates traction stresses. (C) Distribution of preadipocyte migration speeds at 20-minute intervals; a total of 45 images of single cells were included in the distribution. The distribution was fit to a lognormal (*solid red line*) with a mean speed of 0.27 ± 0.09 μm/min. (D) The cell morphologies are estimated with a bounding ellipse, where the minor axis defines left/right (L/R) sides of the cell on its main axis; the total applied traction forces (nondirectional magnitudes) in each side of the cell or within the entire bounded area may be calculated. (E) The total traction force versus the asymmetry index, calculated by the percent difference between the left-sum and the right-sum of the traction force magnitudes divided by the total traction force applied by the whole cell. The preadipocytes applied a wide range of total traction forces; however, the asymmetry index was bounded to ±33%. *(Reprinted with permission Abuhattum S, Weihs D. Asymmetry in traction forces produced by migrating preadipocytes is bounded to 33. Med Eng Phys 2016;38:834–838. https://doi.org/10.1016/j.medengphy.2016.05.013).*

2.8 kPa PAM gels [72]. No preferred cell orientation was observed on the gels, and in most cells the motion was not directly aligned with their initial major axis between the pairs of evaluated time frames (Fig. 9.4E), that is likely because the cells repositioned between imaging times.

Differentiation into adipocytes

Previous research has shown effects of mechanobiology on MSCs prior to their commitment to osteogenic or adipocytic cells [73,74]. In this context, the Weihs lab has evaluated the changes in traction forces during differentiation of preadipocytes into adipocytes. The tractions were evaluated during adipogenesis of mouse embryonic 3T3-L1 preadipocytes, between 3 and 14 days after initiation of adipogenesis, and as compared to the preadipocytes. Following initiation of adipogenesis, traction force microscopy experiments were conducted every 2–3 days; cells were seeded on collagen-coated polyacrylamide gels with a Young's modulus of 2.44 kPa 24 h prior to measurements.

Changes with time from differentiation were observed in cell morphology as well as in the mechanical strains and stresses induced in the gels by cell-applied forces. Preadipocytes exhibited a spindle, fibroblast-like morphology and applied adhesive, traction forces mainly at the cell poles. After differentiation, the cells became rounded and applied forces in different places along the cell's contact area with the gel. After 8 days of differentiation, LDs were observed by microscopy, and those initially increased in number and then merged and increased in size, respectively, demonstrating hyperplasia and hypertrophy. It was observed that although both the cell area and applied traction forces are widely distributed, the ratio between the total traction force to the projected area remains the same for each cell at all evaluated differentiation times. The constant ratio of forces to cell area implies a constant "stress output" (or mechanical energy output) of the adipocytes during all stages of differentiation. As force production in cells is achieved by focal adhesions connected to the actomyosin intracellular network, correlations exist between the cell-applied traction forces, the cell contact area with the gel, and the number of focal adhesions [75].

Effects of externally applied mechanical stresses on adipogenesis

In the sedentary lifestyle of many individuals, and in diseases such as diabetes and obesity, it is important to determine the effects of mechanical tensile strains. Adipocytes in adipose tissues are subjected to substantial mechanical tensile strains in vivo [76,77]. Those strains may influence the mechanoresponsiveness and mechanosensitivity of the cells. Cyclic stretching, vibration, and dynamic loading were found to suppress differentiation in adipocytes during the different development stages [26,78–85].

Sustained tensile strains promote adipogenesis. Mouse embryonic 3T3-L1 were seeded and differentiated and subjected to 12% stretch and nonstretched control. The cells were monitored for 18–28 days. By using a nondestructive adipocyte micrograph image processing method that provided various measures of LD amounts and sizes [86], the adipogenesis was monitored between stretched and nonstretched conditions [47]. In addition, the effects of mitogen-activated protein kinase (MEK/MAPK) inhibitor and peroxisome proliferator–activated receptor gamma (PPARγ) antagonist effects on adipogenesis were examined in the stretched and nonstretched cultures [15].

Stretched cells exhibited a greater number of LDs, which increased the cell size, and also accelerated differentiation as compared to unstretched cells [47]. In addition, it was found that inhibiting the MEK/MAPK pathway has a significant influence on suppressing the accelerated adipogenesis in the stretched cultures [15]. The PPARγ transcription factor, which is involved in the initial phase of cell differentiation, was also evaluated. The PPARγ antagonist did not suppress the accelerated lipid production in stretched cultures, but reduced the percent of lipid area per field of view in the stretched and unstretched cultures.

Adipocyte stiffness increases with accumulation of lipid droplets

Stiffness changes on the cell level will directly affect the micromechanical environment of adjacent cells, particularly the strains and stresses exerted between cells. Hence, it is important to evaluate the stiffness changes occurring in differentiating adipocytes caused by accumulation of intracytoplasmic LDs. Understanding causes for increase or decrease in cell stiffness can reveal aspects of the structure–function–adaptation loop for fat tissues (Fig. 9.5).

The goal of the study performed at the Gefen lab [47] was to compare the stiffness of the cytoplasm and of LDs in maturing adipocytes, i.e., during increased LD volume in the cell; concurrently, the cytoplasm content/volume decreases. The properties in the subcellular level were experimentally evaluated by AFM (Fig. 9.6A) and interferometric phase microscopy [87] and further analyzed and verified using finite element (FE) modeling and simulations (Fig. 9.6B–C).

During initial stages of adipogenesis of 3T3-L1 preadipocytes, the number of LDs increases, but their size remains small (i.e., hyperplasia). As the preadipocytes mature and differentiate, the LDs fuse, and the result is fewer but larger LDs (i.e., hypertrophy). Eventually, mature adipocytes contain many LDs [15,19,86]. Eleven and nineteen days postdifferentiation induction, the effective apparent, localized shear stiffness (the "effective stiffness") was calculated for the lipid-rich and nucleus regions (Fig. 9.6A). The Hertz model was applied to extract the effective stiffness values. The effective stiffness values of the nuclei and LDs, G, were evaluated by fitting the plots of force, F, against the displacement of the cantilever, δ [88]:

$$F = \frac{G}{1-\nu}\tan(\alpha)\delta^2 \tag{9.6}$$

where ν, the Poisson's ratio, was set to 0.45 and $\alpha = 20$ is the edge angle of the pyramid indenter.

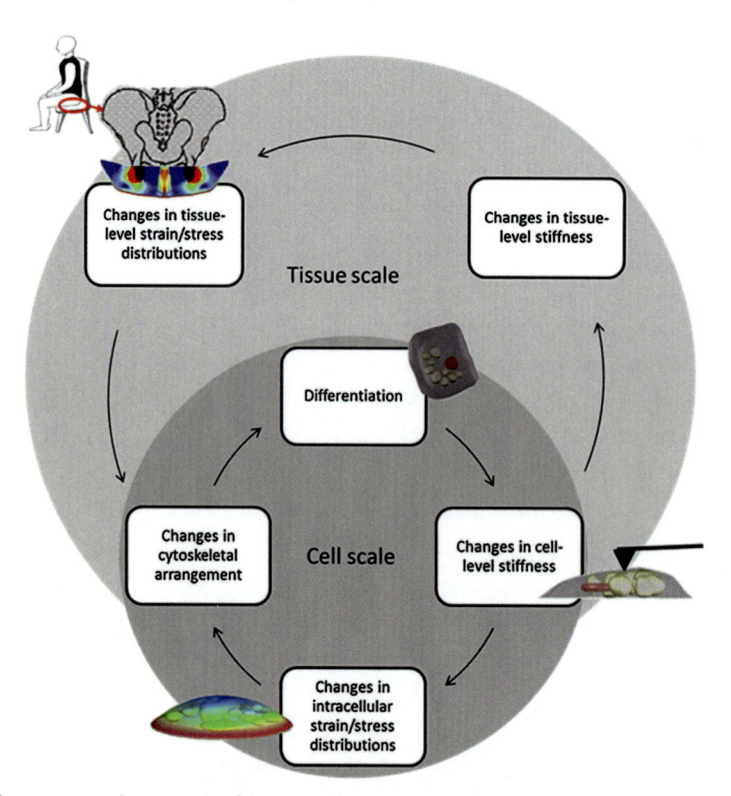

Figure 9.5 The suggested structure—function relationships in fat tissues: the closed-loop coupling between mechanical loads that are being developed in weight-bearing adipose tissues and the adipogenic differentiation process. When adipocytes differentiate, their stiffness gradually changes. Hence, the distributions of strains and stresses in and around the cells change. If the process involves a large number of cells, these stiffness changes also reflect to the tissue scale. At the cell level, such stiffness changes appear to relate to the increasing contents of lipid droplets (adipogenesis) and to rearrangement of the cytoskeleton, which are regulated by activation of different mechanotransduction pathways [14,15,19]. *(Reprinted with permission Shoham N, Girshovitz P, Katzengold R, Shaked NT, Benayahu D, Gefen A. Adipocyte stiffness increases with accumulation of lipid droplets. Biophys J 2014;106: 1421–1431. https://doi.org/10.1016/j.bpj.2014.01.045).*

The mechanics of the lipids and nuclei did not change significantly between 11 and 19 days after differentiation induction. The ratio of the G of the two sites, $\frac{G_{LD}}{G_{nuclei}}$, indicates which part of each cell is stiffer—the LDs or the nucleus. To verify the experimentally obtained ratios, FE modeling was performed to simulate the nanoindentation experiments (Fig. 9.6B–C). The LDs, cytoplasm, nucleus, and plasma membrane were all assumed to be isotropic compressible materials, behaving according to Neo-Hookean strain energy density function (W) [89]:

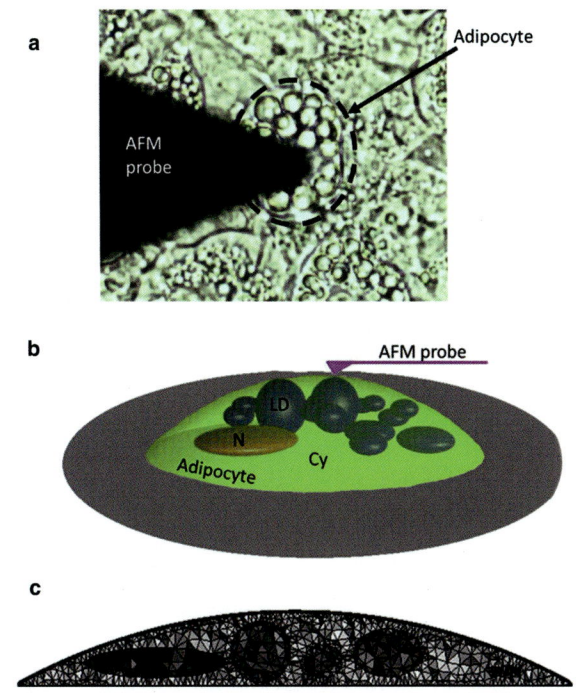

Figure 9.6 (A) The atomic force microscopy (AFM) nanoindentation experiments; and (B) a finite element (FE) model simulating these experiments. *Cy*, cytoplasm; *LD*, lipid droplet; *N*, nucleus, (C) Cross-sectional view of the FE mesh in the model shown in (B). *(Reprinted with permission Shoham N, Girshovitz P, Katzengold R, Shaked NT, Benayahu D, Gefen A. Adipocyte stiffness increases with accumulation of lipid droplets. Biophys J 2014;106:1421−1431. https://doi.org/10.1016/j.bpj.2014.01.045).*

$$W = \frac{G}{2}(\bar{I}_1 - 3) + \frac{\kappa}{2}(J - 1)^2 \tag{9.7}$$

where G is the instantaneous shear modulus, I_1 is the deviatoric part of the left Cauchy−Green deformation tensor, κ is the bulk modulus, and $J = \det(F)$, where F is the deformation gradient tensor.

The study showed that with the progress of the differentiation, the adipocytes stiffen. The effective stiffness of accumulating LDs during adipogenesis was 2.5- to 8.3-fold larger than the cytoplasm effective stiffness. The effective stiffness over LDs was significantly lower than over the nucleus, indicating the local heterogeneity of the cell mechanics. By comparing the cytoplasm effective stiffness which combines all the cellular compartments except the nucleus and LDs [90−96] and the LDs' effective stiffness, the researchers were able to determine the cell stiffness changes that are characteristic for the differentiation of adipocytes, as those changes depend on the volume of the LDs that increases over time [55]; the stiffness of the LDs is unchanged, but their volumetric contribution is more significant as they increase in size.

In the context of structure—function relationships in fat tissue (Fig. 9.5), the results at the cell scale are also affected by the macroscopic scale and vice versa. The differentiation process, specifically the stiffening of adipocytes and the strain and stress distributions caused by the stiffening, can lead to changes at the tissue level, and at the cell level such as rearrangement of the cytoskeleton and activation of different mechanotransduction pathways (Fig. 9.5) [14,15,19—22]. Feedback loops that involve cell- and tissue-level responses in weight-bearing tissues are likely to be involved in many pathologies and diseases such as diabetes and obesity.

References

[1] American Diabetes Association. Diagnosis and classification of diabetes mellitus. Diabetes Care 2014; 37:S81—90. https://doi.org/10.2337/dc14-S081.
[2] Chatterjee S, Khunti K, Davies MJ. Type 2 diabetes. Lancet 2017;389:2239—51. https://doi.org/10.1016/S0140-6736(17)30058-2.
[3] Al-Goblan AS, Al-Alfi MA, Khan MZ. Mechanism linking diabetes mellitus and obesity. Diabetes Metab Syndrome Obes Targets Ther 2014;7:587—91. https://doi.org/10.2147/DMSO.S67400.
[4] Hausman DB, DiGirolamo M, Bartness TJ, Hausman GJ, Martin RJ. The biology of white adipocyte proliferation. Obes Rev 2001;2:239—54. https://doi.org/10.1046/j.1467-789X.2001.00042.x.
[5] Saely CH, Geiger K, Drexel H. Brown versus white adipose tissue: a mini-review. Gerontology 2012; 58:15—23. https://doi.org/10.1159/000321319.
[6] MacDougald OA, Mandrup S. Adipogenesis: forces that tip the scales. Trends Endocrinol Metabol 2002;13:5—11. https://doi.org/10.1016/S1043-2760(01)00517-3.
[7] Trujillo ME, Scherer PE. Adipose tissue-derived factors: impact on health and disease. Endocr Rev 2006;27:762—78. https://doi.org/10.1210/er.2006-0033.
[8] Kershaw EE, Flier JS. Adipose tissue as an endocrine organ. J Clin Endocrinol Metab 2004;89: 2548—56. https://doi.org/10.1210/jc.2004-0395.
[9] Won Park K, Halperin DS, Tontonoz P. Before they were fat: adipocyte progenitors. Cell Metabol 2008;8:454—7. https://doi.org/10.1016/j.cmet.2008.11.001.
[10] Garg A, Agarwal AK. Lipodystrophies: disorders of adipose tissue biology. Biochim Biophys Acta Mol Cell Biol Lipids 2009;1791:507—13. https://doi.org/10.1016/j.bbalip.2008.12.014.
[11] Murphy S, Martin S, Parton RG. Lipid droplet-organelle interactions; sharing the fats. Biochim Biophys Acta Mol Cell Biol Lipids 2009;1791:441—7. https://doi.org/10.1016/j.bbalip.2008.07.004.
[12] Chen Q, Shou P, Zheng C, Jiang M, Cao G, Yang Q, et al. Fate decision of mesenchymal stem cells: adipocytes or osteoblasts? Cell Death Differ 2016;23:1128—39. https://doi.org/10.1038/cdd.2015.168.
[13] Pittenger MF, Mackay AM, Beck SC, Jaiswal RK, Douglas R, Mosca JD, et al. Multilineage potential of adult human mesenchymal stem cells. Science 1999;284:143—7. https://doi.org/10.1126/science.284.5411.143.
[14] Shoham N, Gefen A. Mechanotransduction in adipocytes. J Biomech 2012;45:1—8.
[15] Shoham N, Gottlieb R, Sharabani-Yosef O, Zaretsky U, Benayahu D, Gefen A. Static mechanical stretching accelerates lipid production in 3T3-L1 adipocytes by activating the MEK signaling pathway. Am J Physiol Physiol 2012;302:C429—41. https://doi.org/10.1152/ajpcell.00167.2011.
[16] Adams M, Montague CT, Prins JB, Holder JC, Smith SA, Sanders L, et al. Activators of peroxisome proliferator-activated receptor γ have depot- specific effects on human preadipocyte differentiation. J Clin Invest 1997;100:3149—53. https://doi.org/10.1172/JCI119870.
[17] Coppack SW. Adipose tissue changes in obesity. Biochem Soc Trans 2005;33:1049—52. https://doi.org/10.1042/bst0331049.
[18] Frayn KN, Karpe F, Fielding BA, Macdonald IA, Coppack SW. Integrative physiology of human adipose tissue. Int J Obes 2003;27:875—88. https://doi.org/10.1038/sj.ijo.0802326.

[19] Levy A, Enzer S, Shoham N, Zaretsky U, Gefen A. Large, but not small sustained tensile strains stimulate adipogenesis in culture. Ann Biomed Eng 2012;40:1052–60. https://doi.org/10.1007/s10439-011-0496-x.
[20] Verstraeten VLRM, Renes J, Ramaekers FCS, Kamps M, Kuijpers HJ, Verheyen F, et al. Reorganization of the nuclear lamina and cytoskeleton in adipogenesis. Histochem Cell Biol 2011;135:251–61. https://doi.org/10.1007/s00418-011-0792-4.
[21] Mathieu PS, Loboa EG. Cytoskeletal and focal adhesion influences on mesenchymal stem cell shape, mechanical properties, and differentiation down osteogenic, adipogenic, and chondrogenic pathways. Tissue Eng B Rev 2012;18:436–44. https://doi.org/10.1089/ten.teb.2012.0014.
[22] Feng T, Szabo E, Dziak E, Opas M. Cytoskeletal disassembly and cell rounding promotes adipogenesis from ES cells. Stem Cell Rev Reports 2010;6:74–85. https://doi.org/10.1007/s12015-010-9115-8.
[23] Mizrahi N, Zhou EHH, Lenormand G, Krishnan R, Weihs D, Butler JP, et al. Low intensity ultrasound perturbs cytoskeleton dynamics. Soft Matter 2012;8:2438–43. https://doi.org/10.1039/C2sm07246g.
[24] Gefen A, Weihs D. Cytoskeleton and plasma-membrane damage resulting from exposure to sustained deformations: a review of the mechanobiology of chronic wounds. Med Eng Phys 2016;38:828–33. https://doi.org/10.1016/j.medengphy.2016.05.014.
[25] Tanabe Y, Koga M, Saito M, Matsunaga Y, Nakayama K. Inhibition of adipocyte differentiation by mechanical stretching through ERK-mediated downregulation of PPARγ2. J Cell Sci 2004;117:3605–14. https://doi.org/10.1242/jcs.01207.
[26] Tanabe Y, Matsunaga Y, Saito M, Nakayama K. Involvement of cyclooxygenase-2 in synergistic effect of cyclic stretching and eicosapentaenoic acid on adipocyte differentiation. J Pharmacol Sci 2008;106:478–84. https://doi.org/10.1254/jphs.FP0071886.
[27] Saeed M, Weihs D. Finite element analysis reveals an important role for cell morphology in response to mechanical compression. Biomech Model Mechanobiol 2020;19:1155–64. https://doi.org/10.1007/s10237-019-01276-5.
[28] Vermolen FJ, Harrevelt SD, Gefen A, Weihs D. A particle finite element-based framework for differentiation paths of stem cells to myocytes and adipocytes: hybrid cell-based and finite element modeling. Numer Methods Adv Simul Biomech Biol Process 2018:171–85. https://doi.org/10.1016/B978-0-12-811718-7.00009-5. Academic Press.
[29] Darling EM, Topel M, Zauscher S, Vail TP, Guilak F. Viscoelastic properties of human mesenchymally-derived stem cells and primary osteoblasts, chondrocytes, and adipocytes. J Biomech 2008;41:454–64. https://doi.org/10.1016/j.jbiomech.2007.06.019.
[30] Kwon YN, Kim WK, Lee SH, Kim K, Kim EY, Ha TH, et al. Monitoring of adipogenic differentiation at the single-cell level using atomic force microscopic analysis. Spectroscopy 2011;26:329–35. https://doi.org/10.3233/SPE-2012-0566.
[31] Yu H, Tay CY, Leong WS, Tan SCW, Liao K, Tan LP. Mechanical behavior of human mesenchymal stem cells during adipogenic and osteogenic differentiation. Biochem Biophys Res Commun 2010;393:150–5. https://doi.org/10.1016/j.bbrc.2010.01.107.
[32] Gal N, Lechtman-Goldstein D, Weihs D. Particle tracking in living cells: a review of the mean square displacement method and beyond. Rheol Acta 2013;52:425–43. https://doi.org/10.1007/s00397-013-0694-6.
[33] Gal N, Weihs D. Intracellular mechanics and activity of breast cancer cells correlate with metastatic potential. Cell Biochem Biophys 2012;63:199–209.
[34] Goldstein D, Elhanan T, Aronovitch M, Weihs D. Origin of active transport in breast-cancer cells. Soft Matter 2013;9:7167–73. https://doi.org/10.1039/c3sm50172h.
[35] Butler JP, Tolic-Norrelykke IM, Fabry B, Fredberg JJ. Traction fields, moments, and strain energy that cells exert on their surroundings. Am J Physiol Physiol 2002;282:C595–605.
[36] Massalha S, Weihs D. Metastatic breast cancer cells adhere strongly on varying stiffness substrates, initially without adjusting their morphology. Biomech Model Mechanobiol 2017;16:961–70. https://doi.org/10.1007/s10237-016-0864-4.
[37] Abuhattum S, Weihs D. Asymmetry in traction forces produced by migrating preadipocytes is bounded to 33. Med Eng Phys 2016;38:834–8. https://doi.org/10.1016/j.medengphy.2016.05.013.

[38] Abuhattum S, Gefen A, Weihs D. Ratio of total traction force to projected cell area is preserved in differentiating adipocytes. Integr Biol 2015;7:1212–7. https://doi.org/10.1039/c5ib00056d.

[39] Kraning-Rush CM, Califano JP, Reinhart-King CA. Cellular traction stresses increase with increasing metastatic potential. PLoS One 2012;7:e32572. https://doi.org/10.1371/journal.pone.0032572.

[40] Mierke CT, Rosel D, Fabry B, Brabek J. Contractile forces in tumor cell migration. Eur J Cell Biol 2008;87:669–76. https://doi.org/10.1016/j.ejcb.2008.01.002.

[41] Teo A, Lim M, Weihs D. Embryonic stem cells growing in 3-dimensions shift from reliance on the substrate to each other for mechanical support. J Biomech 2015;48:1777–81. https://doi.org/10.1016/j.jbiomech.2015.05.009.

[42] Trepat X, Wasserman MR, Angelini TE, Millet E, Weitz DA, Butler JP, et al. Physical forces during collective cell migration. Nat Phys 2009;5:426–30. https://doi.org/10.1038/Nphys1269.

[43] Krishnan R, Klumpers DD, Park CY, Rajendran K, Trepat X, van Bezu J, et al. Substrate stiffening promotes endothelial monolayer disruption through enhanced physical forces. Am J Physiol Physiol 2011;300:C146–54. https://doi.org/10.1152/ajpcell.00195.2010.

[44] Mierke CT, Zitterbart DP, Kollmannsberger P, Raupach C, Schlotzer-Schrehardt U, Goecke TW, et al. Breakdown of the endothelial barrier function in tumor cell transmigration. Biophys J 2008;94:2832–46. https://doi.org/10.1529/biophysj.107.113613.

[45] Lldlemkamp LP. Theory of elasticity. Elsevier; 1986.

[46] Engler AJ, Sen S, Sweeney HL, Discher DE. Matrix elasticity directs stem cell lineage specification. Cell 2006;126:677–89. https://doi.org/10.1016/j.cell.2006.06.044.

[47] Shoham N, Girshovitz P, Katzengold R, Shaked NT, Benayahu D, Gefen A. Adipocyte stiffness increases with accumulation of lipid droplets. Biophys J 2014;106:1421–31. https://doi.org/10.1016/j.bpj.2014.01.045.

[48] Gefen A, Haberman E. Viscoelastic properties of ovine adipose tissue covering the gluteus muscles. J Biomech Eng 2007;129:924–30. https://doi.org/10.1115/1.2800830.

[49] Patel PN, Smith CK, Patrick CW. Rheological and recovery properties of poly(ethylene glycol) diacrylate hydrogels and human adipose tissue. J Biomed Mater Res 2005;73A:313–9.

[50] Kallen CB, Lazar MA. Antidiabetic thiazolidinediones inhibit leptin (ob) gene expression in 3T3-L1 adipocytes. Proc Natl Acad Sci U S A 1996;93:5793–6. https://doi.org/10.1073/pnas.93.12.5793.

[51] Steppan CM, Bailey ST, Bhat S, Brown EJ, Banerjee RR, Wright CM, et al. The hormone resistin links obesity to diabetes. Nature 2001;409:307–12. https://doi.org/10.1038/35053000.

[52] Fasshauer M, Klein J, Neumann S, Eszlinger M, Paschke R. Hormonal regulation of adiponectin gene expression in 3T3-L1 adipocytes. 2002. https://doi.org/10.1006/bbrc.2001.6307.

[53] Fasshauer M, Klein J, Neumann S, Eszlinger M, Paschke R. Isoproterenol inhibits resistin gene expression through a G_S-protein-coupled pathway in 3T3-L1 adipocytes. FEBS Lett 2001;500:60–3. https://doi.org/10.1016/S0014-5793(01)02588-1.

[54] Fasshauer M, Klein J, Neumann S, Eszlinger M, Paschke R. Tumor necrosis factor is a negative regulator of resistin gene expression and secretion in 3T3-L1 adipocytes. 2001. https://doi.org/10.1006/bbrc.2001.5874.

[55] Otto TC, Lane MD. Adipose development: from stem cell to adipocyte. Crit Rev Biochem Mol Biol 2005;40:229–42. https://doi.org/10.1080/10409230591008189.

[56] Tang QQ, Daniel Lane M. Adipogenesis: from stem cell to adipocyte keywords. 2012. https://doi.org/10.1146/annurev-biochem-052110-115718.

[57] Kovsan J, Osnis A, Maissel A, Mazor L, Tarnovscki T, Hollander L, et al. Depot-specific adipocyte cell lines reveal differential drug-induced responses of white adipocytes—relevance for partial lipodystrophy. Am J Physiol Metab 2009;296:E315–22. https://doi.org/10.1152/ajpendo.90486.2008.

[58] Doetschman TC, Eistetter H, Katz M, Schmidt W, Kemler R. The in vitro development of blastocyst-derived embryonic stem cell lines: formation of visceral yolk sac, blood islands and myocardium. J Embryol Exp Morphol 1985;87:27–45.

[59] Siti-Ismail N, Bishop AE, Polak JM, Mantalaris A. The benefit of human embryonic stem cell encapsulation for prolonged feeder-free maintenance. Biomaterials 2008;29:3946–52. https://doi.org/10.1016/j.biomaterials.2008.04.027.

[60] Li L, Bennett SAL, Wang L. Role of E-cadherin and other cell adhesion molecules in survival and differentiation of human pluripotent stem cells. Cell Adhes Migrat 2012;6:59—73. https://doi.org/10.4161/cam.19583.

[61] Zheng XR, Zhang X. Microsystems for cellular force measurement: a review. J Micromech Microeng 2011;21:54003.

[62] Harris AK, Wild P, Stopak D. Silicone rubber substrata: a new wrinkle in the study of cell locomotion. Science 1980;208:177—9. https://doi.org/10.1126/science.6987736.

[63] Schaller MD. Cellular functions of FAK kinases: insight into molecular mechanisms and novel functions. J Cell Sci 2010;123:1007—13. https://doi.org/10.1242/jcs.045112.

[64] Nabi IR. The polarization of the motile cell. J Cell Sci 1999;112:1803 LP—1811.

[65] Cramer LP. Forming the cell rear first: breaking cell symmetry to trigger directed cell migration. Nat Cell Biol 2010;12:628—32. https://doi.org/10.1038/ncb0710-628.

[66] Zhong Y, Ji B. Impact of cell shape on cell migration behavior on elastic substrate. Biofabrication 2013;5:015011. https://doi.org/10.1088/1758-5082/5/1/015011.

[67] Lock JG, Wehrle-Haller B, Strömblad S. Cell-matrix adhesion complexes: master control machinery of cell migration. Semin Canc Biol 2008;18:65—76. https://doi.org/10.1016/j.semcancer.2007.10.001.

[68] Sheetz MP, Felsenfeld D, Galbraith CG, Choquet D. Cell migration as a five-step cycle. Cell Behav Control Mech Motil 1999:233—43.

[69] Vermolen FJ, Meijden RP Van Der, Es M Van, Gefen A, Weihs D. Towards a mathematical formalism for semi-stochastic cell-level computational modeling of tumor initiation. Ann Biomed Eng 2015;43:1680—94. https://doi.org/10.1007/s10439-015-1271-1.

[70] Toume S, Gefen A, Weihs D. Low-level stretching accelerates cell migration into a gap. Int Wound J 2017;14:698—703. https://doi.org/10.1111/iwj.12679.

[71] Marom A, Berkovitch Y, Toume S, Alvarez-Elizondo MB, Weihs D. Non-damaging stretching combined with sodium pyruvate supplement accelerate migration of fibroblasts and myoblasts during gap closure. Clin Biomech 2019;62:96—103. https://doi.org/10.1016/j.clinbiomech.2019.01.009.

[72] Munevar S, Wang YL, Dembo M. Traction force microscopy of migrating normal and H-ras transformed 3T3 fibroblasts. Biophys J 2001;80:1744—57.

[73] Fu J, Wang YK, Yang MT, Desai RA, Yu X, Liu Z, et al. Mechanical regulation of cell function with geometrically modulated elastomeric substrates. Nat Methods 2010;7:733—6. https://doi.org/10.1038/nmeth.1487.

[74] Titushkin I, Cho M. Modulation of cellular mechanics during osteogenic differentiation of human mesenchymal stem cells. Biophys J 2007;93:3693—702. https://doi.org/10.1529/biophysj.107.107797.

[75] Legant WR, Miller JS, Blakely BL, Cohen DM, Genin GM, Chen CS. Measurement of mechanical tractions exerted by cells in three-dimensional matrices. Nat Methods 2010;7:969—71. https://doi.org/10.1038/nmeth.1531.

[76] Linder-Ganz E, Shabshin N, Itzchak Y, Gefen A. Assessment of mechanical conditions in sub-dermal tissues during sitting: a combined experimental-MRI and finite element approach. J Biomech 2007;40:1443—54. https://doi.org/10.1016/j.jbiomech.2006.06.020.

[77] Slomka N, Or-Tzadikario S, Sassun D, Gefen A. Membrane-stretch-induced cell death in deep tissue injury: computer model studies. Cell Mol Bioeng 2009;2:118—32. https://doi.org/10.1007/s12195-009-0046-x.

[78] Case N, Xie Z, Sen B, Styner M, Zou M, O'Conor C, et al. Mechanical activation of β-catenin regulates phenotype in adult murine marrow-derived mesenchymal stem cells. J Orthop Res 2010;28:1531—8. https://doi.org/10.1002/jor.21156.

[79] David V, Martin A, Lafage-Proust M-H, Malaval L, Peyroche S, Jones DB, et al. Mechanical loading down-regulates peroxisome proliferator-activated receptor γ in bone marrow stromal cells and favors osteoblastogenesis at the expense of adipogenesis. Endocrinology 2007;148:2553—62. https://doi.org/10.1210/en.2006-1704.

[80] Huang SC, Wu TC, Yu HC, Chen MR, Liu CM, Chiang WS, et al. Mechanical strain modulates age-related changes in the proliferation and differentiation of mouse adipose-derived stromal cells. BMC Cell Biol 2010;11:1—14. https://doi.org/10.1186/1471-2121-11-18.

[81] Sen B, Styner M, Xie Z, Case N, Rubin CT, Rubin J. Mechanical loading regulates NFATc1 and β-catenin signaling through a GSK3β control node. J Biol Chem 2009;284:34607—17. https://doi.org/10.1074/jbc.M109.039453.

[82] Sen B, Xie Z, Case N, Ma M, Rubin C, Rubin J. Mechanical strain inhibits adipogenesis in mesenchymal stem cells by stimulating a durable β-catenin signal. Endocrinology 2008;149:6065—75. https://doi.org/10.1210/en.2008-0687.

[83] Sen B, Xie Z, Case N, Styner M, Rubin CT, Rubin J. Mechanical signal influence on mesenchymal stem cell fate is enhanced by incorporation of refractory periods into the loading regimen. J Biomech 2011;44:593—9. https://doi.org/10.1016/j.jbiomech.2010.11.022.

[84] Tirkkonen L, Halonen H, Hyttinen J, Kuokkanen H, Sievänen H, Koivisto A-M, et al. The effects of vibration loading on adipose stem cell number, viability and differentiation towards bone-forming cells. J R Soc Interface 2011;8:1736—47. https://doi.org/10.1098/rsif.2011.0211.

[85] Turner NJ, Jones HS, Davies JE, Canfield AE. Cyclic stretch-induced TGFβ1/Smad signaling inhibits adipogenesis in umbilical cord progenitor cells. Biochem Biophys Res Commun 2008;377:1147—51. https://doi.org/10.1016/j.bbrc.2008.10.131.

[86] Or-Tzadikario S, Sopher R, Gefen A. Quantitative monitoring of lipid accumulation over time in cultured adipocytes as function of culture conditions: toward controlled adipose tissue engineering. Tissue Eng C Methods 2010;16:1167—81.

[87] Shaked NT, Satterwhite LL, Bursac N, Wax A. Whole-cell-analysis of live cardiomyocytes using wide-field interferometric phase microscopy. Biomed Optic Express 2010;1:706. https://doi.org/10.1364/boe.1.000706.

[88] Bilodeau GG. Regular pyramid punch problem. J Appl Mech Trans ASME 1992;59:519—23. https://doi.org/10.1115/1.2893754.

[89] Slomka N, Gefen A. Confocal microscopy-based three-dimensional cell-specific modeling for large deformation analyses in cellular mechanics. J Biomech 2010;43:1806—16. https://doi.org/10.1016/j.jbiomech.2010.02.011.

[90] Guilak F, Tedrow JR, Burgkart R. Viscoelastic properties of the cell nucleus. Biochem Biophys Res Commun 2000;269:781—6. https://doi.org/10.1006/bbrc.2000.2360.

[91] Caille N, Thoumine O, Tardy Y, Meister JJ. Contribution of the nucleus to the mechanical properties of endothelial cells. J Biomech 2002;35:177—87. https://doi.org/10.1016/S0021-9290(01)00201-9.

[92] Dong C, Skalak R, Sung KLP. Cytoplasmic rheology of passive neutrophils. Biorheology 1991;28:557—67. https://doi.org/10.3233/BIR-1991-28607.

[93] Maniotis AJ, Chen CS, Ingber DE. Demonstration of mechanical connections between integrins cytoskeletal filaments, and nucleoplasm that stabilize nuclear structure. Proc Natl Acad Sci U S A 1997;94:849—54.

[94] Friedl P, Wolf K, Lammerding J. Nuclear mechanics during cell migration. Curr Opin Cell Biol 2011;23:55—64. https://doi.org/10.1016/j.ceb.2010.10.015.

[95] Lombardi ML, Zwerger M, Lammerding J. Biophysical assays to probe the mechanical properties of the interphase cell nucleus: substrate strain application and microneedle manipulation. J Vis Exp 2011:e3087. https://doi.org/10.3791/3087.

[96] Dahl KN, Ribeiro AJ, Lammerding J. Nuclear shape, mechanics, and mechanotransduction. Circ Res 2008;102:1307—18. https://doi.org/10.1161/CIRCRESAHA.108.173989.

[97] Weihs Daphne. Bioengineering studies of cell migration in wound research. In: Gefen Amit, editor. Innovations and Emerging Technologies in Wound Care. Elsevier Academic Press; 2019. p. 103—22.

CHAPTER 10

Optical Coherence Tomography to determine and visualize pathological skin structure changes caused by diabetes

Raman Maiti[1], Roger Lewis[2], Daniel Parker[3], Matt J. Carré[2]

[1]Loughborough University, Loughborough, United Kingdom; [2]University of Sheffield, Sheffield, United Kingdom; [3]School of Health and Society, University of Salford, Salford, United Kingdom

Background and working principle of OCT

Tomographic imaging techniques such as X-ray computed tomography [1], magnetic resonance imaging [2], and ultrasonic imaging [3] have a wide range of applications in the field of medicine. These applications are driven by the variation and limits of resolution and penetration depth each technique can offer. Optical Coherence Tomography (OCT) is a relatively recent form of biological imaging and works on the principle of photon scattering of optical light. Low-coherence light is split using a 2×2 single-mode fiber-optic coupler (also known as a splitter) as shown in Fig. 10.1. The split light is transmitted to a reference arm and sample arm. The reflection from the object of interest, obtained from the sample arm, is recombined with the reference arm at the coupler and they interfere with each other to form an image after signal processing is carried out. A three-dimensional image is obtained by using a galvanometer present at the sample arm to capture lateral scans.

Capabilities for skin imaging

The unique nature of noninvasive penetration makes OCT suitable for cross-sectional imaging of internal structures in biological tissues. The penetration depth of OCT ranges from 10 to 1000 μm and image resolution ranges from 1 to 25 μm. This places the capabilities of OCT between those of confocal microscopy and magnetic resonance/ultrasonic imaging. Two-dimensional OCT scans can be used for measuring the structural properties of the skin whereas three-dimensional scans are mostly used for angiographical analysis. Angiography scans, performed using swept-source systems, are useful to obtain the position and activity of vasculature through the detection of blood flow. Examples of both two- and three-dimensional scans are provided later in this chapter.

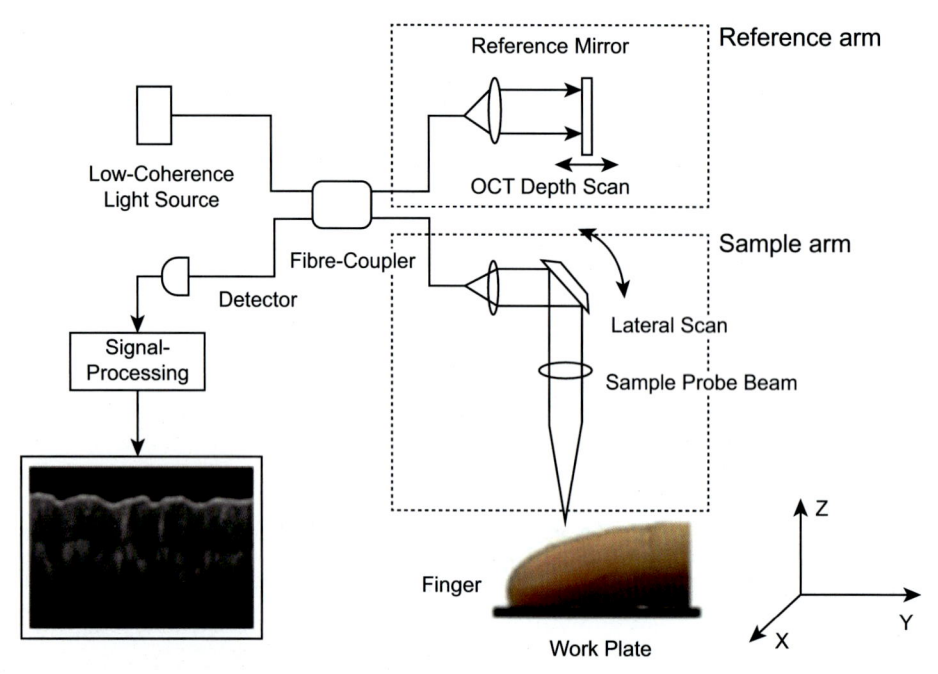

Figure 10.1 Schematic setup diagram of an Optical Coherence Tomography [4].

Common clinical and lab-based applications

To date, OCT has had the largest clinical impact in ophthalmology [5] especially for investigating macular degeneration, glaucoma, and diabetic macular edema. New advances in OCT have made it possible for it to be used in other medical applications. For example, in oncological applications, OCT angiography has been used to image tumors in mice in vivo [6]. Polarized OCT has been used to image dental enamels [7]. High-resolution OCT has also been used in gastroenterology and dermatology. This chapter, however, will focus on the applications of OCT in dermatology through a series of case studies.

OCT setups can be defined as "free-space" and "fiber-based" optical systems. A free-space design uses dynamic focusing to obtain high-resolution images, whereas a fiber-based design reduces the loss of lateral resolution by fusing tomograms that are at the same lateral location, but have different depths [8,9]. A free-space setup can avoid the excessive loss and imbalanced dispersion in the fiber couplers, thereby having better imaging performance than a fiber-based system. However, a fiber-based system has advantages of compactness, easy alignment, and low maintenance, making it more suitable for commercial and clinical uses. In this chapter, two different types of fiber-based

Table 10.1 Details of VivoSight® and in-house Swept-Source OCT systems.

System specification	VivoSight®	In-house swept-source
Center wavelength (nm)	1305	1305
Bandwidth (nm)	147	110
Axial resolution (µm)	5	10
Lateral resolution (µm)	7.5	25
A-scan rate (kHz)	20	20
Imaging depth (mm) in tissue	1.5	1

systems are used, namely the clinical VivoSight (Michelson Diagnostics, Kent, UK) and an in-house swept-source lab-based system (as illustrated schematically in Fig. 10.1). The specifications of both OCT systems are presented in Table 10.1.

Case study 1: use of OCT for skin biomechanics measurements

Early OCT systems were too slow to generate accurate images of tissue structures in vivo with distortions stemming from motion artifacts. This issue has now been overcome with an increased scanning speed of the reference arm [10], allowing detailed images of skin structures to be taken for examination of the morphology of different layers.

Morphological properties of human skin from a range of anatomical locations were reported in a study by Maiti et al. (see Fig. 10.2) [11]. The paper reports a variation of skin thickness, roughness, and undulation which is useful for the development of skin models and grafts. It was observed that skin from the chest (thoracis) region was significantly similar to the volar forearm (antebrachium) in terms of epidermal thickness (93 ± 5 µm), surface roughness (3 ± 0 µm), and dermal–epidermal undulation (3 ± 0 µm) [11]. This means that the design of skin grafts for the chest region can be investigated based on the structure of the volar forearm. In other locations such as the foot (first metatarsal head) and heel (calcaneus) regions, the stratum corneum (SC) layer is significantly thicker due to the requirement of these regions to withstand loading. In these regions, it is the SC thickness and stratum corneum–stratum lucidum (SC-SL) junction that is measured, due to the depth imaging capabilities of the VivoSight system, as can be seen in Fig. 10.3.

OCT can be combined with mechanical test rigs to conduct experiments on the skin's response to loading, or skin can be loaded due to the natural consequence of postural changes. Examples where the VivoSight OCT system has been used to measure structural morphology response due to a change in posture and interactions with surfaces are shown in Fig. 10.4. Fig. 10.4A shows the measurement of volar forearm skin structural morphological changes for different arm postures. The images obtained were subsequently

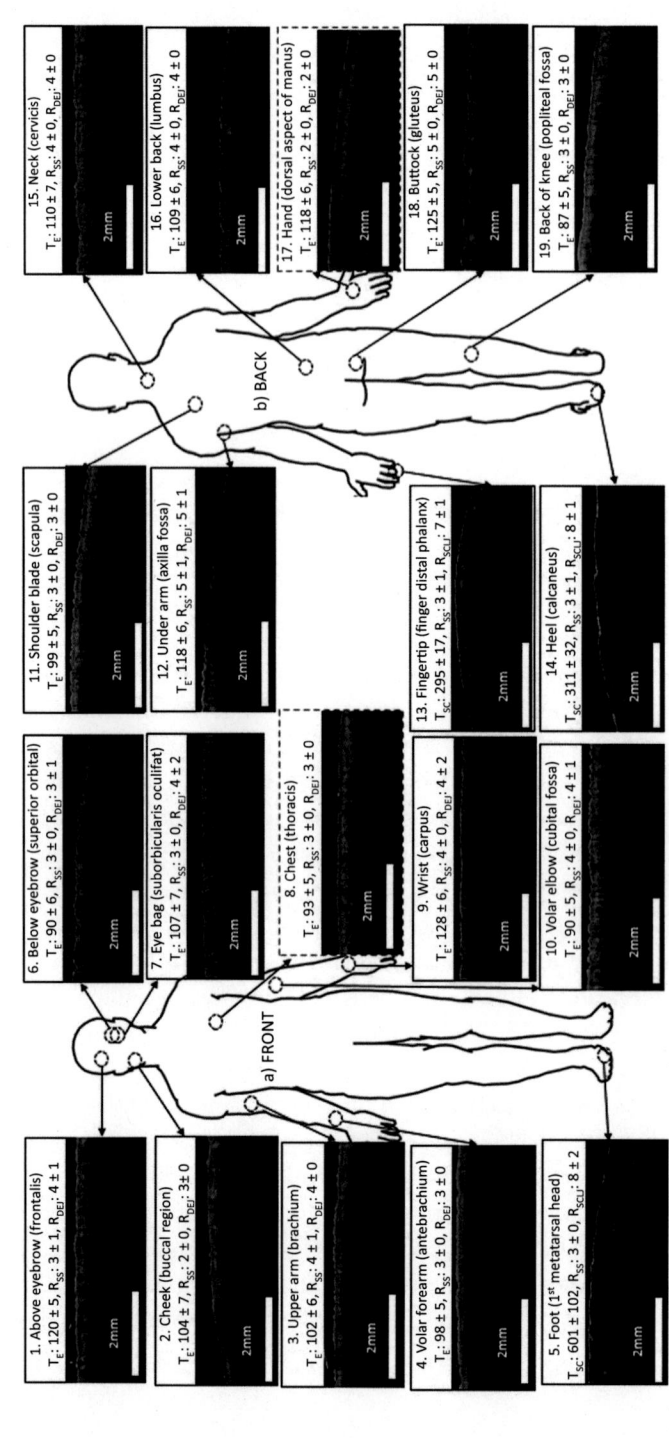

Figure 10.2 Representation of the thickness (T_{SC}: stratum corneum thickness and T_E: epidermal thickness), skin surface roughness (R_{SS}), and undulation at the stratum corneum–stratum lucidum (R_{SCLU}) or dermal–epidermal (R_{DEJ}) junctions. Load was provided either by overextending (nonfacial sites) or manually (facial sites). Values given are mean ± SEM. A *red dotted* border box represents dorsal skin sites [11].

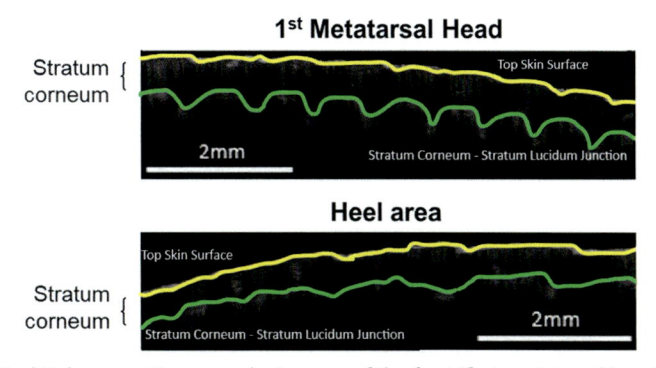

Optical Coherence Tomography to determine and visualize pathological skin structure changes caused by diabetes 165

Figure 10.3 Optical Coherence Tomography images of the foot (first metatarsal head) and heel (calcaneus) regions [11].

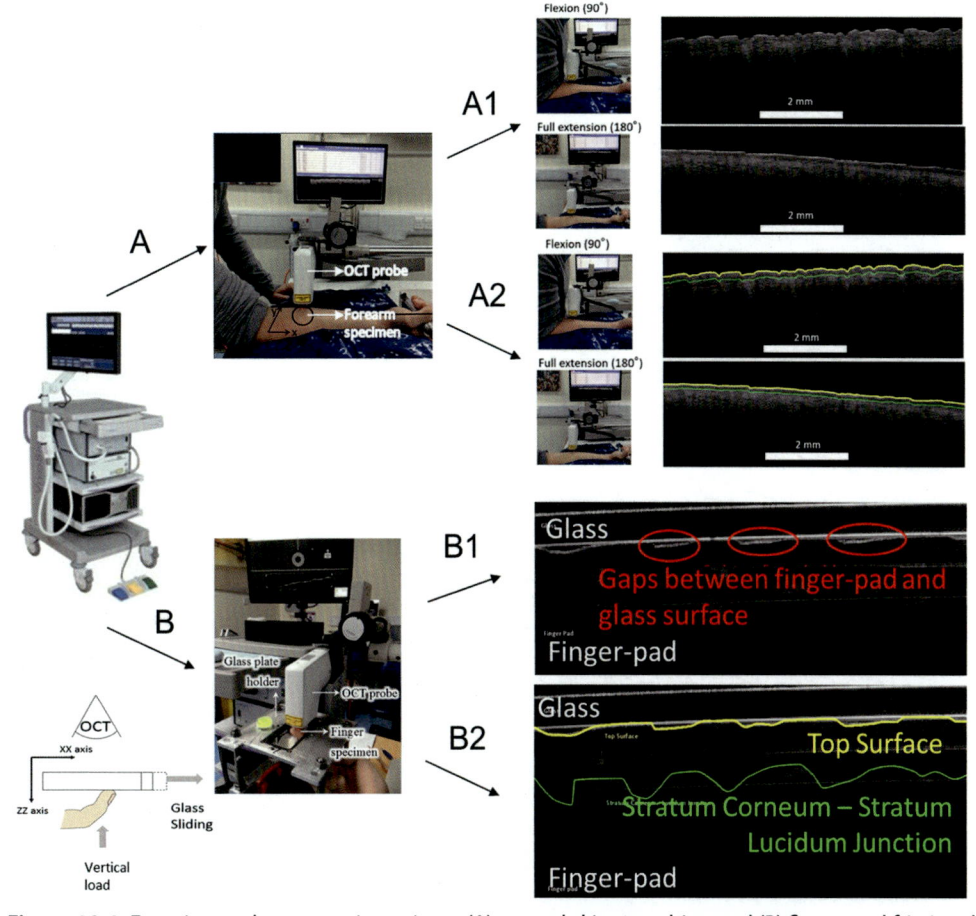

Figure 10.4 Experimental setups to investigate (A) natural skin stretching and (B) finger-pad frictional interactions. The skin layer morphologies are plotted using in-house Matlab detection algorithm [12,13]. *Reprinted from Biotribology, 21, Lee ZS, Maiti R, Carré MJ, Lewis R, Morphology of a human finger-pad during sliding against a grooved plate: A pilot study, 2020, with permission from Elsevier.*

analyzed for the thickness of skin layer, the roughness of top skin surface, and undulation of junction layer using in-house Matlab detection algorithm as shown in Fig. 10.4A2 (details about this algorithm are reported in Ref. [12]).

The VivoSight OCT system was combined with an AMTI force plate (Advanced Mechanical Technology Inc., Massachusetts, USA) as shown in Fig. 10.4B in studies to understand the effect on finger-pad skin surface contact when interacting with a glass plate [4,13]. Fig. 10.4B1 shows images captured from a study where finger loads were applied, normal to the glass plate, between 2 and 6 N ($\pm$0.5 N) and the resistive shear loads were measured when sliding was induced at speeds from 2 to 10 mm/s ($\pm$20%). Captured images were analyzed to predict the contact area at two stages: firstly during the action of normal load; and secondly under the action of normal loading and shear, once sliding had occurred. The changes to the skin morphology were also recorded to investigate the effect of loads on the sublayers of the skin (Fig. 10.4B2). The observed changes in skin layer morphology and contact aid in the understanding of the frictional behavior of finger-pads.

In the study of natural stretching due to posture changes [12], it was found that the epidermal thickness of volar forearm skin reduced when the angle of the arm was changed from 90 degrees flexion to a 180 degrees extension (Fig. 10.5). The reduction of thickness resulted in smoothening of the top surface roughness and junction layer undulation by 40%. A similar effect was observed in morphological properties of the foot and heel when the posture of the metatarsophalangeal and ankle joints was, respectively, changed from flexion to extension. For example, the thickness of the SC at the first metatarsal head on the plantar surface of the foot significantly decreased (450—347 μm; *p = 0.02) when the metatarsophalangeal joint was overextended [11], whereas the thickness of the SC on the plantar surface of the heel increased on overextension. This can be attributed to the orientation of Langer lines on the body [14].

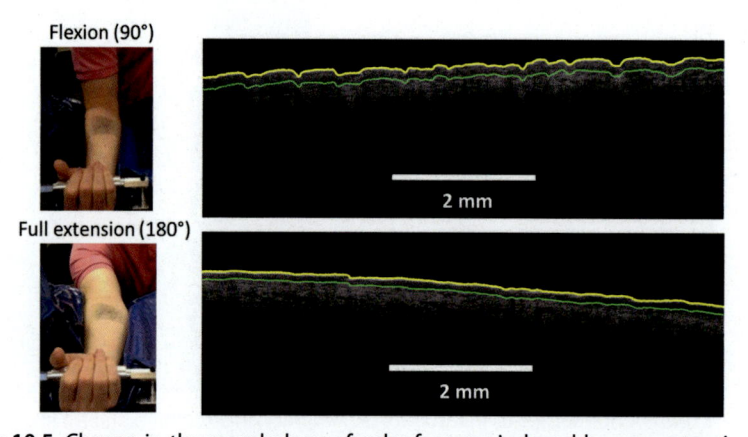

Figure 10.5 Change in the morphology of volar forearm induced by arm extension [12].

In an early finger-pad—surface interaction study, a finger-pad was imaged using an in-house swept-source OCT system while it was pressed statically and then made to slide over a smooth glass surface (roughness less than 0.01 µm), as shown in Fig. 10.6 [4]. The "static" phase is defined as the interaction observed when only loading normal to the surface is applied. In Fig. 10.6, gaps between the finger-pad and glass surface are generated due to the presence of ridges of the finger that define the fingerprint. However, as the shear loading is applied, and continues during sliding of the interface, with the normal loading maintained, the number of gaps reduces drastically. This phase is defined as the "sliding" phase. The contact perimeter measured at the static phase from a series of single 2D OCT images was 50% of the total available contact length. This perimeter increased from 50% to 88% during the sliding phase. It was also noted that the SC-SL junction undulation smoothened in the sliding phase. The finger-pad skin is attached to the underlying tissue and when this is subjected to shear loading during the sliding of glass over the finger-pad, the skin structure responds by flattening the SC-SL junction.

Similar experiments used the VivoSight OCT system to measure morphological changes in a finger-pad when interacting with a grooved surface during sliding [13]. The average roughness of the top surface decreased and SC-SL junction undulation

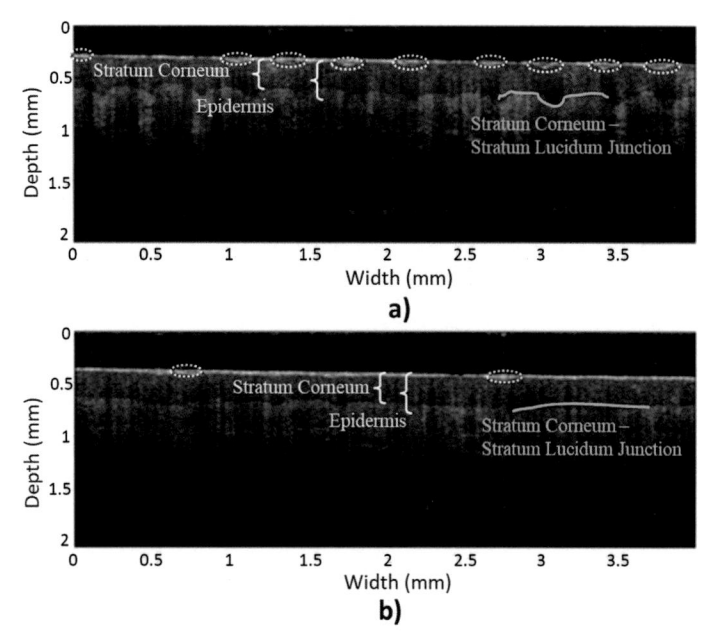

Figure 10.6 Finger-pad interaction of smooth glass surface. (A) Static phase: normal loading only and (B) Sliding phase: normal and shear loading. The *dotted white/red line* represents the gaps in the contact [4].

increased significantly by a factor of 50% during sliding. The top surface of SC gets deformed due to the presence of rough edges on the groove in the plate specimen (represented by white arrows in Fig. 10.7). This interaction resulted in interlocking of the finger-pad ridges (represented by the yellow line) with the rougher edges of the groove surface that might influence friction (as shown in Fig. 10.7) [13].

Case study 2: use of OCT for the diagnosis of atopic dermatitis

OCT has been used for the diagnosis of atopic dermatitis in a clinical and subclinical skin study. The aim of the study was to understand how the blood vascular depth can be used for the diagnosis of therapies related to atopic dermatitis. Angiographic OCT images consisting of the four-dimensional volume of $4 \times 4 \times 2$ mm were captured at 40 frames per second. It was observed that vascular layers were deeper, represented by capillary loop depth (CLD) and superficial plexus depth (SPD), with an increase in the clinically diagnosed severity (using the "EASI" scoring system). The inflammation in the skin pushes the vascular vessels deeper as shown in Fig. 10.8. Also, the dermal—epidermal junction layer was largely undulated for atopic dermatitis volunteers (EASI score 7.5) compared to healthy volunteers [15]. It was interesting to observe that uninvolved skin sites of atopic dermatitis volunteers (EASI score 0.5) were thicker than healthy volunteers (EASI score 0) even though eczema was not evident in those sites.

Future potential of OCT for the diagnosis and management of diabetes-related complications

A serious complication of diabetes which has implications for tissue breakdown is the formation of pathological cross-links within collagen molecules and this can result in mechanically inferior collagen matrix and less mechanically stable skin [16]. The ability to combine OCT with dynamic normal and shear loading offers new approaches to the assessment of the skin's mechanical integrity and stability allowing for a greater understanding of tissue capacity and physiological limits.

The natural response of the skin to increased loading is hyperkeratosis in which the SC becomes thickened [19]. In the diabetic foot when neuropathy is present this process can be further exasperated by unequal loading and insecure gait [17]. This is of concern for patients with diabetes as ulceration risk is increased in the presence of hyperkeratosis [18]. Current clinical approaches to stratify or grade hyperkeratosis are limited and subjective, and do not adequately account for changes to skin texture, thickness, or hydration; this further prevents mechanical characterization [19]. OCT provides an improved ability to characterize surface roughness across a lesion using the approaches outlined above and perhaps more importantly to map the total thickness of lesions of hyperkeratosis allowing for more accurate clinical and mechanical classification.

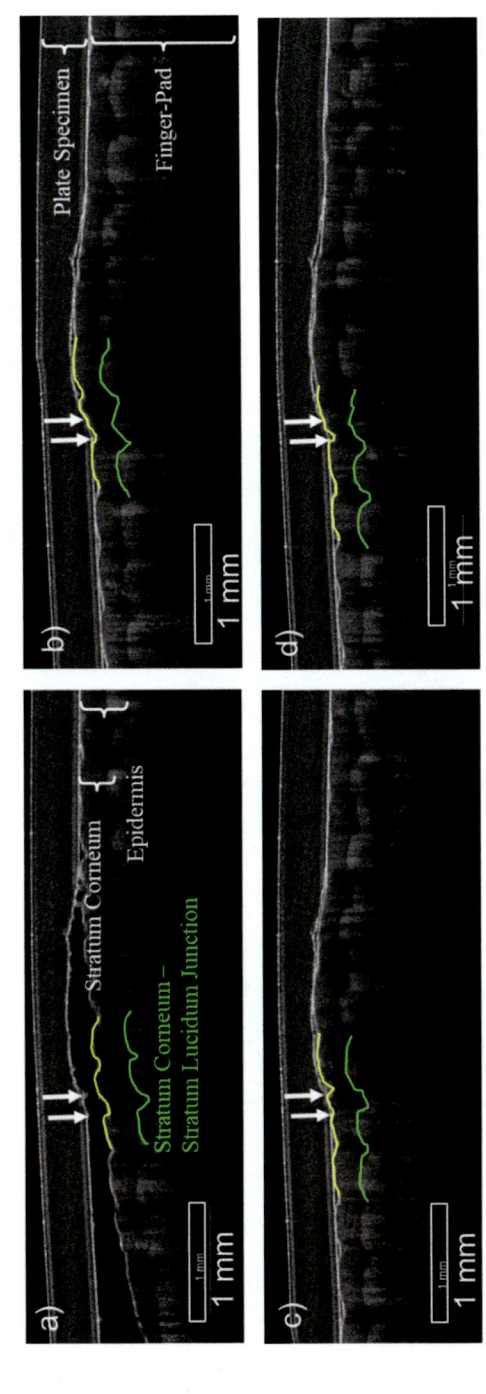

Figure 10.7 Interaction between finger-pad and plate specimen during (a) initial contact, (b) the action of normal load, (c) and (d) sliding under the action of normal and shear loads. The white arrow represents the rough edges of the groove. The interaction of a segment of stratum corneum is highlighted on the top surface in yellow and the stratum corneum—stratum lucidum junction in green. A video file is available in the supplementary files of Ref. [13]. *Reprinted from Biotribology, 21, Lee ZS, Maiti R, Carré MJ, Lewis R, Morphology of a human finger-pad during sliding against a grooved plate: A pilot study, 2020, with permission from Elsevier.*

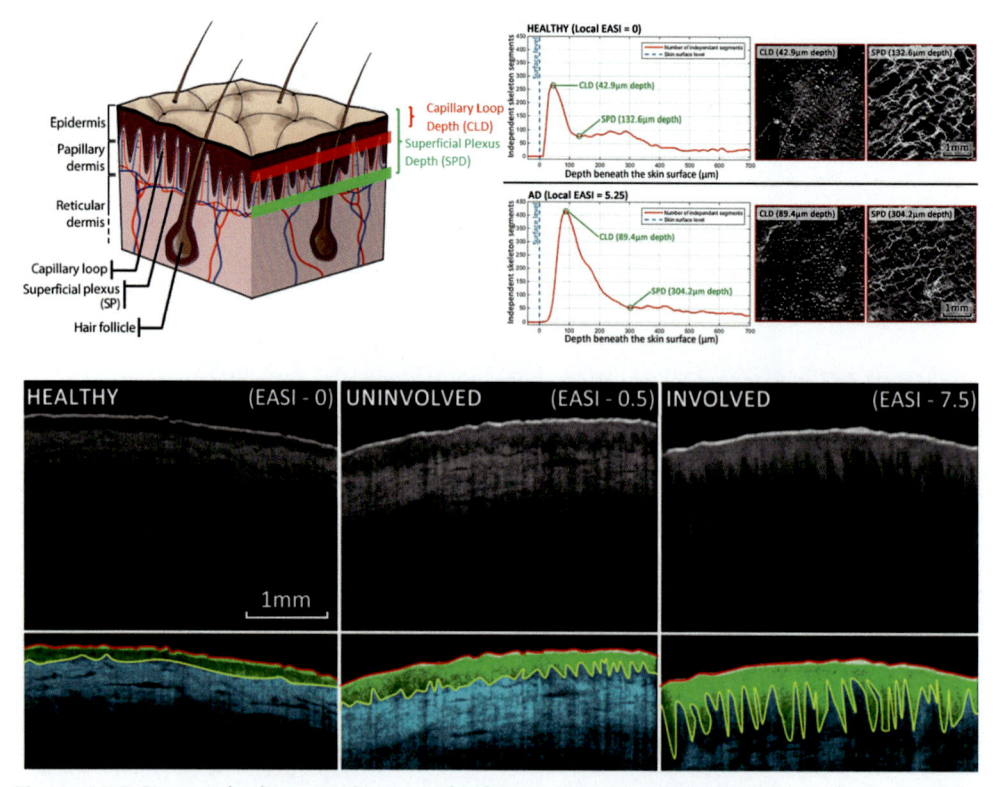

Figure 10.8 Structural schematic diagram of inflammatory skin with vascularized depth signifying capillary loop depth and superficial plexus depth. Analyzed OCT images from the back of the knee (popliteal fossa) for healthy, uninvolved, and involved sites of atopic dermatitis volunteers with relevant EASI score [*adapted with permission from* [15]. © *The Optical Society*].

References

[1] Kinney JH, Johnson QC, Bonse U, Nichols MC, Saroyan RA, Nusshardt R, Pahl R, Brase JM. Three dimensional X-ray computed tomography in materials science. MRS Bull 1988;13:13–8.

[2] Crespo-Facorro B, Kim J-J, Andreasen NC, Spinks R, O'Leary DS, Bockholt HJ, Harris G, Magnotta VA. Cerebral cortex: a topographic segmentation method using magnetic resonance imaging. Physiol Res Neuroimaging Sect 2000;100:97–126.

[3] Cusumano A, Coleman J, Silverman RH, Reinstein DZ, Rondeau MJ, Ursea R, Daly SM, Lloyd HO. Three-dimensional ultrasound imaging: clinical applications. Ophthalmology 1998;105:300–6.

[4] Liu X, Maiti R, Lu ZH, Carré MJ, Matcher SJ, Lewis R. New non-invasive techniques to quantify skin surface strain and sub-surface layer deformation of finger-pad during sliding. Biotribology 2017;12:52–8.

[5] Fercher AF, Hitzenberger CK, Drexler W, Kamp G, Sattmann H. In-vivo optical coherence tomography. Am J Ophthalmol 1993;116:113–4.

[6] Byers RA, Fisher M, Brown NJ, Tozer GM, Matcher SJ. Vascular patterning of subcutaneous mouse fibrosarcomas expressing individual VEGF isoforms can be differentiated using angiographic optical coherence tomography. Biomed Opt Express 2017;8:4551–67.

[7] Colston BW, Sathyam US, DaSilva LB, Everett MJ, Stroeve P, Otis LL, Dental OCT. Opt Express 1998;3:230—8.
[8] Gramatikov BI. Modern technologies for retinal scanning and imaging: an introduction for the biomedical engineer. Biomed Eng Online 2014;13. https://doi.org/10.1186/1475-925X-13-52.
[9] Fercher AF, Drexler W, Hitzenberger CK, Lasser T. Optical coherence tomography — principles and applications. Rep Prog Phys 2003;66:239—303.
[10] Tearney GJ, Bouma BE, Fujimoto JG. High-speed phase- and group-delay scanning with a grating-based phase control delay line. Opt Lett 1997;22:1181—3.
[11] Maiti R, Duan M, Danby SG, Lewis R, Matcher SJ, Carré MJ. Morphological parametric mapping of 21 skin sites throughout the body using optical coherence tomography. J Mech Behav Biomed Mater 2020;102:103501.
[12] Maiti R, Gerhardt L-C, Lee ZS, Byers RA, Woods D, Sanz-Herrera JA, Franklin SE, Lewis R, Matcher SJ, Carré MJ. In-vivo measurement of skin surface strain and sub-surface layer deformation induced by natural tissue stretching. J Mech Behav Biomed Mater 2016;62:556—69.
[13] Lee ZS, Maiti R, Carré MJ, Lewis R. Morphology of a human finger-pad during sliding against a grooved plate: a pilot study. Biotribology 2020;21:100114.
[14] Langer K. On the anatomy and physiology of the skin: the cleavability of the cutis. Br J Plast Surg 1861;31:3—8.
[15] Byers RA, Maiti R, Danby SG, Pang EJ, Mitchel B, Carré MJ, Lewis R, Cork MJ, Matcher SJ. Sub-clinical assessment of atopic dermatitis severity using angiographic optical coherence tomography. Biomed Opt Express 2018;9:2001—17.
[16] Quondamatteo F. Skin and diabetes mellitus: what do we know? Cell Tissue Res 2014;355(1):1—21.
[17] Volmer-Thole M, Lobmann R. Neuropathy and diabetic foot syndrome. Int J Mol Sci 2016;17(6):917.
[18] Murray HJ, Young MJ, Hollis S, Boulton AJ. The association between callus formation, high pressures and neuropathy in diabetic foot ulceration. Diabet Med 1996;13(11):979—82.
[19] Hashmi F, Nester C, Wright C, Newton V, Lam S. Characterising the biophysical properties of normal and hyperkeratotic foot skin. J Foot Ankle Res 2015;8(1):35.

CHAPTER 11

Effects of hyperglycemia and mechanical stimulations on differentiation fate of mesenchymal stem cells

Tasneem Bouzid[1,2], Jung Yul Lim[1,2]
[1]Department of Mechanical and Materials Engineering, University of Nebraska-Lincoln, Lincoln, NE, United States;
[2]Nebraska Center for the Prevention of Obesity Diseases (NPOD), University of Nebraska-Lincoln, Lincoln, NE, United States

Introduction

The prevalence of obesity and diabetes has reached global pandemic levels in recent decades, and continues to have an enormous associated cost, in terms of healthcare and human morbidity. Obesity and diabetes are closely correlated (obese people are more likely to develop type 2 diabetes), and adipose tissue behaves as a probable link between the two. Excess adipose tissue formation results from both increased adipocyte number (hyperplasia) and increased adipocyte size (hypertrophy). The recruitment of new adipocytes from mesenchymal sources such as mesenchymal stem cells (MSCs) is one of the major ways the body increases adipocyte cell number and adipose tissue formation. MSCs are multipotent stem cells that are capable of differentiating into cells in many tissues: bone, fat, muscle, ligament, tendon, etc. The fate decision and terminal differentiation of MSCs can be influenced by soluble and mechanical cues found in the microenvironments surrounding MSCs. Over the last few decades, significant improvements have been made to identify what soluble biochemical factors direct MSC fate into specific phenotypes. Also, attempts have been made to investigate the effect that mechanical environments have on MSC differentiation. In reality, MSCs experience various types of mechanical stimulations including, but not limited to, tensile strain, fluid shear, and compression. There are studies exploiting how mechanical stimulations affect MSC differentiation, even showing that when combined with soluble factors, a synergistic or antagonist effect can be observed. By carefully adjusting the parameters of the biochemical and mechanical cues, one may tune MSC lineage commitment, which can be a helpful tool for developing regenerative and therapeutic strategies.

Interestingly, literatures point to a mutual antagonistic relationship between MSC adipogenesis and osteogenesis in response to external cues, where under certain culture conditions favorable for enhanced osteogenesis, adipogenesis is decreased, and vice versa.

Furthermore, it has been reported that there is a delicate balance between MSC osteogenesis and adipogenesis that involves shared signaling pathways. In this chapter, we will review focusing on MSC osteogenesis and adipogenesis and governing mechanisms under extracellular conditions related to obesity, diabetes, and exercise. Considering hyperglycemia is a defining characteristic of diabetes, and even obesity, the effect of high glucose (HG) levels on MSC behavior will be introduced. Also, the effect of mechanical loading signals on MSC fate will be described. Then, the potential coordination of hyperglycemia and mechanical loading in directing MSC fate decision will be presented.

Effect of hyperglycemia on MSC function and fate

Effect of hyperglycemia on MSC proliferation

Before reviewing the effect of hyperglycemia on MSC lineage commitment, it is notable that the effect of hyperglycemia on the proliferation of MSCs continues to be a topic due to some contradicting results. The inconsistency may depend on the heterogeneity of MSC population, experimental condition, and applied glucose level. In one study, rat bone marrow—derived MSCs (BMSCs) exhibited decreased proliferation when exposed to increased levels of glucose [1]. In another study, HG had a dose-dependent impact on proliferation: BMSCs exposed to 16.5 and 25 mmol/L showed increased proliferation, while those exposed to 35 mmol/L glucose impeded proliferation and increased apoptosis [2]. It was further suggested that hyperglycemia may induce replicative senescence in MSCs [3–5], for instance, exposure to high levels of glucose induces apoptosis via the activation of caspase and leads to a reduction in cell proliferation and multipotency [3]. Also, hyperglycemia induced autophagosome formation in human BMSCs (hBMSCs), which was correlated with senescent behavior; inhibiting autophagosome formation prevented senescence [6].

While most studies used MSCs derived from the perivascular stem cell niche, a study using MSCs derived from the endosteal niche of compact bone [7] found that HG had limited impact on MSC proliferation and did not induce premature replicative senescence. Another study looked at adipose-derived MSCs (ASCs) from diabetic (dASCs) and nondiabetic (nASCs) patients: HG reduced cell replication in both dASCs and nASCs while increasing senescence in both cell types (Fig. 11.1) [8]. Senescence was elevated with increasing glucose level, more in dASCs compared to nASCs, and this effect was reversed with insulin treatment. Thus, one of the reasons that may be behind inconsistency in the effect of HG on MSC proliferation and senescence could be related to the source of MSCs.

Effect of hyperglycemia on MSC osteogenesis

Human and animal studies have shown that diabetes, both type 1 (T1D) and 2 (T2D), are associated with decreased bone strength, and in turn, increased risk of bone fracture

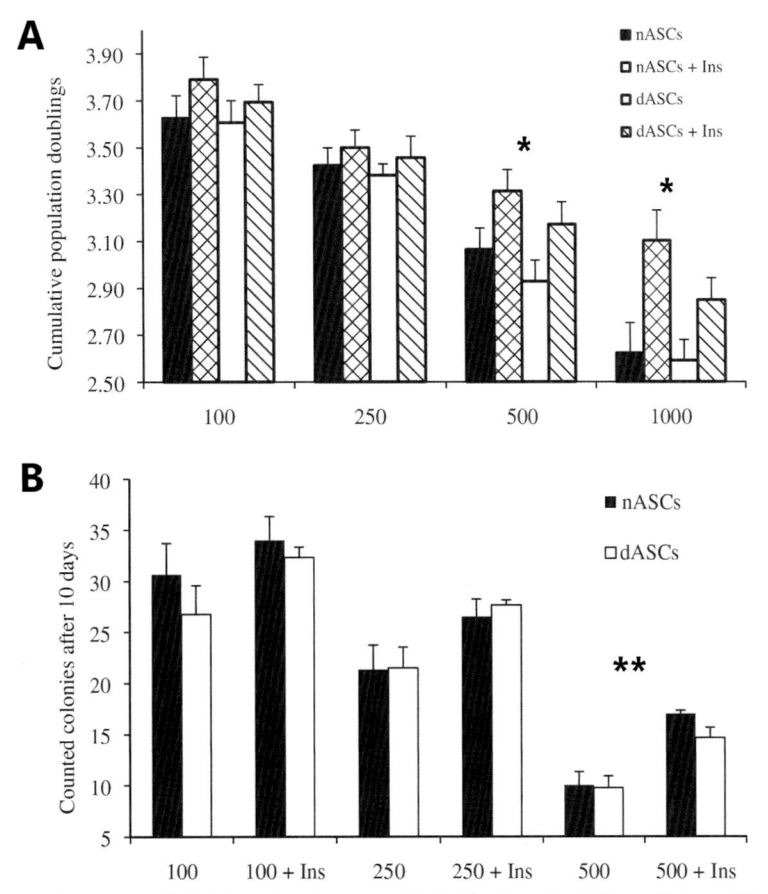

Figure 11.1 Both nASC and dASC replications are diminished under HG conditions. (A) As glucose concentration increased, cell replication was reduced (*: $P < .05$). (B) nASCs/dASC colonies grown were counted after 10 days (**: $P < .01$). Insulin treatment increased population doublings in both cultures. dASC, diabetic adipose-tissue-derived MSC; Ins, insulin; nASC, nondiabetic adipose-tissue-derived MSC. *(Reprinted with permission from Cramer C, Freisinger E, Jones RK, Slakey DP, Dupin CL, Newsome ER, Alt EU, Izadpanah R. Persistent high glucose concentrations alter the regenerative potential of mesenchymal stem cells. Stem Cells Dev. 2010;19:1875–1884.)*

[9–12]. However, it is not fully understood, as studies have different conclusions due to the variety of variables that can impact bone quality. Several factors independent of serum glucose and insulin contents, such as inflammation and the extent to which the patient has diabetic neuropathy or angiopathy, may impact bone density and the risk for falling and subsequent bone fracture. Moreover, diabetes may be associated with a change in bone strength that may not be reflected in bone mass density (BMD) measurements.

In a long-term human study, retarded bone loss was observed in patients with T2D [13]. Since the rate of bone loss directly correlates with the rate of bone turnover, it was determined that T2D is accompanied by a low rate of bone turnover and, in turn, bone accumulation during the period of skeletal growth before epiphyseal closure. Moreover, older patients with a long duration of T2D experienced a protective effect related to low bone turnover. However, studies showed that diabetes remains independently associated with higher fracture risk [13,14]. Patients with T1D showed an insignificant change in BMD, and it was postulated that the retarded rate of bone turnover was offset by the adverse metabolic effects such as acidosis and hypercalciuria [9]. However, in a study of 90 patients with T1D, 34% of the patients exhibited a significantly reduced BMD, and BMD reduction was positively correlated with mean glucose levels, duration of the disease, and insulin dosage [15]. These results were further reinforced with a meta-analysis study on the hip fracture risk and concluded that even though the fracture risk increased for patients with T1D and T2D, BMD was increased in T2D but decreased in T1D patients [16]. T2D thus continues to present a paradox of increased fracture risk but higher observed bone density.

Diabetes affects bone via a range of mechanisms, making it difficult to directly associate the disease with a single metric such as BMD (that does not necessarily encompass overall bone health). Thus, research has focused on elucidating the underlying mechanisms by which diabetes pathophysiology impacts bone quality. Of great recent interest is the effect of hyperglycemia on osteogenesis and bone formation. The amount of MSCs, and the proportion of those cells that differentiate into osteoblasts, influences bone deposition and resorption. Several studies looked at the role high levels of glucose have on lineage commitment and terminal osteogenesis both in vivo and in vitro.

While there remains much to be understood about how hyperglycemia impacts MSC proliferation (the previous section), there is a general consensus that exposure to hyperglycemic conditions inhibits MSC osteogenesis. Chronically high levels of glucose have been shown to decrease several bone differentiation markers and regulators, including alkaline phosphatase (ALP) activity, osteocalcin, matrix metalloproteinase-13 (MMP13), and vascular endothelial growth factor (VEGF). For example, in streptozotocin (STZ)-induced diabetic rats, markers of bone formation such as the number of osteoblasts, ALP activity, and serum osteocalcin were reduced, while markers for bone resorption including the osteoclast number and cathepsin K activity were increased [17]. These changes were reversed with insulin treatment. Researchers have consistently found that HG indeed impacts MSC lineage commitment to bone phenotypes. One study showed that HG impacted BMSC proliferation, ALP activity, and mineralization [1]. In this study, rat BMSCs were treated with concentrations of glucose corresponding to levels often recorded in healthy individuals (5.5 mmol/L), individuals with poorly controlled diabetes (16.5 mmol/L), and individuals with severe hyperglycemia (49.5 mmol/L). As higher levels of glucose were administered, the cultures showed reduced ALP activity and smaller and poorly mineralized nodules. Treatment with insulin or estradiol reversed such changes, e.g., both estradiol and insulin improved nodule formation, mineralization, and matrix deposition. A similar study on murine BMSCs showed that HG treatment (at 25 mM) inhibited osteogenesis by reducing the expression of CXCL13, a consequence mediated by long noncoding RNA (lncRNA) AK028326 (Fig. 11.2) [18]. Another study treated hBMSCs with blood serum obtained from healthy and diabetic patients with a range of glycemic control [19]. Indeed, there was a direct correlation with serum hemoglobin A1c (HbA1c) levels and the reduction in osteogenic potential.

Bone morphogenic proteins (BMPs) play a central regulatory role in the proliferation, adhesion, and fate decision of MSCs. BMPs are from the TGF-β superfamily with at least 14 members found in human [20]. Among these, BMP-2, BMP-4, BMP-6, BMP-7, and BMP-9 are important players in the osteogenesis of MSCs, largely by activating osteogenic expression pathways driven by bone-specific transcription factors [21]. BMP-2 is in particular an important osteoinductive regulator, upstream of transcription factors such as Runx2. It was shown that HG (>25 mmol/L) inhibited BMP-2 activity, in turn, impairing osteogenesis in mouse BMSCs [2]. Indeed, ALP activity, mineralization, and expression of both early and mature osteoblast markers were reduced under HG conditions, which were rescued by the treatment with BMP-2. Interestingly, protein kinase B (PKB also known as Akt) enhanced BMP-2-mediated osteogenesis of MSCs [22]. Furthermore, Runx2 counteracted osteogenic suppression caused by HG via the PI3K/AKT/GSK-3β/β-catenin pathway [23]. Cross talk between BMP-2 and the insulin growth factor (IGF)-activated PI3K-Akt pathway may thus be critical for osteoblast differentiation, maturation, and bone growth. Runx2 could indirectly activate PI3K/Akt signaling, which led to the phosphorylation and subsequent deactivation of GSK3β and nuclear localization of β-catenin [23]. HG significantly reduced the expressions of phosphorylated Akt (pAkt), pGSK3β, and β-catenin compared with MSCs exposed to normal levels of glucose; Runx2 overexpression, on the other hand, increased levels of pAkt, pGSK3β, and β-catenin in MSCs exposed to HG.

Hyperosmotic shock due to HG can lead to increased production of reactive oxygen species (ROS) and DNA fragmentation, leading to increased apoptosis [6,12]. Hyperglycemia leads to increased nonenzymatic protein glycation and, in turn, the formation of proteins known as advanced glycation end products (AGEs) [24]. It has been determined that AGEs accumulate more quickly in the tissues of aging diabetic patients compared to nondiabetic aging patients, and many of the complications can be attributed to the accumulated AGEs [25]. AGEs and their receptors (RAGEs) have been implicated with various cellular processes, including protein modification, nitric oxide (NO) production, and ROS. MSCs exposed to AGEs showed increased ROS generation in a dose-dependent fashion and exhibited inhibited differentiation into bone-forming osteoblasts [4,26]. It is notable that metformin has been revealed to have antioxidative properties, reducing cellular ROS production and reversing the detrimental impact of hyperglycemia on osteoblastic function [27]. Moreover, metformin increased proliferation,

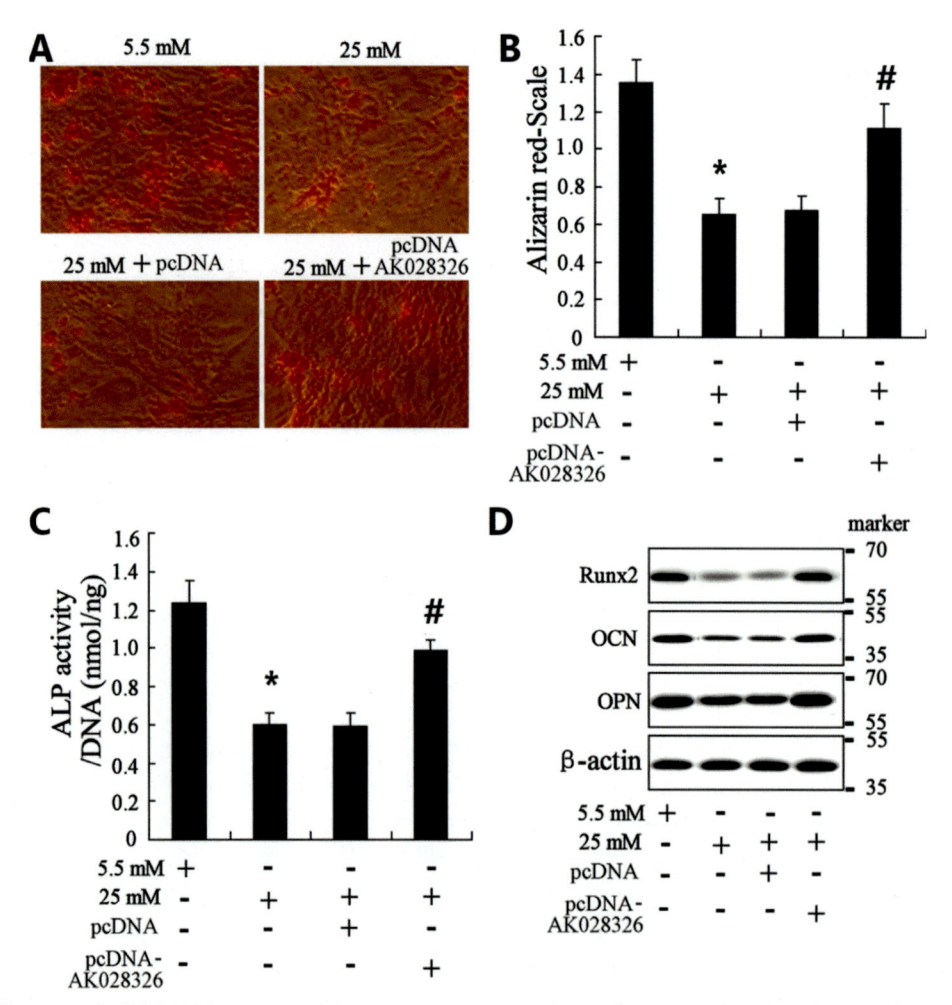

Figure 11.2 AK028326 overexpression restores osteogenic marker expression in HG-treated MSCs. MSCs were transfected with a plasmid for overexpression (pcDNA-AK028326) or an empty control (pcDNA). (A) Mineralized bone nodules are visualized using Alizarin red staining. (B) Semiquantification of staining. (C) ALP activity indicated by the formation of p-nitrophenol. Results are normalized to total protein. (D) Expression of osteogenic markers observed using western blotting. *: $P < .05$ compared with MSCs under normal (5.5 mM) glucose incubation; #: $P < .05$ compared with pcDNA-transfected MSCs under 25 mM glucose incubation. OPN, osteopontin; OCN, osteocalcin. *(Reprinted with permission from Cao B, Liu N, Wang W. High glucose prevents osteogenic differentiation of mesenchymal stem cells via lncRNA AK028326/CXCL13 pathway. Biomed Pharmacother. 2016;84: 544–551.)*

differentiation, ALP activity, and number of mineralized nodules in rat osteoblast culture [28]. Studies on rat bone marrow progenitor cells [29] and human induced pluripotent stem cells [30] showed similar trends, for example, in vivo administration of metformin significantly increased ex vivo ALP activity, type I collagen production, osteocalcin expression, and mineral deposition of cultured BMSCs [29]. Further, for rat BMSCs, metformin treatment induced osteogenesis while inhibiting adipogenesis [31]. In fact, metformin can reverse the proadipogenic effect of rosiglitazone in favor of an osteoblastic phenotype in rat BMSCs [29]. Similarly, oral administration of metformin in mice significantly increased the osteogenic differentiation potential of ASCs, leading to reduced levels of ROS but higher levels of BMP-2, osteocalcin, and osteopontin expressions [32].

Effect of hyperglycemia on MSC adipogenesis

Researches strongly establish an inverse relationship between osteogenic and adipogenic programming. Many of the signaling pathways that are known to induce osteogenesis often do so at the expense of adipogenesis, and vice versa [33]. The same inverse relationship is seen under HG conditions as well. When BMSCs derived from alloxan-induced diabetic and normal rats were exposed to increasing levels of glucose (5.5 mM, 20 mM, 33 mM) with the beginning of differentiation period, the level of triacylglycerol accumulation increased with increasing glucose concentration in both normal and diabetic BMSCs (Fig. 11.3) [12]. Moreover, cells treated with 33 mM D-glucose showed significantly increased mRNA expression of the adipogenic transcription factor, peroxisome proliferator—activated receptor gamma (PPARγ). Meanwhile, cells exposed to 33 mM glucose displayed significant reductions in ALP activity and Runx2 expression. Thus, the conclusion is hyperglycemic conditions improve adipogenesis at the expense of osteogenesis. If treated with antioxidants such as Vitamin C or metformin, this effect could be reversed and as a result, osteogenesis is enhanced and adipogenesis is suppressed [12,33]. In a study using ASCs from dASCs and nASCs patients [8], dASCs showed a greater tendency to differentiate into adipogenesis compared with nASCs. Also, while prolonged exposure to HG increased adipogenesis in both cell types, the change was more significant in dASCs. Under prolonged exposure of hBMSCs to HG, while adipogenesis is increased, not only osteogenesis but also chondrogenesis was suppressed [34].

One of the most prominent molecular pathways that have been implicated in HG regulation of MSC adipogenesis and lipid accumulation is the PPARγ-PI3K/Akt signaling pathway. Murine MSCs exposed to HG-adipogenic medium displayed enhanced adipogenesis, and treatment with GW9662, an irreversible inhibitor of PPARγ, did not impact the level of adipogenesis but significantly reduced the adipogenic induction of lipid accumulation [35]. HG-induced PPARγ expression requires the activation of ERK1/2, the upstream of the PI3K/Akt signaling pathway. Thiazolidinediones such as rosiglitazone are an effective class of antidiabetic agents that bind and activate PPARγ to improve insulin resistance. Interestingly, rosiglitazone has been shown to inhibit osteoblastogenesis and bone formation, while increasing marrow adipocyte count and bone adiposity [36].

Another pathway shown to be involved in HG-dependent MSC adipogenesis is the protein kinase Cβ (PKCβ)-PPARγ pathway. Both rat muscle-derived stem cells

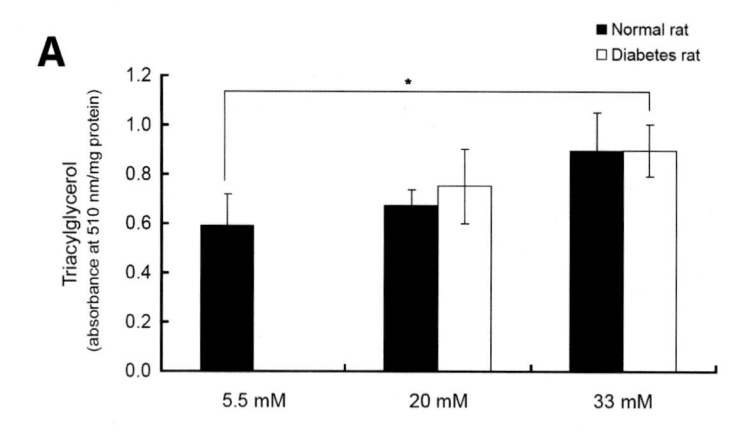

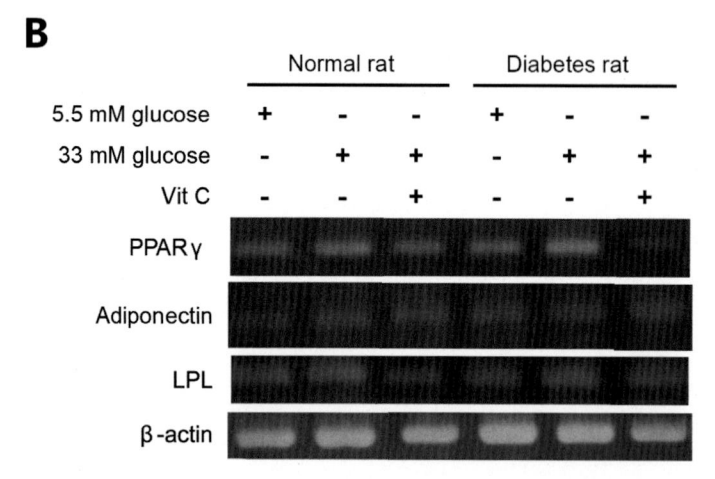

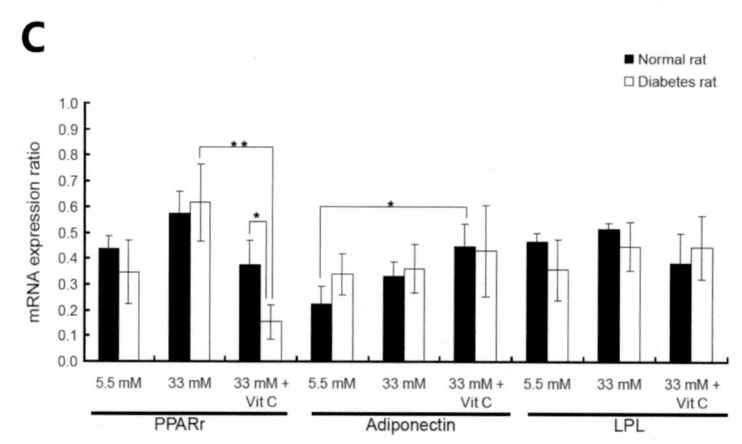

Figure 11.3 HG conditions increase adipogenesis of rat BMSCs. (A) Triacylglycerol levels were quantified from oil red O stained deposits at day 7, and normalized to the total protein amount in each group. (B) The expression of adipogenic markers was determined using RT-PCR at the end of the experimental period. (C) The level of mRNA expression, relative to the loading control. BMSCs were also treated with Vit C. *: $P < .05$; **: $P < .01$. *LPL*, lipoprotein lipase; *PPARγ*, peroxisome proliferator–activated receptor gamma; *Vit C*, vitamin C. *(Reprinted with permission from Li WT, Hu WK, Ho FM. High glucose induced bone loss via attenuating the proliferation and osteoblastogenesis and enhancing adipogenesis of bone marrow mesenchymal stem cells. Biomed Eng-App Bas C. 2013;25:1340010.)*

(MDSCs) and ASCs exposed to HG showed significantly increased adipogenic marker expression and lipid accumulation [37]. ROS produced in response to HG led to the activation of PKCβ, and eventually increased adipogenesis. Direct treatment of cells with oxidizing agents could induce adipogenesis in low glucose (LG) conditions, while silencing of PKCβ with siRNA inhibited HG-enhanced adipogenesis. Additionally, promyelocytic leukemia protein (PML), a tumor suppressor, may play a role in HG-dependent MSC adipogenesis via mediating the process by regulating PKCβ and autophagy [38].

Effect of mechanical loading cues on MSC fate

As described above, the key component of MSC differentiation protocol is the addition of soluble biochemical factors that have been proven to assist in MSC commitment to a specific lineage. While biochemical cues can sufficiently differentiate MSCs, temporal and concentration gradients are difficult to accurately control. Lately, researchers have been exploiting the intrinsic cellular system responsive to mechanical cues. Our bodies are subjected to various mechanical forces, and these forces play a vital role in tissue growth and regeneration. When applied in vitro, mechanical loading such as stretching, compression, and shear stress may provide biomimetic and easily controllable cues for MSC differentiation. In addition to dynamic mechanical loading, MSC lineage commitment and differentiation can also be regulated by static mechanophysical information obtained from the culture substrate. In tissues, cells exist surrounded and supported by various extracellular matrix (ECM) milieus that exhibit a range of micro- and nanotextures, rigidities, and geometries. In vitro, these substrate parameters have been shown to have the potential to affect MSC growth and differentiation. From a tissue engineering standpoint, improving MSC osteogenic and chondrogenic differentiation potential (while suppressing adipogenesis) via dynamic mechanical loading and/or substrate mechanical property is advantageous to better design in vitro and in vivo tissue engineering protocols. Researchers in the fields of obesity and related diseases could be interested in how dynamic mechanical loading cues influence MSC adipogenesis and osteogenesis in the context of widely recognized correlation among obesity, diabetes, and exercise. For example, understanding what factors inhibit MSC adipogenesis (while promoting osteogenesis) may be meaningful for the prevention and treatment of obesity and related diseases. This section will review focusing on the dynamic mechanical loading control of MSC osteogenesis and adipogenesis and the mechanisms by which MSCs sense external forces and translate mechanical signals into differentiation cues.

Effect of mechanical loading cues on MSC osteogenesis
Tensile loading
In the body, tissues are exposed to different levels of mechanical loading, and the cellular response to these mechanical cues is important in healthy tissue homeostasis. MSCs reside

also in bone, which is subjected to extensive mechanical conditioning which plays a central role in bone formation. In recent years, it has been well established that tensile stretch loading has a positive effect on MSC osteogenesis. In fact, orthopedic surgeons have long taken advantage of a process known as "distraction osteogenesis" to stimulate bone regeneration via the mechanical pulling apart of bone segments. This mechanically induced regenerative process was simulated in vitro where rat MSCs were exposed to a short period of cyclic stretch (2000 microstrains in 40 min), and ALP activity was found to increase as a result [39].

In a recent study, equiaxial stretching of human ASCs led to increased osteogenesis, accompanied by intensified cytoskeleton and focal adhesion formation [40]. Focal adhesions are membrane-associated macromolecular complexes through which external mechanical regulatory signals from the ECM are transmitted across the membrane and translated into biochemical signals. The transduction of the mechanical signal through focal adhesion occurs via several mechanosensory proteins including focal adhesion kinase (FAK) followed by a series of downstream pathways [41]. Such pathways include, but are not limited to, PI3K/Akt, mitogen-activated protein kinases (MAPKs), and GTPases of the Rho family. Of particular relevance is the large GTPase RhoA, and its downstream regulator RhoA kinase (ROCK), which have been shown to play a critical role in cytoskeletal tension and MSC differentiation [42].

Numerous key transcription factors and genes have been discovered to be vital players in mediating stretch regulation of MSC osteogenesis. The expression of core-binding factor alpha-1 (Cbfα1), which leads to the activation of many osteoblast-specific genes and proteins including osteocalcin, osteopontin, type I collagen, bone sialoprotein, BMP-2, etc., is one of the key events under stretch [43]. BMP-9 is another powerful osteogenic factor in MSCs. A synergetic effect was observed for BMP-9 and mechanical stretch in the osteogenesis of C3H10T1/2 MSCs (Fig. 11.4) [44]: Stretch alone was incapable of inducing osteogenesis, but when used in conjunction with BMP-9 treatment, a significantly augmented osteogenic response was observed.

Mechanical tensile loading affects the expression of the FOS family of transcription factors, which are vital for bone cell proliferation, differentiation, and apoptosis. Activator Protein-1 (AP-1) transcription factors are dimers of Fos (c-Fos, FosB, DFosB, Fra-1, Fra-2) and Jun (c-Jun, JunB, JunD) proteins, and members of the AP-1 family interact with promoters for several genes essential for osteogenesis, such as ALP, osteocalcin, osteopontin, type I collagen, and runx2. Human BMSCs exposed to 3 days of cyclic tensile loading (2% or 8% strain at 1 Hz for 2 h/day) displayed a time- and intensity-dependent increase in FosB expression, accompanied with increased runx2 and type I collagen [45]. Similarly, cyclic stretch (2%–8% strain at 1 Hz for 15–60 min) induced rat MSC proliferation and upregulation of c-FOS gene expression [46].

Another player in the stretch regulation of MSC function is silent information regulator type 1 (SIRT1). It is a member of nicotinamide adenine dinucleotide (NAD+)-dependent

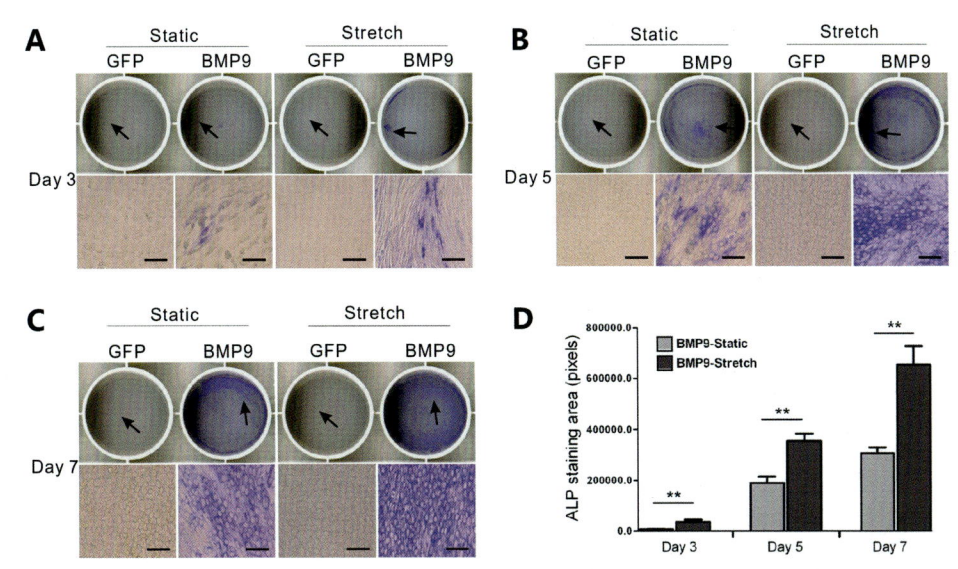

Figure 11.4 BMP-9 overexpression has a synergistic effect on mechanical stretch—induced MSC osteogenesis. C3H10T1/2 MSCs were treated with or without mechanical stretch (6% strain, 1 Hz, 12 h). (A—C) ALP staining, in purple, of the wells (top row) and microscopic images (bottom row) are shown for Days 3, 5, and 7. (D) Quantification of ALP staining area, in pixels, was obtained from the images of the BMP-9-treated samples. Scale bar = 100 μm **: $P < .01$. ALP, alkaline phosphatase; BMP-9, bone morphogenic protein-9; GFP, green fluorescent protein. *(Reprinted with permission from Song Y, Tang Y, Song J, Lei M, Liang P, Fu T, Su X, Zhou P, Yang L, Huang E. Cyclic mechanical stretch enhances BMP9-induced osteogenic differentiation of mesenchymal stem cells. Int Orthop. 2018;42: 947—955.)*

deacetylases that regulate metabolism, senescence, differentiation, and antioxidant responses in MSCs. It was reported for skeletal muscle cells that mechanical stretch resulted in an increase in ROS and suggested that transient increase of SIRT1 by stretch could be protective against stretch-induced oxidative stress [47]. However, when mechanical stretch (2.5% and 5% strain for 2 h/day) was applied to BMSCs, a reduction in intracellular ROS was observed after 3 days [48]. A previous study reported that elevated levels of intracellular H_2O_2 contributed to the decline of hMSC osteogenic differentiation capacity, and moreover, both mRNA and protein expression of SIRT1 were upregulated under stretch via AMPK [49]. Combined data indicate the need of more study on the stretch effect on SIRT1, oxidative stress, and MSC osteogenesis.

The temporality of osteogenic marker expression suggests a strong relationship between stretching parameters and the differentiation potential of the stretched cells. The magnitude and duration of mechanical stretch, as well as frequency and repetition of stretching bouts, all play a role in directing MSC differentiation. Experimental setups differ from study to study, leading to diverging results. For instance, MSCs subjected to cyclic stretch at 1 Hz and strain ranging from 0.8% to 15% showed differential responses in proliferation and differentiation. At 5%, 10%, and 15% strain, MSCs showed a significant increase in proliferation compared with unstretched controls [50]. ALP, on the other hand, increased in activity under 0.8% and 5% strain, but decreased under 10% and 15% strain. Similarly, BMSCs exposed to 10% cyclic stretch showed significantly lower ALP compared with the 5% strain group, suggesting that excessive stretch may actually have an adverse effect on osteogenesis [48]. However, one must take into account that this experiment applied stretch in an intermittent fashion for 2 h daily. One study reported that continuous cyclic stretch actually inhibited osteogenesis [51]. Thus, while many studies looked at the effect of different stretch parameters on MSC differentiation, there lacks coherency in the stretch experimental setups, preventing objective comparisons. There remains a need for more thorough and systematic studies on the effects of various stretch parameters on MSC osteogenesis and on what mechanotransduction pathways dominate under which conditions.

While there is strong evidence that mechanical stretch enhances MSC osteogenesis, it is not the case for all MSCs. In fact, MSCs derived from tooth dental pulp (DPSCs) exhibit the opposite response. DPSCs exposed to cyclic stretch displayed reduced expression of osteogenic markers and proteins such as BMP-2 and ALP, as well as reduced expression of odontogenic markers, DSPP, DSP, and BSP [52]. Apart from DPSCs, however, the effect of mechanical stretch on MSC osteogenesis is well supported in that mechanical stretch consistently leads to increased expression of key osteogenic markers and enhanced bone formation.

Compressive loading

The effect of compressive pressure on MSC osteogenesis has also been studied and taken advantage of in bone tissue engineering. One study applied a range of pressures (10–40 kPa) to osteoblasts, fibroblasts, and endothelial cells at 1 or 0.25 Hz for 1 h/day for 5 days [53]. The cyclic pressure on osteoblasts resulted in increased ALP, but had no effect on osteopontin expression. Fibroblast proliferation increased under cyclic pressure, while endothelial cells remained unaffected. A similar study showed that cyclic pressure led to reduced osteoclast differentiation from precursors, and subsequently, reduced bone resorption [54]. This was supported by the downregulation of key cytokines often secreted in a paracrine fashion by osteoblasts and nonosteoclastic cells (e.g., IL-1α, IL-1β, and TNF-α). Another study from the same group furthered the narrative: while cyclic pressure reduces bone resorption, cyclic pressure may also enhance osteoblast functions pertinent to new bone formation as assessed by enhanced collagen synthesis, calcium accumulation, and osteocalcin expression [55].

Several other studies also reported that compression does induce osteogenesis of MSCs. For instance, dynamic hydraulic compression (1 psi at 1 Hz) delivered to both hASCs and hMSCs led to increased osteogenic gene expression, with hMSCs more receptive than hASCs [56]. Considering that BMSCs, like mature osteoblasts, are local targets of estrogen action, the potential role of estrogen and its receptor (ER) in BMSC mechanotransduction was explored [57]. BMSCs were treated with compression loading under additional estrogen or ER antagonist treatment. It was observed that compression stimulated BMSC proliferation and osteogenesis via F-actin and estrogen further enhanced this effect, suggesting ER may be a key mediator of the BMSC biochemical response to compression.

Interestingly, it was suggested that MSCs are capable of distinguishing between compressive loading and dynamic tensile stretching [58]. In the study, dynamic tensile loading of MSCs upregulated fibroblastic and osteogenic genes and inhibited genes associated with chondrogenesis; on the other hand, dynamic compression enhanced the expression of chondrogenic genes (α-catenin, cathepsin B, aggrecan, BMP-6). As a mechanism, it was proposed that the difference may be due to altered catenin signaling: compression upregulated α-catenin which is responsible for inactivating β-catenin required for osteogenesis; to the contrary, tensile stretch upregulated β-catenin to induce osteogenesis but not chondrogenesis.

While studies showed dynamic compression loading leading to osteogenic terminal differentiation of MSCs, chondrogenic differentiation has been the major target in many of the compression studies (this chapter will not focus on chondrogenesis). It is notable while some stretch studies demonstrated that stretch may overcome any regulation by soluble factors [59], the current body of compression research does not support a similar phenomenon. For example, combining compression and shear was needed for MSC chondrogenesis [60], hypoxia was more stimulatory to chondrogenesis than compression [61], etc.

Fluid shear

Fluid flow inside the bone has been demonstrated to be involved in bone formation and remodeling. Mechanical loading induces interstitial fluid flow in bone, and the resultant shear stress can be a powerful regulator of cell migration, communication, and MSC fate decision by affecting biochemical transport, ECM deposition, cytoskeletal straining, and cell deformation [62,63]. In fact, it was demonstrated that fluid flow—induced shear stress alone had the capability to direct osteoblast and MSC migrations [64,65] and stimulate MSC osteogenic differentiation [66].

Fluid shear stimulation initiates a wide range of signaling cascades, including transient cytosolic accumulation of Ca^{2+}, release of prostaglandins and NO, phosphorylation of MAPKs such as ERK and p38, etc. The intracellular accumulation of Ca^{2+} is one of the most upstream events occurring in bone cells and MSCs exposed to fluid shear, and the Ca^{2+} responsiveness to fluid shear may alter depending on the type of flow (steady, oscillatory, or pulsatile) [67] or on the cell interaction with the cultured substrate [68]. More downstream effectors of fluid shear mechanotransduction include MAPKs, a family of kinases that respond to various extracellular stimuli including environmental stress and inflammatory cytokine and are capable of regulating proliferation, differentiation, apoptosis, and metabolism. ERK1/2 and p38 are important focal points of MAPK cascades and, as research supports, mechanotransduction pathways. Fluid shear stress activates ERK1/2 with focal adhesion signaling, FAK, working as an upstream of ERK1/2 phosphorylation [69]. As related pathway, actin cytoskeletal contraction also contributes to the fluid shear–induced osteogenesis in MSCs [70]: C3H10T1/2 MSCs exposed to fluid shear for 1 h increased the activity of RhoA/ROCK that acts as an upstream effector of ERK1/2. As in Fig. 11.5 [70], interrupting cytoskeletal tension, polymerization, and dynamics by various inhibitors greatly abrogated fluid shear–induced upregulation of Runx2 expression. Thus, the literature establishes both focal adhesion and cytoskeletal tension as important regulators of fluid shear–induced MSC osteogenesis.

Similar to the case of tensile stretch loading, gene expression and terminal differentiation can be affected by both shear stress magnitude and duration. In one study, MSCs displayed shear stress (0.015, 0.030, 0.045, or 0.060 Pa)-dependent increases in type I collagen and osteopontin gene expression [71]. Also, intermittent fluid shear stress applied for 11 h to hBMSCs resulted in significantly elevated osteogenic gene expression (RUNX2, ALP, osteocalcin, and collagen) and ALP activity [72]. Moreover, MSCs exposed to different magnitudes and duration of fluid shears exhibited marked

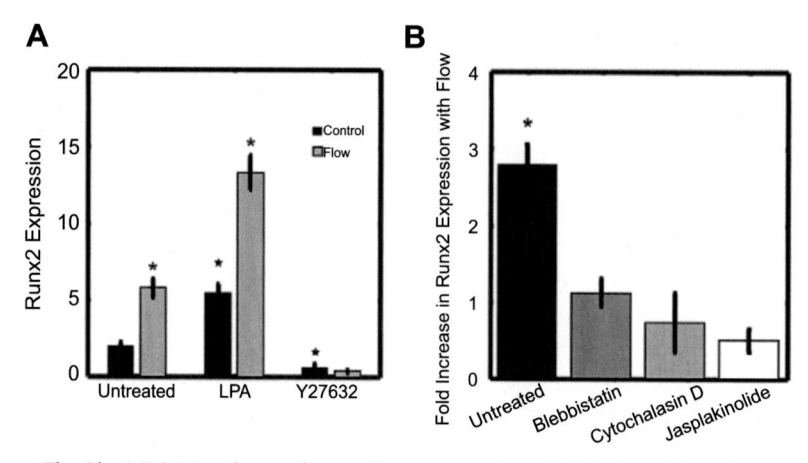

Figure 11.5 The RhoA/ROCK pathway plays an important role in fluid-induced Runx2 upregulation in C3H10T1/2 MSCs. (A) RhoA activator LPA increased Runx2 expression, and the application of fluid flow further increased Runx2 expression. Incubation with ROCK inhibitor Y27632 greatly abrogated Runx2 expression in all cells. *: $P < .01$ compared with untreated control. (B) Inhibition of myosin II activity with blebbistatin, or impacting cytoskeletal polymerization or dynamics with cytochalasin D and jasplakinolide, respectively, all significantly reduced fluid flow induction of Runx2 expression. *: $P < .01$ compared with the treated data. *(Reprinted with permission from Arnsdorf EJ, Tummala P, Kwon RY, Jacobs CR. Mechanically induced osteogenic differentiation—the role of RhoA, ROCKII and cytoskeletal dynamics. J Cell Sci. 2009;122;546–553.)*

upregulation of MAP3K8 and IL-1 [73]. Of those two, MAP3K8 is directly upstream of the ERK1/2 (MEK1 and MEK2), p38 (MEK3 and MEK6), and JNK (MECH4 and MECH7) signaling.

As briefly noted above on flow-induced Ca^{2+} response, the type of flow may determine the extent to which MSCs respond to fluid shear. Human fetal osteoblasts showed much less response in Ca^{2+} under oscillating flow relative to either steady or pulsing flow [67]. On the other hand, MSCs responded strongly to oscillatory flows showing significant increases in intracellular Ca^{2+}, osteopontin and osteocalcin gene expressions, and proliferation as well [74]. The general consensus in cell proliferation is that cell cycle arrest precedes osteogenic differentiation. It was proposed that the heterogeneity of the MSC population used may lead to seemingly paradoxical finding. In a study using BMSCs [75], it was observed that pulsatile flow, but not steady flow, can modulate BMP-2, BMP-7, and TGF-β1 expressions and via that mechanism, may enhance the expression of bone-like ECM proteins in MSCs. Reported data indicate the need to investigate more about the potential different effect of steady, pulsatile, and oscillatory flows on MSC functions.

Effect of mechanical loading cues on MSC adipogenesis
Tensile loading
Given that a wide range of cell types have been established to be mechanoresponsive, it is an interesting question whether adipocytes are mechanoresponsive as well. In the body, adipose tissue is subjected to compound mechanical loading conditions including compressive, tensile, and shear strains largely as a result of bodyweight loads and weight-bearing [76]. Adipocytes are thus physiologically exposed to a wide range of body load strains, and the loading profile heavily depends on lifestyle, age, injury, and disease in the individual [77].

In accordance with the previously described observations on hyperglycemia, studies on mechanobiology have also established a reciprocal relationship between bone and fat formation under mechanical loading. So, there also exists a general inverse relationship in the mechanical stretch regulation of adipogenesis and osteogenesis. For in vivo, PPARγ-deficient mice displayed decreased adipogenesis but increased bone mass [78]. In an in vitro study with ASCs cultured in both normal and adipogenic medium, cyclic stretch stimulated osteogenesis while inhibiting adipogenesis [79], in accordance with the study cited above [59].

The adipogenesis/osteogenesis relationship relies on canonical Wnt/β-catenin signaling, which represses PPARγ expression and subsequent adipogenesis, while activating runx2 and therefore enhancing osteogenesis [80,81]. Such regulation has already been established in the absence of mechanical stimulus. Cyclic stretch loading was found to trigger β-catenin nuclear translocation, leading to an osteogenic fate induction rather than the lesser favored adipogenic fate [82]. As a mechanism, a recent study proposed that

PI3K/Akt/GSK-3β pathway may be responsible for the mechanically induced β-catenin signaling and related osteogenesis versus adipogenesis behavior [83].

The stretch control of MSC adipogenesis may also depend on what stage of MSC lineage commitment that the loading is delivered. In our study, we examined the effect of mechanical stretch on MSC adipogenesis by applying stretch during the MSC commitment period [84]. BMPs have been traditionally known to promote osteogenesis and bone formation, but BMP-4 has been evidenced to play a vital role in inducing MSC commitment to adipocyte [85]. Using BMP-4-induced MSC adipogenesis model, we showed that cyclic stretch (10% strain, 1 Hz, 2 h/day) applied during the BMP-4 pretreatment (commitment) period effectively suppressed MSC adipogenesis (Fig. 11.6) [84]. It is known that BMP-4 activation of MSC adipogenesis can be achieved through Smad and p38 MAPK [85]. In our study, however, stretch suppression of BMP-4 induction of MSC adipogenesis may be accomplished via upregulation of ERK1/2 but not through downregulating Smad1/5/8 or p38. We also observed that silencing FAK with small hairpin RNA (shRNA) resulted in significantly suppressed MSC adipogenesis under static condition [86]. Considering that FAK is an upstream effector of ERK, our combined data may suggest that focal adhesion–based mechanosensing may be an important mediator for MSC adipogenesis under BMP-4 for both static and dynamic conditions. In a study with MSC without BMP-4, stretch inhibition of MSC adipogenesis during the differentiation stage was achieved through the activation of TGF-β1/Smad2 signaling pathway [87].

Strain magnitude has also been revealed to play an important role in MSC differentiation. When equiaxial cyclic tensile loading (10% strain, 0.5 Hz for 48 h) was applied to murine ASCs, an expected reduction in adipogenesis was observed [88]. However, when the peak strain was reduced to 2%, the suppressive effect of stretch was significantly diminished; at 0.5% strain, stretch had almost no suppressive effect on MSC adipogenesis. The existing body of literature supports the existence of strain thresholds beyond which MSCs respond to stretch. Moreover, the observed strain magnitude dependency of MSC differentiation upkeeps the postulation that there may exist different signaling pathways at play in MSC mechanotransduction depending on the strain magnitude.

A widely accepted concept is that cells exposed to persistent mechanical strain for a long period of time experience strain tolerance, or a ceasing of the cellular response to applied stretch. Studies have turned focus to understanding how singular or repetitive stretch may differentially impact MSC differentiation. While one study sought to overcome the strain tolerance by applying a gradually increasing strain scheme [89], many other studies incorporated refractory periods between loads to ameliorate the effects of strain tolerance and sustain, or in some cases even enhance, the initial stretch effects for the entirety of the period. In a study that investigated the effects of low-intensity vibration (<10 microstrain, 90 Hz) and high-magnitude strain (20,000 microstrain, 0.17 Hz) on MSC adipogenesis, it was observed that adipogenesis was inhibited when

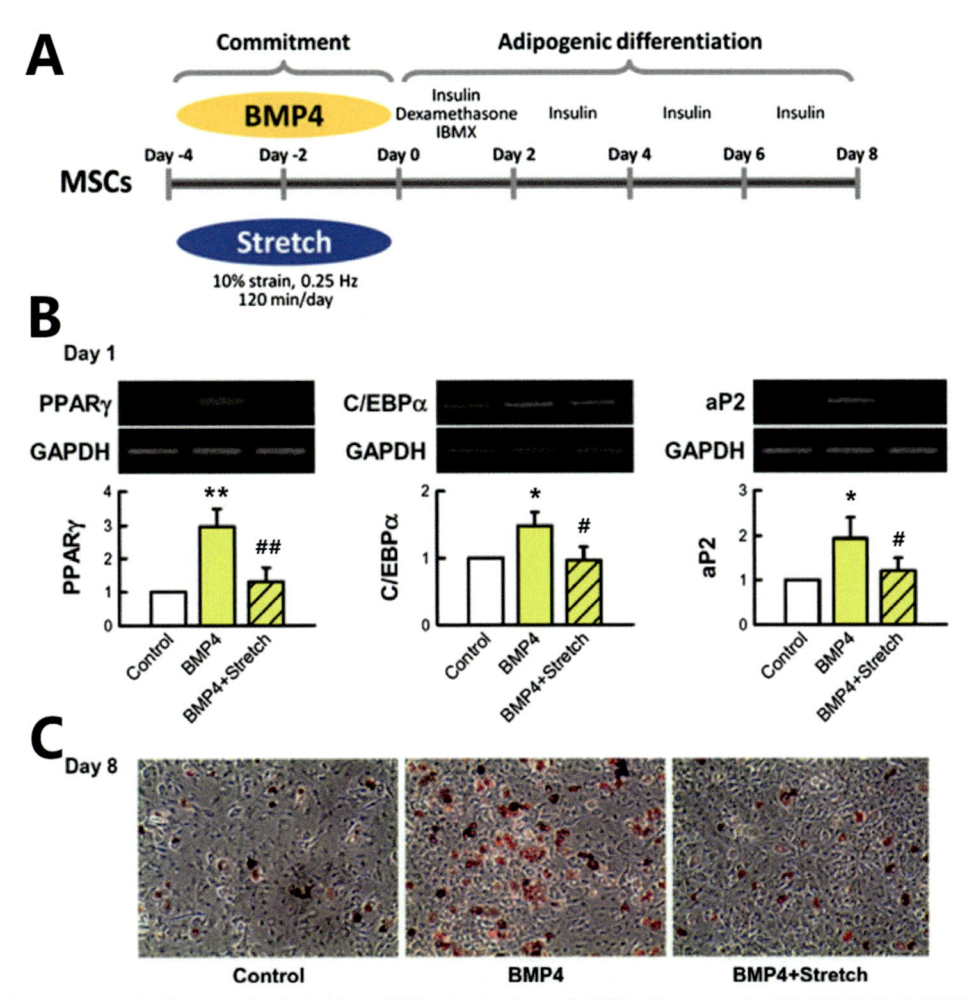

Figure 11.6 Cyclic stretch diminishes BMP-4 induction of MSC adipogenesis. (A) C3H10T1/2 MSCs were cultured for 4 days prior to the induction of adipogenesis for 8 days. Cyclic stretch (10% strain, 0.25 Hz, 120 min/day) was applied in conjunction with BMP-4 during the commitment stage. (B) Stretch reversed the observed BMP-4-induced increase in adipogenic gene expressions, measured using RT-PCR. *: $P < .05$, **: $P < .01$ compared with control; #: $P < .05$, ##: $P < .01$ compared with BMP-4. (C) Oil red O staining indicated an increase in lipid synthesis with BMP-4 treatment, reversed with the application of stretch. *Ap2*, adipocyte protein 2; *BMP-4*, bone morphogenic protein-4; *C/EBPα*, CCAAT/enhancer-binding protein α; *PPARγ*, peroxisome proliferator–activated receptor gamma. *(Reprinted with permission from Lee JS, Ha L, Park JH, Lim JY. Mechanical stretch suppresses BMP4 induction of stem cell adipogenesis via upregulating ERK but not through downregulating Smad or p38. Biochem Biophys Res Commun. 2012;418;278–283.)*

the cells were subjected to 22 min bouts of either low-intensity vibration or high-magnitude strain with at least 1 h break between bouts [90]. Interestingly, when the refractory period was extended to 3 h, adipogenesis was suppressed even more. Therefore, it can be concluded that cyclic stretch suppression of MSC adipogenesis also depends on the duration of any incorporated refractory periods between loads.

As described above, research consistently establishes cyclic mechanical stretch as a positive instigator for MSC osteogenesis while inhibiting adipogenesis. On the other hand, studies utilizing noncyclic (static) stretching, i.e., elongation to a certain strain and keeping the strain for prolonged time, reported that static stretch may provide an opposite effect [91]. It was shown that static tensile strains at 6%, 9%, and 12% applied to 3T3-L1 preadipocytes produced more adipogenesis relative to cultures subjected to 3% strain or nonstretched control, indicating that relatively large levels of static strain seem to enhance adipogenesis of preadipocytes [92]. In another study by the same group, 12% static stretching applied to 3T3-L1 preadipocytes enhanced adipogenesis via the MEK/MAPK pathway (Fig. 11.7) [93]. When an MEK/MAPK inhibitor (PD98059) was used, the observed increase in adipogenesis by static stretch was notably reduced. These suggest that more systematic studies are required to thoroughly compare the mechanical stretch mediation of differentiation in MSCs and 3T3-L1 preadipocytes, along with the other loading parameters (strain, duration, refractory period, etc.).

Fluid shear

Similar to bone, mechanical forces applied to adipose tissue can provide compression and fluid shear. The existing literature, however, remains lacking about the compression control of MSC adipogenesis. For fluid flows, some studies did investigate the fluid shear effect on MSC adipogenesis. One study looked at the effects of oscillatory fluid flow on MSC osteogenesis, chondrogenesis, and adipogenesis by assessing the expression of Runx2, Sox9, and PPARγ, respectively [70]. In the absence of fluid flow, a treatment with RhoA activator LPA, which increases cytoskeletal tension, upregulated Runx2 expression but downregulated PPARγ. Meanwhile, treatment with cytochalasin, an actin disruptor, increased the expression of PPARγ and the chondrogenic marker Sox9. Interestingly, fluid flow—induced shear upregulated all three transcription factors (Runx2, Sox9, and PPARγ). Together, it was proposed that in static culture, RhoA-mediated cytoskeletal tension is a negative regulator of MSC adipogenesis, whereas an intact actin cytoskeletal under tension is required for flow-induced signaling and the resultant PPARγ and Sox9 expression.

Yes-associated proteins (YAP), an important player in stem cell fate decision, has also been revealed to be activated in MSCs stimulated by fluid shear. YAP is a transcriptional cofactor that, with its paralogue PDZ-binding motif (TAZ), shuttles between the nucleus and cytoplasm and binds to transcription factors, regulating their activity [94]. In a study using a multishear microfluidic device, MSCs exposed to fluid flows displayed shear

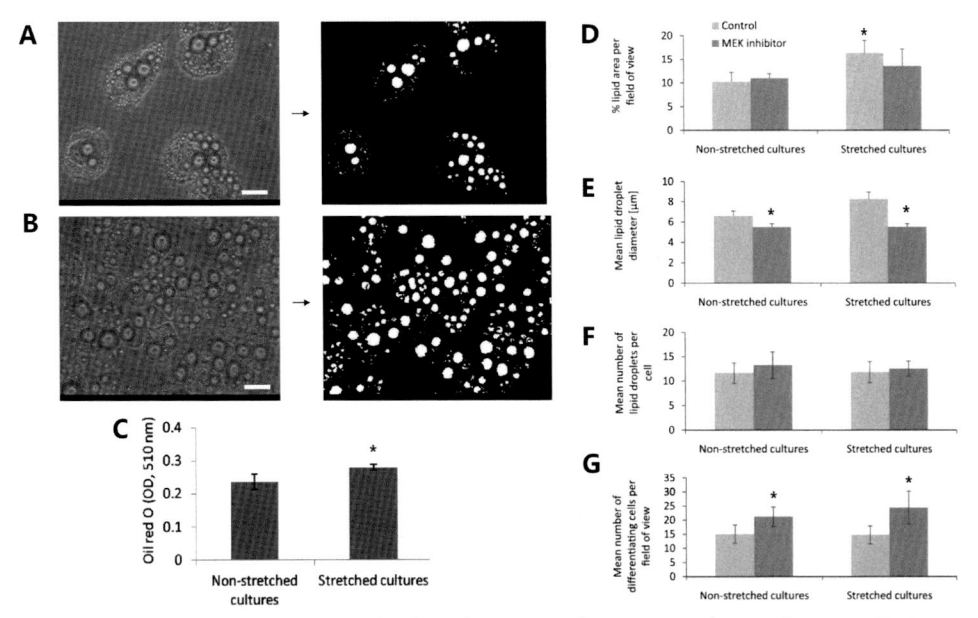

Figure 11.7 Static stretch accelerates lipid production via the MEK signaling pathway in 3T3-L1 preadipocytes. Oil red O staining of nonstretched (A) and stretched (12% static strain) (B) cultures, 17 days postinduction of differentiation. The original micrograph (left) was converted into a black/white image (right). (C) Lipid accumulation at day 17 was measured with spectrophotometry. Scale bar = 30 μm. Effects of MEK inhibitor (PD98059) on lipid production were evaluated to show (D) percentage lipid area per FOV, (E) mean LD diameter, (F) mean number of LDs per cell, and (E) mean number of differentiated cells per FOV. FOV, field of view; LD, lipid droplet; OD, optical density. *: $P < .05$ compared with nonstretched cultures (C) and with nonstretched cultures without MEK inhibitor (D–G). *(Reprinted with permission from Shoham N, Gottlieb R, Sharabani-Yosef O, Zaretsky U, Benayahu D, Gefen A. Static mechanical stretching accelerates lipid production in 3T3-L1 adipocytes by activating the MEK signaling pathway. Am J Physiol Cell Physiol. 2012;302;C429–C441.)*

stress—mediated regulation of YAP (Fig. 11.8) [95]. Increasing the magnitude of fluid shear stress led to an increase in YAP, and subsequently, decreased adipogenesis, increased osteogenesis, and initiated the dedifferentiation of chondrocytes. The results suggest that, as with mechanical stretch, the effect of fluid shear on MSC adipogenesis appears to be magnitude dependent as well. Additionally, from a study using 3T3-L1 preadipocytes, it was proposed that the effect of fluid shear on adipogenesis may also depend on what stage of differentiation the stimulus was applied [96].

Another study with a microfluidic system reported contrary data on the fluid shear control of MSC adipogenesis [97]. hMSCs were maintained in suspension on a microfluidic platform, and terminal differentiation was assessed by adipogenic markers (SSP1, CFL1, and LPL1) and lipid accumulation. The microfluidic device applied flow rates comparable to the accelerated rates found in the arterial system (15 dyne/cm^2),

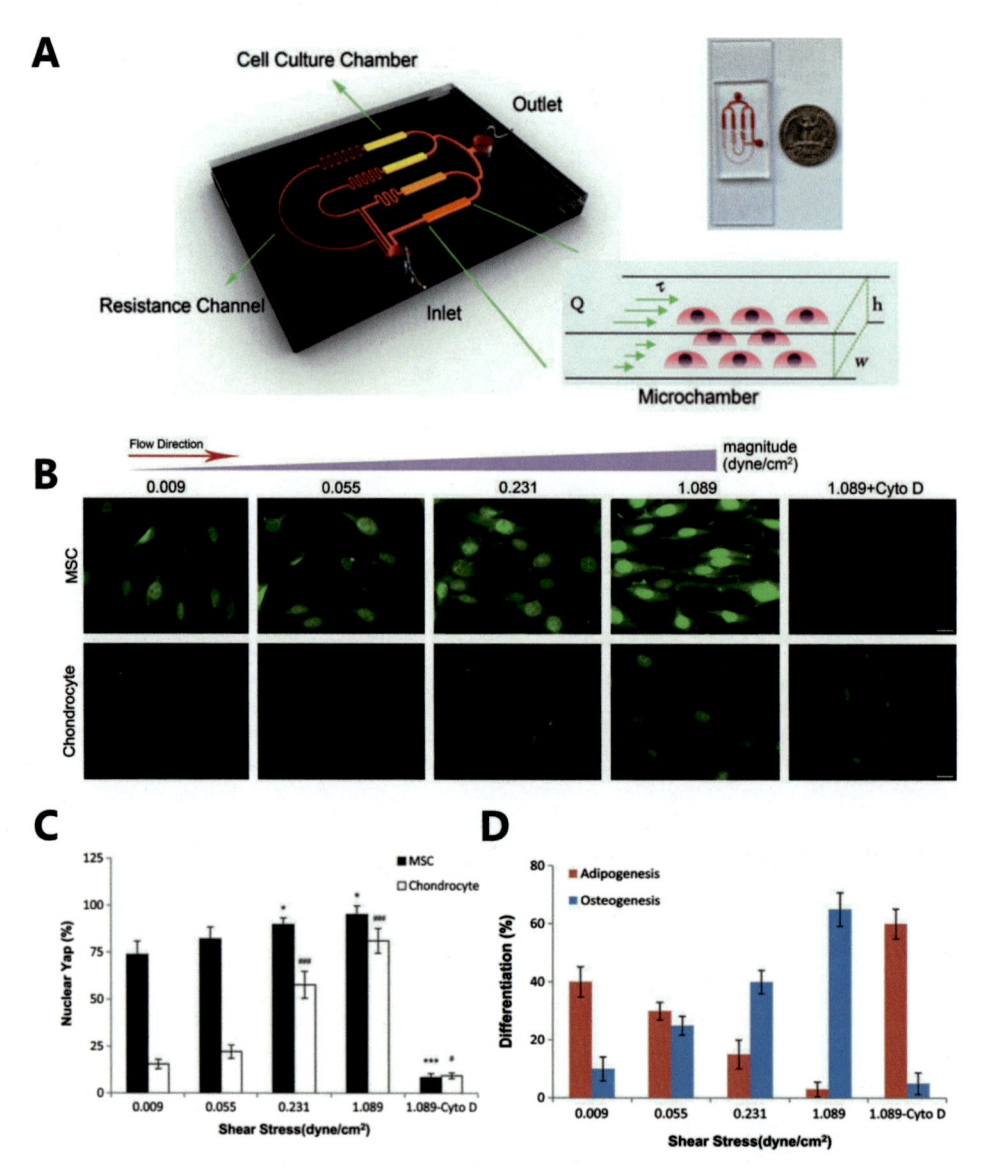

Figure 11.8 Fluid shear may regulate MSC fate via controlling YAP. (A) A schematic diagram of the multishear microfluidic device (left) and a photograph image of a prototype (right). (B) MSCs and chondrocytes were immunostained for YAP (green) and exposed to increasing levels of fluid shear strain. (C) The percentage of cells with nuclear YAP for each condition was quantified. (D) Percentage of osteogenesis and adipogenesis of MSCs after 5 days of exposure to a mixed induction medium. Q, flow rate in the microchamber; τ, shear stress; h, height of the microchamber; w, width of the microchamber; Cyto D, cytochalasin D. Scale bar = 20 μm. * and #: $P < .05$; *** and ###: $P < .0001$ compared with 0.009 dyn/cm^2. *(Reprinted with permission from Zhong W, Tian K, Zheng X, Li L, Zhang W, Wang S, Qin J. Mesenchymal stem cell and chondrocyte fates in a multishear microdevice are regulated by Yes-associated protein. Stem Cells Dev. 2013;22:2083−2093.)*

and hMSCs exhibited sustained viability and visible lipid accumulation. Interestingly, shear stimulation induced hMSCs to differentiate into brown adipose tissue once placed into adipogenic culture media. Combined with a significant increase in the expression of adipogenic markers, this study supports that fluid shear may help accelerate the adipogenesis of hMSCs in culture.

Unfortunately, microfluidic systems have limitations. The two-dimensionality and small-scale nature of microfluidic devices sometimes create a physiologically inaccurate model, and thus may not enable effective clinical translation. Researchers began implementing innovative cell culture systems, such as 3D perfusion-based systems, to better mimic in vivo environments by allowing controlled nutrient transport and waste clearance while also controlling the level of applied shear stress. However, there remain challenges when designing perfusion systems. For instance, adipocytes can easily be washed away by fluid flow. The issue may be resolved with a recently developed double-layered perfusion device, in which adipocytes are cultured on the bottom layer and the adipogenic culture media flows in a chamber above perfusing through a porous membrane into the bottom layer [98]. This device may allow for not only the natural exchange of nutrients but also the maintenance of a constant level of mild shear stress on the adipocytes. Interstitial fluid shear stress in vivo, which is caused by fluid flowing through the complex mesh of the ECM, is extremely low (less than 0.01 Pa). Shear stresses applied in the perfusion system used in this study were calculated to be 2.5×10^{-4} dyne/cm^2 (at 5 nL/s) and 7.5×10^{-4} dyne/cm^2 (at 15 nL/s). While studies using physiologically excessive levels of shear stress (1 Pa) showed that fluid flow inhibited adipocyte maturation [96], this perfusion system showed that low levels of applied shear stress can lead to accelerated adipogenesis. Looking at just these data, it can be proposed that low levels of shear stress may have a different impact on MSC adipogenesis compared to higher levels. However, the different types of fluid flow may also impact nutrient exchange, so more research is needed to closely investigate the direct effect of fluid shear as a single factor in MSC adipogenesis.

Perspective

This chapter introduced two factors, hyperglycemia and mechanical loading, involved in the control of MSC fate decision and terminal differentiation focusing on osteogenesis and adipogenesis that are especially relevant for diabetic and obese patients. Considering the data reported for the two parameters, it would be pertinent to model MSC mechanobiological interactions in an obese and hyperglycemic environment in culture. However, there is almost none that investigates MSC fate decision under both HG and mechanical loading. Recently, Lustig et al. [99] exposed 3T3-L1 preadipocytes to both physiologically high level of glucose (450 mg/dL) and sustained tensile stretch (12% static strain) and analyzed adipogenesis and lipid production over the course of

19 days. They observed that HG and static tensile strain accelerated the lipid production by 1.7- and 1.4-fold, respectively. Based on the result, it will be an interesting question whether or not combining HG and tensile stretch has a synergistic effect on lipid production and adipogenesis. The obtained representative micrographs do appear to show somewhat of a synergistic effect with larger lipids found in HG-treated stretched cells (Fig. 11.9) [99]. Since it is early to make a final conclusion with minimal amount of reported data, more research will be required on the combination of HG and mechanical stimulation on MSC adipogenesis (and osteogenesis). Other forms of mechanical loading (cyclic stretch, fluid shear, compression, etc.) and various physiological parameters of mechanical loading should also be considered. Furthermore, the relationship between HG, mechanotransduction, and inflammatory signaling pathways needs to be further elucidated. For instance, the mechanism study may focus on the PI3K/AKT/GSK-3β/β-catenin pathways as they appear to overlap between HG treatment and mechanotransduction and possibly dictate the delicate balance between MSC osteogenesis and adipogenesis (as described in previous sections). Moreover, in addition to relatively well-established MSC mechanosensation through focal adhesion (e.g., FAK) and cytoskeletal tension (e.g., ROCK) signaling, recent recognition of MSC mechanotransduction through the Linker of Nucleoskeleton and Cytoskeleton (LINC) complex [100] may be investigated in MSC fate decision to osteogenesis versus adipogenesis without or with hyperglycemia condition.

In conclusion, hyperglycemia and mechanical loading are powerful regulators of MSC fate commitment into the osteogenic and adipogenic phenotypes. By controlling glucose and mechanical parameters, fate decision between both lineages can be controlled. Literatures support the supposition that HG enhances MSC adipogenesis, while inhibiting osteogenesis. A somewhat similar inverse relationship between the two commitment pathways is observed in the mechanical stretch control of MSC differentiation, i.e., mechanical induction of osteogenesis at the expense of adipogenesis. Further, the addition of soluble factors may alter MSC sensitivity to the applied biochemical or mechanical cue. For instance, the addition of BMP-2 rescued MSC osteogenic suppression caused by HG; BMP-9 augmented mechanical stretch induction of osteogenesis; BMP-4 enhanced cyclic stretch inhibition of adipogenesis. The potential coeffect that simultaneous mechanical and soluble stimulation (especially glucose) has on MSC fate decision remains an open area to be further explored. Moreover, while researchers have assembled an impressive body of knowledge regarding MSC fate regulation via soluble and biophysical cues, there remains strong needs for systematic molecular signaling mechanism studies and the utilization of cell-scale model systems that more accurately mimic the diseased physical and biochemical microenvironments.

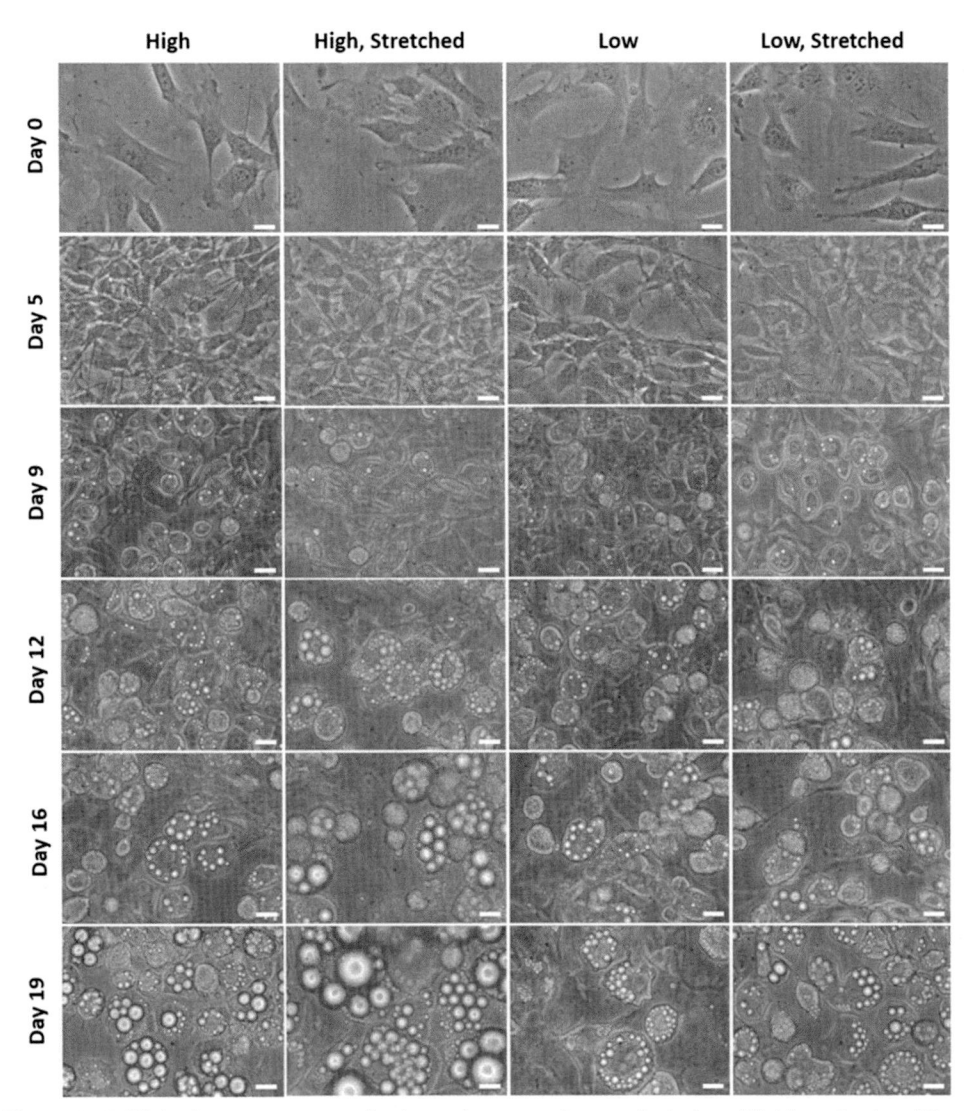

Figure 11.9 High glucose appears to further enhance static stretch—induced lipid production. Micrographs of 3T3-L1 cell cultures at days 0, 5, 9, 12, 16, and 19 postinduction of differentiation. Scale bar = 20 μm. *(Reprinted with permission from Lustig M, Gefen A, Benayahu D. Adipogenesis and lipid production in adipocytes subjected to sustained tensile deformations and elevated glucose concentration: a living cell-scale model system of diabesity. Biomech Model Mechanobiol. 2018;17;903—913.)*

Acknowledgments

We thank the support from NSF GRFP 1610400 given to Tasneem Bouzid; NIH/NIGMS COBRE NPOD (P20GM104320, PI: Zempleni), NIH/NIGMS GP IDeA-CTR (1U54GM115458-01, PI: Rizzo), AHA Scientist Development Grant 17SDG33680170 (PI: Duan), NSF Grant 1826135 (PI: Yang), and University of Nebraska Collaboration Initiative grants (PI: Lim; PI: Yang) all given to Jung Yul Lim.

References

[1] Gopalakrishnan V, Vignesh RC, Arunakaran J, Aruldhas MM, Srinivasan N. Effects of glucose and its modulation by insulin and estradiol on BMSC differentiation into osteoblastic lineages. Biochem Cell Biol 2006;84:93—101.

[2] Wang J, Wang B, Li Y, Wang D, Lingling E, Bai Y, Liu H. High glucose inhibits osteogenic differentiation through the BMP signaling pathway in bone mesenchymal stem cells in mice. EXCLI J 2013;12:584—97.

[3] Stolzing A, Coleman N, Scutt A. Glucose-induced replicative senescence in mesenchymal stem cells. Rejuvenation Res 2006;9:31—5.

[4] Stolzing A, Sellers D, Llewelyn O, Scutt A. Diabetes induced changes in rat mesenchymal stem cells. Cell Tissue Organ 2010;19:453—65.

[5] Stolzing A, Bauer E, Scutt A. Suspension cultures of bone-marrow-derived mesenchymal stem cells: effects of donor age and glucose level. Stem Cell Dev 2012;21:2718—23.

[6] Chang TC, Hsu MF, Wu KK. High glucose induces bone marrow-derived mesenchymal stem cell senescence by upregulating autophagy. PloS One 2015;10. e0126537.

[7] Al-Qarakhli AM, Yusop N, Waddington RJ, Moseley R. Effects of high glucose conditions on the expansion and differentiation capabilities of mesenchymal stromal cells derived from rat endosteal niche. BMC Mol Cell Biol 2019;20:1—8.

[8] Cramer C, Freisinger E, Jones RK, Slakey DP, Dupin CL, Newsome ER, Alt EU, Izadpanah R. Persistent high glucose concentrations alter the regenerative potential of mesenchymal stem cells. Stem Cell Dev 2010;19:1875—84.

[9] Krakauer JC, McKenna MJ, Buderer NF, Rao DS, Whitehouse FW, Parfitt AM. Bone loss and bone turnover in diabetes. Diabetes 1995;44:775—82.

[10] De Molon RS, Morais-Camilo JA, Verzola MH, Faeda RS, Pepato MT, Marcantonio Jr E. Impact of diabetes mellitus and metabolic control on bone healing around osseointegrated implants: removal torque and histomorphometric analysis in rats. Clin Oral Implants Res 2013;24:831—7.

[11] Jiao H, Xiao E, Graves DT. Diabetes and its effect on bone and fracture healing. Curr Osteoporos Rep 2015;13:327—35.

[12] Li WT, Hu WK, Ho FM. High glucose induced bone loss via attenuating the proliferation and osteoblastogenesis and enhancing adipogenesis of bone marrow mesenchymal stem cells. Biomed Eng-App Bas C. 2013;25:1340010.

[13] Strotmeyer ES, Cauley JA, Schwartz AV, Nevitt MC, Resnick HE, Zmuda JM, Bauer DC, Tylavsky FA, de Rekeneire N, Harris TB, Newman AB. Diabetes is associated independently of body composition with BMD and bone volume in older white and black men and women: the Health, Aging, and Body Composition Study. J Bone Miner Res 2004;19:1084—91.

[14] Bonds DE, Larson JC, Schwartz AV, Strotmeyer ES, Robbins J, Rodriguez BL, Johnson KC, Margolis KL. Risk of fracture in women with type 2 diabetes: the women's health initiative observational study. J Clin Endocrinol Metab 2006;91:3404—10.

[15] Kayath MJ, Dib SA, Vieira JH. Prevalence and magnitude of osteopenia associated with insulin-dependent diabetes mellitus. J Diabetes Complicat 1994;8:97—104.

[16] Vestergaard P. Discrepancies in bone mineral density and fracture risk in patients with type 1 and type 2 diabetes—a meta-analysis. Osteoporos Int 2007;18:427—44.

[17] Hie M, Tsukamoto I. Increased expression of the receptor for activation of NF-κB and decreased runt-related transcription factor 2 expression in bone of rats with streptozotocin-induced diabetes. Int J Mol Med 2010;26:611—8.
[18] Cao B, Liu N, Wang W. High glucose prevents osteogenic differentiation of mesenchymal stem cells via lncRNA AK028326/CXCL13 pathway. Biomed Pharmacother 2016;84:544—51.
[19] Deng X, Xu M, Shen M, Cheng J. Effects of type 2 diabetic serum on proliferation and osteogenic differentiation of mesenchymal stem cells. J Diabetes Res 2018;2018:5765478.
[20] Hanna A, Frangogiannis NG. The role of the TGF-beta superfamily in myocardial infarction. Fron Cardiovasc Med 2019;6:140.
[21] Li X, Cao XU. BMP signaling and skeletogenesis. Ann Ny Acad Sci 2006;1068:26—40.
[22] Mukherjee A, Rotwein P. Akt promotes BMP2-mediated osteoblast differentiation and bone development. J Cell Sci 2009;122:716—26.
[23] Chen Y, Hu Y, Yang L, Zhou J, Tang Y, Zheng L, Qin P. Runx2 alleviates high glucose-suppressed osteogenic differentiation via PI3K/AKT/GSK3β/β-catenin pathway. Cell Biol Int 2017;41:822—32.
[24] Jakus V, Bauerova K, Michalkova D, Carsky J. Serum levels of advanced glycation end products in poorly metabolically controlled children with diabetes mellitus: relation to HbA1c. Diabetes Nutr Metab 2001;14:207—11.
[25] Vlassara H. Recent progress in advanced glycation end products and diabetic complications. Diabetes 1997;46:S19—25.
[26] Kume S, Kato S, Yamagishi SI, Inagaki Y, Ueda S, Arima N, Okawa T, Kojiro M, Nagata K. Advanced glycation end-products attenuate human mesenchymal stem cells and prevent cognate differentiation into adipose tissue, cartilage, and bone. J Bone Miner Res 2005;20:1647—58.
[27] Zhen D, Chen Y, Tang X. Metformin reverses the deleterious effects of high glucose on osteoblast function. J Diabetes Complicat 2010;24:334—44.
[28] Cortizo AM, Sedlinsky C, McCarthy AD, Blanco A, Schurman L. Osteogenic actions of the antidiabetic drug metformin on osteoblasts in culture. Eur J Pharmacol 2006;536:38—46.
[29] Molinuevo MS, Schurman L, McCarthy AD, Cortizo AM, Tolosa MJ, Gangoiti MV, Arnol V, Sedlinsky C. Effect of metformin on bone marrow progenitor cell differentiation: in vivo and in vitro studies. J Bone Miner Res 2010;25:211—21.
[30] Wang P, Ma T, Guo D, Hu K, Shu Y, Xu HH, Schneider A. Metformin induces osteoblastic differentiation of human induced pluripotent stem cell-derived mesenchymal stem cells. J Tissue Eng Regen Med 2018;12:437—46.
[31] Gao Y, Xue J, Li X, Jia Y, Hu J. Metformin regulates osteoblast and adipocyte differentiation of rat mesenchymal stem cells. J Pharm Pharmacol 2008;60:1695—700.
[32] Marycz K, Tomaszewski KA, Kornicka K, Henry BM, Wroński S, Tarasiuk J, Maredziak M. Metformin decreases reactive oxygen species, enhances osteogenic properties of adipose-derived multipotent mesenchymal stem cells in vitro, and increases bone density in vivo. Oxid Med Cell Longev 2016;2016:9785890.
[33] Gu Q, Gu Y, Yang H, Shi Q. Metformin enhances osteogenesis and suppresses adipogenesis of human chorionic villous mesenchymal stem cells. Tohoku J Exp Med 2017;241:13—9.
[34] Keats E, Khan ZA. Unique responses of stem cell-derived vascular endothelial and mesenchymal cells to high levels of glucose. PloS One 2012;7. e38752.
[35] Chuang CC, Yang RS, Tsai KS, Ho FM, Liu SH. Hyperglycemia enhances adipogenic induction of lipid accumulation: involvement of extracellular signal-regulated protein kinase 1/2, phosphoinositide 3-kinase/Akt, and peroxisome proliferator-activated receptor γ signaling. Endocrinology 2007;148:4267—75.
[36] Ali AA, Weinstein RS, Stewart SA, Parfitt AM, Manolagas SC, Jilka RL. Rosiglitazone causes bone loss in mice by suppressing osteoblast differentiation and bone formation. Endocrinology 2005;146:1226—35.
[37] Aguiari P, Leo S, Zavan B, Vindigni V, Rimessi A, Bianchi K, Franzin C, Cortivo R, Rossato M, Vettor R, Abatangelo G. High glucose induces adipogenic differentiation of muscle-derived stem cells. Proc Natl Acad Sci Unit States Am 2008;105:1226—31.

[38] Morganti C, Missiroli S, Lebiedzinska-Arciszewska M, Ferroni L, Morganti L, Perrone M, Ramaccini D, Occhionorelli S, Zavan B, Wieckowski MR, Giorgi C. Regulation of PKCβ levels and autophagy by PML is essential for high-glucose-dependent mesenchymal stem cell adipogenesis. Int J Obes 2019;43:963—73.

[39] Qi MC, Hu J, Zou SJ, Chen HQ, Zhou HX, Han LC. Mechanical strain induces osteogenic differentiation: cbfa1 and Ets-1 expression in stretched rat mesenchymal stem cells. Int J Oral Maxillofac Surg 2008;37:453—8.

[40] Virjula S, Zhao F, Leivo J, Vanhatupa S, Kreutzer J, Vaughan TJ, Honkala AM, Viehrig M, Mullen CA, Kallio P, McNamara LM. The effect of equiaxial stretching on the osteogenic differentiation and mechanical properties of human adipose stem cells. J Mech Behav Biomed 2017;72: 38—48.

[41] Andalib MN, Lee JS, Ha L, Dzenis Y, Lim JY. Focal adhesion kinase regulation in stem cell alignment and spreading on nanofibers. Biochem Biophys Res Commun 2016;473. 920-5.

[42] Andalib MN, Lee JS, Ha L, Dzenis Y, Lim JY. The role of RhoA kinase (ROCK) in cell alignment on nanofibers. Acta Biomater 2013;9. 7737-45.

[43] Kearney EM, Farrell E, Prendergast PJ, Campbell VA. Tensile strain as a regulator of mesenchymal stem cell osteogenesis. Ann Biomed Eng 2010;38:1767—79.

[44] Song Y, Tang Y, Song J, Lei M, Liang P, Fu T, Su X, Zhou P, Yang L, Huang E. Cyclic mechanical stretch enhances BMP9-induced osteogenic differentiation of mesenchymal stem cells. Int Orthop 2018;42:947—55.

[45] Haasper C, Jagodzinski M, Drescher M, Meller R, Wehmeier M, Krettek C, Hesse E. Cyclic strain induces FosB and initiates osteogenic differentiation of mesenchymal cells. Exp Toxicol Pathol 2008; 59:355—63.

[46] Song G, Ju Y, Shen X, Luo Q, Shi Y, Qin J. Mechanical stretch promotes proliferation of rat bone marrow mesenchymal stem cells. Colloid Surf. B 2007;58:271—7.

[47] Pardo PS, Mohamed JS, Lopez MA, Boriek AM. Induction of Sirt1 by mechanical stretch of skeletal muscle through the early response factor EGR1 triggers an antioxidative response. J Biol Chem 2011; 286:2559—66.

[48] Chen X, Yan J, He F, Zhong D, Yang H, Pei M, Luo ZP. Mechanical stretch induces antioxidant responses and osteogenic differentiation in human mesenchymal stem cells through activation of the AMPK-SIRT1 signaling pathway. Free Radic Biol Med 2018;126:187—201.

[49] Ho PJ, Yen ML, Tang BC, Chen CT, Yen BL. H_2O_2 accumulation mediates differentiation capacity alteration, but not proliferative decline, in senescent human fetal mesenchymal stem cells. Antioxidants Redox Signal 2013;18:1895—905.

[50] Koike M, Shimokawa H, Kanno Z, Ohya K, Soma K. Effects of mechanical strain on proliferation and differentiation of bone marrow stromal cell line ST2. J Bone Miner Metabol 2005;23:219—25.

[51] Shi Y, Li H, Zhang X, Fu Y, Huang Y, Lui PP, Tang T, Dai K. Continuous cyclic mechanical tension inhibited Runx2 expression in mesenchymal stem cells through RhoA-ERK1/2 pathway. J Cell Physiol 2011;226:2159—69.

[52] Cai X, Zhang Y, Yang X, Grottkau BE, Lin Y. Uniaxial cyclic tensile stretch inhibits osteogenic and odontogenic differentiation of human dental pulp stem cells. J Tissue Eng Regen M 2011;5:347—53.

[53] Nagatomi J, Arulanandam BP, Metzger DW, Meunier A, Bizios R. Frequency-and duration-dependent effects of cyclic pressure on select bone cell functions. Tissue Eng 2001;7:717—28.

[54] Nagatomi J, Arulanandam BP, Metzger DW, Meunier A, Bizios R. Effects of cyclic pressure on bone marrow cell cultures. J Biomech Eng 2002;124. 308-14.

[55] Nagatomi J, Arulanandam BP, Metzger DW, Meunier A, Bizios R. Cyclic pressure affects osteoblast functions pertinent to osteogenesis. Ann Biomed Eng 2003;31:917—23.

[56] Park SH, Sim WY, Min BH, Yang SS, Khademhosseini A, Kaplan DL. Chip-based comparison of the osteogenesis of human bone marrow-and adipose tissue-derived mesenchymal stem cells under mechanical stimulation. PloS One 2012;7. e46689.

[57] Zhang M, Chen FM, Wang AH, Chen YJ, Lv X, Wu S, Zhao RN. Estrogen and its receptor enhance mechanobiological effects in compressed bone mesenchymal stem cells. Cell Tissue Organ 2012;195:400—13.

[58] Haudenschild AK, Hsieh AH, Kapila S, Lotz JC. Pressure and distortion regulate human mesenchymal stem cell gene expression. Ann Biomed Eng 2009;37:492−502.
[59] David V, Martin A, Lafage-Proust MH, Malaval L, Peyroche S, Jones DB, Vico L, Guignandon A. Mechanical loading down-regulates peroxisome proliferator-activated receptor γ in bone marrow stromal cells and favors osteoblastogenesis at the expense of adipogenesis. Endocrinology 2007; 148:2553−62.
[60] Schätti O, Grad S, Goldhahn J, Salzmann G, Li Z, Alini M, Stoddart MJ. A combination of shear and dynamic compression leads to mechanically induced chondrogenesis of human mesenchymal stem cells. Eur Cell Mater 2011;22:b97.
[61] Meyer EG, Buckley CT, Thorpe SD, Kelly DJ. Low oxygen tension is a more potent promoter of chondrogenic differentiation than dynamic compression. J Biomech 2010;43:2516−23.
[62] Riehl BD, Lim JY. Macro and microfluidic flows for skeletal regenerative medicine. Cells 2012;1. 1225-45.
[63] Salvi JD, Lim JY, Donahue HJ. Finite element analyses of fluid flow conditions in cell culture. Tissue Eng C Method 2010;16. 661-70.
[64] Riehl BD, Lee JS, Ha L, Kwon IK, Lim JY. Flowtaxis of osteoblast migration under fluid shear and the effect of RhoA kinase silencing. PloS One 2017;12. e0171857.
[65] Riehl BD, Lee JS, Ha L, Lim JY. Fluid-flow-induced mesenchymal stem cell migration: role of focal adhesion kinase and RhoA kinase sensors. J R Soc Interface 2015;12:20141351.
[66] Yourek G, McCormick SM, Mao JJ, Reilly GC. Shear stress induces osteogenic differentiation of human mesenchymal stem cells. Regen Med 2010;5:713−24.
[67] Jacobs CR, Yellowley CE, Davis BR, Zhou Z, Cimbala JM, Donahue HJ. Differential effect of steady versus oscillating flow on bone cells. J Biomech 1998;31. 969-76.
[68] Salvi JD, Lim JY, Donahue HJ. Increased mechanosensitivity of cells cultured on nanotopographies. J Biomech 2010;43. 3058-62.
[69] Young SR, Gerard-O'Riley R, Kim JB, Pavalko FM. Focal adhesion kinase is important for fluid shear stress-induced mechanotransduction in osteoblasts. J Bone Miner Res 2009;24:411−24.
[70] Arnsdorf EJ, Tummala P, Kwon RY, Jacobs CR. Mechanically induced osteogenic differentiation− the role of RhoA, ROCKII and cytoskeletal dynamics. J Cell Sci 2009;122:546−53.
[71] Zhang H, Kay A, Forsyth NR, Liu KK, El Haj AJ. Gene expression of single human mesenchymal stem cell in response to fluid shear. J Tissue Eng 2012;3:1.
[72] Liu L, Shao L, Li B, Zong C, Li J, Zheng Q, Tong X, Gao C, Wang J. Extracellular signal-regulated kinase1/2 activated by fluid shear stress promotes osteogenic differentiation of human bone marrow-derived mesenchymal stem cells through novel signaling pathways. Int J Biochem Cell Biol 2011;43: 591−601.
[73] Glossop JR, Cartmell SH. Effect of fluid flow-induced shear stress on human mesenchymal stem cells: differential gene expression of IL1B and MAP3K8 in MAPK signaling. Gene Expr Patterns 2009;9: 381−8.
[74] Li YJ, Batra NN, You L, Meier SC, Coe IA, Yellowley CE, Jacobs CR. Oscillatory fluid flow affects human marrow stromal cell proliferation and differentiation. J Orthop Res 2004;22:1283−9.
[75] Sharp LA, Lee YW, Goldstein AS. Effect of low-frequency pulsatile flow on expression of osteoblastic genes by bone marrow stromal cells. Ann Biomed Eng 2009;37:445−53.
[76] Shoham N, Gefen A. Mechanotransduction in adipocytes. J Biomech 2012;45:1−8.
[77] Linder-Ganz E, Shabshin N, Itzchak Y, Gefen A. Assessment of mechanical conditions in sub-dermal tissues during sitting: a combined experimental-MRI and finite element approach. J Biomech 2007; 40:1443−54.
[78] Akune T, Ohba S, Kamekura S, Yamaguchi M, Chung UI, Kubota N, Terauchi Y, Harada Y, Azuma Y, Nakamura K, Kadowaki T. PPAR γ insufficiency enhances osteogenesis through osteoblast formation from bone marrow progenitors. J Clin Invest 2004;113:846−55.
[79] Yang X, Cai X, Wang J, Tang H, Yuan Q, Gong P, Lin Y. Mechanical stretch inhibits adipogenesis and stimulates osteogenesis of adipose stem cells. Cell Prolif 2012;45:158−66.
[80] Takada I, Kouzmenko AP, Kato S. Wnt and PPARγ signaling in osteoblastogenesis and adipogenesis. Nat Rev Rheumatol 2009;5:442.

[81] Baron R, Kneissel M. WNT signaling in bone homeostasis and disease: from human mutations to treatments. Nat Med 2013;19:179—92.

[82] Sen B, Xie Z, Case N, Ma M, Rubin C, Rubin J. Mechanical strain inhibits adipogenesis in mesenchymal stem cells by stimulating a durable β-catenin signal. Endocrinology 2008;149:6065—75.

[83] Song F, Jiang D, Wang T, Wang Y, Lou Y, Zhang Y, Ma H, Kang Y. Mechanical stress regulates osteogenesis and adipogenesis of rat mesenchymal stem cells through PI3K/Akt/GSK-3β/β-catenin signaling pathway. BioMed Res Int 2017;2017:6027402.

[84] Lee JS, Ha L, Park JH, Lim JY. Mechanical stretch suppresses BMP4 induction of stem cell adipogenesis via upregulating ERK but not through downregulating Smad or p38. Biochem Biophys Res Commun 2012;418:278—83.

[85] Huang H, Song TJ, Li X, Hu L, He Q, Liu M, Lane MD, Tang QQ. BMP signaling pathway is required for commitment of C3H10T1/2 pluripotent stem cells to the adipocyte lineage. Proc Natl Acad Sci Unit States Am 2009;106:12670—5.

[86] Lee JS, Ha L, Kwon IK, Lim JY. The role of focal adhesion kinase in BMP4 induction of mesenchymal stem cell adipogenesis. Biochem Biophys Res Commun 2013;435:696—701.

[87] Li R, Liang L, Dou Y, Huang Z, Mo H, Wang Y, Yu B. Mechanical stretch inhibits mesenchymal stem cell adipogenic differentiation through TGFβ1/Smad2 signaling. J Biomech 2015;48:3665—71.

[88] Huang SC, Wu TC, Yu HC, Chen MR, Liu CM, Chiang WS, Lin KM. Mechanical strain modulates age-related changes in the proliferation and differentiation of mouse adipose-derived stromal cells. BMC Cell Biol 2010;11:18.

[89] Diederichs S, Böhm S, Peterbauer A, Kasper C, Scheper T, Van Griensven M. Application of different strain regimes in two-dimensional and three-dimensional adipose tissue—derived stem cell cultures induces osteogenesis: implications for bone tissue engineering. J Biomed Mater Res 2010;94:927—36.

[90] Sen B, Xie Z, Case N, Styner M, Rubin CT, Rubin J. Mechanical signal influence on mesenchymal stem cell fate is enhanced by incorporation of refractory periods into the loading regimen. J Biomech 2011;44:593—9.

[91] Hara Y, Wakino S, Tanabe Y, Saito M, Tokuyama H, Washida N, Tatematsu S, Yoshioka K, Homma K, Hasegawa K, Minakuchi H. Rho and Rho-kinase activity in adipocytes contributes to a vicious cycle in obesity that may involve mechanical stretch. Sci Signal 2011;4:ra3.

[92] Levy A, Enzer S, Shoham N, Zaretsky U, Gefen A. Large, but not small sustained tensile strains stimulate adipogenesis in culture. Ann Biomed Eng 2012;40. 1052-60.

[93] Shoham N, Gottlieb R, Sharabani-Yosef O, Zaretsky U, Benayahu D, Gefen A. Static mechanical stretching accelerates lipid production in 3T3-L1 adipocytes by activating the MEK signaling pathway. Am J Physiol Cell Physiol 2012;302:C429—41.

[94] Zhao B, Tumaneng K, Guan KL. The Hippo pathway in organ size control, tissue regeneration and stem cell self-renewal. Nat Cell Biol 2011;13:877—83.

[95] Zhong W, Tian K, Zheng X, Li L, Zhang W, Wang S, Qin J. Mesenchymal stem cell and chondrocyte fates in a multishear microdevice are regulated by Yes-associated protein. Stem Cell Dev 2013; 22:2083—93.

[96] Choi J, Lee SY, Yoo YM, Kim CH. Maturation of adipocytes is suppressed by fluid shear stress. Cell Biochem Biophys 2017;75:87—94.

[97] Adeniran-Catlett AE, Weinstock LD, Bozal FK, Beguin E, Caraballo AT, Murthy SK. Accelerated adipogenic differentiation of hMSC s in a microfluidic shear stimulation platform. Biotechnol Prog 2016;32:440—6.

[98] Liu Y, Kongsuphol P, Gourikutty SB, Ramadan Q. Human adipocyte differentiation and characterization in a perfusion-based cell culture device. Biomed Microdevices 2017;19:8.

[99] Lustig M, Gefen A, Benayahu D. Adipogenesis and lipid production in adipocytes subjected to sustained tensile deformations and elevated glucose concentration: a living cell-scale model system of diabesity. Biomech Model Mechanobiol 2018;17:903—13.

[100] Bouzid T, Kim E, Riehl BD, Esfahani AM, Rosenbohm J, Yang R, Duan B, Lim JY. The LINC complex, mechanotransduction, and mesenchymal stem cell function and fate. J Biol Eng 2019; 13:68.

CHAPTER 12

Clinical complications of tendon tissue mechanics due to collagen cross-linking in diabetes

Jennifer A. Zellers[1], Jeremy D. Eekhoff[2], Simon Y. Tang[1], Mary K. Hastings[1], Spencer P. Lake[2]

[1]Washington University School of Medicine in St. Louis, St. Louis, MO, United States; [2]Washington University in St. Louis, St. Louis, MO, United States

Clinical problem

Diabetes mellitus (DM) affects 422 million adults worldwide, which is ~8.5% of the world's population [1]. DM is the result of endocrine dysfunction and chronic hyperglycemia and is associated with impairment of the vascular, immune, neurological, and musculoskeletal systems. There are a host of effects on the musculoskeletal system, including complications specific to tendon tissue that lead to tendon-limited joint range of motion and three times higher rates of tendon injury [2] relative to individuals without DM. Accumulation of advanced glycation end products (AGEs) within the tendon, which often accompanies DM, is thought to be a contributor to joint range of motion and higher tendon injury rates associated with DM. Since collagen turnover is slower in tendon than other collagenous tissues [3], tendons are thus particularly susceptible to nonenzymatic cross-linking and the resulting accumulation of AGEs. This mechanism has been thought to contribute to the clinical observations of shortening and stiffening in tendon tissue that limits joint excursion [4–6] and ultimately contributes to increased plantar pressure [6–9], ulceration [7,10–12], and amputation [13].

Elevated AGEs in the tendons of individuals with DM could also alter the tendon's ability to maintain homeostasis and heal from injury because of effects on collagen structure, collagen–collagen binding, and activation of inflammatory pathways [15]. DM impairs homeostasis and has been associated with higher rates of asymptomatic tendon changes, including impaired collagen organization [16–18] and calcific changes in the enthesis of the Achilles tendon [19], and sonographic abnormalities in rotator cuff tendons [20]. In the context of tendon injury, data from animal models indicate impaired tendon healing in tendons from diabetic animals, with altered tendon biochemistry and poor recovery of mechanical properties [21–23]. Concerns for inadequate healing are multiplied when considering the risk of infection and other complications with surgical intervention. Future investigation into the mechanisms of tendon homeostatic dysregulation could identify more specific treatment targets.

Additional study of the response to tendon-specific treatments in people with DM is warranted to allow clinicians to make more informed, patient-centered treatment decisions, recovery timelines, and prognoses.

Tendons from individuals with DM have been suggested to have altered morphology [2,24], mechanics [14,25–30], composition [14], and gliding capacity [26]. Appreciating that DM has implications for tendon healing and homeostasis without a clear understanding of underlying mechanisms makes for a clinical challenge. There may be alterations in tendon homeostasis due to a host of mechanical and biochemical factors. This chapter will provide an overview of many of these proposed mechanisms in an effort to assimilate what is known about DM-associated tendon changes encompassing information within the literature and viewpoints from multiple disciplines.

Clinical syndromes of tendon dysfunction in individuals with DM

Tendons transfer muscle-generated forces to bone; however, the specific functional requirements vary greatly by tendon size, orientation, and anatomical location. This variability in function is matched with differences in tendon composition, structure, and surrounding structures. To highlight the impact of DM on the variety of clinical presentations of tendon disorders, we discuss three pathologies of particular relevance to individuals with DM: trigger finger, rotator cuff pathology, and Achilles tendon dysfunction.

Trigger finger

The finger flexors of the hand are critical for the fine motor tasks of daily living. These tendons are long, originating in the muscle bellies of the proximal forearm, traveling within the carpal tunnel and finger pulley systems, and inserting into the distal phalanges. The finger flexors are encased in a synovial sheath that aids their ability to efficiently glide within adjacent tissues.

Trigger finger (Fig. 12.1) is a condition in which dysfunction of the long flexor tendons in the hand limit the ability to extend a digit. This condition is more prevalent in individuals with DM [31], with up to 20% prevalence, compared to 2% or less in the general population [32,33]. Trigger finger also occurs bilaterally and impacts multiple digits more frequently in individuals with compared to those without DM [32,34]. This condition is thought to be a precursor to cheiroarthropathy ("stiff hand syndrome") in the hand and is also associated with carpal tunnel syndrome [32]. From a pathophysiology standpoint, trigger finger is characterized by tendon changes consistent with tendinosis (i.e., fusiform tendon thickening), causing the tendon to increase in size and limit movement through the palmar pulley system, specifically the proximal or A1 pulley [33]. Additionally, it has been suggested that the accumulation of AGEs limits

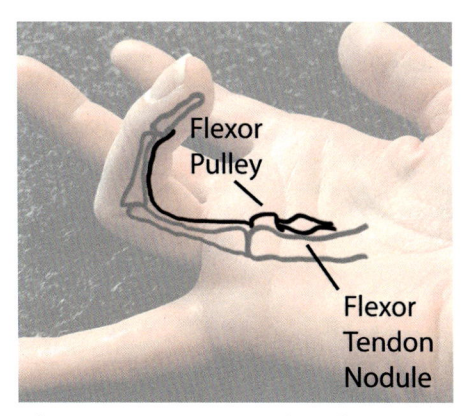

Figure 12.1 Trigger finger. When attempting to extend the finger, the nodule of the affected flexor tendon catches in the fibrous pulley, limiting range of motion.

gliding between the tendon and its synovial sheath, thereby restricting movement of the digit [32]. It may be that alterations in tendon homeostasis and healing from microinjury associated with overuse in the presence of DM drive the presence of trigger finger.

Treatment for trigger finger is based on severity. More mild cases are typically treated with a limited course of corticosteroid injections, whereas severe cases are treated surgically with the release of the A1 pulley [32]. Short-term response to treatment appears to be fairly similar in individuals with DM; however, individuals with DM are more likely to have a recurrence of trigger finger in the long-term compared to those without DM [32,35]. There are also some additional risks associated with treatment in diabetic individuals. Corticosteroid injection is associated with transient hyperglycemia (up to a few days in duration), though this risk likely does not outweigh the benefit associated with treatment [32]. There is a high rate of recurrence following corticosteroid injection that appears more pronounced in individuals with DM [36], so moving more quickly toward surgery has been proposed as a more cost-effective approach [37,38]. Surgery comes with its own risks, however, which may be of greater concern in individuals with DM [39].

Rotator cuff pathology

The rotator cuff plays a critical role in the biomechanics and stability of the glenohumeral joint [40]. Atraumatic rotator cuff pathology, including tearing, tends to occur with advancing age [41]. Pathology typically develops first in the supraspinatus at the superior aspect of the shoulder and may progress to other tendons [41]. Unlike rupture in a tendon like the Achilles, rotator cuff tears can lack an acute, inciting trauma [41]. The chronicity of this problem likely contributes to the tendency for the tendon ends to retract, making repair challenging with high retear rates relative to other tendon repairs [42].

Individuals with DM are at greater risk of rotator cuff injury compared to the general population [43] [hazard ratio of 2.11 (95% CI = 2.02−2.20) [44]]. DM has been associated with calcific changes in the rotator cuff tendons [2] as well as increased risk of requiring rotator cuff surgical repair [45]. Additionally, healing from rotator cuff repair is inferior in individuals with DM compared to those without [46−48]. Individuals with DM have been found to have inferior functional outcomes [47], particularly in those individuals who receive hemodialysis [46]. Individuals with DM and poor glycemic control are also at increased risk of poor healing and retearing [49,50].

Achilles dysfunction

The Achilles tendon is a power-generating tendon with substantial contributions to physical tasks that include walking, stair climbing, running, and jumping. The Achilles is covered in a paratenon rather than a synovial sheath [51], and does not have the space constraints of the finger flexors or rotator cuff. Also, unlike the rotator cuff, Achilles tendon pain does not typically precede tendon rupture. In fact, the majority of individuals with a history of Achilles rupture report no tendon symptoms prior to rupture [52,53]. The Achilles can become adherent to surrounding tissues, but this has primarily been described with recovery from rupture [54,55].

In the context of DM, the Achilles tendon has been thought to become short and stiff leading to abnormal gait mechanics and plantar pressures increasing the risk of foot ulceration [4−6,8,9]. Achilles tendon lengthening procedures are used when more conservative measures are unsuccessful in healing plantar foot ulcers. Achilles tendon lengthening restores ankle joint dorsiflexion mobility in the short term [12]; however, this comes at the expense of ankle power [7] and quality of life [56]. While largely successful in healing the initial wound, these procedures do carry risks associated with surgery as well as the risk of developing ulceration in other locations due to changes in gait mechanics associated with gastrocsoleus insufficiency [12].

Advanced glycation end products in collagenous tissues
Introduction to advanced glycation end products

AGEs are the heterogeneous by-product of the Maillard reaction [57−60]. In this reaction, glycation of a protein occurs as an amino group binds with a reducing sugar such as glucose, forming a Schiff base. The Schiff base is relatively unstable and undergoes rearrangement to form a more stable Amadori product [60]. Over time, metabolic by-products are formed which can interact with proteins to form additional AGEs [58]. In a collagen-rich extracellular matrix, AGEs can form a cross-link between two collagen fibrils, affecting fiber mechanics, thermal stability, enzymatic degradation, and collagen packing [57,60,61]. Adducts can also form, in which AGEs bind to a single collagen molecule affecting the charge profile of the molecule and cell—collagen interaction [57,60,61]. Cross-links between molecules can last the lifetime of the protein to which it is bound. Due to the low turnover of collagen particularly in the core of the tendon compared to other tissues such as skeletal muscle [3], AGEs accumulation is believed to be high in tendon. AGEs accumulation is of particular concern in the tendons of people with DM given the chronic hyperglycemic environment and high levels of endogenous AGEs production.

Structural consequences of advanced glycation end products on collagen

Glycation of collagen is associated with alterations in fibril diameter [14,26] (Table 12.1), stiffness [62], and periodicity [59]. Fibril diameter has been reported to decrease [14,26] and stiffness to increase [62] with glycation. It appears, though, that increases in collagen fibril stiffness are attenuated at higher levels of tendon architecture [62]. In addition to changes in collagen fibril structure, alterations in collagen fibril alignment can occur with increased glycation. Collagen fibrils are aligned with a characteristic banding (i.e., D-period), and glycation has been associated with decreased D-period in collagen fibrils at rest, impacting tensile behavior [59]. Gautieri et al. [59] report this decrease in D-period in length associated with glycation and advancing age in methylglyoxal-treated rat tail tendon and in cadaveric human tendon. With strain, this study found that glycated fibrils deformed more than control tendons (80% compared to ~60% deformation), though this has yet to be replicated in human tendon specimens, perhaps due to heterogeneity in a limited number of specimens [59]. These results suggest that bonding between collagen and the absence of adequate sliding mechanics [59,62,63] may result in a larger contribution of deformation of the collagen to overall tendon deformation [59].

Signaling role of advanced glycation end products

The impact of AGEs is not only limited to direct binding of AGEs to collagen but also to its signaling role. AGEs are associated with several signaling cascades that promote a proinflammatory environment; however, the cascade associated with AGEs binding to their receptor (RAGE) is best described in the literature [58]. RAGE is expressed in a variety of cell types including macrophages, endothelial cells, and smooth muscle cells, and expression is upregulated in the context of diabetes [61]. RAGE can bind to a variety of ligands, resulting in the activation of multiple signaling pathways. Combined signaling from RAGE and TLR2/4 can further magnify the inflammatory response [64]. RAGE binding stimulates the NF-κB pathway, producing reactive oxygen species and cytokines that form a positive feedback loop that promotes a chronic, proinflammatory environment [61,64,65].

Table 12.1 Studies reporting collagen fibrillar and organizational characteristics, in vivo models.

Author	Model of diabetes	n	Collagen fibril size	Collagen organization	Other/notes
Achilles tendon					
Batista 2008 [16]	Human, type unspecified	DM = 70 Control = 10		↓	Assessed via ultrasound.
Abate 2012 [62]	Human, T2DM	DM = 136 Controls = 273		[a]	Higher rates of asymptomatic abnormalities and changes at the enthesis on ultrasound in individuals with DM compared to controls.
De Jonge 2015 [17]	Human, T2DM	DM, T2 = 24 DM, T1 = 24 Controls = 44		↓[b]	Assessed via ultrasound tissue characterization.
Guney 2015 [28] Volper 2015 [63]	Human, T2DM Rat, streptozotocin-induced T1DM	DM = 21 Control = 21 n = 40 divided among DM (acute), DM (chronic), DM + insulin, control groups		↓ =	Assessed via histology. Assessed via histology.
Wong 2015 [64]	Human, T1DM	DM = 7 Control = 10		=	No statistically significant differences in tendon structure on ultrasound tissue characterization.
Couppé 2016 [14]	Human, T1 and T2 DM	DM = 40 Control = 10	↓[c]		Collagen density ↑ Assessed via transmission electron microscopy.
Flexor digitorum longus tendon					
Studentsova 2018 [26]	C57Bl/6J mice, high fat diet-induced T2DM	DM = 3 Control = 3	↓	=	Lipid deposits in midsubstance of tendon. Assessed via transmission electron microscopy.

Upward arrow indicates value higher in DM group, downward arrow indicates value lower in DM group, and equal sign indicates no group differences relative to controls.
[a] Higher rates of abnormalities noted in the midportion observed in controls compared to individuals with DM.
[b] Between–group differences noted between individuals with T2DM and their matched controls; nonstatistically significant trend observed in individuals with T1DM and their matched controls.
[c] Noted nonstatistically significant trend toward smaller collagen fibril diameter in individuals with DM compared to controls.

Clinical aspects of advanced glycation end products in tendon

There is currently no established way of directly quantifying AGEs in vivo in tendon tissue. AGEs can be quantified in vivo using blood tests as well as with skin intrinsic fluorescence [66]. It is unclear if there is a relationship between AGEs concentrations present in tendons and AGEs present in other tissues [14]. This inability to directly measure AGEs accumulation in skeletal tissues in vivo highlights the need for surrogate measures that will allow the in vivo characterization of tendon tissue and the functional impact due to AGEs accumulation. Quantitative magnetic resonance imaging (MRI) has been used to characterize other nontendinous, collagenous tissues. It has been observed that AGEs accumulation results in loss of tissue hydration, which can be assessed using quantitative MRI [67]. MRI-based evaluation of tendon poses additional challenges due to the lower signal-to-noise ratio; however, with advances in spin sequences, the Achilles and patellar tendons are beginning to be imaged using diffusion tensor MRI [68–70]. In the future, diffusion tensor imaging may allow for quantitative assessment of collagen organization in tendon at the subfascicle level. Advanced ultrasound imaging (e.g., shear wave elastography) has been used to assess DM-related changes in tendon tissue quality in vivo [29]. Given the resolution of ultrasound imaging, these changes likely reflect differences at the fascicle level.

Another area of interest has been in the development of pharmaceutical interventions to inhibit the formation of AGEs or promote the breakdown of AGEs. There have been a number of proposed interventions that have had some success in vitro and preclinical models, including the AGE breakers, aminoguanidine and alagebrium, and AGE inhibitors such as vitamin B [71,72]. However, the translation of these interventions into the clinic has been limited due to concerns regarding toxicity and efficacy, respectively [71,72]. In addition to the development of new interventions, there has been investigation into the effect of existing therapeutics on AGEs, including aldose reductase inhibitors, medications to treat hyperglycemia, and medications to lower blood pressure and lipids [71]. Discerning the direct effect on AGEs accumulation rather than indirect effects on AGEs from improved glycemic or lipid control is challenging, and investigations into the direct effect on AGEs are ongoing [71]. Investigation into therapeutics has been in the context of addressing vascular and kidney disease, so the effect of therapeutics aimed at inhibiting or breaking AGEs in tissues with slow collagen turnover, like tendons, is not known.

Collagen sliding and homeostasis
Microstructural mechanics of tendon

The aligned and hierarchical structure of tendon is crucial for interactions between and within the collagenous substructures at varying length scales. Beginning at larger length scales, collagen fascicles are the largest substructure of tendons. The matrix between fascicles, or the interfascicular matrix (IFM), is rich in elastic fibers and lubricin and contains a distinct cell population from that which is located within fascicles [73–76]. The composition of the IFM seems to be uniquely suited to regulate structural interactions between fascicles: While an abundance of lubricin creates a low gliding resistance environment to allow fascicles to slide along one another, the meshlike network of elastic fibers may link adjacent fascicles and prevent excessive fascicular motion in order to retain the tendon structure and provide some resistance to nonaxial loading [74,77–80]. Fascicles are, in turn, composed of collagen fibers. Experiments using photobleached lines or grids onto regions of tendon have clearly demonstrated that fibers slide past one another when tendons are strained, consequently decreasing the strain experienced by individual fibers [81–83]. Likewise, rotation of collagen fibers may similarly attenuate the strain on individual collagen fibers [82,84]. The next smallest substructure is the collagen fibril. Interactions within fibrils are largely dominated by enzymatic cross-linking between collagen molecules within the fibril [85]. Still, by measuring fibril strain within tendon using small-angle X-ray scattering (SAXS) further strain attenuation has been demonstrated at this length scale due to sliding between individual collagen molecules within fibrils (Fig. 12.2) [86,87].

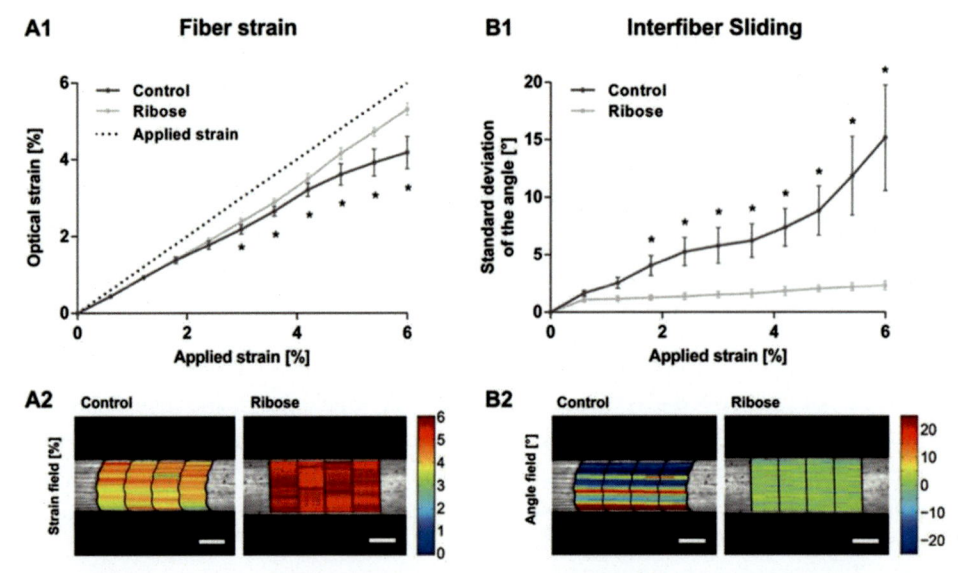

Figure 12.2 Contributions of fiber strain and interfiber sliding to tendon kinematics and the effect of ribose-induced glycation. (A1) Fiber strain and (B1) interfiber sliding between strain increments were determined from the displacements of the photobleached lines on the extracellular matrix ($n = 8$). (A2) Representative strain and (B2) angle fields measured at 6% applied strain. Photobleached lines of the nonglycated fascicle display a typical waviness, representing interfiber sliding. In sharp contrast, the photobleached lines of the ribose-treated (glycated) fascicle remain nearly straight throughout the entire stretching experiment (*$P < .05$; scale bars = 100 m) *(Used with permission from Gautieri A, Passini FS, Silván U, et al. Advanced glycation end-products: mechanics of aged collagen from molecule to tissue. Matrix Biol. 2017;59:95–108. https://doi.org/10.1016/j.matbio.2016.09.001).*

Generally, each tendon substructure experiences slightly less strain than the next largest substructure. This strain attenuation at each length scale of tendon is posited to reduce the accumulation of microdamage while also increasing the maximum strain which a tendon can be subjected to before failure. Presumably, any inhibition of strain attenuation would increase the mechanical damage that would form and accumulate at smaller length scales given an equivalent load applied to the whole tendon.

Impact of AGEs on tendon microstructural interactions

Research on the effects of AGEs on the microstructural mechanics of tendon has focused on using agents such as glucose, ribose, and methylglyoxal to induce glycation in harvested tendons. Most notably, AGEs were shown to severely inhibit collagen sliding at the fiber and fibril length scales [59,62,63]. Even sliding between collagen molecules was similarly diminished [88]. As one consequence of decreased collagen sliding, there was less strain attenuation in glycated tendon and therefore greater strain in individual fibers and fibrils compared to strains for the full tendon [59]. As stated above, this may lead to an increase in microdamage at these smaller length scales during normal tendon loading. Moreover, because cellular mechanotransduction in tendon is thought to operate in part by sensing shearing between collagen fibers, another possible consequence of diminished collagen sliding from glycation is a decrease in the cellular response to mechanical load [89]. This loss of mechanical signaling could disrupt the homeostasis where the cellular activity to produce and repair extracellular matrix proteins meets the demands created by accumulated microdamage from repeated loading [57].

The precise mechanism by which glycation inhibits collagen sliding is not fully understood. At smaller length scales, specific AGEs such as glucosepane and pentosidine form covalent cross-links both within and between collagen molecules [85,90–92]. These cross-links add excessive stability to the molecular network within fibrils, thus preventing motion between collagen molecules within that network [85,88]. However, the mechanisms at larger length scales are less clear. The short, subnanometer length of the individual cross-links make it unlikely that covalent bonds are formed between fibers, but there is still diminished collagen sliding between fibers with accumulation of AGEs. Instead, it has been proposed that glycation on the surface of collagen fibers changes the interactions between fibers, and interfiber sliding is consequently inhibited through those changes [93,94]. No research to date has directly evaluated the effect of glycation on the IFM, although there may be accumulation of AGEs in IFM proteins such as elastin which could alter collagen sliding or other structural interactions between fascicles. Further work that focuses on clarifying the effects and mechanisms of AGEs at different length scales in tendon as well as continued research identifying how increased glycation interacts with other complications of DM will improve our understanding of the microstructural mechanics under diabetic conditions.

Noncollagenous proteins in the tendon in diabetic conditions

The implications of DM and collagen mechanics has been the most studied DM-related tendon complication. The role of noncollagenous tendon components has only recently been described in the literature and is primarily characterized in the absence of disease; very little is known about the effect of DM on these other tissue constituents. The impact of hyperglycemia and DM on elastin and proteoglycans has been characterized more extensively in other collagenous tissues, which may be informative to investigating tendon tissue in the future.

Elastin

Elastin is a hydrophobic, fibrillar protein that comprises between 1% and 4% of the dry mass of tendon [80,95]. Elastin has low stiffness but is highly extensible, demonstrates nearly perfect spring mechanics with regard to hysteresis, and withstands extremely repetitive loads [96,97]. Elastin is found in vascular tissues, lungs, skin, ligaments, and tendon due to its resilience to tensile loading. In tendons, elastin contributes to the mechanical behavior of the tissue [98] and has been proposed to potentially support the mechanical environment for tenocytes [99]. Tendons from elastin haploinsufficient mice have been found to have a 38% reduction in elastin content accompanied with a 14% increase in linear stiffness compared to tendons from wild-type mice [98]. Depletion of elastin with elastase in rat Achilles has been associated with tendon changes similar to those observed with tendinosis, including collagen disorganization, tendon thickening, and hypercellularity, though it is unclear whether the changes in cellular response were mediated through loss of elastin versus directly from elastase treatment [100]. Depletion of elastin in human supraspinatus tendon with elastase similarly demonstrated decreased resistance to shear loading [80].

Diabetes, vascular disease, and aging have all been associated with higher amounts of glycation of elastin [101,102]. While there are similarities between tendon behavior reported in DM and elastin depletion, whether or not elastin depletion is a mechanism for aberrant tendon behavior in individuals with DM is unknown. AGEs, as well as the binding of AGE with its receptor (RAGE), have been found to increase tissue stiffness in lung [103] and vascular [104,105] tissues. AGEs accumulation in tendon tissue could influence elastin content and function in the context of DM via several potential mechanisms, given that adequate hydration is required to maintain the mechanical behavior of elastin [95], elastase occurs with inflammatory processes [100,106], and elastin synthesis decreases with age [107].

Proteoglycans

Proteoglycans contribute up to 3.5% of the wet weight of tendon [108]. Proteoglycans consist of a central protein core with glycosaminoglycan (highly sulfated, polysaccharide,

GAG) side chains. The role of proteoglycans in tendon tissue depends on the tendon, region, and anatomical location. GAG depletion in suprasinatus [109], patellar [110], and rat tail tendons [111] has been reported to have minimal to no effect on tendon shear, compressive, and tensile behavior, respectively. In the Achilles tendon, however, the presence of the GAG decorin has been associated with tendon changes improving elasticity and ultimate strength [112–114]. GAG content is particularly concentrated at the distal third of the Achilles tendon, near the insertion, with diminished ultimate load and elastic modulus in this area in particular with GAG depletion [112]. Regional differences in proteoglycan content has also been reported in the supraspinatus tendon, specifically in the distribution of aggrecan and biglycan content [115]. In addition to tendon mechanics, proteoglycans have been suggested to play a role in fibril structure [114], fiber alignment [113,116], cell bonding [116], and prevention of calcification [117].

Proteoglycans appear to be negatively impacted by both a hyperglycemic environment [108] and accumulation of AGEs [116]. Burner et al. found proteoglycan content (i.e., biglycan, decorin, fibromodulin, lumican, versican) to decrease in tendons incubated in a hyperglycemic relative to a normoglycemic environment [108]. This study further identified diminished sulfation levels, which alter the tissue's affinity for water and could impact tissue material properties and potentially alter tissue stiffness [108]. Taken together, the loss of GAG content, which may be most substantial in the insertional portion of the tendon, could potentially contribute to the high frequency of insertional Achilles pathology in individuals with DM [19]. However, this mechanism has not been investigated to the authors' knowledge.

Tendon fibroblasts

There is growing evidence that DM has not only a mechanical but also a biochemical effect on tendon fibroblasts. Independent of a mechanical effect of AGEs, accumulation of AGEs has been reported to affect cell proliferation [118], adenosine triphosphate (ATP) production [118], ossification [119], and viability [119,120]. The effects of AGEs accumulation in tendon are complex and multifaceted, extending beyond the proposed mechanical effects that have been described.

Tendon mechanics

Diabetes is associated with altered tendon mechanics. Clinical treatment tends to be based on the theoretical framework that tendon stiffness increases with DM due to the accumulation of advanced glycation end products. Despite this clinical mindset, experimental data are conflicting as to whether or not Achilles tendon stiffness increases with DM (Table 12.2). The lack of consistent findings may be largely due to differences in the model of DM, methodology, and data interpretation.

Table 12.2 Reported impact of diabetes on tendon mechanics.

Author	Model of diabetes	n	CSA/ Volume	Length	Stiffness	Young's modulus	Maximum load at failure	Other/notes
Achilles tendon								
Reddy 2003 [124]	Rabbit tendon incubated in ribose	Glycated = 6 Control = 6				↑	↑	Maximum load ↑ Maximum stress ↑ Maximum strain ↑ Energy absorption ↑ Toughness ↑
De Oliveira 2012 [125]	Rat, streptozotocin-induced T1DM	DM = 7 DM + intervention = 11 Control = 11 Control + intervention = 10 (treadmill intervention)	↓ / = [b]			↓		Maximum strain ↑ Energy/tendon area ↑ Displayed is the nonintervention group comparison unless otherwise noted.
Cheing 2013 [24]	Human, T2DM	DM = 23 DM + neuropathy = 9 Control = 32	↑			=		Mechanics assessed with ultrasound indentation system.
Connizzo 2014 [27]	db/db mouse	DM = 8 Control = 10	↓		↓	↓		Transition strain ↓ Differences only observed at the insertion; no differences in any tendon at the midsubstance.

Study	Model	Sample sizes						Notes
Evranos 2015 [29]	Human, T2DM	DM + foot ulcer = 35; DM = 43; Control = 33	↑[d]					Mechanics assessed with ultrasound shear wave elastography.
Guney 2015 [28]	Human, T2DM	DM = 21; Control = 21			↓	↓		Toughness ↓; Maximum load ↓; Energy at break point ↓; Study reports additional outcomes not listed here.
Volper 2015 [63]	Rat, streptozotocin-induced T1DM	8–12 per group (DM-acute, DM-chronic, DM + insulin, control)		=				
Couppé 2016 [14]	Human, T1 and T2 DM	DM = 34–41; Control = 9	=		=	=	= / ↑[e]	Mechanics assessed with dynamometry and ultrasound.
Silva 2017 [126]	Rat, streptozotocin-induced T1DM	DM = 15; DM + intervention = 15; Control = 15; Control + intervention = 14 (Resistive jumping intervention)	↓			↓	↑	Displayed is the nonintervention group comparison unless otherwise noted.
Petrovic 2018 [30]	Human, T1 and T2 DM	DM = 20; DM + neuropathy = 13; Control = 23			↑			Hysteresis ↑; Mechanics assessed with ultrasound during walking.
Svensson 2018 [127]	Rat, MGO					↑		Fibril mechanics. Failure strain ↑; Failure stress ↑; Failure energy ↑

Continued

Table 12.2 Reported impact of diabetes on tendon mechanics.—cont'd

Author	Model of diabetes	n	CSA/Volume	Length	Stiffness	Young's modulus	Maximum load at failure	Other/notes
Flexor digitorum longus tendon								
Studentsova 2018 [26]	C57Bl/6J mice, high fat diet–induced T2DM	Groups: High fat diet/High-to-low fat diet/Low fat diet N = 6–10 per diet per time point			=/↑[c]		=/↓[c]	
Patellar tendon								
Lancaster 1994 [25]	Canine (naturally occurring T1DM)	DM = 7 Control = 27	=		↑/=[a]	=	=	Creep =
Connizzo 2014 [27]	db/db mouse	DM = 8 Control = 10	↓	↓	↓	=		Transition strain ↓
Supraspinatus tendon								
Connizzo 2014 [27]	db/db mouse	DM = 8 Control = 10	↓		↓	=		Transition strain ↓
Thomas 2014 [128]	Rat, streptozotocin-induced T1DM	DM = 8 Control = 10	=		=	=		
Tail tendon								
Andreassen 1981 [129]	Rat, streptozotocin-induced T1DM	DM(10 days) = 8 DM(30 days) = 8 DM + insulin = 8 Control = 17			↑		↑	
Li 2013 [66]	Rat, MGO	MGO = 6 Control = 6						Tangential moduli =

Study	Model	Sample sizes					Notes
Fessel 2014 [65]	Rat, MGO	MGO (6 h incubation) = 9 MGO (24 h incubation) = 5 MGO (96 h incubation) = 5 Control = 10			↑		Ultimate stress ↑ Ultimate strain = Yield stress ↑ Yield strain ↑ Energy to yield ↑ Toughness = Reported results are for fascicle-level mechanics. Study also reports fiber mechanics.
Gonzalez 2014 [130]	Zucker diabetic Sprague–Dawley rats	DM = 4 Control = 5		=	↑		Ultimate tensile strength ↑ Tissue relaxation ↓ Reported results are for fascicle-level mechanics. Study also reports fibril mechanics.
Volper 2015 [63]	Rat, streptozotocin-induced T1DM	8–12 per group (DM-acute, DM-chronic, DM + insulin, control)			=		Ultimate stress = Testing performed to individual fascicles.

Continued

Table 12.2 Reported impact of diabetes on tendon mechanics.—cont'd

Author	Model of diabetes	n	CSA/Volume	Length	Stiffness	Young's modulus	Maximum load at failure	Other/notes
Gautieri 2017 [59]	Rat, ribose incubated	Treated = 8, Untreated = 8				=		Ultimate stress ↑, Failure strain ↓, Stress relaxation ↓. Reported results are for fascicle-level mechanics. Study also reports fibril and fiber mechanics.
Svensson 2018 [127]	Rat, MGO					↑		Fascicle mechanics. Failure strain ↑, Failure stress ↑, Failure energy ↑

Upward arrow indicates value higher in DM group, downward arrow indicates value lower in DM group, and equal sign indicates no group differences relative to controls. To be included, studies needed to compare a DM group with a non-DM control group. Human studies reporting only CSA or volume were not included as these are largely reported in the cited systematic review. MGO, Methylglyoxal-treated tendon.

[a] Differences between groups were not statistically significant when correcting for age.
[b] Statistically significant lower CSA in treadmill DM group but not in sedentary DM group.
[c] Differences were statistically significant and some, but not all, time points.
[d] Differences were statistically significant between DM + ulcer group and DM/control groups. Between-group differences in stiffness observed at middle and distal locations, no differences between groups at proximal tendon.
[e] Statistically significant increase in DM group at common force, no between group differences at maximum force.

For example, human studies have demonstrated that the Achilles tendon thickens in humans with DM, and this assertion has been supported by systematic review-level evidence [2]. Increased tendon thickness has led to speculation that tendon stiffness increases in humans. It is important to appreciate that alterations in tendon geometry (cross-sectional area) will only increase stiffness of the tissue if Young's modulus remains unchanged. If, for example, there is fatty infiltration of the tissue, the cross-sectional area could increase but the modulus could simultaneously decrease. A simultaneous increase in cross-sectional area and decrease in modulus would not necessarily result in a net increase in tendon stiffness. The concern becomes that tendon material properties were not assessed in the majority of these studies, so the assertion that tendon stiffness increases is based on tissue geometry alone, without accounting for alterations in tissue quality.

Other human studies have reported increases in Achilles tendon stiffness with DM using real-time ultrasound during walking or with muscle contraction to estimate tendon stiffness [14,30]. These studies are very informative in understanding tendon movement as part of the musculotendinous unit but are limited by a lack of assessment of isolated tendon function. Because DM is associated with fibrosis, fatty infiltration, and weakening of the muscle, it is challenging to make assertions about the tendon when performance is linked so closely to muscle function. Two studies have investigated tendon mechanics in humans using techniques that do not require muscle contraction. Evranos et al. compared Achilles tendon stiffness assessed with shear wave elastography in individuals with DM and history of ulceration, DM without history of ulceration, and a nondiabetic control group [29]. This study found that individuals with DM and a history of ulceration had lower stiffness in their Achilles tendons. Similarly, Guney et al. directly measured tendon stiffness in cadaveric tendon from donors with and without DM and reported a 21% reduction in Achilles tendon stiffness in diabetic compared to nondiabetic tendon [28].

Several studies have characterized whole tendon mechanics in animal models of DM (Table 12.2), again with inconsistent findings. Stiffness and Young's modulus changes with DM are commonly reported and of clinical interest in tendon given the implications of tendons limiting joint range of motion. Studies using ribose incubation or methylglyoxal-treated tendon have consistently found modulus to be higher in treated compared to untreated tendons [62,94,121]. Models of type 1 DM, such as streptozotocin-induced DM, have demonstrated inconsistent findings with regard to changes in modulus [25,122–126]. Models of type 2 DM have reported decreased or no difference in modulus in DM tendon, which differed based on which tendon was investigated [27]. However, the combined effect of modulus and cross-sectional area has resulted in conflicting findings for tendon stiffness in animal models of type 2 DM [26,27]. These contradictory findings relating to tendon stiffness could be due to increases in tendon cross-sectional area and stiffness isolated to specific portions of the tendon [27]. Collectively, these findings raise the question of the independent effects of DM versus body mass index on tendon mechanical properties.

Achilles tendons in rodent models of DM have been found to be responsive to tendon loading to restore more typical tendon mechanics [122,123], though the direction of these changes is inconsistent and needing additional investigation. Additional concerns, such as impaired tendon gliding in the Achilles tendon of a mouse model of DM [26], have been reported which may be important to consider given the clinical importance of maintaining joint range of motion.

Increases in tendon stiffness are often attributed to accumulation of AGEs and nonenzymatic collagen binding. As discussed above, nonenzymatic binding impairs collagen sliding. However, it is unclear how strongly this contributes to increased tensile stiffness at the level of the whole tendon [62,63]. Additional research is warranted to identify changes in tendon composition and organization at larger length scales of tendon architecture as well as the extent to which these changes attenuate impaired tendon mechanics at smaller length scales.

There are several potential factors that may contribute to altered tendon function at the whole tendon level. It is possible that impaired sliding alters cell signaling in the tendon, leading to homeostatic dysregulation and changes to whole tendon mechanics over time. Fatty infiltration and weakness of the muscle along with sedentary behavior may chronically decrease loading on the Achilles tendon over time. Decreased loads have been associated with lower tendon stiffness in healthy tendons [131,132]. Additionally, infiltration of fat into the tendon has been described in the context of tendon degeneration [53,133] and has been reported in one study of a mouse model of DM [26]. There are potentially a variety of other mechanisms and it is possible that multiple changes may occur simultaneously and vary between individuals. Needless to say, identifying mechanisms for altered tendon function and potential treatment targets in individuals with DM is a complex problem and one that necessitates additional research.

Summary

Diabetes is characterized by a hyperglycemic environment, which promotes AGEs accumulation particularly in slow-turnover tissues like tendon. AGEs accumulation has a host of downstream effects on tendon tissue, including alterations in collagen structure and binding as well as in the noncollagenous components of tendon tissue. These changes are believed to result in stiffening of the tendon tissue; however, experimental data suggest that the problem may be more complex. For example, alterations in tendon biochemistry and sliding mechanics may lead to impaired homeostasis and consequent degenerative changes. Additional investigation translating the effect of AGEs in tendon tissue to specific tendon-related clinical syndromes would be beneficial to identify specific treatment targets and better tailor medical and rehabilitative intervention for individuals with DM.

References

[1] World Health Organization. Global report on diabetes. 2016. http://www.who.int/about/licensing/.

[2] Ranger TA, Wong AMY, Cook JL, Gaida JE. Is there an association between tendinopathy and diabetes mellitus? A systematic review with meta-analysis. Br J Sports Med 2016;50(16):982—9. https://doi.org/10.1136/bjsports-2015-094735.

[3] Heinemeier KM, Schjerling P, Heinemeier J, Magnusson SP, Kjaer M. Lack of tissue renewal in human adult Achilles tendon is revealed by nuclear bomb 14C. FASEB J 2013;27:2074—9. https://doi.org/10.1096/fj.12-225599.

[4] Giacomozzi C, D'Ambrogi E, Uccioli L, MacEllari V. Does the thickening of Achilles tendon and plantar fascia contribute to the alteration of diabetic foot loading? Clin Biomech 2005;20(5):532—9. https://doi.org/10.1016/j.clinbiomech.2005.01.011.

[5] Cronin NJ, Peltonen J, Ishikawa M, et al. Achilles tendon length changes during walking in long-term diabetes patients. Clin Biomech 2010;25(5):476—82. https://doi.org/10.1016/j.clinbiomech.2010.01.018.

[6] D'Ambrogi E, Giacomozzi C, Macellari V, Uccioli L. Abnormal foot function in diabetic patients: the altered onset of windlass mechanism. Diabet Med 2005;22(12):1713—9. https://doi.org/10.1111/j.1464-5491.2005.01699.x.

[7] Maluf KS, Mueller MJ, Strube MJ, Engsberg JR, Johnson JE. Tendon Achilles lengthening for the treatment of neuropathic ulcers causes a temporary reduction in forefoot pressure associated with changes in plantar flexor power rather than ankle motion during gait. J Biomech 2004;37(6):897—906. https://doi.org/10.1016/j.jbiomech.2003.10.009.

[8] Zou D, Mueller MJ, Lott DJ. Effect of peak pressure and pressure gradient on subsurface shear stresses in the neuropathic foot. J Biomech 2007;40(4):883—90. https://doi.org/10.1016/j.jbiomech.2006.03.005.

[9] Lott DJ, Zou D, Mueller MJ. Pressure gradient and subsurface shear stress on the neuropathic forefoot. Clin Biomech 2008;23(3):342—8. https://doi.org/10.1016/j.clinbiomech.2007.10.005.

[10] Laborde JM. Neuropathic plantar forefoot ulcers treated with tendon lengthenings. Foot Ankle Int 2008;29(4):378—84. https://doi.org/10.3113/FAI.2008.0378.

[11] Van Bael K, Van Der Tempel G, Claus I, et al. Gastrocnemius fascia release under local anaesthesia as a treatment for neuropathic foot ulcers in diabetic patients: a short series. Acta Chir Belg 2016;116(6):367—71. https://doi.org/10.1080/00015458.2016.1192378.

[12] Mueller MJ, Sinacore DR, Hastings MK, Strube MJ, Johnson JE. Effect of Achilles tendon lengthening on neuropathic plantar ulcers. J Bone Jt Surg 2003;85-A(8):1436—45. https://doi.org/10.1016/j.legalmed.2014.08.005.

[13] Adler AI, Boyko EJ, Ahroni JH, Smith DG. Lower-extremity amputation in diabetes. Diabetes Care 1999;22(7):1029—35.

[14] Couppé C, Svensson RB, Kongsgaard M, et al. Human Achilles tendon glycation and function in diabetes. J Appl Physiol 2016;120(2):130—7. https://doi.org/10.1152/japplphysiol.00547.2015.

[15] Hudson BI, Lippman ME. Targeting RAGE signaling in inflammatory disease. Annu Rev Med 2018;69(1):349—64. https://doi.org/10.1146/annurev-med-041316-085215.

[16] Batista F, Nery C, Pinzur M, et al. Achilles tendinopathy in diabetes mellitus. Foot Ankle Int 2008;29(5):498—501. https://doi.org/10.3113/FAI.2008.0498.

[17] De Jonge S, Rozenberg R, Vieyra B, et al. Achilles tendons in people with type 2 diabetes show mildly compromised structure: an ultrasound tissue characterisation study. Br J Sports Med 2015;49(15):995—9. https://doi.org/10.1136/bjsports-2014-093696.

[18] Abate M, Salini V, Antinolfi P, Schiavone C. Ultrasound morphology of the Achilles in asymptomatic patients with and without diabetes. Foot Ankle Int 2014;35(1):44—9. https://doi.org/10.1177/1071100713510496.

[19] Ursini F, Arturi F, D'Angelo S, et al. High prevalence of Achilles tendon enthesopathic changes in patients with type 2 diabetes without peripheral neuropathy. J Am Podiatr Med Assoc 2017;107(2):99—105. https://doi.org/10.7547/16-059.

[20] Abate M, Schiavone C, Salini V. Sonographic evaluation of the shoulder in asymptomatic elderly subjects with diabetes. BMC Muscoskel Disord 2010;11. https://doi.org/10.1186/1471-2474-11-278.

[21] Ahmed AS, Schizas N, Li J, et al. Type 2 diabetes impairs tendon repair after injury in a rat model. J Appl Physiol 2012;113(11):1784—91. https://doi.org/10.1152/japplphysiol.00767.2012.

[22] David MA, Jones KH, Inzana JA, Zuscik MJ, Awad HA, Mooney RA. Tendon repair is compromised in a high fat diet-induced mouse model of obesity and type 2 diabetes. PLoS One 2014;9(3):1—8. https://doi.org/10.1371/journal.pone.0091234.

[23] Bedi A, Fox AJS, Harris PE, et al. Diabetes mellitus impairs tendon-bone healing after rotator cuff repair. J Shoulder Elbow Surg 2010;19(7):978—88. https://doi.org/10.1016/j.jse.2009.11.045.

[24] Cheing GLY, Chau RMW, Kwan RLC, Choi C, Zheng Y. Clinical biomechanics do the biomechanical properties of the ankle — foot complex in fluence postural control for people with type 2 diabetes? JCLB 2013;28(1):88—92. https://doi.org/10.1016/j.clinbiomech.2012.09.001.

[25] Lancaster RL, Haut RC, Decampi CE. Changes in the mechanical properties of patellar tendon preparations of spontaneously diabetic dogs under long-term insulin therapy. J Biomech 1994;27(8):1105—8.

[26] Studentsova V, Mora KM, Glasner MF, Buckley MR, Loiselle AE. Obesity/type II diabetes promotes function-limiting changes in murine tendons that are not reversed by restoring normal metabolic function. Sci Rep 2018;8(1):1—10. https://doi.org/10.1038/s41598-018-27634-4.

[27] Connizzo BK, Bhatt PR, Liechty KW, Soslowsky LJ. Diabetes alters mechanical properties and collagen fiber re-alignment in multiple mouse tendons. Ann Biomed Eng 2014;42(9):1880—8. https://doi.org/10.1007/s10439-014-1031-7.

[28] Guney A, Vatansever F, Karaman I, Kafadar I, Oner M, Turk C. Biomechanical properties of Achilles tendon in diabetic vs. non-diabetic patients. Exp Clin Endocrinol Diabetes 2015;123(7):428—32. https://doi.org/10.1055/s-0035-1549889.

[29] Evranos B, Idilman I, Ipek A, Polat SB, Cakir B, Ersoy R. Real-time sonoelastography and ultrasound evaluation of the Achilles tendon in patients with diabetes with or without foot ulcers: a cross sectional study. J Diabet Complicat 2015;29(8):1124—9. https://doi.org/10.1016/j.jdiacomp.2015.08.012.

[30] Petrovic M, Deschamps K, Verschueren SM, et al. Altered leverage around the ankle in people with diabetes: a natural strategy to modify the muscular contribution during walking? Gait Posture 2017;57:85—90. https://doi.org/10.1016/j.gaitpost.2017.05.016.

[31] Junot HSN, Anderson Hertz AFL, Gustavo Vasconcelos GR, da Silveira DCEC, Paulo Nelson B, Almeida SF. Epidemiology of trigger finger: metabolic syndrome as a new perspective of associated disease. Hand 2019. https://doi.org/10.1177/1558944719867135.

[32] Kuczmarski AS, Harris AP, Gil JA, Weiss APC. Management of diabetic trigger finger. J Hand Surg Am 2019;44(2):150—3. https://doi.org/10.1016/j.jhsa.2018.03.045.

[33] Abate M, Schiavone C, Salini V, Andia I. Management of limited joint mobility in diabetic patients. Diabetes Metab Syndrome Obes Targets Ther 2013;6:197—207. https://doi.org/10.1016/j.gexplo.2017.01.008.

[34] Berlanga-de-Mingo D, Lobo-Escolar L, López-Moreno I, Bosch-Aguilá M. Association between multiple trigger fingers, systemic diseases and carpal tunnel syndrome: a multivariate analysis. Rev Española Cirugía Ortopédica y Traumatol (English Ed.) 2019;63(4):307—12. https://doi.org/10.1016/j.recote.2018.12.008.

[35] Ashour A, Alfattni A, Hamdi A. Functional outcome of open surgical A1 pulley release in diabetic and nondiabetic patients. J Orthop Surg 2018;26(1):1—4. https://doi.org/10.1177/2309499018758069.

[36] Chang CJ, Chang SP, Kao LT, Tai TW, Jou IM. A meta-analysis of corticosteroid injection for trigger digits among patients with diabetes. Orthopedics 2018;41(1):e8—14. https://doi.org/10.3928/01477447-20170727-02.

[37] Luther GA, Murthy P, Blazar PE. Cost of immediate surgery versus non-operative treatment for trigger finger in diabetic patients. J Hand Surg Am 2016;41(11):1056—63. https://doi.org/10.1016/j.jhsa.2016.08.007.

[38] Brozovich N, Agrawal D, Reddy G. A critical appraisal of adult trigger finger: pathophysiology, treatment, and future outlook. Plast Reconstr Surg Glob Open. 2019;7(8):1–6. https://doi.org/10.1097/GOX.0000000000002360.

[39] Werner BC, Boatright JD, Chhabra AB, Dacus AR. Trigger digit release: rates of surgery and complications as indicated by a United States medicare database. J Hand Surg Eur 2016;41(9):970–6. https://doi.org/10.1177/1753193416653707.

[40] Huri G, Kaymakoglu M, Garbis N. Rotator cable and rotator interval: anatomy, biomechanics and clinical importance. EFORT Open Rev 2018;4(2):56–62. https://doi.org/10.1302/2058-5241.4.170071.

[41] Mall NA, Lee AS, Chahal J, et al. An evidenced-based examination of the epidemiology and outcomes of traumatic rotator cuff tears. Arthrosc J Arthrosc Relat Surg 2013;29(2):366–76. https://doi.org/10.1016/j.arthro.2012.06.024.

[42] Narvani A, Imam M, Godenèche A, et al. Degenerative rotator cuff tear, repair or not repair? A review of current evidence. Ann R Coll Surg Engl 2020;102(4):248–55. https://doi.org/10.1308/rcsann.2019.0173.

[43] Park HB, Gwark JY, Im JH, Jung J, Na JB, Yoon CH. Factors associated with atraumatic posterosuperior rotator cuff tears. J Bone Jt Surg 2018;100(16):1397–405. https://doi.org/10.2106/JBJS.16.01592.

[44] Lin TTL, Lin CH, Chang CL, Chi CH, Chang ST, Sheu WHH. The effect of diabetes, hyperlipidemia, and statins on the development of rotator cuff disease: a nationwide, 11-year, longitudinal, opulation-based follow-up study. Am J Sports Med 2015;43(9):2126–32. https://doi.org/10.1177/0363546515588173.

[45] Huang SW, Wang WT, Chou LC, Liou TH, Chen YW, Lin HW. Diabetes mellitus increases the risk of rotator cuff tear repair surgery: a population-based cohort study. J Diabet Complicat 2016;30(8):1473–7. https://doi.org/10.1016/j.jdiacomp.2016.07.015.

[46] Wu KT, Chou WY, Ko JY, Siu KK, Yang YJ. Inferior outcome of rotator cuff repair in chronic hemodialytic patients. BMC Muscoskel Disord 2019;20(1):1–8. https://doi.org/10.1186/s12891-019-2597-x.

[47] Clement ND, Hallett A, MacDonald D, Howie C, McBirnie J. Does diabetes affect outcome after arthroscopic repair of the rotator cuff? J Bone Jt Surg 2010;92(8):1112–7. https://doi.org/10.1302/0301-620X.92B8.23571.

[48] Fermont AJM, Wolterbeek N, Wessel RN, Baeyens JP, De Bie RA. Prognostic factors for successful recovery after arthroscopic rotator cuff repair: a systematic literature review. J Orthop Sports Phys Ther 2014;44(3):153–63. https://doi.org/10.2519/jospt.2014.4832.

[49] Cho NS, Moon SC, Jeon JW, Rhee YG. The influence of diabetes mellitus on clinical and structural outcomes after arthroscopic rotator cuff repair. Am J Sports Med 2015;43(4):991–7. https://doi.org/10.1177/0363546514565097.

[50] Kim YK, Jung KH, Kim JW, Kim US, Hwang DH. Factors affecting rotator cuff integrity after arthroscopic repair for medium-sized or larger cuff tears: a retrospective cohort study. J Shoulder Elbow Surg 2018;27(6):1012–20. https://doi.org/10.1016/j.jse.2017.11.016.

[51] Pierre-Jerome C, Moncayo V, Terk MR. MRI of the Achilles tendon: a comprehensive review of the anatomy, biomechanics, and imaging of overuse tendinopathies. Acta radiol 2010;51:438–54. https://doi.org/10.3109/02841851003627809.

[52] Kader D, Mosconi M, Benazzo F, Maffulli N. Achilles tendon rupture. In: Tendon injuries: basic science and clinical medicine; 2005. p. 187–200.

[53] Kannus P, Józsa L. Histopathological preceding changes rupture of a tendon. J Bone Joint Surg Am 1991;73(10):1507–25. http://europepmc.org/abstract/MED/1748700.

[54] Lui TH. Endoscopic adhesiolysis for extensive tibialis posterior tendon and Achilles tendon adhesions following compound tendon rupture. BMJ Case Rep 2013:1–4. https://doi.org/10.1136/bcr-2013-200824.

[55] Ahn JH, Choy WS. Tendon adhesion after percutaneous repair of the Achilles tendon: a case report. J Foot Ankle Surg 2011;50(1):93–5. https://doi.org/10.1053/j.jfas.2010.07.008.

[56] Mueller MJ, Sinacore DR, Hastings MK, Lott DJ, Strube MJ, Johnson JE. Impact of Achilles tendon lengthening on functional limitations and perceived disability in people with a neuropathic plantar ulcer. Diabetes Care 2004;27(7):1559−64. https://doi.org/10.2337/diacare.27.7.1559.

[57] Snedeker JG. How high glucose levels affect tendon homeostasis. In: Ackermann PW, Hart DA, editors. Metabolic influences on risk for tendon disorders. Springer International Publishing; 2016. p. 191−8. https://doi.org/10.1007/978-3-319-33943-6.

[58] Ahmed N. Advanced glycation endproducts - role in pathology of diabetic complications. Diabetes Res Clin Pract 2005;67(1):3−21. https://doi.org/10.1016/j.diabres.2004.09.004.

[59] Gautieri A, Passini FS, Silván U, et al. Advanced glycation end-products: mechanics of aged collagen from molecule to tissue. Matrix Biol 2017;59:95−108. https://doi.org/10.1016/j.matbio.2016.09.001.

[60] Bailey AJ, Paul RG, Knott L. Mechanisms of maturation and ageing of collagen. Mech Ageing Dev 1998;106(1−2):1−56. https://doi.org/10.1016/S0047-6374(98)00119-5.

[61] Avery NC, Bailey AJ. The effects of the Maillard reaction on the physical properties and cell interactions of collagen. Pathol Biol 2006;54(7):387−95. https://doi.org/10.1016/j.patbio.2006.07.005.

[62] Abate M, Schiavone C, Di Carlo L, Salini V. Achilles tendon and plantar fascia in recently diagnosed type II diabetes: role of body mass index. Clin Rheumatol 2012;31(7):1109−13. https://doi.org/10.1007/s10067-012-1955-y.

[63] Volper BD, Huynh RT, Arthur KA, et al. Influence of acute and chronic streptozotocin-induced diabetes on the rat tendon extracellular matrix and mechanical properties. Am J Physiol Regul Integr Comp Physiol 2015;309(9):R1135−43. https://doi.org/10.1152/ajpregu.00189.2015.

[64] Wong AMY, Docking SI, Cook JL, Gaida JE. Does type 1 diabetes mellitus affect Achilles tendon response to a 10 km run? A case control study. BMC Muscoskel Disord 2015;16(1):345. https://doi.org/10.1186/s12891-015-0803-z.

[65] Fessel G, Li Y, Diederich V, et al. Advanced glycation end-products reduce collagen molecular sliding to affect collagen fibril damage mechanisms but not stiffness. PLoS One 2014;9(11). https://doi.org/10.1371/journal.pone.0110948.

[66] Li Y, Fessel G, Georgiadis M, Snedeker JG. Advanced glycation end-products diminish tendon collagen fiber sliding. Matrix Biol 2013;32(3−4):169−77. https://doi.org/10.1016/j.matbio.2013.01.003.

[67] Rojas A, Delgado-López F, González I, Pérez-Castro R, Romero J, Rojas I. The receptor for advanced glycation end-products: a complex signaling scenario for a promiscuous receptor. Cell Signal 2013;25(3):609−14. https://doi.org/10.1016/j.cellsig.2012.11.022.

[68] Lin L. RAGE on the toll road? Cell Mol Immunol 2006;3(5):351−8.

[69] Ediger MN, Olson BP, Maynard JD. Noninvasive optical screening for diabetes. J Diabetes Sci Technol 2009;3(4):776−80. https://doi.org/10.1177/193229680900300426.

[70] Jazini E, Sharan AD, Morse LJ, et al. Alterations in T2 relaxation magnetic resonance imaging of the ovine intervertebral disc due to nonenzymatic glycation. Spine 2012;37(4):209−15. https://doi.org/10.1097/BRS.0b013e31822ce81f.

[71] Wengler K, Tank D, Fukuda T, et al. Diffusion tensor imaging of human Achilles tendon by stimulated echo readout-segmented EPI (ste-RS-EPI). Magn Reson Med 2018;80(6):2464−74. https://doi.org/10.1002/mrm.27220.

[72] Wengler K, Fukuda T, Tank D, et al. In vivo evaluation of human patellar tendon microstructure and microcirculation with diffusion MRI. J Magn Reson Imag 2019. https://doi.org/10.1002/jmri.26898.

[73] Sarman H, Atmaca H, Cakir O, et al. Assessment of postoperative tendon quality in patients with Achilles tendon rupture using diffusion tensor imaging and tendon fiber tracking. J Foot Ankle Surg 2015;54(5):782−6. https://doi.org/10.1053/j.jfas.2014.12.025.

[74] Engelen L, Stehouwer CDA, Schalkwijk CG. Current therapeutic interventions in the glycation pathway: evidence from clinical studies. Diabetes Obes Metabol 2013;15(8):677−89. https://doi.org/10.1111/dom.12058.

[75] Abate M, Schiavone C, Pelotti P, Salini V. Limited joint mobility in diabetes and ageing: recent advances in pathogenesis and therapy. Int J Immunopathol Pharmacol 2010;23(4):997—1003. https://doi.org/10.1177/039463201002300404.
[76] Grant TM, Thompson MS, Urban J, Yu J. Elastic fibres are broadly distributed in tendon and highly localized around tenocytes. J Anat 2013;222(6):573—9. https://doi.org/10.1111/joa.12048.
[77] Godinho MSC, Thorpe CT, Greenwald SE, Screen HRC. Elastin is localised to the interfascicular matrix of energy storing tendons and becomes increasingly disorganised with ageing. Sci Rep 2017;7(1):1—11. https://doi.org/10.1038/s41598-017-09995-4.
[78] Thorpe CT, Peffers MJ, Simpson D, Halliwell E, Screen HRC, Clegg PD. Anatomical heterogeneity of tendon: fascicular and interfascicular tendon compartments have distinct proteomic composition. Sci Rep February 2016;6:1—12. https://doi.org/10.1038/srep20455.
[79] Thorpe CT, Karunaseelan KJ, Ng Chieng Hin J, et al. Distribution of proteins within different compartments of tendon varies according to tendon type. J Anat 2016;229(3):450—8. https://doi.org/10.1111/joa.12485.
[80] Kohrs RT, Zhao C, Sun YL, et al. Tendon fascicle gliding in wild type, heterozygous, and lubricin knockout mice. J Orthop Res 2011;29(3):384—9. https://doi.org/10.1002/jor.21247.
[81] Alberti KA, Sun JY, Illeperuma WR, Suo Z, Xu Q. Laminar tendon composites with enhanced mechanical properties. J Mater Sci 2015;50(6):2616—25. https://doi.org/10.1007/s10853-015-8842-2.
[82] Henninger HB, Valdez WR, Scott SA, Weiss JA. Elastin governs the mechanical response of medial collateral ligament under shear and transverse tensile loading. Acta Biomater 2015;25:304—12. https://doi.org/10.1016/j.actbio.2015.07.011.
[83] Fang F, Lake SP. Multiscale mechanical integrity of human supraspinatus tendon in shear after elastin depletion. J Mech Behav Biomed Mater 2016;63:443—55. https://doi.org/10.1016/j.jmbbm.2016.06.032.
[84] Lee AH, Elliott DM. Multi-scale loading and damage mechanisms of plantaris and rat tail tendons. J Orthop Res 2019;37(8):1827—37. https://doi.org/10.1002/jor.24309.
[85] Fang F, Lake SP. Multiscale strain analysis of tendon subjected to shear and compression demonstrates strain attenuation, fiber sliding, and reorganization. J Orthop Res 2015;33(11):1704—12. https://doi.org/10.1002/jor.22955.
[86] Szczesny SE, Elliott DM. Interfibrillar shear stress is the loading mechanism of collagen fibrils in tendon. Acta Biomater 2014;10(6):2582—90. https://doi.org/10.1016/j.actbio.2014.01.032.
[87] Shearer T, Thorpe CT, Screen HRC. The relative compliance of energy-storing tendons may be due to the helical fibril arrangement of their fascicles. J R Soc Interface 2017;14(133):2—8. https://doi.org/10.1098/rsif.2017.0261.
[88] Eekhoff JD, Fang F, Lake SP. Multiscale mechanical effects of native collagen cross-linking in tendon. Connect Tissue Res 2018. https://doi.org/10.1080/03008207.2018.1449837.
[89] Sasaki N, Odajima S. Elongation mechanisms of collagen fibrils and force-strain relations of tendons at each level of structural hierarchy. J Biomech 1996;29(9):1131—6. https://doi.org/10.1016/0021-9290(96)00024-3.
[90] Bianchi F, Hofmann F, Smith AJ, Thompson MS. Probing multi-scale mechanical damage in connective tissues using X-ray diffraction. Acta Biomater 2016;45:321—7. https://doi.org/10.1016/j.actbio.2016.08.027.
[91] Collier TA, Nash A, Birch HL, de Leeuw NH. Effect on the mechanical properties of type I collagen of intra-molecular lysine-arginine derived advanced glycation end-product cross-linking. J Biomech 2018;67:55—61. https://doi.org/10.1016/j.jbiomech.2017.11.021.
[92] Patel D, Sharma S, Bryant SJ, Screen HRC. Recapitulating the micromechanical behavior of tension and shear in a biomimetic hydrogel for controlling tenocyte response. Adv Healthc Mater 2017;6(4):1—7. https://doi.org/10.1002/adhm.201601095.
[93] Sell DR, Biemel KM, Reihl O, Lederer MO, Strauch CM, Monnier VM. Glucosepane is a major protein cross-link of the senescent human extracellular matrix: relationship with diabetes. J Biol Chem 2005;280(13):12310—5. https://doi.org/10.1074/jbc.M500733200.

[94] Monnier VM, Mustata GT, Biemel KL, et al. Cross-linking of the extracellular matrix by the Maillard reaction in aging and diabetes: an update on "a puzzle nearing resolution.". Ann N Y Acad Sci 2005;1043:533−44. https://doi.org/10.1196/annals.1333.061.

[95] Snedeker JG, Gautieri A. The role of collagen crosslinks in ageing and diabetes - the good, the bad, and the ugly. Muscles Ligaments Tendons J 2014;4(3):303−8. https://doi.org/10.11138/mltj/2014.4.3.303.

[96] Collier TA, Nash A, Birch HL, de Leeuw NH. Preferential sites for intramolecular glucosepane cross-link formation in type I collagen: a thermodynamic study. Matrix Biol 2015;48:78−88. https://doi.org/10.1016/j.matbio.2015.06.001.

[97] Reddy GK. Glucose-mediated in vitro gycation modulates biomechanical integrity of the soft tissues but not hard tissues. J Orthop Res 2003;21:738−43.

[98] Vrhovski B, Weiss AS. Biochemistry of tropoelastin. Eur J Biochem 1998;258:1−18.

[99] Gosline J, Lillie M, Carrington E, Guerette P, Ortlepp C, Savage K. Elastic proteins: biological roles and mechanical properties. Philos Trans R Soc B Biol Sci 2002;357(1418):121−32. https://doi.org/10.1098/rstb.2001.1022.

[100] Baldock C, Oberhauser AF, Ma L, et al. Shape of tropoelastin, the highly extensible protein that controls human tissue elasticity. Proc Natl Acad Sci U S A 2011;108(11):4322−7. https://doi.org/10.1073/pnas.1014280108.

[101] Eekhoff JD, Fang F, Kahan LG, et al. Functionally distinct tendons from elastin haploinsufficient mice exhibit mild stiffening and tendon-specific structural alteration. J Biomech Eng November 2017;139:1−9. https://doi.org/10.1115/1.4037932.

[102] Wu YT, Su WR, Wu PT, Shen PC, Jou IM. Degradation of elastic fiber and elevated elastase expression in long head of biceps tendinopathy. J Orthop Res 2017;35(9):1919−26. https://doi.org/10.1002/jor.23500.

[103] Wu YT, Wu PT, Jou IM. Peritendinous elastase treatment induces tendon degeneration in rats: a potential model of tendinopathy in vivo. J Orthop Res 2016;34(3):471−7. https://doi.org/10.1002/jor.23030.

[104] Konova E, Baydanoff S, Atanasova M, Velkova A. Age-related changes in the glycation of human aortic elastin. Exp Gerontol 2004;39(2):249−54. https://doi.org/10.1016/j.exger.2003.10.003.

[105] Nicoloff G, Nikolov A, Dekov D. Serum AGE-elastin derived peptides among diabetic children. Vasc Pharmacol 2005;43(4):193−7. https://doi.org/10.1016/j.vph.2005.03.007.

[106] Al-Robaiy S, Weber B, Simm A, et al. The receptor for advanced glycation end-products supports lung tissue biomechanics. Am J Physiol Lung Cell Mol Physiol 2013;305(7). https://doi.org/10.1152/ajplung.00090.2013.

[107] Gu Q, Wang B, Zhang XF, Ma YP, Liu JD, Wang XZ. Contribution of receptor for advanced glycation end products to vasculature-protecting effects of exercise training in aged rats. Eur J Pharmacol 2014;741:186−94. https://doi.org/10.1016/j.ejphar.2014.08.017.

[108] Brüel A, Oxlund H. Changes in biomechanical properties, composition of collagen and elastin, and advanced glycation endproducts of the rat aorta in relation to age. Atherosclerosis 1996;127(2):155−65. https://doi.org/10.1016/S0021-9150(96)05947-3.

[109] Mania-Pramanik J, Potdar SS, Vadigoppula A, Elastase SS. A predictive marker of inflammation and/or infection. J Clin Lab Anal 2004;18(3):153−8. https://doi.org/10.1002/jcla.20015.

[110] Kostrominova TY, Brooks SV. Age-related changes in structure and extracellular matrix protein expression levels in rat tendons. Age (Omaha). 2013;35(6):2203−14. https://doi.org/10.1007/s11357-013-9514-2.

[111] Burner T, Gohr C, Mitton-Fitzgerald E, Rosenthal AK. Hyperglycemia reduces proteoglycan levels in tendons. Connect Tissue Res 2012;53(6):535−41. https://doi.org/10.3109/03008207.2012.710670.

[112] Fang F, Lake SP. Multiscale mechanical evaluation of human supraspinatus tendon under shear loading after glycosaminoglycan reduction. J Biomech Eng 2017;139(7):071013. https://doi.org/10.1115/1.4036602.

[113] Dourte LM, Pathmanathan L, Jawad AF, et al. Influence of decorin on the mechanical, compositional, and structural properties of the mouse patellar tendon. J Biomech Eng 2012;134(3):1–8. https://doi.org/10.1115/1.4006200.

[114] Fessel G, Snedeker JG. Equivalent stiffness after glycosaminoglycan depletion in tendon - an ultrastructural finite element model and corresponding experiments. J Theor Biol 2011;268(1):77–83. https://doi.org/10.1016/j.jtbi.2010.10.007.

[115] Rigozzi S, Müller R, Snedeker JG. Local strain measurement reveals a varied regional dependence of tensile tendon mechanics on glycosaminoglycan content. J Biomech 2009;42(10):1547–52. https://doi.org/10.1016/j.jbiomech.2009.03.031.

[116] Gordon JA, Freedman BR, Zuskov A, Iozzo RV, Birk DE, Soslowsky LJ. Achilles tendons from decorin- and biglycan-null mouse models have inferior mechanical and structural properties predicted by an image-based empirical damage model. J Biomech 2015;48(10):2110–5. https://doi.org/10.1016/j.jbiomech.2015.02.058.

[117] Robinson KA, Sun M, Barnum CE, et al. Decorin and biglycan are necessary for maintaining collagen fibril structure, fiber realignment, and mechanical properties of mature tendons. Matrix Biol 2017;64(class I):81–93. https://doi.org/10.1016/j.matbio.2017.08.004.

[118] Matuszewski PE, Chen YL, Szczesny SE, et al. Regional variation in human supraspinatus tendon proteoglycans: decorin, biglycan, and aggrecan. Connect Tissue Res 2012;53(5):343–8. https://doi.org/10.3109/03008207.2012.654866.

[119] Reigle KL, Di Lullo G, Turner KR, et al. Non-enzymatic glycation of type I collagen diminishes collagen-proteoglycan binding and weakens cell adhesion. J Cell Biochem 2008;104(5):1684–98. https://doi.org/10.1002/jcb.21735.

[120] Gohr CM, Fahey M, Rosenthal AK. Calcific tendonitis: a model. Connect Tissue Res 2007;48(6):286–91. https://doi.org/10.1080/03008200701692362.

[121] Patel SH, Yue F, Saw SK, et al. Advanced glycation end-products suppress mitochondrial function and proliferative capacity of achilles tendon-derived fibroblasts. Sci Rep 2019;9:12614. https://doi.org/10.3997/2214-4609.201404048.

[122] Xu L, Xu K, Wu Z, et al. Pioglitazone attenuates advanced glycation end products- induced apoptosis and calcification by modulating autophagy in tendon-derived stem cells. J Cell Mol Med 2020:1–12. https://doi.org/10.1111/jcmm.14901.

[123] Mifune Y, Inui A, Muto T, et al. Influence of advanced glycation end products on rotator cuff. J Shoulder Elbow Surg 2019;28(8):1490–6. https://doi.org/10.1016/j.jse.2019.01.022.

[124] Reddy GK. Cross-linking in collagen by nonenzymatic glycation increases the matrix stiffness in rabbit Achilles tendon. Exp Diabesity Res 2004;5(2):143–53. https://doi.org/10.1080/15438600490277860.

[125] De Oliveira RR, Bezerra MA, De Lira KDS, et al. Aerobic physical training restores biomechanical properties of Achilles tendon in rats chemically induced to diabetes mellitus. J Diabet Complicat 2012;26(3):163–8. https://doi.org/10.1016/j.jdiacomp.2012.03.017.

[126] Silva RTB, Castro PV de, Coutinho MPG, Brito ACN de L, Bezerra MA, Moraes SRA de. Resistance jump training may reverse the weakened biomechanical behavior of tendons of diabetic Wistar rats. Fisioter e Pesqui 2017;24(4):399–405. https://doi.org/10.1590/1809-2950/17198024042017.

[127] Svensson RB, Smith ST, Moyer PJ, Magnusson SP. Effects of maturation and advanced glycation on tensile mechanics of collagen fibrils from rat tail and Achilles tendons. Acta Biomater 2018;70:270–80. https://doi.org/10.1016/j.actbio.2018.02.005.

[128] Thomas SJ, Sarver JJ, Yannascoli SM, et al. The effect of isolated hyperglycemia on native mechanical and biologic shoulder joint properties in a rat model hyperglycemia and shoulder properties. J Orthop Res 2014;32(11):1464–70. https://doi.org/10.1002/jor.22695.

[129] Andreassen TT, Seyer-Hansen K, Bailey AJ. Thermal stability, mechanical properties and reducible cross-links of rat tail tendon in experimental diabetes. BBA Gen Subj 1981;677(2):313–7. https://doi.org/10.1016/0304-4165(81)90101-X.

[130] Gonzalez AD, Gallant MA, Burr DB, Wallace JM. Multiscale analysis of morphology and mechanics in tail tendon from the ZDSD rat model of type 2 diabetes. J Biomech 2014;47(3):681–6. https://doi.org/10.1016/j.jbiomech.2013.11.045.

[131] Maganaris CN, Reeves ND, Rittweger J, et al. Adaptive response of human tendon to paralysis. Muscle Nerve January 2006;33:85—92. https://doi.org/10.1002/mus.20441.
[132] Epro G, Mierau A, Doerner J, et al. The Achilles tendon is mechanosensitive in older adults: adaptations following 14 weeks versus 1.5 years of cyclic strain exercise. J Exp Biol 2017;220(5):1008—18. https://doi.org/10.1242/jeb.146407.
[133] Józsa L, Kannus P. Histopathological findings in spontaneous tendon ruptures. Scand J Med Sci Sports 1997;7(9):113—8.

CHAPTER 13

A phenomenological dashpot model for morphoelasticity for the contraction of scars

F.j. Vermolen
Computational Mathematics Group (CMAT), Division of Mathematics and Statistics, Faculty of Sciences, University of Hasselt, Diepenbeek, Belgium

Introduction

Deep tissue injury is characterized by damage on deeper dermal layers such as the dermis and subcutaneous tissue. These layers are constructed by collagen and contain various cell types such as immune cells, endothelial cells (from blood vessels), and fibroblasts. Fibroblasts are responsible for the expression of collagen, but also exert pulling (contractile) forces on their direct environment. Numerous mathematical models exist that describe the evolution of skin after an injury. These models focus on different aspects of wound healing. These aspects include angiogenesis, wound closure, wound contraction, hemostasis, or remodeling. In this manuscript, we will focus on mechanical issues. Modern surveys regarding treatment of burns and contractions can be found in Refs. [1,2].

In a great variety of applications in computational mechanics, the deformations are small, and hence standard Hooke's law provides a reasonable approximation for the computation of displacements and stresses. However, there are also various cases, in particular in soft materials, where it is necessary to include plasticity, that is, deformations are permanent. The reason is that either deformations are so large that the associated stresses exceed the von Mises stress, or the structure of the tissue has undergone some microstructural changes so that the material properties and dimensions have changed. Many classical mechanical frameworks for wound healing do not incorporate plasticity either due to large stresses or as a result of growth phenomena. Examples of such established mechanical frameworks for wound contraction are by Refs. [3,4], where the latter considers stress-determined differentiation of fibroblasts to myofibroblasts. Next to entirely partial differential equations based models, there exist (hybrid) agent-based models where cells are treated as separate entities that migrate though the tissue. Here cellular processes like cell migration, cell division, death, and differentiation are treated as stochastic processes. See the work by Refs. [5,6] as examples of agent-based models applied to tumor development and wound contraction, respectively. Morphoelasticity constitutes a framework in which deformations are combined with growth and shrinkage phenomena. Morphoelastic frameworks have been used successfully to model growth of tumors [7] or to model permanent contractures that occur as a result of deep tissue injury [8]. Most mathematical models for tissue growth and contraction are constructed on the basis of the assumption that superficial processes determine morphological changes. Morphoelastic models treat the interplay between growth and shrinkage of tissues as a result of biophysical processes, such as cellular growth or large deformations in the interior, hence in the bulk, of the tissue. Hence morphoelastic models are capable of simulating growth phenomena in which growth occurs as a result of internal (bulk) phenomena, rather than by processes that occur on the boundary. For cases of tumor growth and wound contraction, this is a much more realistic point of view. In the case of contraction of wounds, the (large) strains are the main reason of the plastic deformation, and, here, morphoelasticity merely models a materialistic change of the material.

The model for morphoelasticity involves a set of complicated partial differential equations that follow from a balance of momentum, and an evolution equation for the effective Eulerian strain. A derivation of the equations can be found in Ref. [9], and the equations are as follows:

$$\frac{D(\rho \mathbf{v})}{Dt} + \rho \mathbf{v} \nabla \cdot \mathbf{v} - \nabla \cdot \sigma(\varepsilon, \mathbf{L}) = \rho \mathbf{f}, \tag{13.1}$$

$$\frac{D\varepsilon}{Dt} + \varepsilon \, \text{skw}(\mathbf{L}) - \text{skw}(\mathbf{L}) \, \varepsilon + (tr(\varepsilon) - 1) \, \text{sym}(\mathbf{L}) = -\mathbf{G}, \tag{13.2}$$

where

$$\mathbf{L} = \nabla \mathbf{v}, \text{ and sym}(\mathbf{L}) = \frac{\mathbf{L} + \mathbf{L}^T}{2}, \text{ and skw}(\mathbf{L}) = \frac{\mathbf{L} - \mathbf{L}^T}{2}.$$

Here $\mathbf{v}$ and ε, respectively, denote the displacement velocity and effective Eulerian strain, and these parameters are solved for directly in these model equations. Furthermore, ρ and σ denote the density of the tissue and stress in the tissue. The first equation entails a momentum balance, and $\mathbf{f}$ represents a body force in the tissue that results by pulling forces that the cells exert on their environment. Further, the time derivative in the above partial differential equations represents the material derivative, which is commonly used in problems from fluid mechanics and structural mechanics, and this derivative entails a derivative with respect to time in a moving (Lagrangian) framework. This derivative entails temporal changes when the observer travels alongside with a particle, and it is given by

$$\frac{Df}{Dt} = \frac{\partial f}{\partial t} + \mathbf{v} \cdot \nabla f. \tag{13.3}$$

For the relation between the stress tensor σ, the strain tensor ε, and the displacement velocity gradient tensor $\mathbf{L}$, one can use a viscoelastic model, such as the one by Kelvin–Voigt. As mentioned earlier, differential Eqs. (13.1) and (13.2) are solved for the strain ε

and displacement velocity **v**. The displacement vector **u** follows from integrating the displacement velocity over time. This set of equations was used for the simulation of the evolution of skin grafts over time by Ref. [8]. The computed results were validated by experiments done by Ref. [10]. Next to the solution of these partial differential equations, the solution methods involve the movement of the computational mesh. From an analytic point of view, it is possible to demonstrate various mathematical properties of the solution to the morphoelasticity equations, such as symmetry of the strain tensor in general dimensions, and stability of the solution for the one-dimensional case. Another issue is the determination of the cell forces, which involves the computation of the cell densities (number of cells per unit of volume). This cell density typically follows from a partial differentiation that models a cell balance. Hence these equations provide a complicated problem in which the solution is not tractable and requires sufficient computational resources. It is the aim of the current manuscript to provide a phenomenological (hence less physics-based) model that is able to reproduce the most important trends.

In order to make morphoelasticity a more tractable problem, we propose a heavily simplified formalism in which we consider an ordinary differential equation that is inspired on a balance of momentum, and the evolution of plastic deformations. These plastic deformations can be a result of large deformations or as a result of growth or shrinkage phenomena. This makes it possible to simulate growth and shrinkage problems in combinations with a mechanical balance. An advantage of this simplified model is the quick access of parameters and the straightforward analysis of the model. Further, one can link the current formalism through fitting input parameters to reproduce the results that have been obtained by using the more complicated models. A drawback is that geometrical effects cannot be modeled using the current simplified formalism.

The manuscript continues with a quick motivation of stability of the model, presenting the differential equations, analysis of the system, and subsequently some numerical simulation case studies will be presented. In this manuscript, we will also analyze the impact of uncertainty on the results. Finally some conclusions are drawn.

The mathematical model

Before we introduce the phenomenological model for morphoelasticity, we consider the one-dimensional counterpart of the morphoelasticity model and we consider its stability.

Stability of the one-dimensional model

Though the formal assessment of the multidimensional model is not very hard from a conceptual point of view, it is laborious as a result of the cross-term involved, and therefore, we will limit ourselves to assessing stability for the one-dimensional case. As far as we know, this has not been done in earlier studies. For the one-dimensional case, the morphoelasticity equations are given by:

$$\begin{cases} \rho\left(\dfrac{\partial v}{\partial t} + 2v\dfrac{\partial v}{\partial x}\right) - E\dfrac{\partial \varepsilon}{\partial x} - \mu\dfrac{\partial^2 v}{\partial x^2} = \rho f \\ \dfrac{\partial \varepsilon}{\partial t} + \varepsilon\dfrac{\partial v}{\partial x} + (\varepsilon - 1)\dfrac{\partial v}{\partial x} + \alpha\,\varepsilon = 0. \end{cases} \quad (13.4)$$

This system is supplemented with initial conditions for v and ε, as well as with boundary conditions for the case of a finite domain. Note that $\varepsilon = 0$ and $v = 0$ represents an equilibrium if $f = 0$. We are interested in the case of small strains and small displacement velocities, and we analyze the behavior of the solution around $v = 0$ and $\varepsilon = 0$, then a linearization in v and ε around $(0, 0)$ gives

$$\begin{cases} \rho\dfrac{\partial v}{\partial t} - E\dfrac{\partial \varepsilon}{\partial x} - \mu\dfrac{\partial^2 v}{\partial x^2} = \rho f, \\ \dfrac{\partial \varepsilon}{\partial t} - \dfrac{\partial v}{\partial x} + \alpha\,\varepsilon = 0. \end{cases} \quad (13.5)$$

For $f = 0$, it is immediately clear that $(v, \varepsilon) = (0, 0)$ represents an equilibrium. We will motivate that this equilibrium is stable, that is, it will always be possible to choose initial perturbations such that the impact for all $t > 0$ is as small as one wishes (we omit the technical, formal ε, δ definition of Lyapunov stability). Formally, one should distinguish the notation for the solution to the linearized problem from the solution of the original equation. Upon considering a large domain of computation, we apply a Fourier Transform over the spatial coordinate, which we choose as

$$\widehat{f}(t, \omega) = \mathscr{F}(f(t, x)) = \int_{-\infty}^{\infty} e^{-i\omega x} f(t, x) dx, \text{ and}$$
$$f(t, x) = \mathscr{F}^{-1}\left(\widehat{f}(t, \omega)\right) = \dfrac{1}{2\pi}\int_{-\infty}^{\infty} e^{i\omega x} \widehat{f}(t, \omega) d\omega. \quad (13.6)$$

Here i denotes the imaginary unit, and note that one can make different choices for the constants in front of the integral as long as their product is equal to $\dfrac{1}{2\pi}$ for the one-dimensional case. The Fourier transformation of $f(t, x)$ is denoted by $\widehat{f}(t, \omega)$ where ω denotes the frequency. A Fourier transform of a signal in the current contents is interpreted as the decomposition of the signal into cycles (harmonics) or periodic signals such as sines and cosines with respect to the spatial position (location). The alternative of having a Fourier series over a finite (periodic) domain would give the same conclusions; however, one needs more algebraic steps then. Application of the Fourier transform, and using $F\left(\frac{\partial f}{\partial x}\right) = i\omega \widehat{f}$, gives the following set of ordinary differential equations of the Fourier transformed solution (which are in the frequency domain):

$$\begin{cases} \rho \dfrac{\partial \widehat{v}}{\partial t} - iE\omega \widehat{\varepsilon} + \mu \omega^2 \widehat{v} = \rho \widehat{f}, \\ \dfrac{\partial \widehat{\varepsilon}}{\partial t} - i\omega \widehat{v} + \alpha \widehat{\varepsilon} = 0. \end{cases} \quad (13.7)$$

It is easy to write the above system in the form of $\mathbf{y}' + A\mathbf{y} = \mathbf{f}$, where A represents the system matrix. Linear stability of the equilibrium $\widehat{v} = 0$ and $\widehat{\varepsilon} = 0$ for $f = 0$ is warranted if the real part of the eigenvalues of the system matrix is nonnegative. The eigenvalues are given by

$$\lambda_{\pm} = \frac{1}{2}\left(\alpha + \omega^2 \frac{\mu}{\rho}\right) \pm \frac{1}{2}\sqrt{\left(\alpha + \omega^2 \frac{\mu}{\rho}\right)^2 - 4\frac{\omega^2}{\rho}(\mu\alpha + E)}.$$

Furthermore, if the argument in the square root is positive then convergence to the equilibrium proceeds monotonically, else an oscillatory behavior is to be expected as a result of complex eigenvalues. Since the parameters are nonnegative, linear stability is guaranteed for all choices. This can also be extended to nonzero forces by the use of a homogenization technique (or a particular solution) and hence this stability implies that the current model yields convergence to a steady-state solution as $t \to \infty$ if the forcing possesses a limit as $t \to \infty$. Since $\widehat{v}$ and $\widehat{\varepsilon}$ tend to zero as $t \to 0$ it can be concluded from linearity of the (inverse) Fourier transform that v and ε also tend to zero as $t \to 0$, and hence it can be concluded that the morphoelasticity model is stable. This important observation is recalled during the assessment of the phenomenological model.

The phenomenological model

Next, we elaborate on what morphoelasticity actually is, and on the implications of the formalism. The treatment is such that we simplify the morphoelastic formulation, as outlined in the introduction, to a simple mass–spring–dashpot system. For the reader with basic knowledge on mass–spring systems, this section should be easily accessible. For the reader who furthermore possesses basic knowledge on (numerical methods of) dynamical systems, the results should even be reproducible. First we consider a model for the contraction of skin or a scar, and subsequently we consider the expansion of an organ or a tumor.

To this end, we consider a mass m, on a spring with stiffness, k, and damping coefficient, b. Furthermore, the displacement of the mass is denoted by u. On the mass, a force F is exerted. The equilibrium position changes over time. The initial position of the mass is denoted by X_0. The main idea of morphoelasticity is that the equilibrium positions over time. The current equilibrium position at time t is denoted by $X(t)$. Note that $(X, u) = (X, u)(t)$. Further the current position of the mass is denoted by x, which also changes over time, hence $x = x(t)$. The current displacement, current equilibrium, and current position are related via

$$u(t) = x(t) - X(t) \Leftrightarrow x(t) = X(t) + u(t). \tag{13.8}$$

The force balance follows via Newton's second law, and this is given by

$$mu'' + bu' + ku = F(t). \tag{13.9}$$

The first, second, and third terms in the left-hand side, respectively, denote the inertia, damping, and stiffness. The above equation is used to model to the position of the wound edge. The forcing $F(t)$ represents forces that are directed in the scar area, and the stiffness represents the forces that are exerted on the wound edge to move the wound edge toward the (current) equilibrium position. As initial conditions, we consider

$$u(0) = u'(0) = 0, \quad X(0) = X_0, \tag{13.10}$$

which says that there is no initial displacement or deformation as well as displacement velocity and that the initial state equilibrium state of the material is prescribed. The above model mimics the case that has been sketched in Fig. 13.1, in which the mass movement is only determined by the forcing F that is exerted in the scar region and by the spring force that is exerted by the undamaged portion of the tissue on the other side.

The formal set of equations to be solved is

$$\begin{cases} mv' + bv + ku = F(t), & t > 0, \\ u' = v, & t > 0 \\ X' = G, & t \geq 0, \\ x = X + u, & t > 0. \end{cases} \tag{13.11}$$

Here G is a general function that incorporates the evolution of permanent changes. This term can be compared to the α—term in the full morphoelastic model from the introduction. The nature of this function depends on the application that is considered. For the case of wound contraction, this term can be compared to the α—term in the full morphoelastic model from the introduction. This matter will be clarified in the next paragraphs. Further, note that in general $X(t)$ does not represent a physical position,

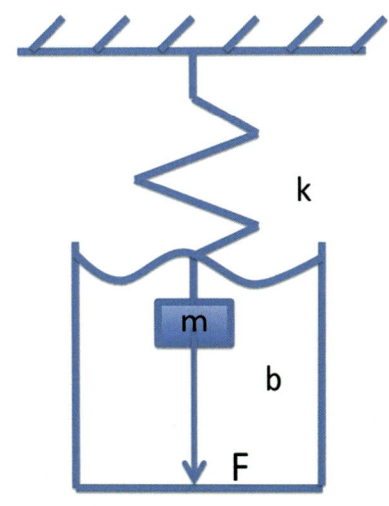

Figure 13.1 Schematic of a spring–dashpot system, of a mass that is subject to vibrations with damping. The spring models the reactive force that is exerted on the wound edge by the undamaged skin, the liquid models the viscous, damping behavior of skin, and the forcing F models the cell traction forces that are exerted by the (myo)fibroblasts in the scar.

but a virtual position. The actual physical coordinate or physical position is $x(t)$, which is determined by the current force $F(t)$, the reaction force with reaction constant k_2, and the equilibrium position $X(t)$. We consider the case of skin tissue and propose that the equilibrium position changes over time according to a linear relation with the displacement u, that is,

$$X'(t) = G(u) = \alpha u(t), \quad \alpha \geq 0. \tag{13.12}$$

This choice mimics the case that the degree of displacement determines how the tissue obtains its permanently (plastic) changed structure. Since we only follow the spatial position of the rigid mass, there are no spatial variations, and hence the current displacement is closest to the current (Eulerian) strain in the morphoelasticity formulation in the introduction. Note that if $\alpha = 0$, then the equilibrium position $X(t)$ remains constant, and hence the inflicted deformations are not causing any permanent changes.

Analysis and computational methodologies

First we consider the simplest case in which there is no damping and no inertia. Subsequently, we deal with damping.

Zero inertia and zero damping

First we consider the simplest case in which inertia and damping are neglected, that is, we treat the case that $m = 0$ and $b = 0$. Then upon using Eq. (13.12), we get

$$u(t) = \frac{F(t)}{k}, \quad X' = \alpha u = \frac{\alpha}{k} F(t). \tag{13.13}$$

This implies that

$$X(t) = X_0 + \frac{\alpha}{k} \int_0^t F(s) \, ds. \tag{13.14}$$

Using $x(t) = X(t) + u(t)$, we get for the current position:

$$x(t) = X_0 + \frac{\alpha}{k} \int_0^t F(s) \, ds + \frac{F(t)}{k} = X_0 + \frac{1}{k} \left(\int_0^t \alpha F(s) \, ds + F(t) \right). \tag{13.15}$$

It is noted that the second and third terms in the right-hand side of the above equation correspond to permanent and temporary deformation, respectively. The second term contains the *history* of the imposed forces (due to the integral over time t), whereas the third term contains the effects from the *current* force. In this sense, the second term, that is, the integral over t, represents a *hysteresis* term.

The improper integral $\int_0^\infty F(t) \, dt$ can only exist if $F(t) \to 0$ as $t \to \infty$ (hence this is a necessary condition), one sees that $X(t)$ and $x(t)$ have the same limits as $t \to \infty$, which allows to determine the final wound edge position

$$X_f = x_f = X_0 + \frac{\alpha}{k} \int_0^\infty F(s) \, ds. \tag{13.16}$$

The maximum contraction of the wound is determined whenever $x'(\theta) = 0$, and this gives the solution of

$$x'(\theta) = \frac{1}{k}(\alpha F(\theta) + F'(\theta)) = 0. \tag{13.17}$$

which gives the interesting observation that the morphoelasticity rate parameter α not only determines the extent of plasticity, but also shifts (delays) the maximum contraction.

Zero inertia with damping

Using the initial condition for u, and solving the first-order differential equation for u, gives

$$u(t) = \frac{1}{b} \int_0^t e^{\frac{k}{b}(s-t)} F(s) \, ds, \tag{13.18}$$

and formally for the equilibrium, we get the following expression:

$$X(t) = X_0 + \alpha \int_0^t u(s)ds = X_0 + \frac{\alpha}{b} \int_0^t \int_0^s e^{\frac{k}{b}(s'-s)} F(s')ds'ds. \tag{13.19}$$

This gives the following closed form for the current position:

$$x(t) = X_0 + \frac{1}{b}\left[\int_0^t e^{\frac{k}{b}(s-t)} F(s)ds + \alpha \int_0^t \int_0^s e^{\frac{k}{b}(s'-s)} F(s')ds'ds\right]. \tag{13.20}$$

Here the α-term accounts for hysteresis. The case that inertia is not negligible can be dealt with analogously using variation of parameters or Laplace transforms, one has to distinguish between the cases $b^2 - 4mk$ is positive, negative, or zero. The expressions to be obtained involve the Wronskian determinant, and will amount to lengthy integral expressions. Since this does not contribute to any insight, these expressions are not evaluated further. For this case, one can also derive a time at which the contraction is maximal. The expression looks rather awkward, and therefore this is omitted.

Numerical results

The simple model enables fast and easy access to simulations. First, we present some basic runs, and then we show some parameter sensitivity analysis in terms of Monte Carlo simulations. The basic input parameters are listed in Table 13.1.

Basic runs

The equations are solved using data that are as close as possible to skin properties. Table 13.1 lists the values of the input parameters. The simulations that we present represent the position of the point mass as a function of time. The point mass represents the edge of the wound that is contracted as a result of forces that are exerted in the interior of the wound. These forces are predominantly exerted by myofibroblasts, or differentiated fibroblasts, with the potential of exerting larger forces. We apply a temporary forcing

$$F = F_0 \exp\left(-\frac{(t-\tau)^2}{2\sigma^2}\right), \tag{13.21}$$

Table 13.1 Input values for the basic runs.

m	1200	kg/m^3
b	50	$Pa\,s$
k	0.5	MPa
τ	70	$days$
σ	21	$days$
F_0	$-0.5 \cdot 10^4$	N/m^3
α	10^{-7}	s^{-1}

which gives a peak force of F_0 centered at $t = \tau$ with an interval size determined by σ. We motivate this formula for the forcing as follows. As soon as wounding occurs, then the immune cells are called to migrate toward the wound site in order to clear up the debris, and to neutralize pathogens. Furthermore, the immune cells secrete chemokines that attract the fibroblasts. The fibroblasts are responsible for the expression of collagen (in particular a provisional collagen type III is secreted), which builds the integrity of the skin. Fibroblasts can differentiate to myofibroblasts, who continue expressing collagen, but also exert large contractile forces on their immediate environment. This causes contraction of the scar. Subsequently, after some time, the myofibroblasts lose their activity either as a result of apoptosis or as a result of other processes as soon as the collagen matrix has been developed. Then the forcing disappears. Hence this force mimics the temporary presence of myofibroblasts in the wound area. This temporary force is modeled by a Gaussian, and here τ, which is the time of onset of the force determined by processes like cell migration speeds, and the working of the immune system. If the immune system is weak, then τ will be larger since fewer chemokines will have been secreted by the lower counts of immune cells. This implies that there will be fewer fibroblasts migrating toward the wound and hence the onset of the contractile force will be delayed. Furthermore, since the density of (myo)fibroblasts will only develop over a longer time period, it will follow that the contraction forcing has to be smeared out over a longer time and thereby σ could possibly be larger as well for a weaker immune system. A weaker immune system could also imply the reduction of the maximal forcing F_0. This reduction is possibly caused by the presence of lower (myo)fibroblasts counts.

In Fig. 13.2, we plot the displacement u, equilibrium wound edge position $X(t)$, and actual wound edge position $x(t)$ for a morphoelasticity parameter $\alpha = 10^{-7} s^{-1} = 8.64 \cdot 10^{-3}/\text{day}$. It can be seen that the displacement exhibits a peak at time θ, which coincides with the maximum of the force. Despite the decrease of the displacement, the equilibrium position has changed over time and therefore, despite the decrease of the displacement toward zero, the actual wound edge tends to a remaining displacement as a result of the new equilibrium. The equilibrium should be interpreted as a dynamic limit, which coincides with the new asymptotic position of the displacement as a result of the permanent material change. Next, we show how the morphoelasticity parameter α influences the actual $x(t)$ and the equilibrium $X(t)$ displacement of the wound. Larger α—values make the equilibrium change more rapidly and easily and herewith each displacement causes an immediate permanent material change. Hence it becomes easier to deform the tissue material, and hence larger wound edge displacements are obtained as well as larger wound edge positions.

We finally note that combining Eq. (13.16) with the forcing (Eq. 13.21), and for the case of no damping and no inertia, we get the following expression for the final position of the wound edge:

$$X_f = x_f = X_0 + F_0 \frac{\alpha \sigma}{k} \sqrt{\frac{\pi}{2}} \left(1 + \text{erf}\left(\frac{\tau}{\sigma\sqrt{2}}\right)\right), \qquad (13.22)$$

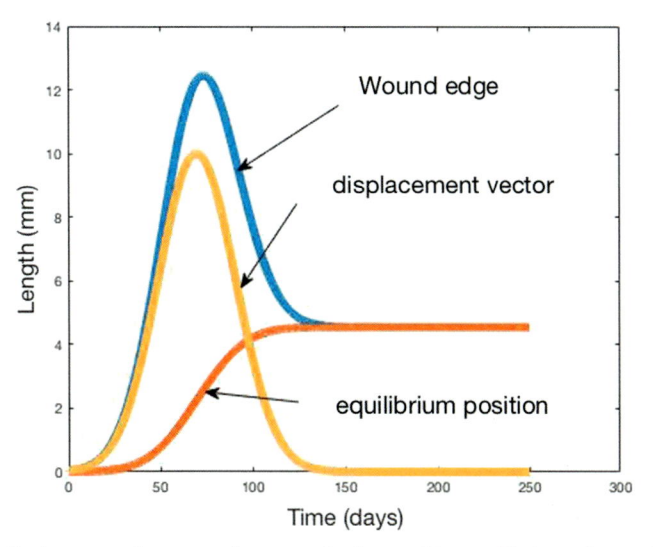

Figure 13.2 The displacement between the wound edge and its equilibrium, the equilibrium, and the actual wound edge position as a function of time for $\alpha = 10^{-7} s^{-1} = 8.64 \cdot 10^{-3}$/day. The input parameters have been listed in Table 13.1.

where the error function is defined as

$$\operatorname{erf}(x) = \frac{2}{\sqrt{\pi}} \int_0^x e^{-t^2} dt.$$

Since $b > 0$, it can be shown that this limit also holds for the cases with damping and using the current set of parameters, it follows that $X_f - X_0 = 4.5461$, which matches very well with the asymptotic behavior in Fig. 13.2. We can derive a similar equation for the time, τ, at which the contraction is maximal:

$$\theta = \tau + \alpha \sigma^2, \tag{13.23}$$

which shows that the delay of the maximum occurred contraction is proportional to the morphoelasticity rate constant and the duration of the forcing. Substituting $\tau = 70$ days, $\sigma = 21$ days, and $\alpha = 10^7$ s $= 0.864 \cdot 10^{-2}$/days gave $\theta - \tau = 3.81$ days, and $\alpha = 5 \cdot 10^{-7}$ s^{-1} and $\alpha = 10^{-6}$ s^{-1}, respectively, give $\theta - \tau = 19.05$ days and $\theta - \tau = 38.10$ days. These values match very well with Fig. 13.3. Further, the maximum contraction is given by

$$x(\theta) = X_0 + \frac{F_0}{k}\left(\alpha\sigma\sqrt{\frac{\pi}{2}}\operatorname{erf}\left(\frac{\alpha\sigma}{\sqrt{2}}\right) + e^{-\frac{\alpha}{2}}\right). \tag{13.24}$$

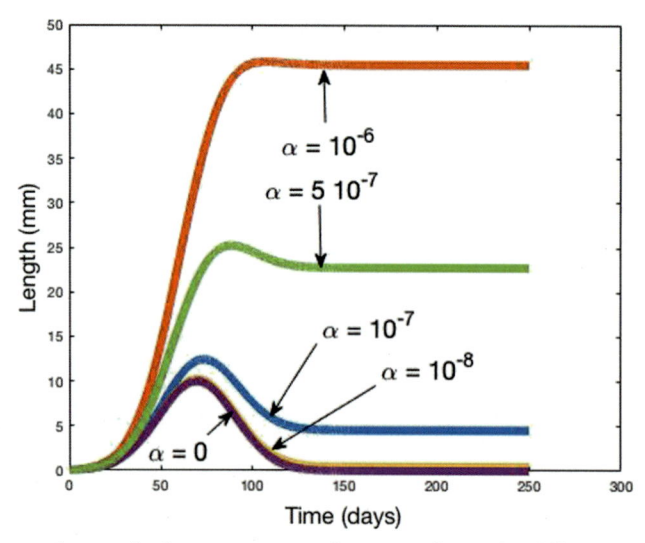

Figure 13.3 The actual wound edge position as a function of time for different values of morphoelasticity parameter α. The input parameters have been listed in Table 13.1.

Monte Carlo simulations

We carry out a parameter sensitivity analysis on the basis of sampling of input parameters from statistical distributions. We use the parameters in Table 13.1 as basis input parameters, except that the average value of the α—value of $5 \cdot 10^{-7}$ s^{-1}. We sample from normal distributions, where the mean is given by the values from Table 13.1, and the variance is given by 0.25 times the mean. Hence if μ_M denotes the mean of an input parameter M, then parameter M is sampled from

$$M \sim \mu_M + 0.25\, \mu_M\, \mathcal{N}(0,1).$$

This sampling is applied for the parameters k, τ, σ, F_0, and α, where the means follow from Table 13.1. The number of Monte Carlo samples was 100,000, where the actual computation took a couple of seconds. All samples are statistically independent. The results have been presented in Figs. 13.4 and 13.5. The resulting sample correlation coefficients are presented in Table 13.2.

Fig. 13.4 shows scatter plots for the final wound edge position, X_f, as a function of the input parameters. Regarding the elasticity k, it is reflected that smaller values of k represent smaller stiffness and hence weaker tissues. This implies that the same force applied on a weaker tissue will have a larger displacement, and hence a larger shift of the wound edge toward the center of the wound. Further, the time of onset of the forcing is varied, and

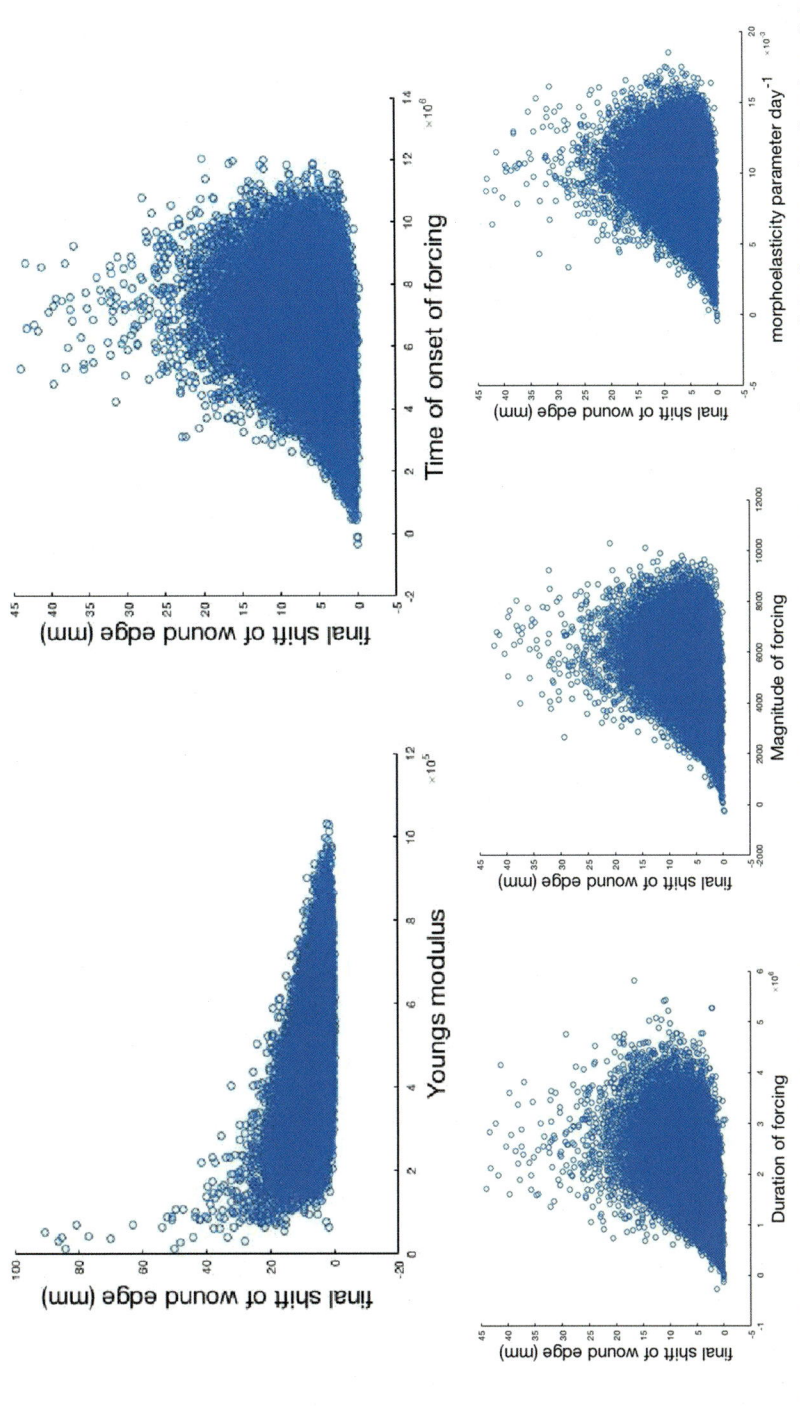

Figure 13.4 Scatter plots of the final wound edge position with the Young's modulus k, the forcing F_0, and the morphoelasticity parameter α. The number of Monte Carlo samples was 100,000; outliers beyond the 99% confidence interval have been removed.

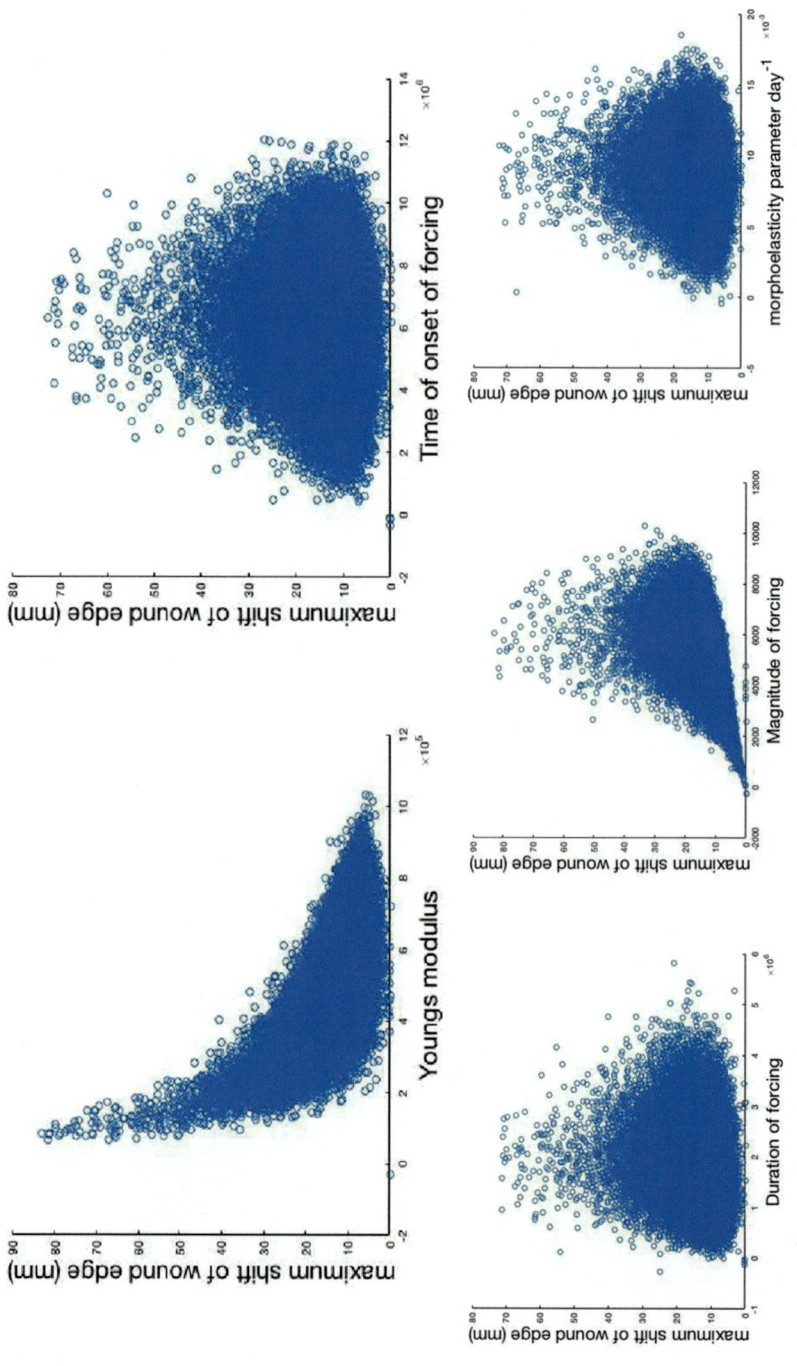

Figure 13.5 Scatter plots of the final wound edge position with the Young's modulus k, the time of onset of the forcing σ, forcing F_0, and the morphoelasticity parameter α. The number of Monte Carlo samples was 100,000, outliers beyond the 99% confidence interval have been removed.

Table 13.2 Correlations between the input and output parameters on the basis of the full differential equations. Here X_f and x_m, respectively, represent the final wound edge position and the extremal wound edge position.

	k	τ	σ	$|F_0|$	α
X_f	−0.3618	0.2886	0.4092	0.2887	0.2929
x_m	−0.4159	0.0702	0.0953	0.3310	0.0716

this parameter models the progress of the migration of fibroblasts, as well as the differentiation rate of fibroblasts to myofibroblasts. A higher rate of fibroblasts migration and its differentiation rate are modeled by the smaller onsets, τ of the forcing. It can be seen that the forcing sets in later, then the time integral over the forcing becomes a little larger, though this effect is not significant for the current parameter choice. However, we observe that the later onset of the forcing gives a larger shift of the final wound edge position, which reflects a larger contraction of the wound. Herewith a slower migration rate of fibroblasts is disadvantageous for the amount of contraction of the wound since the contraction is larger. The duration of the forcing, σ, gives a larger integral over the forcing over time, and hence the contraction becomes larger. This is clearly visible in Fig. 13.4 and in Table 13.2, though the correlations are not as strong as for the maximum shift of the wound edge, x_m. The same can be said regarding the magnitude of the forcing. The value of α reflects the adaptivity of the tissue to the straining, and reflects the ease at which plastic deformations are made. It can be seen that the alpha parameter has a significant impact on the plastic deformation, which is quantified by X_f.

Fig. 13.5 shows scatter plots for the maximum shift of the wound edge, x_m, as a function of the input parameters. Similar observations to Fig. 13.4 are seen; however, the correlations between the onset and duration, τ and σ, of the forcing, as well as the morphoelasticity parameter, α, are weaker since x_m depends less on the time integral over the forcing than the final wound position, X_f. Fig. 13.6 shows histograms of the maximum shift of the wound edge and of the final shift of the wound edge. It can be seen that the distributions look like a lognormal distribution. Furthermore, for the larger α−value, the difference between the maximum shift of the wound edge and the final shift of the wound edge is smaller than for the case of smaller α−values. This can also be seen in Fig. 13.3.

Discussion and conclusions

Of course, the current model represents a heavy simplification of reality, but in fact, all models are representations and simplifications of reality. The art of modeling is based on to what extent one can simplify. Simplified models have the merit of being characterized

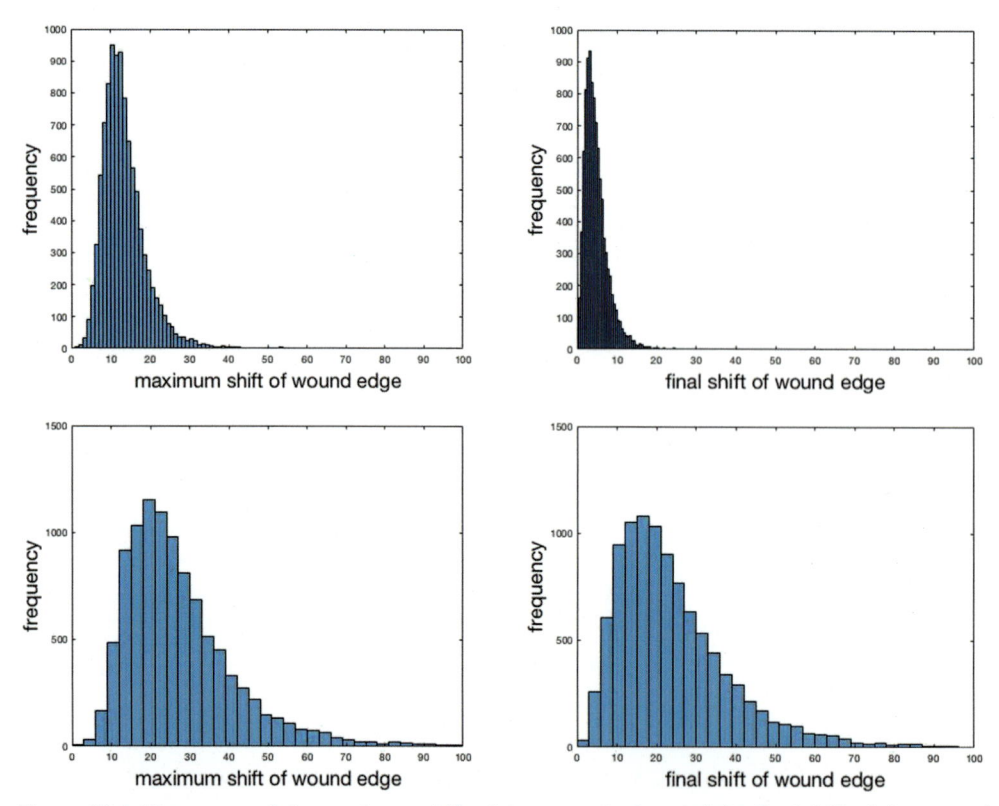

Figure 13.6 Histograms of the maximum shift of the wound edge and the final shift of the wound edge for the case that average α-value of 10^{-7} and $5 \cdot 10^{-7}$

by a small parameter space and by a simple, easy, and cheap simulation of the process one is interested in. These merits compromise the physics behind the processes. It is impossible to incorporate all the physical mechanisms in modeling a process. One important reason is that not all the physical processes are known and if they are known there are often multiple ways of dealing with these processes from a mathematical point of view. Once one has constructed the ultimate mathematical model, then one has to find the input parameters, which are hard to get since many of the parameters that are invented by modelers have never been measured or there is a large discrepancy in the literature about them. Hence more realistic and herewith models that are more physics-based are not necessarily better. Therefore, we investigate the performance of this extremely simplified way of modeling permanent changes of tissues in the wake of postburn scars. We also note that the current heavily simplified model generates extremely fast results and that one could investigate the possibility of combining this model with more complicated models through parameter fitting so that results from complicated models can be reproduced using this type of simplified models.

In Fig. 13.7, the wound area has been plotted as a function of time by the use of a complete morphoelasticity model. The results from Fig. 13.7 can be compared to the results in Fig. 13.2, where one could analyze the curve for $1 - x(t)$. It is obvious that the wound area tends to a new equilibrium. Regarding the Monte Carlo simulations, it can be seen that the permanent wound contraction increases for increase of α, τ, and σ. The increase of τ and σ may be linked to a later onset of contractile forces as a result of slower cell migration speeds and a weaker immune reaction. Weakening of the immune system can be linked to diabetes [11], in which hyperglycemia (excessively high blood sugar (glucose) levels) causes a reduced expression of β-defensins, which are antimicrobial peptides that enhance the resistance of epithelial areas. This weakening of the immune reaction increases the sensitivity of infections. This is observed among patients with type 2 diabetes mellitus, as well as among patients that are subject to aging and chronic diseases. Since the actual wound healing starts after the immune response, the frustration of the immune system contributes to delayed wound healing. Unfortunately, this is not the entire story. In Ref. [12], cellular dysfunctioning of diabetic fibroblasts was experimentally observed in mice. It is well known that fibroblasts are an essential cell phenotype for

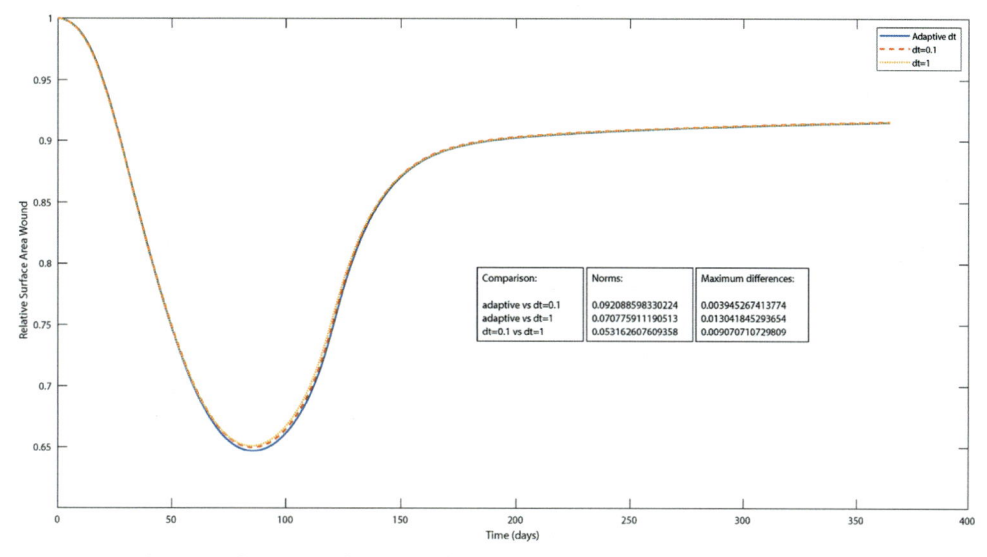

Figure 13.7 The wound area as a function of time for a complete morphoelasticity model. The result was generated, thanks to Ms Ginger Egberts.

wound repair. It has been evidenced in Ref. [12] that normal (wild-type) fibroblasts migrate 32% and 77% more than diabetic fibroblasts in response of fibronectin and collagen I, respectively. This implies that, in general, diabetic fibroblasts will arrive later in the wound site than normal fibroblasts and hence the onset of collagen expression, as well as the onset of contractile forces will be delayed as a result of diabetes. Other side effects are the diminished expression of VEGF by diabetic fibroblasts and herewith the development of a vascular network in the wound/ scar region is delayed, which gives rise to hypoxia and a further delayed and weakened of wounds, and herewith the time interval during which tissue is being regenerated is longer. Another side effect of diabetes is the reduction of migration speed of diabetic keratinocytes [13], which are an essential cell phenotype for the closure of the denuded epidermis. This decrease of keratinocyte migration speed reduces the wound closure rate. Wound closure is crucial for further wound healing and scar evolution. Both lower cell migration speeds and weaker immune systems can be linked to diabetes [11–13]. Furthermore, lower cell migration rates can be linked to a larger period of time over which the forces act, and hence this increases the magnitude of the permanent contraction. This can also be seen from Eq. (13.22).

Acknowledgment

The author thanks his PhD students Qiyao Peng and Ginger Egberts from the Delft University of Technology in the Netherlands for the fruitful discussions on morphoelasticity and for Fig. 13.6 in this manuscript. Furthermore, financial support from the Dutch Burns Foundation (project 17.105) and the China Scholarship Council is gratefully acknowledged.

References

[1] Hop MJ, Langenberg LC, Hiddingh J, Stekelenburg CM, van der Wal MBA, Hoogewerf CJ, van Koppen MLJ, Polinder S, van Zuijlen PPM, van Baar ME, Middelkoop E. Reconstructive surgery after burns: a 10 year follow-up study. Burns 2014;40:1544–51.

[2] Stekelenburg CM, Marck RE, Tuinebreijer WE, de Vet HC, Ogawa R, van Zuijlen PP. A systematic review on burn scar contracture treatment: searching for evidence. J Burn Care Res 2015;36(3): e153–61.

[3] Olsen L, Sherratt JA, Maini PK. A mechanochemical model for adult dermal wound closure and the permanence of the contracted tissue displacement role. J Theor Biol 1995;177:113–28.

[4] Javierre E, Moreo P, Doblare M, Garcia–Aznar JM. Numerical modeling of a mechano–chemical theory for wound contraction analysis. Int J Solid Struct 2009;46(20):3597–606.

[5] Byrne H, Drasdo D. Individual–based and continuum models of growing cell populations: a comparison. J Math Biol 2009;58:657–87.

[6] Boon WM, Koppenol DC, Vermolen FJ. A multi-agent cell-based model for wound contraction. J Biomech 2016;49(8):1388–401.

[7] Goriely A, Moulton D. Morphoelasticity: a theory of elastic growth. In: Ben Amar M, Goriely A, Michael Müller M, Cugliandolo L, editors. New trends in the physics and mechanics of biological systems: lecture notes of the Les Houches Summer School, vol. 92; July 2009. https://doi.org/10.1093/acprof:oso/9780199605835.003.0006.

[8] Koppenol DC, Vermolen FJ. Biomedical implications from a morphoelastic continuum model for the simulation of contracture formation in skin grafts that cover excised burns. Biomech Model Mechanobiol 2017;16(4):1187—206.

[9] Hall CL. reportModelling of some biological materials using continuum mechanics [Ph.D. thesis]. Australia: Queensland University of Technology; n.d.

[10] El Hadidy M, Tesauro P, Cavallini M, Colonna M, Rizzo F, Signorini M. Contraction and growth of deep burn wounds covered by non-meshed and meshed split thickness skin grafts in humans. Burns 1994;20:226—8.

[11] Kiselar JG, Wang X, Dubyak GR, El Sanadi C, Ghosh SK, Lundberg K, Williams WM. Modification of β-defensin-2 by dicarbonyls methylglyoxal and glyoxal inhibits antibacterial and chemotactic function in vitro. PLoS One 2015;10(8):e0130533. https://doi.org/10.1371/journal.pone.0130533.

[12] Lerman OZ, Galiano RD, Armour M, Levine JP, Gurtner GC. Cellular dysfunction in the diabetic fibroblasts: impairment in migration, vascular endothelial growth factor production, and response to hypoxia. Am J Pathol 2003;162(1):303—12. https://doi.org/10.1016/S0002-9440(10)63821-7.

[13] Li L, Zhang J, Zhang Q, Xiang F, Jia J, Wei P, Zhang J, Hu J, Huang Y. High glucose suppresses keratinocyte migration through the inhibition of p38 MAPK/Autophagy pathway. Front Physiol 2019;10:24. https://doi.org/10.3389/fphys.2019.00024.

CHAPTER 14

Mechanobiology of diabetes and its complications: from mechanisms to effective mechanotherapies

Chenyu Huang[1], Rei Ogawa[2]

[1]Department of Dermatology, Beijing Tsinghua Changgung Hospital, School of Clinical Medicine, Tsinghua University, Beijing, China; [2]Department of Plastic, Reconstructive and Aesthetic Surgery, Nippon Medical School, Tokyo, Japan

Diabetes mellitus (DM) is a chronic disease that is prevalent in nearly all countries regardless of their income. The most recent data from the International Diabetes Federation indicate that in 2019, DM affected 463 million adults: thus, the global prevalence is 9.3%. The vast majority of these patients (90%) have DM Type 2. The total number of DM patients is estimated to grow to 578 and 700 million by 2030 and 2045, respectively [1]. It is now widely accepted that the complications of DM are far more important than the high blood glucose itself since these complications are very hard to cure and associate with severe outcomes if they are ignored. Of particular concern are diabetic wounds, which require significant medical attention and intervention and thus associate with considerable economic expenditure. This is due in part to the profound intractability of these wounds along with their high prevalence: according to the database recording until September 2015, the global prevalence of the most common diabetic wounds, namely, diabetic foot ulcer (DFU), was 6.3% (95%CI = 5.4%−7.3%) [2].

Currently, there is increasing interest in the roles that mechanical forces play in both health and disease: Mechanobiological research has shown that many cell types sense and transduce mechanical forces and that these signals alter their intracellular biochemistry, gene expression, and the following cell behavior [3,4]. Moreover, it is now being increasingly recognized that these cellular responses to mechanical force can be exploited by mechanotherapy, namely, the application of various mechanical forces, to reduce or reverse injury to damaged tissues or promote the homeostasis of healthy tissues [5]. In this chapter, we will focus on the mechanobiology of diabetes and its complications, particularly diabetic wounds, and how the mechanotherapeutic interventions of negative pressure wound therapy (NPWT) and extracorporeal shock wave therapy (ESWT) can improve the outcomes of this disease. Our intention is to increase understanding of the involvement of mechanical forces in diabetes and its complications, with the hope that this will promote basic research into the underlying mechanisms and thereby facilitate the development of novel clinical approaches that can restore structure and function and aid the management of DM and its complications.

Mechanobiological changes in DM that affect wound healing

Normal wound healing involves a number of phases that progress in a timely and orderly fashion and eventually result in the restoration of the structural and functional integrity of the skin. Chronic diabetic wounds are characterized clinically by delays and disruption of this healing process [6]. There are multiple lines of evidence showing that DM associates with mechanobiological changes at various levels, all of which contribute significantly to the development and impaired healing of diabetic wounds. An example of **macro-level changes** is the stiffness of the joint and thickness of the skin and in diabetes: This impairs joint mobility and gait, which in turn results in abnormally high plantar foot pressure. When this pressure is combined with neuropathy, it significantly increases the risk of plantar ulcers [7]. Thus, prophylactic or therapeutic interventions that relieve, reduce, or redistribute the plantar pressure may play important roles in plantar ulcer management [8,9]. In terms of **cell-level changes**, multiple studies have shown that biomechanical cellular processes in wound healing are impaired in diabetes. One of these processes is wound contraction, which is driven by the contraction of myofibroblasts. Wound contraction is reduced in diabetes. For example, the study of Yu et al. with a diabetic wound model showed that the rate of wound contraction in diabetic mice decreased twofold compared with that observed in control mice [10]. Similarly, diabetes associates with impaired migration of cells that participate in wound healing. Cell migration is strongly shaped by mechanical cues such as extracellular matrix (ECM) stiffness and tensile/compressive stress from other cells. In particular, tensile stress arises from cells pulling on each other via cell—cell adhesion [11]. In diabetes, endothelial cells at the margins of DFUs express greater levels of adhesion molecules such as endothelial cell adhesion molecule. However, this change does not associate with the expected increased infiltration of inflammatory cells such as macrophages and $CD3^+$ T cells. This suggests that inflammatory cell infiltration has been disrupted despite the greater extravasation potential of the endothelial cells [12]. In addition, keratinocytes at the margins of diabetic ulcers exhibit impaired migration, as shown by reduced LM-3A32 expression. This change associates with high keratinocyte proliferation ($K67^+$) and activation ($K16^+$) yet differentiated phenotype ($K10^-/K2^-$). It is possible that the inability of keratinocytes in diabetic ulcers to adopt a migratory phenotype is at least partly responsible for the delayed wound closure in diabetes [13].

Diabetes also associates with changes in neurosensitivity to mechanical forces. For example, patients with non-insulin-dependent DM of moderate duration often exhibit somatic hypersensitivity in the eyes, including feelings of a foreign body in the eyes. Cui et al. showed that the corneal nerve terminals of these patients exhibit changes in the location, structure, and function of the mechanosensitive channel called pannexin1. Specifically, compared to pannexin1 from normal synaptosomes, pannexin1 in diabetic synaptosomes shows a strong localization to the membrane, upregulated glycosylation, less S-nitrosylation of the glyco-pannexin1, and greater release of ATP after potassium chloride stimulation. This study also showed that S-nitrosylation of pannexin1 may be an important mechanism for dictating the mechanochemical activity of pannexin1. Since sensory nerves play important roles in wound healing, including in the eye [14], these above defects in cornal nerve mechanosensation are likely to weaken corneal epithelium healing and thereby impair corneal functions [15].

Many studies also show that the diabetic wound dermis exhibits mechanobiological changes at the **molecular level**, specifically in mechanotransduction pathways. For example, when normal dermal fibroblasts are cultured with high glucose concentrations, their expression of focal adhesion kinase (FAK) is downregulated. This downregulation is also observed in dermal fibroblasts from diabetic animals. FAK integrates external mechanical stimuli from integrins and growth factor receptors and plays an important role in wound healing by regulating cell migration, proliferation, and differentiation and by promoting collagen deposition. The downregulation of FAK in diabetic fibroblasts was found to be due to the increased activity of calpain-1, which is induced by hyperglycemia: this high enzymatic activity degrades FAK. As shown by studies on diabetic mice with a splinted excisional skin wound, this downregulation of FAK significantly reduces collagen deposition in the dermis as well as a key mechanical property of the skin, namely, stiffness. The fact that intraperitoneal injections of an inhibitor of calpain-1 normalized the mechanical properties of the diabetic skin and improved wound healing suggests that targeting this pathway therapeutically could improve wound healing in diabetic patients [16].

The study of Yu et al. mentioned above that showed diminished wound contraction in diabetic mice also demonstrates derangement of a mechanotransduction pathway. Specifically, they found that the slow wound contraction in diabetes reflects aberrantly low mRNA expression by dermal fibroblasts of Yes-associated protein (YAP), which is a protein that mediates mechanotransduction. The downregulation of YAP in turn reduces the expression of transforming growth factor (TGF)-β and connective tissue growth factor, both of which play important roles in wound healing [10].

Mechanotherapy for diabetic wounds

At present, there are two main mechanotherapies that promote soft tissue healing in diabetic wounds, namely, NPWT and ESWT. They are medical devices that respectively use negative pressure and shock wave to reduce diabetic wound size. Both operate at the tissue level via mechanobiological signaling pathways.

NPWT (also called microdeformation wound therapy or vacuum-assisted closure) involves packing a polyurethane sponge into the wound bed, applying an occlusive dressing over the sponge, and then applying negative pressure on the wound via an outlet tube that is connected to a vacuum pump [17]. It mainly improves wound healing by

(i) removing extracellular fluid and (ii) inducing wound shrinkage (macrodeformation) and wound surface undulation (microdeformation), thereby activating cellular/molecular wound healing responses and promoting an optimal wound environment [18]. The molecular mechanisms that mediate the positive outcomes of these NPWT-induced changes in the wound remain to be completely clarified. However, studies on animal models and human diabetics are gradually revealing some of these mechanisms. In terms of animal models, Ma et al. showed that when wounds in the streptozotocin-induced diabetic rat model are exposed to 7—10 days of NPWT (−125 mmHg), their expression of angiogenin-1, Tie-2, and collagen type IV are upregulated. These changes are concomitant with the proliferation of pericytes, which are cells that wrap themselves around capillary endothelium and support both structural integrity and functional stability of microvasculature. These changes facilitate new vessel integrity/maturation and thereby promote blood flow perfusion [19]. In addition, the study of Chen et al. in diabetic mice showed that NPWT mobilizes fibrocytes [20]. Fibrocytes are a unique leukocyte subpopulation in the peripheral blood that migrate to wound bed soon after injury, differentiate into collagen- and α-smooth muscle actin (SMA)-producing fibroblast- and myofibroblast-like cells, and regulate normal wound healing by stimulating reepithelialization, wound contraction, and angiogenesis [21]. Indeed, Chen et al. also showed that NPWT promoted collagen-1 and α-SMA production in the wound and augmented angiogenesis [20]. In humans, multiple studies on continuous NPWT (−125 mmHg) for 7—10 days show that the therapy not only promotes DFU healing, it also changes the expression in the ulcer of multiple factors that are important for proper wound healing, as follows. (i) Thus, 7-day NPWT increases the protein and mRNA expression of cellular fibronectin and TGF-β1, which play central roles in wound healing [22]. (ii) The same regimen also lowers the expression of IκB-α and NF-κB P65, which are markers of a proinflammatory state [23]. This antiinflammatory effect of NPWT is supported by the fact that NPWT inhibits IL-6, tumor necrosis factor-α, and inducible nitric oxide synthase (NOS) expression, in part by downregulating the MAPK-JNK signaling pathway [24]. (iii) Ten-day NPWT decreases interleukin (IL)-1β, matrix metalloproteinase (MMP)-1, and MMP-9 mRNA levels and increases tissue inhibitor of metalloproteinases (TIMP)-1 mRNA level. The effects on the MMPs and TIMP suggest that NPWT promotes switching from ECM degradation to ECM deposition [25]. (iv) Seven-day NPWT significantly increased the number of circulating endothelial progenitor cells and the levels of VEGF and stromal cell—derived factor 1α (SDF1-α) in the serum and granulation tissues. Since endothelial progenitor cells are mobilized from the bone marrow by VEGF and SDF1-α to the peripheral blood and finally to ischemic wounds to repair damaged blood vessels, these findings suggest that NPWT promotes angiogenesis [26]. (v) Eight-day NPWT of neuropathic, noninfected DFUs reduced plasma angiopoietin-2 levels. This cytokine is higher in diabetic patients, associates with chronic diabetic complications, and, in certain contexts, acts as an antiangiogenic factor [27]. In summary, NPWT may improve diabetic wound healing by normalizing the angiogenic milieu, promoting angiogenesis, downregulating the proinflammatory status of the chronic wound, and promoting ECM deposition over ECM degradation.

In ESWT (also called shock wave therapy), biphasic high-energy acoustic waves are applied to the wound. These waves can be generated via electromagnetic, electrohydraulic, or piezoelectric technologies [28]. Both short- and long-term studies show that ESWT improves the healing of diabetic wounds. For example, when DFUs were treated twice a week for 3 weeks with ESWT at an energy flux density of 0.2 mJ/mm^2 and 250 shocks/cm^2 plus 500 shocks on the arterial beds supplying the ulcer, the ulcer area dropped significantly. Moreover, this was shown to associate with increased tissue oxygenation of the skin adjacent to the ulcer ($P = .044$) [29]. Increased local tissue oxygenation is known to promote angiogenesis [30]. The beneficial effects of ESWT on DFU healing were also shown by a clinical study with a 5-year follow-up period. Thus, 40 nonhealing (>3 months) DFUs from 38 patients and 32 foot ulcers from 29 nondiabetic patients were treated twice a week for 3 weeks with ESWT delivered at an ulcer size-dependent dosage: The number of pulses that were delivered was calculated by multiplying treatment area (cm^2) by 8. At least 500 shocks at E2 at 4 Hz (equivalent to 0.11 mJ/mm^2 energy flux density) were administered. Three months, one year, and five years later, the complete plus ≥50% healing rates of the diabetic ulcers were 42.5%, 73%, and 46%, respectively. However, these rates were still significantly lower than the rates of the nondiabetic ulcers (71.9%, 97%, and 77%, respectively). ESWT also improved local blood flow perfusion in both the diabetic and nondiabetic groups: This effect was already observed at 6 weeks and was further improved at 1 year. However, by 5 years, this effect of ESWT had waned, particularly in the diabetic group [31]. Similarly, a meta-analysis reported that ESWT of DFUs is superior to standard wound care in terms of complete wound healing (Odds Ratio = 2.66, 95%CI = 1.03−6.87, $I^2 = 0$%) and time to healing (64.5 ± 8.06 vs. 81.17 ± 4.35 days) [32]. The molecular mechanisms that underlie the positive effect of ESWT on diabetic wound are still being explored but a number of studies show that like NPWT, ESWT promotes angiogenesis and downregulates proinflammatory activity in the wound. Thus, when rats with streptozotocin-induced diabetes incur a dorsal skin defect and then undergo ESWT, the wound size improves in association with increased local neoangiogenesis (as shown by greater wound blood perfusion and increased expression of VEGF and endothelial NOS expression), cell regeneration (as shown by greater PCNA expression), and an earlier antiinflammatory responses (as shown by less leukocyte infiltration at the wound edge) [33]. Similarly, Chen et al. showed that in rats with streptozotocin-induced diabetes, the enhanced wound healing after ESWT (800 impulses, 0.09 mJ/mm^2, twice within a week) associates with VEGF and the MAPK pathway. Thus, 3 and 10 days after ESWT, the expression of VEGF, endothelial NOS, and Ki-67 in the wound rose. Anti-VEGF monoclonal antibody treatments blocked these changes as well as reducing the ability of ESWT to promote wound healing. The upregulated VEGF protein expression after ESWT associated with increased mRNA expressions on day 3 of the MAPK pathway related genes of Kras, Raf1, Mek1, Jnkk, Jnk, and Jun [34]. Similarly, Hayashi et al. showed that low-energy ESWT (energy flow density/shot = 0.25 mJ/mm^2, frequency = 4 Hz) accelerated the wound healing in mice with streptozotocin-induced diabetes by inducing endothelial NOS, which in turn promoted VEGF expression and neovascularization [35].

Diabetic wounds are highly intractable and common and thus have long been the focus of global efforts to identify effective therapies and preventive strategies. Recent progress in our understanding of the mechanobiology that underlies normal and aberrant wound healing suggests that targeting pathological mechanobiological processes and their underlying molecular mechanisms may be a fruitful therapeutic avenue for diabetic wounds. This is supported by the well-documented effectiveness of NPWT and ESWT, two well-known mechanotherapies, in diabetic wounds. It is likely that as our understanding of wound healing mechanobiology improves, further mechanotherapeutic strategies that act at the molecular, cellular, and/or tissue levels will evolve. This research may also help identify mechanobiological biomarkers that can serve to diagnose potentially problematic wound healing at an early stage, thus promoting administration of interventions that will help sidestep the development of large, unmanageable, infected wounds that require drastic measures such as amputation.

References

[1] International Diabetes Federation Atlas. 9th ed. The International Diabetes Federation Diabetes Atlas; 2019.
[2] Zhang P, Lu J, Jing Y, Tang S, Zhu D, Bi Y. Global epidemiology of diabetic foot ulceration: a systematic review and meta-analysis. Ann Med 2017;49:106—16.
[3] Ingber DE. Tensegrity-based mechanosensing from macro to micro. Prog Biophys Mol Biol 2008;97: 163—79.
[4] Ingber DE. Mechanobiology and diseases of mechanotransduction. Ann Med 2003;35:564—77.
[5] Huang C, Holfeld J, Schaden W, Orgill D, Ogawa R. Mechanotherapy: revisiting physical therapy and recruiting mechanobiology for a new era in medicine. Trends Mol Med 2013;9:555—64.
[6] Lazarus GS, Cooper DM, Knighton DR, Margolis DJ, Pecoraro RE, Rodeheaver G, et al. Definitions and guidelines for assessment of wounds and evaluation of healing. Arch Dermatol 1994;130:489—93.
[7] Fernando DJS, Masson EA, Veves A, Boulton AJM. Relationship of limited joint mobility to abnormal foot pressures and diabetic foot ulceration. Diabetes Care 1991;14:8—11.
[8] Boulton AJM. Pressure and the diabetic foot: clinical science and offloading techniques. Am J Surg 2004;187:17S—24S.
[9] Kato H, Takada T, Kawamura T, Hotta N, Torii S. The reduction and redistribution of plantar pressures using foot orthoses in diabetic patients. Diabetes Res Clin Pract 1996;31:115—8.
[10] Yu J, Choi S, Um J, Park KS. Reduced expression of YAP in dermal fibroblasts is associated with impaired wound healing in type 2 diabetic mice. Tissue Eng Regen Med 2017;14:49—55.
[11] Saw TB, Jain S, Ladoux B, Lim CT. Mechanobiology of collective cell migration. Cell Mol Bioeng 2015;8:3—13.

[12] Galkowska H, Wojewodzka U, Olszewski WL. Low recruitment of immune cells with increased expression of endothelial adhesion molecules in margins of the chronic diabetic foot ulcers. Wound Repair Regen 2005;13:248—54.

[13] Usui ML, Mansbridge JN, Carter WG, Fujita M, Olerud JE. Keratinocyte migration, proliferation, and differentiation in chronic ulcers from patients with diabetes and normal wounds. Histochem Cytochem 2008;56:687—96.

[14] Yu FS, Yin J, Lee P, Hwang FS, McDermott M. Sensory nerve regeneration after epithelium wounding in normal and diabetic cornea. Expet Rev Ophthalmol 2015;10:383—92.

[15] Cui H, Liu Y, Qin L, Wang L, Huang Y. Increased membrane localization of pannexin1 in human corneal synaptosomes causes enhanced stimulated ATP release in chronic diabetes mellitus. Medicine (Baltim) 2016;95:e5084.

[16] Liu W, Ma K, Kwon SH, Garg R, Patta YR, Fujiwara T, Gurtner GC. The abnormal architecture of healed diabetic ulcers is the result of FAK degradation by Calpain 1. J Invest Dermatol 2017;137: 1155—65.

[17] Saxena V, Orgill D, Kohane I. A set of genes previously implicated in the hypoxia response might be an important modulator in the rat ear tissue response to mechanical stretch. BMC Genom 2007;8:430.

[18] Huang C, Leavitt T, Bayer LR, Orgill DP. Effect of negative pressure wound therapy on wound healing. Curr Probl Surg 2014;51:301—31.

[19] Ma Z, Li Z, Shou K, Jian C, Li P, Niu Y, et al. Negative pressure wound therapy: regulating blood flow perfusion and microvessel maturation through microvascular pericytes. Int J Mol Med 2017;40: 1415—25.

[20] Chen D, Zhao Y, Li Z, Shou K, Zheng X, Li P, et al. Circulating fibrocyte mobilization in negative pressure wound therapy. J Cell Mol Med 2017;21:1513—22.

[21] Kao HK, Chen B, Murphy GF, Li Q, Orgill DP, Guo L. Peripheral blood fibrocytes: enhancement of wound healing by cell proliferation, re-epithelialization, contraction, and angiogenesis. Ann Surg 2011;254:1066—74.

[22] Yang S, Zhu L, Han R, Sun L, Dou J. Effect of negative pressure wound therapy on cellular fibronectin and transforming growth factor-β1 expression in diabetic foot wounds. Foot Ankle Int 2017;38:893—900.

[23] Wang T, He R, Zhao J, Mei JC, Shao MZ, Pan Y, et al. Negative pressure wound therapy inhibits inflammation and upregulates activating transcription factor-3 and downregulates nuclear factor-κB in diabetic patients with foot ulcerations. Diabetes Metab Res Rev 2017;33. https://doi.org/10.1002/dmrr.2871.

[24] Wang T, Li X, Fan L, Chen B, Liu J, Tao Y, et al. Negative pressure wound therapy promoted wound healing by suppressing inflammation via downregulating MAPK-JNK signaling pathway in diabetic foot patients. Diabetes Res Clin Pract 2019;150:81—9.

[25] Karam RA, Rezk NA, Rahman TMA, Saeed MA. Effect of negative pressure wound therapy on molecular markers in diabetic foot ulcers. Gene 2018;667:56—61.

[26] Mu S, Hua Q, Jia Y, Chen MW, Tang Y, Deng D, et al. Effect of negative-pressure wound therapy on the circulating number of peripheral endothelial progenitor cells in diabetic patients with mild to moderate degrees of ischaemic foot ulcer. Vascular 2019;27:381—9.

[27] Hohendorff J, Drozdz A, Borys S, Ludwig-Slomczynska AH, Kiec-Wilk B, Stepien EL. Effects of negative pressure wound therapy on levels of angiopoetin-2 and other selected circulating signaling molecules in patients with diabetic foot ulcer. J Diabetes Res 2019;2019:1756798.

[28] Wang CJ. Extracorporeal shockwave therapy in musculoskeletal disorders. J Orthop Surg Res 2012;7: 11.

[29] Jeppesen SM, Yderstraede KB, Rasmussen BSB, Hanna M, Lund L. Extracorporeal shockwave therapy in the treatment of chronic diabetic foot ulcers: a prospective randomised trial. J Wound Care 2016;25:641—9.

[30] Schugart RC, Friedman A, Zhao R, Sen CK. Wound angiogenesis as a function of tissue oxygen tension: a mathematical model. Proc Natl Acad Sci U S A 2008;105:2628—33.

[31] Wang CJ, T Wu C, Yang YJ, Liu RT, Kuo YR. Long-term outcomes of extracorporeal shockwave therapy for chronic foot ulcers. J Surg Res 2014;189:366—72.

[32] Hitchman LH, Totty JP, Raza A, Cai P, Smith GE, Carradice D, et al. Extracorporeal shockwave therapy for diabetic foot ulcers: a systematic review and meta-analysis. Ann Vasc Surg 2019;56:330—9.

[33] Kuo YR, Wang CT, Wang FS, Chiang YC, Wang CJ. Extracorporeal shock-wave therapy enhanced wound healing via increasing topical blood perfusion and tissue regeneration in a rat model of STZ-induced diabetes. Wound Repair Regen 2009;17:522—30.

[34] Chen RF, Chang CH, Wang CT, Yang MY, Wang CJ, Kuo YR. Modulation of VEGF and MAPK-related pathway involved in extracorporeal shockwave therapy accelerate diabetic wound healing. Wound Repair Regen 2019;27:69—79.

[35] Hayashi D, Kawakami K, Ito K, Ishii K, Tanno H, Imai Y, et al. Low-energy extracorporeal shock wave therapy enhances skin wound healing in diabetic mice: a critical role of endothelial nitric oxide synthase. Wound Repair Regen 2012;20:887—95.

CHAPTER 15

Application of tissue mechanics to clinical management of risk in the diabetic foot

Daniel Parker, Farina Hashmi
School of Health and Society, University of Salford, Salford, United Kingdom

Plantar soft tissue

The plantar soft tissue is composed of the skin and the plantar fat pads which lie on the outer surface of the sole. Together these tissues provide an interface between the load-bearing skeletal system and the ground during movement.

These tissues are exposed to a wide variety of loads associated with daily activities such as prolonged static standing, cyclic long-distance walking, and acute impacts of running and jumping. Loading can also occur in multiple directions due to the complex musculoskeletal arrangement within the foot allowing for both extrinsic and intrinsic control across multiple joints to produce a large range of movement in multiple planes.

To enable movement while protecting the skeletal system from excessive or damaging loads these tissues have a specialized multilayered structure which combines advantageous properties of viscous fat and elastic-fibrous matrices [1].

Plantar skin

The plantar skin is a multilayered structure divided as a superficial epidermis and deep dermis and hypodermis (or subcutis) [2]. Within the epidermis, regions are subdivided further with cells at different stages of differentiation on their migration from the stratum basale to the stratum corneum [3]. Throughout this process, the cells degrade and become highly keratinized producing a thick protective layer [4]. Keratin is a composite of two protein types (high and low sulfur content) produced within the cell and as a result fiber composition and arrangement can be highly controlled leading to multiple keratin variants [5,6]. The stratum corneum is the thickest component of the epidermis and has a specific keratinous arrangement which provides resistance to physical stress [7,8]. The dermoepidermal junction is the point where plantar dermis is tightly bound to the epidermis by the presence of rete ridges which reduce the impact of shear and prevent horizontal displacement [9]. This close adhesion of the skin layers improves the grip which can be generated on the plantar surface [9]. Keratin 9 is unique to the palmar and plantar epidermis and is situated on the peaks of the dermoepidermal junction, suggesting the function of reinforcing the stress-bearing portions of the epidermis [10]. Keratin 16 is found in the troughs of the dermoepidermal junction and provides plasticity between the ridges. Keratins 6 and 16, at the expense of K1 and K10, are expressed as a result of wounding or perturbation of the epidermis [11]. There are other distinguishing characteristics of plantar skin, as a whole, compared to non-ridged skin, including greater epidermal cell density and collagen density [12]. The unique nature of plantar skin enables it to resist tearing at the surface while also distributing throughout the tissue the high compressive and shear stresses experienced during weight-bearing motion.

Despite these structural qualities, plantar hyperkeratosis can occur in response to increased loading, resulting in increased stratum corneum thickness [13]. An increase in the degree of hyperkeratosis results in a reduction in stratum corneum hydration and an increased stiffness [14]. Plantar skin sensitivity can also be affected as transmission of stimuli occurs through multiple tissue layers before reaching a receptor [15]. The dermis and hypodermis contain a large number of sensory receptors, which provide feedback crucial to the control of motion during gait and protective sensation to detect potential harm [16].

The dermis contains elastin and collagen fibers which provide mechanical strength [12,17]. Elastin is a compliant biological rubber with an ordered and coiled fibrous structure which when stretched is brought into tension with increasing stiffness [5]. When tension is released the coiled structure allows elastic recoil and enables the tissue to restore its original shape [17]. Collagen fibers are arranged in random orientations within the dermis and have a folded structure which creates an increasing stiffness within the tissue as it is stretched [5,12,17].

Plantar fat pads

Multiple distinct fibroadipose layers lie deep to the plantar skin within the soft tissue; these layers are further specialized under regions of the heel and metatarsal heads with defined fat pads. The internal arrangement of these fat pads follows a regular honeycomb pattern in which chambers are divided into globules by fibroelastic bundles of collagen and elastin, forming a closed cell honeycomb structure [1,18]. The adipose enclosed within this structure differs from storage fat by inclusion of more unsaturated fatty acids which produce a softer and less viscous tissue which has been suggested to increase the efficacy as a viscous damper [19]. Although it is a highly effective shock-absorbing structure when compared to other regions of the foot it is still possible to damage the plantar soft tissue by high repetitive loading [20]. This suggests that, like most biological structures, the plantar soft tissue has a functional level at which it is most effective.

Structure of the heel pad

Within the heel pad, deep to the plantar skin are two fibroadipose layers which provide protective cushioning for loads transmitted through the calcaneus. These distinct layers described through histological studies can be easily observed using B–mode ultrasound due to the varied concentration of adipose and thus altered echogenicity (Fig. 15.1).

The most superficial fibroadipose layer can be described as microchambers due to the densely packed fibroadipose structure composed primarily of collagen matrix with small and regularly spaced globules of adipose [21]. The microchambers are fixed through fibrous ties to the reticular dermis and form a natural heel cup which resists excessive tissue bulging during compression [18,22]. The second deeper fibroadipose layer consists of macrochambers with large volumes of adipose bound by thick fibrous septa arranged in an overlapping spiral formation around the tuberosity of the calcaneus [21]. The arrangement of these chambers and the specialized honeycomb internal structure described above allow this layer to act as a viscous damper, dissipating the large impact energy of ground contact during loading. Between the micro- and macrochamber layers lies a subcutaneous and vestigial muscle layer the panniculus carnosus [18]. Although this muscle has no direct function in the heel its metabolic requirement has been proposed as a risk for tissue injury and atrophy leading to ulceration when blood flow is restricted [23]. The blood supply to this region is provided by the posterior tibial artery and the anterior and posterior branches of the peroneal artery with vascular trees branching along the septa of the fat pads before supplying the dermis [1]. A venous plexus formed within this region

The Heel Pad

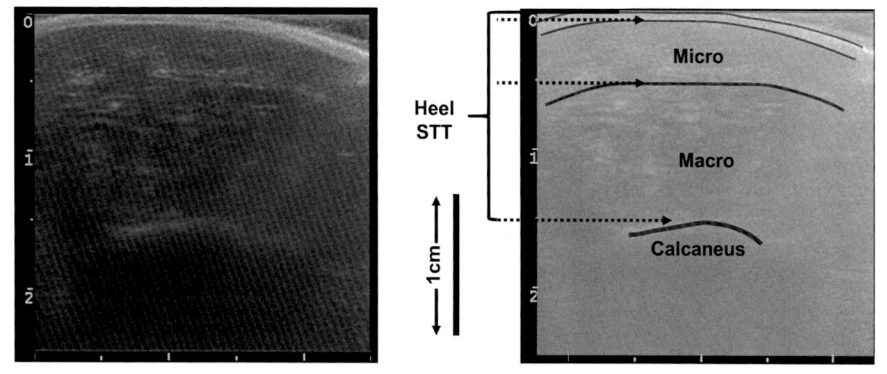

Figure 15.1 B-mode ultrasound coronal plane image of the plantar tissue structure in the heel pad in which the following layers are identified: *Macro*, macrochamber layer; *Micro*, microchamber layer, and the Heel Soft Tissue Thickness (STT) spans between the outer surface of the skin and the apex of the tuberosity of the calcaneus.

has been proposed to act as a secondary damper within the heel pad due to the energy use to displace blood which pools within it during gait [24]. The perfusion pressure which fills this region may also play a role in the rebound of the tissue between cycles.

Structure of the metatarsal pads

The forefoot is also adapted to provide cushioning during gait with subdermal fibroadipose structures positioned under loadbearing metatarsal heads. As with the heel pad on examination of ultrasound imaging these can be observed clearly and divided into two layers (Fig. 15.2). Cadaveric dissection has demonstrated distinct submetatarsal cushions which support the heads of the metatarsals [25]. Histomorphological studies for this region are limited; however, Wang et al. [18] described a similar basic structure to that found in the heel with epidermis, dermis, and two subcutaneous fat layers (superficial and deep) divided by the presence of the panniculus carnosus muscle. The subcutaneous layers of the metatarsal region also retain a dense microchamber structure superficially and a thicker macrochamber layer with large compartments of adipocytes [18]. Crossing the metatarsophalangeal joints lie tendons of the flexor muscles which are maintained in position by a flexor sheath and the transverse metatarsal ligament [25]. These tendon and ligamentous structures are resistant to compression and thus care should be taken to ensure they are not considered within the assessment of soft tissue. Computed topography studies have also identified changes to the skeletal structure with increased

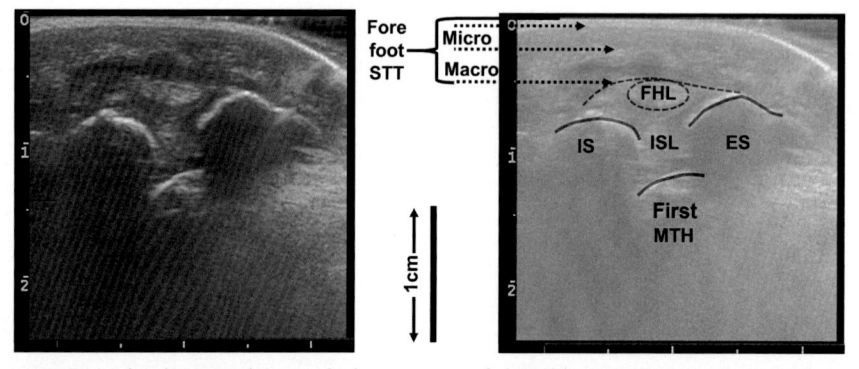

Figure 15.2 B-mode ultrasound coronal plane image of the plantar tissue structure in the forefoot pad at the first metatarsal head in which the following structures are identified: *ES*, external sesamoid; *FHL*, flexor hallucis longus; *first MTH*, first metatarsal head; *IS*, internal sesamoid; *ISL*, intersesamoidal ligament; *Macro*, macrochamber layer; *Micro*, microchamber layer, and the Forefoot Soft Tissue Thickness (STT) spans between the outer surface of the skin and the flexor sheath for the flexor hallucis longus tendon.

extension of the metatarsophalangeal joint and changes to joint condition for patients with diabetes; this weakens the joint, offloads the toe, and thus exposes the plantar tissue to increased loads [26].

The reduced size of the individual forefoot pads when compared to the heel is in part due to their reduced role in impact loading and also due to the other mechanisms within the forefoot which allow redistribution and dissipation of energy. The plantar aponeurosis forms strong fibrous ties with the dermal and subdermal tissue in the metatarsal region. This strong fibrous band allows force imposed on the metatarsal region to be transmitted to the calcaneus; this when combined with the transverse metatarsal ligament which ties the five metatarsals together ensures that the metatarsal region is loaded evenly [25]. The flexibility and large range of motion of the metatarsophalangeal joints also act to prevent overloading of the skeletal system; however, this motion poses a challenge to the positioning of the metatarsal pads and displacement commonly occurs in the presence of skeletal deformity [27].

Effect of diabetes on plantar soft tissue structure

It is well established that changes to the loadbearing soft tissues of the foot occur throughout the aging process leading to disruption of this specialized structure [1,28]. Advanced glycation end products (AGEs) are the main contributor to this disruption causing pathological cross-links to form within the collagen and elastin structures of the skin and fat pads [29]. The primary mechanical function of the collagenous tissue is to prevent the sliding of molecules under loads, which is achieved by the binding of intermolecular cross-links [30]. When glucose binds to collagen and elastin molecules it prevents long multiprotein chain formation and allows for irregular cross-link formation leading to nodular masses or plaques within the tissue [28] (Fig. 15.3).

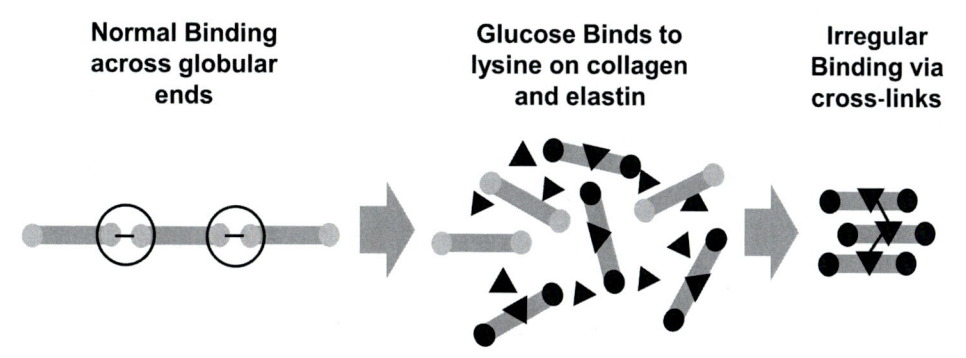

Figure 15.3 Mechanism of advanced glycation end products (AGEs) in the disruption of protein binding.

Hyperglycemia associated with diabetes can result in acceleration of these pathways with the accumulation of AGEs in collagenous structures causing an increase in diameter and closer packing of the fibers, in addition to irregular fiber morphology that may lead to increased stiffness [29]. This leads to a generally reduced capacity to handle loading at even body weight levels when exposed for periods of prolonged walking [31].

Differences in tissue thickness for the skin and fat pads have been reported by comparative studies. In a systematic review conducted by Morrison et al., metatarsal skin thickness and submetatarsal pad thickness were found to be reduced in diabetes participants while heel pad thickness displayed mixed results across studies but generally did not vary greatly ($\sim$5%) [32]. Analysis of tissue sublayers has also identified specific changes with diabetes. At the heel ultrasound studies showed increased tissue thickness of both microchambers (+4%) and macrochambers (+6%) [33]. Skin dermis thickness was found to be increased while epidermis thickness is decreased for people with diabetes when compared to age-matched controls [18].

A reduced capacity and inherent susceptibility to breakdown is a key component of ulceration risk in the diabetic foot [34]. In the presence of peripheral neuropathy where protective sensation is lost, risk is increased as the individual can often sustain damaging loading for longer [31]. Other key factors associated with tissue damage and ulceration risk include the level of vascular supply to the region, the local environment and capacity to clear toxins from the region, the duration for which a load is applied and held, and the magnitude of the load applied [35,36].

Although risk classification strategies for management of diabetes vary by locality, diabetes care is predominantly focused on vascular supply, nerve supply, and clinically observed tissue breakdown [37,38]. Measurement of these factors in a clinical setting is done to ensure the risk to which the patient is exposed is reduced in line with the susceptibility of the tissue. These approaches typically measure the level of irreparable damage which has occurred to a tissue through: degradation of arterial inflow leading to loss of measurable oxygen supply to tissue, nerve damage, or cell death causing a loss of protective sensation and perception, tissue hardening in response to increased load, or scaring as a result of prior breakdown. As a result, management of conditions in this way considerably limits the ability of clinicians to detect problems within tissue before damage occurs and to ensure this is managed effectively.

Biomechanics of the plantar soft tissue

This review focuses on the properties which can be readily quantified in a clinical setting, without the need for invasive procedures, highly specialized equipment, or the use of time-consuming protocols. Achieving accurate, valid measurement methods will ensure the true application of knowledge to the improvement of foot health outcomes.

In the first instance, to account for variation in the size of anatomical structures under investigation stress and strain should be used in which stress represents the load applied to a fixed cross-sectional area and strain represents the deformation as a percentage of the original tissue thickness [5]. Due to the varied tissue structure, composition and functional role of the plantar soft tissues, there is no single value which can fully explain the tissue's capacity. Further to this to fully understand and monitor the function of the plantar soft tissue it is necessary to characterize the response of the tissue to load using a combination of quasi-static and dynamic measures.

Nature of plantar soft tissue

The plantar soft tissue can be broadly described as exhibiting varied degrees of viscoelastic and hydrostatic response to loading. These behaviors in response to load are governed by the complex structural arrangement of the plantar tissue layers discussed above. Although the tissue structures could be considered individually to be elastic in the fibrous septa and viscous in the fatty globules these structures do not act independently and must be considered as composite structures or functional units.

Viscoelastic structures are highly loading rate dependent, with higher loading rates resulting in greater amounts of energy lost during the loading cycle [5]. For the plantar soft tissues high loading rates seen in impact or rapid compression during weight-bearing will result in the most energy loss while slow controlled loading allows greater energy return. This enables the plantar tissue to function as a natural damper/shock absorber [20].

Hydrostatic structures comprise a fluid-filled chamber surrounded by fibrous walls, compressive load placed on the enclosed fluid forces the walls into tension producing an even distribution of load throughout the structure [5]. Although the fat pads of the plantar tissue do exhibit this arrangement the large compressive strain possible within these layers suggests a delayed hydrostatic response occurs. Effectively within the subcutaneous tissues an initial large strain viscoelastic response occurs until the fibers of the septal walls are brought into tension and the chamber becomes incompressible. The hydrostatic response may act to prevent over compression of the tissue and reduce damage within the tissue due to long durations of compressive load during activities such as prolonged standing; however, studies of this component of the soft tissue response are largely absent from the literature.

The variety of mechanical responses and rate-dependent nature of the plantar soft tissue pose a challenge for measurement in vivo as many standardized approaches adopted from the field of engineering assume a degree of homogeneity which simply does not exist in biological structures.

Approaches to measurement of soft tissue mechanics

At present a range of systems exist within a research environment which are designed to evaluate the plantar soft tissues; however, no system exists in routine clinical practice for this purpose. Devices typically combine force or load measurement with imaging systems such as ultrasound, MRI, or CT to allow tissue deformation to be assessed.

Existing in vivo work can be divided into three groups:
(1) Nonactuated systems which rely on ultrasound imaging and static or quasi-static loading (standing) or freehand compression by the assessor [34,39–43] and those which utilized ultrasound shear wave elasticity with quasi-static loading [44–47].
(2) Devices to apply controlled dynamic loading [31,33,48–50] and including systems which use a calibrated pendulum or gait simulation routine [24,51,52].
(3) Walkway devices and those designed to be used in shoe to capture properties of the tissue during normal walking gait [53–56].

In many cases these systems are not suitable for implementation in clinical practice; however, advances in sensor technology and additive manufacture approaches do offer some routes to development of these systems further [57].

Approaches to superficial tissue characterization
Plantar skin elasticity

Application of vertical negative pressure can be applied using a cutometer to determine the elastic and viscoelastic properties of the plantar epidermis [93]. The method involves the application of constant negative pressure on the skin, which is brought to tension for a specific amount of time and then released. This generates a stress relaxation curve (Fig. 15.4), which allows estimation of the initial elastic response (Ue), the viscous component (U_{v1} and U_{v2}), the immediate retraction (Ur), and the plastic component (U_{v2}).

The variations in biophysical skin properties, related to body region, age, and sex, are known [58]. Elastic deformation reduces with age ($r^2 = 0.57$, $P < .001$, n = 30) [59]. Although there appears to be no difference between the sexes regarding biomechanical properties of skin, there is a significant difference in sebum content (males: 60.39 ± 74.52, females: 42.19 ± 54.10) [59]. The low transepidermal water loss values are reported between 10.00 and 13.16 g/m^2/h (dorsal hand, ventral forearms, scalp, elbow flexure, spinal area, and calf) while the highest numbers are between 41.69 and 46.37 g/m^2/h (ventral side of the hands, right axilla, and arch of the inner right foot). Stratum corneum hydration, using capacitance measures, is lowest on the scalp (27.7 AU) and highest in the spinal C4 area (84.9 AU) [58].

However, until recently, little was known about plantar skin in this context [60]. The validity of specific measurement techniques has been tested and the protocols have been used, not only to demonstrate differences in hydration and elasticity between different

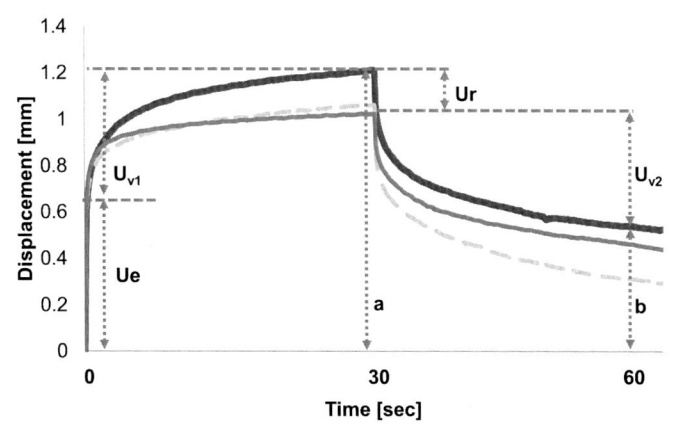

Figure 15.4 Curve illustrating the viscoelastic properties of healthy plantar skin (*thick black curve*) and the center (*dashed gray curve*) and edge (*gray curve*) of plantar callus generated by the application of vertical negative pressure (500 mbar). The healthy skin curve labels describe b/a, plastic component; U_e, immediate traction; U_r, immediate retraction of skin after the removal of negative pressure; U_{v1}, U_{v2} viscous component.

sites across plantar skin, but also to test the effects of treatments of plantar callus [14,59,61]. Fig. 15.4 illustrates the differences in stiffness properties between callus and healthy skin. Positive correlations between hydration and elasticity have also been evaluated [14].

Soft tissue hardness

The use of handheld durometers designed for estimation of the resistance of a material to indentation has been applied to the characterization of the plantar soft tissue due to its convenience of use and minimal preparation needed [62]. A durometer functions by measuring the displacement of an indentation tip under compression; a spring with calibrated stiffness resists compression providing a gauge which represents the work required to displace the tissue [63]. The most appropriate durometer type for the measurement of plantar soft tissue is the type OO, which applies a weak load and is calibrated for measurement of living biological tissue [62].

In many cases studies have applied protocol and definitions from the ASTM D2240 [63]. Application of this test method has some limitations for the analysis of skin hardness, most notably is the need for a minimum depth of 6 mm which is considerably larger than the thickness of the plantar skin. As a result, any measurements made using this method will incorporate hardness of the composite soft tissue layers which limits specificity possible when comparing layers across foot regions.

When the regional tissue is considered as a composite, increased hardness has been observed under loadbearing regions such as the heel, lateral midfoot, and forefoot when compared to nonloadbearing plantar regions such as the medial midfoot

[64–66]. Links between plantar tissue hardness and surface temperature have also been shown by Seixas et al. [66]; however, the mechanism which underpins this relationship is still unclear and it is likely that multiple factors including muscle function (heat generation) and tissue thickness (heat transmission) may play a role.

To date approaches have not assessed the relationship between skin stiffness determined using cutometer measures and skin hardness determined using a durometer.

Approaches to whole tissue characterization

For each of the variables discussed in this section data were collected in vivo for the heel pad soft tissue using an actuator-driven platen with integrated ultrasound imaging and force measurement [51]. Prior to testing the tissue was assessed using ultrasound to identify the location of the calcaneal tuberosity; the participant's foot was then positioned and braced to minimize leg movement and allow maximum compression of the whole tissue to be achieved (Fig. 15.5). Plantar heel pad tissue properties were calculated across three different loading conditions: cyclic quasi-static triangle wave at 6 mm/s velocity, cyclic sine wave with 22 mm/s peak velocity, and cyclic gait simulation with peak velocity of 311 mm/s. The data set has been presented as whole group average (n = 24, Male = 50%, Age = 43.29 ± 18.17, BMI = 23.93 ± 2.29) and also subdivided by age into young (n = 12, Male = 33%, Age = 24.5 ± 2.4, BMI = 22.98 ± 2.33) and elder (n = 12, Male = 67%, Age = 61.08 ± 4.59, BMI = 24.89 ± 1.81).

Compressibility of the plantar soft tissue

Compressibility is assessed using an imaging modality such as ultrasound. An image is captured in an unloaded state and a maximally loaded state such as weight-bearing and a compressibility ratio is determined.

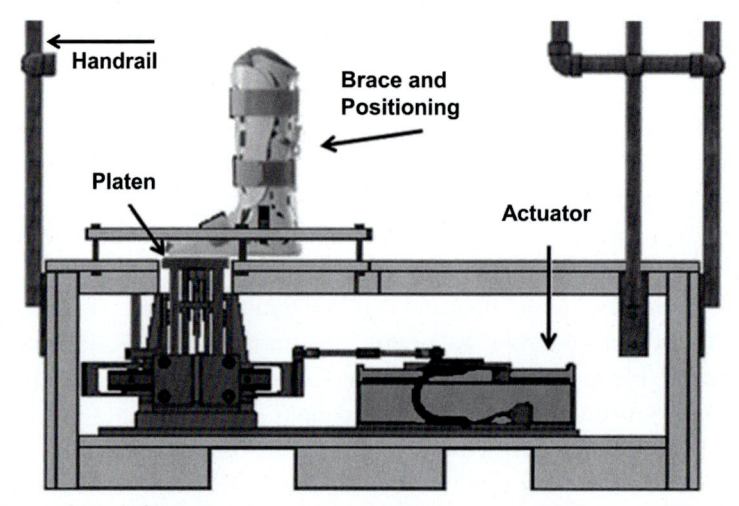

Figure 15.5 Soft tissue response imaging device setup and brace positioning.

$$\text{Compressibility} = \text{Peak Loaded Thickness}/\text{Unloaded Thickness}.$$

Compressibility can be determined for a single layer or for each layer within the tissue imaged. Compressibility varies by region and the nature of the tissue structure under compression, for example, in the heel region the most compressive structure is the macrochamber layer, with minimal compression in other layers.

A benefit of this measure is the relative ease of capture, requiring access to ultrasound and a simple platform. Analysis and interpretation are also quick to perform and intuitive. The challenge with this approach is its quasi-static nature. Given that the plantar soft tissues have viscous and elastic components, assessment based solely on factors like compressibility will be limited in terms of tissue characterization.

The compressibility for the whole group (n = 24) was 34% (±7%) at quasi-static rates, 38% (±4%) at low rates, and 36% (±6%) at high rates of compression (Fig. 15.6). The minimal rate effect observed when foot position and loading conditions are controlled suggests it is suitable to calculate compressibility at any rate of compression. No effect of age was observed on the measured compressibility index. Prior studies have observed increased compressibility with age at quasi-static rates [67]; however, in this case data were collected using an indentation approach, where increased compressibility is possible due to horizontal displacement of tissue under compression. Studies assessing compressibility in the metatarsal head region are limited with indentation studies demonstrating lower compressibility in the elderly population at both low and high velocities [68].

Stiffness of the plantar soft tissue

The plantar soft tissue exhibits a large displacement capacity, a nonlinear response to load, and a high rate dependence. These factors pose challenges for characterization of

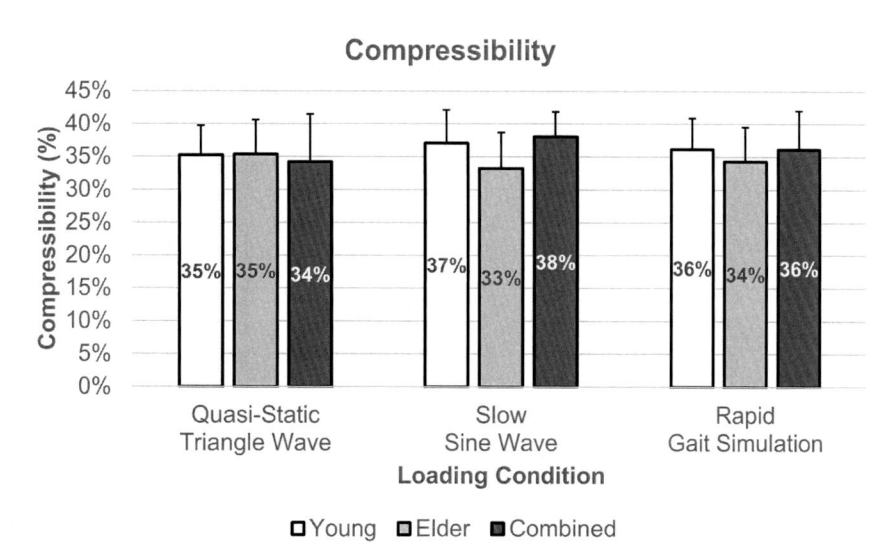

Figure 15.6 Compressibility for the plantar heel region in healthy young (n = 12) and elder (n = 12) groups. Combined all participant data (n = 24) have also been presented.

properties such as stiffness in a clinically meaningful way. As a result, standard testing protocols and analysis must be applied to characterize the stiffness in a way which allows comparison between individuals.

The Young's modulus represents the gradient of the stress—strain curve during the initial linear region in which the tissue could be considered to be purely elastic or Hookean (Fig. 15.7). For biological structures and especially the plantar soft tissue this linear region is often only present over a short strain range; further to this it is only possible to determine at low strain rates due to the viscous components of the tissue and the rate-dependent nature of the tissue response.

$$\text{Initial Elastic Modulus} = \frac{\text{Initial stress}}{\text{Initial strain}}$$

The dependence of the Young's modulus on the initial (low load, low rate) response considerably limits the relevance of this measure for the assessment of tissue function. However, this measure can be used to quantify changes to the elastic components of the plantar soft tissue (septal walls) and thus may be suitable for measurement of diabetes related changes caused by glycation pathways discussed previously.

The elastic modulus is highly rate dependent with increases in modulus observed between slow cyclic compression and loading at rates equivalent to walking gait (Fig. 15.8). An increased elastic modulus was observed in the elder group at all loading rates with a 57% increase at quasi-static rates, a 104% increase at slow rates, and a slight increase (8%) at rapid rates. These results align with prior studies showing linear stiffness of the plantar

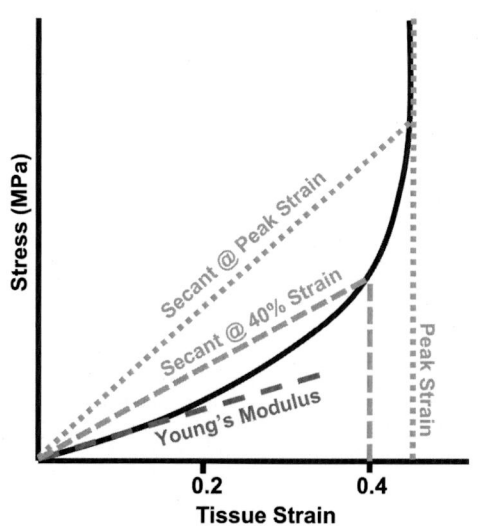

Figure 15.7 Nonlinear response of the plantar soft tissue in loading and the gradient of each estimate modulus.

Initial Elastic Modulus

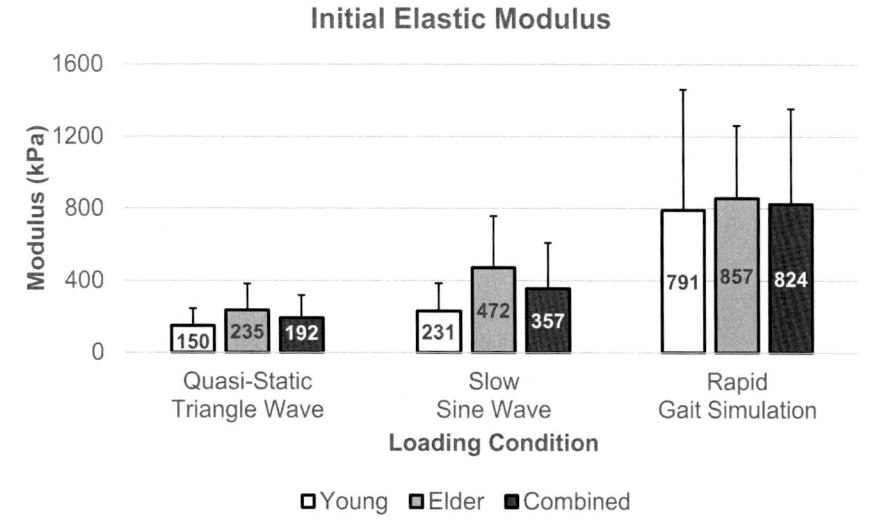

Figure 15.8 Initial elastic modulus for the plantar heel region in healthy young (n = 12) and elder (n = 12) groups. Combined all participant data (n = 24) have also been presented.

soft tissue, assessed at low velocities, to be greater in healthy elderly populations when compared to healthy young populations [67,69]. At other regions of the foot (MTH region) the linear stress–strain behavior has also been shown to be greater in the elderly population [39,70]. These findings suggest that quasi-static and slow rates of compression are most suitable for assessment of the linear elastic response of the plantar soft tissue. The elastic modulus has been shown to be greater in the elderly population when assessed at low velocities [39], and has also been shown to be greater in this population when calculated across a range of loading rates [56,68] suggesting isolation of elastic and viscous components may give a clearer image of the change which has occurred.

An alternative approach is to calculate a secant modulus which represents a gradient between zero and a defined point on the stress–strain curve [71]. The secant modulus is perhaps the simplest measure of tissue stiffness and this can be used to determine stiffness at set strain points (e.g., 10% strain) or at physiologically relevant limit such as peak strain (Fig. 15.7).

$$\text{Secant modulus at peak strain} = \frac{\text{Stress at peak strain}}{\text{Peak strain}}$$

The secant modulus at peak strain explains the force required to reach incompressible state of the tissue and thus the range through which the tissue can function to distribute load and as such provides an insight into the capacity of the tissue. Alternatively applying a standardized protocol with set loading rate and strain points to the measurement of the secant modulus could be used to permit comparison across individuals or to a reference grading system. The latter approach is seen within ISO standards [72] designed to characterize materials which are commonly used in the production of foot orthotics and footwear.

The secant modulus is also found to increase slightly with loading rate; however, it is less sensitive to loading rate than the elastic modulus. At low loading rates increases in the secant modulus are observed in the elder group with a 42% increase at quasi-static rates and a 57% increase at slow rates. A slight reduction (−11%) was observed in the secant modulus for the elder group at rapid rates of compression (Fig. 15.9).

Measurements of the secant modulus display a large range across prior studies [40,56,71,73], with no consistent pattern. Due to the reliance of the secant modulus on peak conditions, the applied loading routine and the nature of that loading (bulk compression or indentation) can have considerable effect. This limits the use of this characteristic within study comparisons unless a standardized approach to assessment is adopted. Percentage differences observed between the secant modulus of different populations may provide a more valid cross study variable.

Although neither of these approaches could be said to explain the response of a tissue to dynamic loading when applied following a standard approach, they do allow assessment of functionally relevant components of the plantar tissue's response.

Other approaches include estimation of the bulk compressive modulus or viscoelastic properties through dynamic material analysis or stress relaxation methodologies [49,74]. These approaches can provide a more standardized and clearly defined mechanical response; however, they are often very difficult to conduct in vivo due to the requirement to fix the test sample and conduct considerable preconditioning.

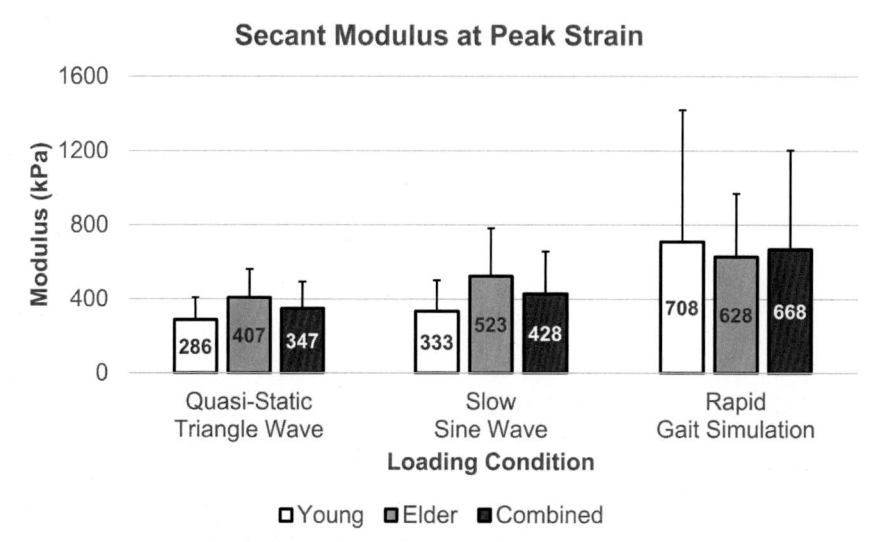

Figure 15.9 Secant modulus for the plantar heel region in healthy young (n = 12) and elder (n = 12) groups. Combined all participant data (n = 24) haves also been presented.

Shear wave elastography, a function of many commercial ultrasound systems, may also provide a convenient clinical assessment of stiffness allowing for the estimated shear and Young's moduli to be generated automatically. This approach has been applied to the assessment of both whole tissue and sublayer structure [47]. Chatzistergos et al. [46] have demonstrated significant linear relationships between the stiffness variables calculated through shear wave elastography and finite element/mechanical approaches; however, the authors recommended a focus on assessing difference in tissue stiffness with shear wave elastography due to the systemic underestimation of absolute stiffness variables.

Percentage energy loss

Assessing the full loading and unloading response of the tissue can allow estimation of the energy used to deform the tissue under compression. For the elastic components shape changes are reversible with energy is stored (when stretched) and released when the structure recoils, however, for the viscous component's energy is dissipated through mechanisms of viscous dampening which produces heat (Fig. 15.10). This dissipation of energy within the tissue acts to reduce the intensity of impact.

$$Percentage\ Energy\ Loss = \frac{Area\ bound\ by\ the\ stress - strain\ curve}{Area\ below\ the\ loading\ curve}$$

As the measured stress—strain curve can be either tissue or region specific it is important to consider the approach to measurement. Approaches which do not use imaging to determine tissue or sublayer deformation may be subject to error by inclusion of energy lost within other structures and joints in line with loading [75]. To perform effective

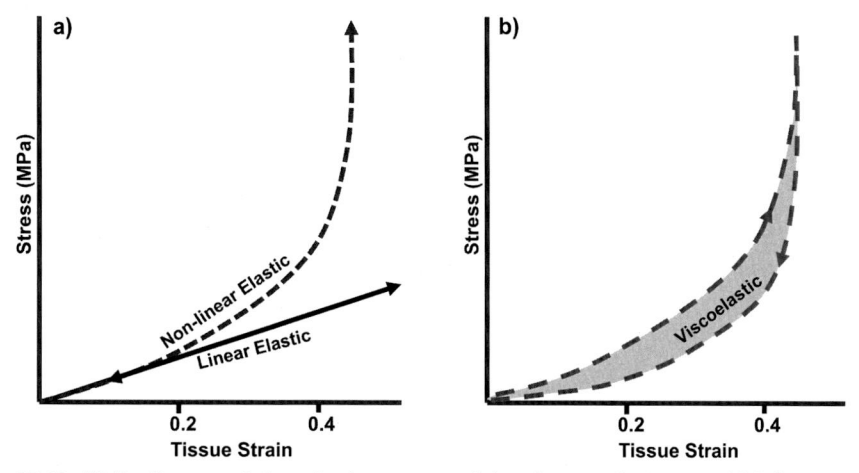

Figure 15.10 (A) Nonlinear and viscoelastic response of the plantar soft tissue and (B) the energy loss, area bound between loading and unloading curves.

assessment of the energy loss within the tissue it is also necessary to conduct preconditioning or multiple loading cycles prior to measurement as the first cycle response has been shown increased magnitude and variability when compared to a subsequent response after multiple loading cycles [75].

The percentage energy loss within the plantar heel pad tissue demonstrates clear increase from ∼30% to ∼60% from quasi-static to slow rates of compression; percentage energy loss did not demonstrate further increases with increased loading rate. It should be noted, however, that the absolute energy observed does continue to increase as a function of increased compression rate with whole group quasi-static $5.08 \pm 3.34 \, J/m^2$, slow $14.61 \pm 7.37 \, J/m^2$, and rapid $21.62 \pm 13.31 \, J/m^2$.

At low loading rates reductions in percentage energy loss were observed in the elder group with a -18.6% reduction at quasi-static rates and a -9.5% reduction at slow rates. No difference was observed at rapid rates of compression. The response observed at low rates does not align to prior studies which demonstrated increased percentage energy loss in an elder population in indentation testing [67]. Prior pendulum impact studies at rates equivalent to gait have also demonstrated no significant differences [52]; this may be the result of the inclusion of whole lower leg response to the pendulum and not specific response within the heel region. At other regions of the foot (MTH region) the percentage energy dissipated has been shown to be greater in the elderly population when tested at both low and high velocities [68].

Effect of diabetes on plantar soft tissue biomechanics

The changes to tissue structure as a result of diabetes have considerable impact on the tissue mechanics and the functional capacity of the plantar soft tissue. This reduced capacity can lead to an increased risk of tissue breakdown during even routine daily activities such as walking [38].

Plantar skin elasticity

Although the alterations in plantar skin structural and mechanical properties are suggested from the research, the direct correlation between the progression of diabetes and changes in mechanical properties is yet to be confirmed. This can only be achieved through longitudinal studies using large sample sizes and more work on measurement protocols. The latter is supported by the dearth of research on plantar skin in this field. Researchers have addressed these pragmatic issues by adopting strategies for collecting data on patients with different secondary complications of diabetes and evaluating data on the duration of diabetes in cross-sectional studies. Patients with and without diabetes, with and without peripheral neuropathy, have demonstrated significantly different plantar tissue mechanical properties, including plantar skin [76].

Based on cross-sectional studies plantar skin elasticity has been shown to decrease in people with diabetes [76,77]. Hashmi et al. [76] measured the presence of epidermal AGEs in patients with diabetes and the elasticity of the skin in the same group of participants. The results revealed a significantly greater amount of AGEs in the diabetic epidermis compared to controls and a significantly reduced plasticity of the plantar skin. Again, this was a cross-sectional study; therefore no direct relationship between AGE cross-linking and the mechanical properties of the affected tissues could be confirmed. Lechner et al. [77] reported a reduced total deformation under fixed load in the diabetic population at the heel (−14%), metatarsal head (−25%), and dorsum (−26%) with significant differences found across the plantar metatarsal and dorsal foot skin.

Plantar tissue hardness

The hardness of plantar skin determined by durometer measurement is increased in people with diabetes and significantly increased when both diabetes and neuropathy are present [64,65,78,79]. A positive correlation between neuropathy and tissue hardness has been shown for vibration perception threshold [64] and for foot loading [78,79]. Given the limitations of this approach it is not possible to determine if the changes identified occur within the plantar skin alone or within the composite of tissues. Despite these limitations a shore value greater than 35 degrees has been proposed to imply a risk for ulcer formation in the feet of people with diabetes [65].

Plantar tissue compressibility

For the heel region an increased compressibility (+15%) has been observed when using a fixed low load (30N) and quasi-static ultrasound indentation approach in vivo [48]. Compressive tests of in vitro tissue also showed a slight (6%) increase in heel compressibility when assessed 1 Hz cycles with loads equivalent to body weight [74]. It is hypothesized that the structural changes to the septal walls within the heel fat pad underpin this increased compressibility. This creates an extended loading curve and, in turn, an increased ability to dissipate energy and distribute load. Although advantageous for normal gait this is likely to result in further damage to the septal walls and reduce function of the hydrostatic skeleton which enables prolonged weight-bearing activities.

The effect of diabetes on the compressibility of the metatarsal fat pads has not been demonstrated in vivo. In vitro testing has demonstrated slight reductions within the region of the first (−4%) and third (−13%) metatarsal heads, with a slight increase (+6%) under the fifth metatarsal head [74].

Plantar tissue stiffness

Many cross-sectional studies have identified a significant increase in the initial elastic modulus for people with diabetes compared to age-matched control participants [34,42,43,74,80]. Increases are found across different plantar regions and for a range of measurement approaches and characterization methods with typical increases in the range of 50%—150%. Hsu et al. [33] applied the secant modulus to assessment of heel pad sublayers in patients with diabetes and demonstrated increased stiffness in the microchamber layer (+32%), yet a reduced stiffness in the combined microchamber and skin layers (−66%) demonstrating the need to asses tissue layers independently.

The presence of an increased stiffness at both low and high displacement rates suggests that changes have occurred which effect both the elastic and viscous responses of the fat pads. This is confirmed by the observations made in histological studies, which displayed significant disruption both to septal wall structure and to the elastin and collagen constituents [18]. Changes to the highly specialized fatty acid composition of the heel have also been observed although the small sample size of this study limits findings [19]. Significant correlations between septal wall thickness and elastic modulus have been demonstrated across multiple foot regions [81]. Further exploration of the fatty acid composition across the foot regions and their relationship to the mechanical response may yield insights into the viscous component.

Percentage energy loss

The percentage energy dissipated at the heel was shown to be greater in subjects with diabetes with increases of 7% for in vitro testing and increases of 29% for in vivo assessment [40,74]. Although the effect of diabetes was similar, due to differing methodologies, in vitro studies observed a considerably higher energy dissipation of 64% while the in vivo approach reported only 36%. These differences are expected to be the result of differing loading rates applied (1 Hz/~30 mm/s in vitro and 6 mm/s in vivo) and are in line with the step change seen from quasi-static to dynamic test approaches (Fig. 15.11). Energy loss was found to further increase from 63% to 75% with increased test frequency from 1 to 10 Hz for diabetic tissue tested in vitro; this relationship was consistent with the change observed in nondiabetic tissue [74]. For the metatarsal region energy dissipation was also observed to increase; however, the observed change was much lower with energy dissipation at the first metatarsal increased by 1% when tested in vivo and 4% in vitro [74,82]. Methodologies and loading rates were more consistent in this comparison with both approaches using rates greater than 30 mm/s and yielding energy dissipations of 60%—70%. Significant correlations have been shown across a range of foot tissue regions between the amount of collagen and elastin within the tissue and the energy loss has been shown [81].

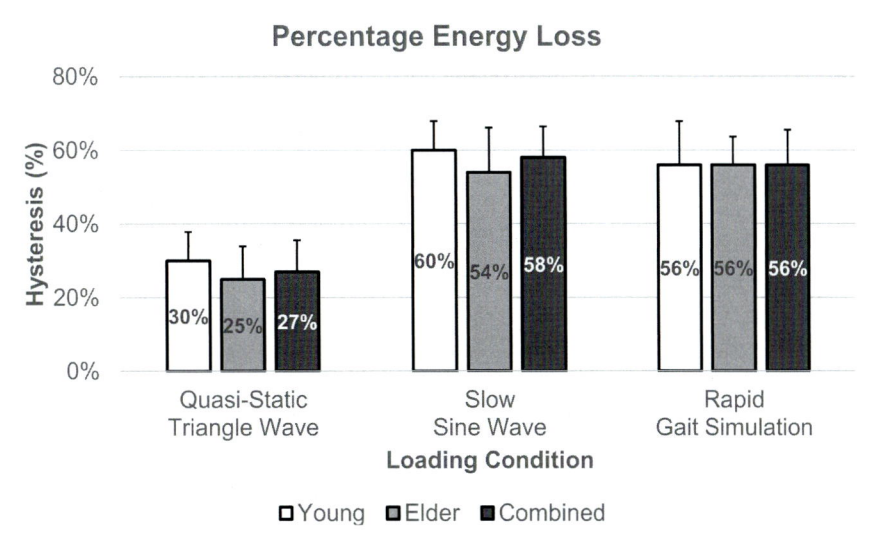

Figure 15.11 Percentage energy loss for the plantar heel region in healthy young (n = 12) and elder (n = 12) groups. Combined all participant data (n = 24) has also been presented.

It is expected that the combination of increased compressibility and increased stiffness observed at the heel results in an increased work within the tissue and thus greater capacity for energy loss. The slightly reduced compressible range at the metatarsals in patients with diabetes may explain the lower change observed in energy dissipation for this region.

Loading and simulation

The behavior of the skin and deeper tissues under shear load is still an active area of investigation and it is expected that outcomes of this work will have impact on the design of clinically relevant measurement approaches and should be considered going forward [36].

The most common foot skin complaints are calluses and blisters, which have a detrimental impact on the functional status of the foot [83]. Although it is known that these conditions form due to high activity levels or ill-fitting footwear and hosiery, a direct link between causal factors and the mechanical behavior of plantar epidermis has not been elucidated. Callus and blisters are thought to develop in response to compression, friction, and shear forces [14]. Although compression forces have been measured successfully in foot tissues, little is known about the contribution of friction and shear forces to the risk of developing foot pathologies.

As a result, clinical therapies are mainly focused on addressing the compression (vertical) forces, with the shear and friction (horizontal) forces often being neglected. A greater understanding of the behavior of the skin and soft tissues in normal and diseased states is required.

Summary of the current evidence base

Diabetes has been shown to clearly change the mechanical response of the plantar soft tissue and thus the inherent capacity of this tissue to protect the skeletal system during activity. An increase in the stiffness of the skin and deeper soft tissues across a range of loading conditions may be the clearest indication of the effects of structural changes which have occurred [81,84]. The disruption of the regular structure within the fat pads of the plantar tissue has impact on the compressive range over which tissue functions, and in turn the potential for the tissue to dissipate energy [81]. Although these changes from a mechanical perspective appear to improve the soft tissue's ability to handle increased loads, they may result in an increase in the overall stress imposed on the calcaneus and increase the risk of deep tissue damage [85].

Relevance for clinical management of diabetes

Currently, there are no clinical therapies that are used to prevent the accumulation of AGEs in collagen tissues. Therapies that can reverse the stiffening process of plantar soft tissues in diabetes in combination with appropriate glycemic control would be beneficial. In the absence of such therapies appropriate measurement and monitoring is essential to prevent tissue damage.

Measuring changes to soft tissue structure, although appealing as quick diagnostic approach, does not always clearly relate to the tissue mechanics [32,81]. The relationships between the complexity of the plantar tissue structure and the range of functional capacity or redundancy within the tissue are still not clearly understood. However, in some cases it is possible to make clinical judgment based on diagnostic imaging alone. For example, there is strong evidence to support that the physical displacement of the forefoot and heel pad tissues plays an important role in the clinical consequences of the foot ulceration [27,86].

Identification of patients where tissue mechanical changes are developing may be possible by assessment of other commonly used measures in the management of diabetes. Chatzistergos (2014) demonstrated a relationship between tissue stiffness and fasting blood sugar and triglyceride measures suggesting a link to hyperglycemia [94]. Further to this the association between increased tissue stiffness and increased age has been demonstrated through multiple studies allowing for some degree of risk stratification to be applied [67,68].

The relationship between internal tissue properties and external stress or plantar pressure has been demonstrated both in vivo and through finite element modeling [43,87]. Jan [43] showed that for the first metatarsal head there were significant correlations between soft tissue thickness and both peak pressure and peak pressure gradient. The effective Young's modulus and initial Young's modulus calculated using a handheld ultrasound indenter correlated with peak pressure gradient. The peak pressure gradient describes the distribution of pressure around a peak and a high peak pressure gradient has been suggested to contribute to skin breakdown due to large shearing stresses generated [88]. Cheung et al. [87] used a finite element modeling approach to vary the plantar soft tissue mechanical properties and found that fivefold increase in soft tissue stiffness resulted in 33% increase in pressure for the rearfoot and a 35% increase for the forefoot.

The increase in stiffness of the plantar soft tissue not only leads to a decrease in shock absorption during weight-bearing activities but also compromises the ability to dissipate internal shear stresses [89]. High shear stress may also contribute to the formation of plantar callus and therefore aggravate the high plantar pressure distributions and further increase the risk of the development of diabetic foot ulcers [90]. Although it is recognized that mechanical dermatoses of plantar epidermis increase the risk of ulceration in diabetes little work has been conducted on effects of these conditions on the mechanical behaviors of plantar skin [91].

An initial prospective study investigating tissue thickness and stiffness assessed using strain elastography has shown some improvement in the ability to predict ulceration within 12 months [92]. Further work of this nature to monitor tissue mechanics and the outcomes of patients at risk of ulceration is crucial to the development of appropriate methodologies.

Outlook in the context of clinical practice

The morbidity, mortality, and additional medical cost are considerably increased for patients with diabetic foot ulceration, yet the current approaches to prevention do not assess the mechanical integrity of the tissues most at risk [37,38]. Given the clear changes to both structure and mechanics of the plantar tissue in diabetes, this lack of routine assessment of these factors prevents a clinician from making a truly informed choice about management for their patient.

To ensure adoption of the measurement of soft tissue mechanics within clinical practice, it is essential to progress development "outside of the lab" and the next generation of devices should focus on enabling use within a clinical setting and without the requirement for specialist training.

There exists a need to bring together techniques and protocols to develop a full picture of plantar anatomical structures, their interaction, and their response to loading (pressure and shear). Introduction of these assessment approaches within longitudinal studies will broaden the knowledge base and clinical relevance. Exploring how and at what rate the plantar tissue mechanics change over time in people with diabetes will provide a better understanding of the capacity for tissues to manage external loads (in vivo) and to function in the dissipation of potentially damaging forces. The insights from this work will have applications to the assessment of patients with diabetes and therefore may enable clinicians to not only manage the progressive high-risk complications of diabetes but also detect and respond to the early signs of diabetic foot disease.

References

[1] Jahss MH, Michelson JD, Desai P, Kaye R, Reich S, Kummer F, et al. Investigations into the fat pads of the sole of the foot: anatomy and histology. Foot Ankle Int 1992;13:233–42. https://doi.org/10.1177/107110079201300502.

[2] Honari G, Maibach H. Skin structure and function. Clin Asp 2014:1–10. https://doi.org/10.1016/B978-0-12-420130-9.00001-3.

[3] Weinstein GD, Van Scott EJ. Autoradiographic analysis of turnover times of normal and psoriatic epidermis. J Invest Dermotol 1965;45:257–62.

[4] Fuchs E. Keratins and the skin. Annu Rev Cell Dev Biol 1995;11:123–53.

[5] Ennos A. Solid biomechanics. Princeton University Press; 2012.

[6] Broekaert D, Cooreman K, Coucke P, Nsabumukunzi S, Reyniers P, Kluyskens P, et al. A quantitative histochemical study of sulphydryl and disulphide content during normal epidermal keratinization. Histochem J 1982;14:573–84. https://doi.org/10.1007/BF01011890.

[7] Swensson O, Langbein L, McMillan J, Stevens H, Leigh I, McLean W, et al. Specialized keratin expression pattern in human ridged skin as an adaptation to high physical stress. Br J Dermatol 1998;139:767.

[8] Thomas SE, Dykes PJ, Marks R. Plantar hyperkeratosis: a study of callosities and normal plantar skin. J Invest Dermatol 1985;85:394–7. https://doi.org/10.1111/1523-1747.ep12277052.

[9] Derler S, Gerhardt L-CC. Tribology of skin: review and analysis of experimental results for the friction coefficient of human skin. Tribol Lett 2012;45:1–27. https://doi.org/10.1007/s11249-011-9854-y.

[10] Görmar FE, Bernd A, Bereiter-Hahn J, Holzmann H. A new model of epidermal differentiation: induction by mechanical stimulation. Arch Dermatol Res 1990;282:22–32. https://doi.org/10.1007/BF00505641.

[11] Paladini RD, Coulombe PA. The functional diversity of epidermal keratins revealed by the partial rescue of the keratin 14 null phenotype by keratin 16. J Cell Biol 1999;146:1185–201. https://doi.org/10.1083/jcb.146.5.1185.

[12] Vela-Romera A, Carriel V, Martín-Piedra MA, Aneiros-Fernández J, Campos F, Chato-Astrain J, et al. Characterization of the human ridged and non-ridged skin: a comprehensive histological, histochemical and immunohistochemical analysis. Histochem Cell Biol 2019;151:57–73. https://doi.org/10.1007/s00418-018-1701-x.

[13] Boyle CJ, Plotczyk M, Villalta SF, Patel S, Hettiaratchy S, Masouros SD, et al. Morphology and composition play distinct and complementary roles in the tolerance of plantar skin to mechanical load. Sci Adv 2019;5. https://doi.org/10.1126/sciadv.aay0244. eaay0244.

[14] Hashmi F, Nester C, Wright C, Newton V, Lam S. Characterising the biophysical properties of normal and hyperkeratotic foot skin. J Foot Ankle Res 2015;8. https://doi.org/10.1186/s13047-015-0092-7.

[15] Strzalkowski NDJ, Triano JJ, Lam CK, Templeton CA, Bent LR. Thresholds of skin sensitivity are partially influenced by mechanical properties of the skin on the foot sole. Phys Rep 2015;3. https://doi.org/10.14814/phy2.12425.

[16] Maurer C, Mergner T, Bolha B, Hlavacka F. Human balance control during cutaneous stimulation of the plantar soles. Neurosci Lett 2001;302:45–8.

[17] Hussain SH, Limthongkul B, Humphreys TR. The biomechanical properties of the skin. Dermatol Surg 2013;39:193–203. https://doi.org/10.1111/dsu.12095.
[18] Wang YN, Lee K, Ledoux WR. Histomorphological evaluation of diabetic and non-diabetic plantar soft tissue. Foot Ankle Int 2011;32:802–10. https://doi.org/10.3113/FAI.2011.0802.
[19] Buschmann WR, Hudgins LC, Kummer F, Desai P, Jahss MH. Fatty acid composition of normal and atrophied heel fat pad. Foot Ankle 1993;14:389–94. https://doi.org/10.1177/107110079301400704.
[20] Jørgensen U. Achillodynia and loss of heel pad shock absorbency. Am J Sports Med 1985;13:128–32. https://doi.org/10.1177/036354658501300209.
[21] Blechschmidt E. The structure of the calcaneal padding. Foot Ankle 1982;2:260–83.
[22] Hsu CC, Tsai WC, Wang CL, Pao SH, Shau YW, Chuan YS. Microchambers and macrochambers in heel pads: are they functionally different? J Appl Physiol 2007;102:2227–31. https://doi.org/10.1152/japplphysiol.01137.2006.
[23] Cichowitz A, Pan WR, Ashton M. The heel. Ann Plast Surg 2009;62:423–9. https://doi.org/10.1097/SAP.0b013e3181851b55.
[24] Weijers RE, Kessels AGH, Kemerink GJ. The damping properties of the venous plexus of the heel region of the foot during simulated heelstrike. J Biomech 2005;38:2423–30. https://doi.org/10.1016/j.jbiomech.2004.10.006.
[25] Bojsen-Moller F, Flagstad KE. Plantar aponeurosis and internal architecture of the ball of the foot. J Anat 1976;121:599–611.
[26] Robertson D, Mueller M, Smith K, Johnson J, Commean P, Pilgram T, et al. Structural changes in the forefoot of individuals with diabetes and a prior plantar ulcer. J Bone Joint Surg Br 2002;84:1395–404.
[27] Stainsby GD. Pathological anatomy and dynamic effect of the displaced plantar plate and the importance of the integrity of the plantar plate-deep transverse metatarsal ligament tie-bar. Ann R Coll Surg Engl 1997;79:58–68.
[28] Brash PD, Foster J, Vennart W, Anthony P, Tooke JE. Magnetic resonance imaging techniques demonstrate soft tissue damage in the diabetic foot. Diabet Med 1999;16:55–61. https://doi.org/10.1046/j.1464-5491.1999.00005.x.
[29] Dalal S, Widgerow AD, Evans GRD. The plantar fat pad and the diabetic foot - a review. Int Wound J 2015;12:636–40. https://doi.org/10.1111/iwj.12173.
[30] Avery NC, Bailey AJ. Enzymic and non-enzymic cross-linking mechanisms in relation to turnover of collagen: relevance to aging and exercise. Scand J Med Sci Sports 2005;15:231–40. https://doi.org/10.1111/j.1600-0838.2005.00464.x. John Wiley & Sons, Ltd.
[31] Kwak Y, Kim J, Lee KM, Koo S. Increase of stiffness in plantar fat tissue in diabetic patients. J Biomech 2020;107:109857. https://doi.org/10.1016/j.jbiomech.2020.109857.
[32] Morrison T, Jones S, Causby RS, Thoirs K. Can ultrasound measures of intrinsic foot muscles and plantar soft tissues predict future diabetes-related foot disease? A systematic review. PLoS One 2018;13. https://doi.org/10.1371/journal.pone.0199055.
[33] Hsu C, Tsai W, Hsiao T, Tseng F, Shau Y, Wang CL, et al. Diabetic effects on microchambers and macrochambers tissue properties in human heel pads. Clin Biomech 2009;24:682–6.
[34] Klaesner JW, Hastings MK, Zou D, Lewis C, Mueller MJ. Plantar tissue stiffness in patients with diabetes mellitus and peripheral neuropathy. Arch Phys Med Rehabil 2002;83:1796–801. https://doi.org/10.1053/apmr.2002.35661.
[35] Linder-Ganz E, Engelberg S, Scheinowitz M, Gefen A. Pressure–time cell death threshold for albino rat skeletal muscles as related to pressure sore biomechanics. J Biomech 2006;39:2725–32. https://doi.org/10.1016/j.jbiomech.2005.08.010.
[36] Bader DL, Worsley PR. Technologies to monitor the health of loaded skin tissues. Biomed Eng Online 2018;17:1–19. https://doi.org/10.1186/s12938-018-0470-z.
[37] NICE. Diabetic foot problems: prevention and management [NG19]. National Institue for Health Care Excellence; 2015.

[38] Bus SA, Lavery LA, Monteiro-Soares M, Rasmussen A, Raspovic A, Sacco ICN, et al. Guidelines on the prevention of foot ulcers in persons with diabetes (IWGDF 2019 update). Diabetes Metab Res Rev 2020;36. https://doi.org/10.1002/dmrr.3269.

[39] Zheng YP, Choi YKCC, Wong K, Chan S, Mak AFTT. Biomechanical assessment of plantar foot tissue in diabetic patients using an ultrasound indentation system. Ultrasound Med Biol 2000;26: 451−6. https://doi.org/10.1016/S0301-5629(99)00163-5.

[40] Hsu TC, Wang CL, Shau YW, Tang FT, Li KL, Chen CY. Altered heel-pad mechanical properties in patients with type 2 diabetes mellitus. Diabet Med 2000;17:854−9. https://doi.org/10.1046/j.1464-5491.2000.00394.x.

[41] Chao CYL, Zheng YP, Cheing GLY. Epidermal thickness and biomechanical properties of plantar tissues in diabetic foot. Ultrasound Med Biol 2011;37:1029−38. https://doi.org/10.1016/j.ultrasmedbio.2011.04.004.

[42] Sun J, Cheng B, Zheng Y, Huang Y. Changes in the thickness and stiffness of plantar soft tissues in people with diabetic peripheral neuropathy. Arch Phys Med Rehabil 2011;92:1484−9.

[43] Jan Y-K, Lung C-W, Cuaderes E, Rong D, Boyce K. Effect of viscoelastic properties of plantar soft tissues on plantar pressures at the first metatarsal head in diabetics with peripheral neuropathy. Physiol Meas 2013;34:53−66. https://doi.org/10.1088/0967-3334/34/1/53.

[44] Naemi R, Chatzistergos P, Sundar L, Chockalingam N, Ramachandran A. Differences in the mechanical characteristics of plantar soft tissue between ulcerated and non-ulcerated foot. J Diabet Complicat 2016;30:1293−9. https://doi.org/10.1016/j.jdiacomp.2016.06.003.

[45] Tas S, Bek N, Ruhi Onur M, Korkusuz F. Effects of body mass index on mechanical properties of the plantar fascia and heel pad in asymptomatic participants. Foot Ankle Int 2017;38:779−84. https://doi.org/10.1177/1071100717702463.

[46] Chatzistergos PE, Behforootan S, Allan D, Naemi R, Chockalingam N. Shear wave elastography can assess the in-vivo nonlinear mechanical behavior of heel-pad. J Biomech 2018;80:144−50. https://doi.org/10.1016/j.jbiomech.2018.09.003.

[47] Wu CH, Lin CY, Hsiao MY, Cheng YH, Chen WS, Wang TG. Altered stiffness of microchamber and macrochamber layers in the aged heel pad: shear wave ultrasound elastography evaluation. J Formos Med Assoc 2018;117:434−9. https://doi.org/10.1016/j.jfma.2017.05.006.

[48] Tong J, Lim CS, Goh OL. Technique to study the biomechanical properties of the human calcaneal heel pad. Foot 2003;13:83−91. https://doi.org/10.1016/S0958-2592(02)00149-9.

[49] Behforootan S, Chatzistergos PE, Chockalingam N, Naemi R. A simulation of the viscoelastic behaviour of heel pad during weight-bearing activities of daily living. Ann Biomed Eng 2017;45:2750−61. https://doi.org/10.1007/s10439-017-1918-1.

[50] Williams ED, Stebbins MJ, Cavanagh PR, Haynor DR, Chu B, Fassbind MJ, et al. A preliminary study of patient-specific mechanical properties of diabetic and healthy plantar soft tissue from gated magnetic resonance imaging. J Eng Med 2017;231:625−33. https://doi.org/10.1177/0954411917695849.

[51] Parker D, Cooper G, Pearson S, Crofts G, Howard D, Busby P, et al. A device for characterising the mechanical properties of the plantar soft tissue of the foot. Med Eng Phys 2015;37:1098−104. https://doi.org/10.1016/j.medengphy.2015.08.008.

[52] Alcántara E, Forner A, Ferrús E, García AC, Ramiro J. Influence of age, gender, and obesity on the mechanical properties of the heel pad under walking impact conditions. J Appl Biomech 2002;18: 345−56. https://doi.org/10.1123/jab.18.4.345.

[53] Cavanagh PR. Plantar soft tissue thickness during ground contact in walking. J Biomech 1999;32: 623−8. https://doi.org/10.1016/S0021-9290(99)00028-7.

[54] Telfer S, Woodburn J, Cavanagh PR. Footwear embedded ultrasonography to determine plantar soft tissue properties for finite element simulations. Footwear Sci 2015;7:9424280. https://doi.org/10.1080/19424280.2015.1036931.

[55] Wearing SC, Smeathers JE. The heel fat pad: mechanical properties and clinical applications. J Foot Ankle Res 2011;4:I14. https://doi.org/10.1186/1757-1146-4-S1-I14.

[56] Gefen A, Megido-Ravid M, Itzchak Y. In vivo biomechanical behavior of the human heel pad during the stance phase of gait. J Biomech 2001;34:1661−5. https://doi.org/10.1016/S0021-9290(01)00143-9.

[57] Schimmoeller T, Colbrunn R, Nagle T, Lobosky M, Neumann EE, Owings TM, et al. Instrumentation of off-the-shelf ultrasound system for measurement of probe forces during freehand imaging. J Biomech 2019;83:117−24. https://doi.org/10.1016/j.jbiomech.2018.11.032.

[58] Kleesz P, Darlenski R, Fluhr JW. Full-body skin mapping for six biophysical parameters: baseline values at 16 anatomical sites in 125 human subjects. Skin Pharmacol Physiol 2011;25:25−33. https://doi.org/10.1159/000330721.

[59] Firooz A, Sadr B, Babakoohi S, Sarraf-Yazdy M, Fanian F, Kazerouni-Timsar A, et al. Variation of biophysical parameters of the skin with age, gender, and body region. Sci World J 2012;2012. https://doi.org/10.1100/2012/386936.

[60] Smalls LK, Randall Wickett R, Visscher MO. Effect of dermal thickness, tissue composition, and body site on skin biomechanical properties. Skin Res Technol 2006;12:43−9. https://doi.org/10.1111/j.0909-725X.2006.00135.x.

[61] Hashmi F, Wright C, Nester C, Lam S. The reliability of non-invasive biophysical outcome measures for evaluating normal and hyperkeratotic foot skin. J Foot Ankle Res 2015;8. https://doi.org/10.1186/s13047-015-0083-8.

[62] Cuaderes E, Khan MM, Azzarello J, Lamb WL. Reliability and limitations of the durometer and pressureStat to measure plantar foot characteristics in native americans with diabetes. J Nurs Meas 2009;17:3−18. https://doi.org/10.1891/1061-3749.17.1.3.

[63] ASTM International. D2240-15 e1. Standard test method for rubber property—durometer hardness, vol. 05; 2010. https://doi.org/10.1520/D2240-15E01.

[64] Piaggesi A, Romanelli M, Schipani E, Campi F, Magliaro A, Baccetti F, et al. Hardness of plantar skin in diabetic neuropathic feet. J Diabet Complicat 1999;13:129−34. https://doi.org/10.1016/S1056-8727(98)00022-1.

[65] Periyasamy R, Anand S, Ammini AC. The effect of aging on the hardness of foot sole skin: a preliminary study. Foot 2012;22:95−9. https://doi.org/10.1016/j.foot.2012.01.003.

[66] Seixas A, Ammer K, Carvalho R, Vilas-Boas JP, Mendes J, Vardasca R. Relationship between skin temperature and soft tissue hardness in diabetic patients: an exploratory study. Physiol Meas 2019;40:0−8. https://doi.org/10.1088/1361-6579/ab2f03.

[67] Hsu TC, Wang CL, Tsai WC, Kuo JK, Tang FT. Comparison of the mechanical properties of the heel pad between young and elderly adults. Arch Phys Med Rehabil 1998;79:1101−4. https://doi.org/10.1016/S0003-9993(98)90178-2.

[68] Hsu CC, Tsai WC, Chen CPC, Shau YW, Wang CL, Chen MJL, et al. Effects of aging on the plantar soft tissue properties under the metatarsal heads at different impact velocities. Ultrasound Med Biol 2005;31:1423−9. https://doi.org/10.1016/j.ultrasmedbio.2005.05.009.

[69] Chao CYL, Zheng YP, Huang YP, Cheing GLY. Biomechanical properties of the forefoot plantar soft tissue as measured by an optical coherence tomography-based air-jet indentation system and tissue ultrasound palpation system. Clin Biomech 2010;25:594−600. https://doi.org/10.1016/j.clinbiomech.2010.03.008.

[70] Kwan RLC, Zheng YP, Cheing GLY. The effect of aging on the biomechanical properties of plantar soft tissues. Clin Biomech 2010;25:601−5. https://doi.org/10.1016/j.clinbiomech.2010.04.003.

[71] Wearing SC, Smeathers JE, Yates B, Urry SR, Dubois P. Bulk compressive properties of the heel fat pad during walking: a pilot investigation in plantar heel pain. Clin Biomech 2009;24:397−402. https://doi.org/10.1016/j.clinbiomech.2009.01.002.

[72] BS ISO. ISO 7743:2011 - Rubber, vulcanized or thermoplastic — determination of compression stress-strain properties. 2011.

[73] Wang CL, Shau YW, Hsu TC, Chen HC, Chien SH. Mechanical properties of heel pads reconstructed with flaps. J Bone Joint Surg Br 1999;81:207—11. https://doi.org/10.1302/0301-620X.81B2.9056.

[74] Pai S, Ledoux WR. The compressive mechanical properties of diabetic and non-diabetic plantar soft tissue. J Biomech 2010;43:1754—60. https://doi.org/10.1016/j.jbiomech.2010.02.021.

[75] Aerts P, Ker RF, De Clercq D, Ilsley DW, Alexander RM. The mechanical properties of the human heel pad: a paradox resolved. J Biomech 1995;28:1299—308. https://doi.org/10.1016/0021-9290(95)00009-7.

[76] Hashmi F, Malone-Lee J, Hounsell E. Plantar skin in type II diabetes: an investigation of protein glycation and biomechanical properties of plantar epidermis. Eur J Dermatol 2006;16:23—32.

[77] Lechner A, Akdeniz M, Tomova-Simitchieva T, Bobbert T, Moga A, Lachmann N, et al. Comparing skin characteristics and molecular markers of xerotic foot skin between diabetic and non-diabetic subjects: an exploratory study. J Tissue Viability 2019;28:200—9. https://doi.org/10.1016/j.jtv.2019.09.004.

[78] Charanya G, Patil KM, Narayanamurthy VB, Parivalavan R, Visvanathan K. Effect of foot sole hardness, thickness and footwear on foot pressure distribution parameters in diabetic neuropathy. Proc Inst Mech Eng Part H J Eng Med 2004;218:431—43. https://doi.org/10.1243/0954411042632117.

[79] Thomas VJ, Patil KM, Radhakrishnan S, Narayanamurthy VB, Parivalavan R. The role of skin hardness, thickness, and sensory loss on standing foot power in the development of plantar ulcers in patients with diabetes mellitus—a preliminary study. Int J Low Extrem Wounds 2003;2:132—9. https://doi.org/10.1177/1534734603258601.

[80] Zheng YP, Choi YKC, Wong K, Mak AFT. Indentation assessment of plantar foot tissue in diabetic patients. In: Proc First Jt BMES/EMBS Conf 1999 IEEE Eng Med Biol 21st Annu Conf 1999 Annu Fall Meet Biomed Eng Soc Cat N. 1; 1999. p. 7803.

[81] Ledoux WR, Pai S, Shofer JB, Wang YN. The association between mechanical and biochemical/histological characteristics in diabetic and non-diabetic plantar soft tissue. J Biomech 2016;49:3328—33. https://doi.org/10.1016/j.jbiomech.2016.08.021.

[82] Hsu CC, Tsai WC, Shau YW, Lee KL, Hu CF. Altered energy dissipation ratio of the plantar soft tissues under the metatarsal heads in patients with type 2 diabetes mellitus: a pilot study. Clin Biomech 2007;22:67—73. https://doi.org/10.1016/j.clinbiomech.2006.06.009.

[83] Dunn JE, Link CL, Felson DT, Crincoli MG, Keysor JJ, McKinlay JB. Prevalence of foot and ankle conditions in a multiethnic community sample of older adults. Am J Epidemiol 2004;159:491—8. https://doi.org/10.1093/aje/kwh071.

[84] Quondamatteo F. Skin and diabetes mellitus: what do we know? Cell Tissue Res 2014;355:1—21. https://doi.org/10.1007/s00441-013-1751-2.

[85] Lin CY, Chuang HJ, Cortes DH. Investigation of the optimum heel pad stiffness: a modeling study. Australas Phys Eng Sci Med 2017;40:585—93. https://doi.org/10.1007/s13246-017-0565-z.

[86] Petre M, Erdemir A, Cavanagh PR. An MRI-compatible foot-loading device for assessment of internal tissue deformation. J Biomech 2008;41:470—4. https://doi.org/10.1016/j.jbiomech.2007.09.018.

[87] Cheung JTM, Zhang M, Leung AKL, Fan YB. Three-dimensional finite element analysis of the foot during standing - a material sensitivity study. J Biomech 2005;38:1045—54. https://doi.org/10.1016/j.jbiomech.2004.05.035.

[88] Mueller MJ, Zou D, Lott DJ. "Pressure gradient" as an indicator of plantar skin injury. Diabetes Care 2005;28:2908—12. https://doi.org/10.2337/diacare.28.12.2908.

[89] Pai S, Ledoux WR. The shear mechanical properties of diabetic and non-diabetic plantar soft tissue. J Biomech 2012;45:364—70. https://doi.org/10.1016/j.jbiomech.2011.10.021.

[90] Yavuz M, Tajaddini A, Botek G, Davis B. Temporal characteristics of plantar shear distribution: relevance to diabetic patients. J Biomech 2008;41:556—9.

[91] Murray HJ, Young MJ, Hollis S, Boulton AJM. The association between callus formation, high pressures and neuropathy in diabetic foot ulceration. Diabet Med 1996;13:979—82. https://doi.org/10.1002/(SICI)1096-9136(199611)13:11<979::AID-DIA267>3.0.CO;2-A.

[92] Naemi R, Chatzistergos P, Suresh S, Sundar L, Chockalingam N, Ramachandran A. Can plantar soft tissue mechanics enhance prognosis of diabetic foot ulcer? Diabetes Res Clin Pract 2017;126:182—91. https://doi.org/10.1016/j.diabres.2017.02.002.

[93] Hashmi F, Malone-Lee J. Measurement of skin elasticity on the foot. Skin Research and Technology 2007;13(3):252—8. https://doi.org/10.1111/j.1600-0846.2007.00216.x. In press.

[94] Chatzistergos PE, Naemi R, Sundar L, Ramachandran A, Chockalingam N. The relationship between the mechanical properties of heel-pad and common clinical measures associated with foot ulcers in patients with diabetes. Journal of Diabetes and its Complications 2014;28(4):488—93. https://doi.org/10.1016/j.jdiacomp.2014.03.011. In press.

CHAPTER 16

Bone carriers in diabetic foot osteomyelitis

Cristian Nicoletti

Diabetic Foot Unit, Pederzoli Hospital, Peschiera del Garda, Italy

Background

Diabetes is one of the most common chronic disease worldwide, and its prevalence is progressively increasing; in 2019 there were about 463 million people with diabetes, and this number will increase to almost 700 million by 2045 all over the world (IDF DIABETES ATLAS Ninth edition 2019) [32].

Diabetic foot ulcerations (DFUs) are a complication of diabetes mellitus caused by external or internal trauma associated with different stages of diabetic neuropathy and peripheral vascular disease [1], and it is estimated that between 19% and 34% of persons with diabetes are likely to develop a foot ulcer at some point of their life [2]. The most serious consequence of DFUs is major or minor amputation [3] leading to a dramatic loss in the life expectancy of these patients, which places them at risk of higher mortality rates than colon, prostate, and breast cancer or Hodgkin's disease [4]. The most frequent causes of amputation in these patients are ischemia and infection [5], and in particular the latter remains the most frequent diabetic complication, affecting up to 60% of DFUs, sometimes requiring hospitalization, and is the most common precipitating event leading to amputations [6].

Diabetic foot osteomyelitis (DFO) is a common complication of patients with diabetic foot infections; it was reported that nearly 20%–60% of patients with diabetic foot might suffer from DFO [7], a very difficult-to-treat situation as its treatment might include the management of chronic ulcers, necrotic soft tissues, gangrenes, and, of course, the infected bones [8], and failure is associated with high rates of amputation.

DFO typically involves the forefoot and develops by contiguous spread from overlying soft tissues and penetration through cortical bone and into the medullary cavity [9].

DFO current diagnosis and treatment

DFO must be promptly and accurately diagnosed, on the basis of clinical presentation and imaging studies (mainly plain X-ray, magnetic resonance imaging, and scintigraphy). Medical and/or surgical treatments are the cornerstones of DFO treatment.

The literature shows that the traditional treatment has been the resection of necrotic and infected bone; however there were some studies that demonstrated highest remission rates after the treatment with antibiotics alone [10].

Treatment involves the use of antibiotic agents with adequate penetration to bone, choice of which is best guided by bone culture; at present, no specific antibiotic regimen has shown superiority for the treatment of DFO, while the optimal treatment duration is still to be defined, even if long antibiotic duration (weeks to months) is to be preferred.

Surgery is required primarily when there are exposed or necrotic bones, large sequestra, and severe gangrenous soft tissue necrosis [11].

Despite the seriousness of this complication, unfortunately there are no agreed-upon guidelines for the management of DFO, and the International Working Group on Diabetic Foot recognized that DFO was an area in which guidelines for diagnosing and treatment were needed [1]. In particular, the choice of surgical versus medical approach needs to be further defined.

Rationale for the use of topical antibiotic therapy

Diabetic foot infections are often polymicrobial, especially in the chronic wound. Recurrence of infection in long-standing ulcers requires repeated courses of antibiotics, but the benefits are frequently hampered by intolerance and adverse effects, especially in frail, diabetic patients with multiple comorbidities.

Despite the progress in systemic antibiotic usage, its efficacy can be impaired by low tissue penetration due to the peripheral artery disease, manifested in the more distal vessels, as well as the presence of impaired microcirculation.

The development of biofilms in chronic wounds represents an additional challenge, as biofilms protect pathogens from host immunity and systematically administered antibiotics. Thus, a multiplicity of issues have meant that targeting antibiotic-resistant organisms has been an increasing problem in recent decades [4].

Furthermore, after surgical removal of infected bone, the risk for biomechanical changes, that increase recurrent ulceration rates by pressure transfer to other foot locations, and the risk for recurrent infection due to remaining infected bone are consistent [1].

For all these reasons, topical delivery of antibiotics represents an attractive emerging modality in this field; it allows to achieve some key advantages:
- High antibiotic concentrations in the affected area (10—100 times higher than serum levels) above the minimum inhibitory concentration, which cannot be achieved with the use of systemic antibiotics. This can be of importance in cases where the penetration of the systemically administered antibiotics in the infected area might be suboptimal because of compromised vascular perfusion and/or the presence of bacterial biofilm.

- Limited systemic absorption of the locally applied antibiotic that reduces considerably the risk of toxicity and avoids many of the adverse drug reaction caused by systemic antibiotics. This could be especially useful for patients who are intolerant to systemic administration or have impaired renal or liver function.
- Reduction of resistant microorganisms, and this could bring significant public health benefits, in view of the current threat of multidrug-resistant microbial strains [4].

Classes of local antibiotic delivery systems

Much of the provenance of topical antibiotic therapies originates from orthopedic-driven interventions. Several devices loaded with antibiotics have been used for the in situ treatment of chronic osteomyelitis, primarily to fill anatomical defects secondary to surgical debridement.

They can be classified as resorbable and nonresorbable antibiotic delivery systems [4].
- Nonresorbable Antibiotic Delivery Systems

Polymethyl methacrylate (PMMA) beads represent the major class of nonbiodegradable carrier systems. Beads can be impregnated with one or a combination of antibiotics such as glycopeptides and aminoglycosides.

Antibiotic release from PMMA is initially high during the first 48—72 h, but quickly falls to lower subtherapeutic levels, eluting for many times (weeks or even years) [5].

Advantages: Longevity and structural support.

Disadvantages: This material requires surgical removal on completion of drug elution, to avoid becoming a focus for biofilm formation [6]. This further intervention can increase the perioperative associated risk, in particular in fragile patients like diabetics are.
- Resorbable Antibiotic Delivery Systems

There are several types, namely, bone substitutes, natural protein-based polymers, synthetic polymers, and composite carriers.

Calcium sulfate can be loaded with water-soluble antibiotics such as aminoglycosides, glycopeptides (such as vancomycin), fluoroquinolones (such as moxifloxacin), or daptomycin [4]. The elution properties have been studied in vitro and in animal models, drawing the conclusion that there is a high initial elution phase, providing about 45% of vancomycin and about 80% of gentamycin and clindamycin within the first 24 h. This was followed by a more gradual second phase of drug delivery over a further 10 days [7].

Collagen, fibrin, and thrombin are naturally occurring protein-based polymers that can be manufactured into a meshlike structure, creating a scaffold that allows direct binding of antibiotics, which are then released as the structure is broken down, usually within days.

Collagen fleece is a widely used biodegradable carrier system, which stimulates osteoblast proliferation, promoting mineralization and production of collagenous tissue; it is usually impregnated with a broad-spectrum antibiotic such as an aminoglycoside. Changing the porosity of collagen or treating it with chemicals can modify and control drug elution rates [8].

Chitosan is a polymerized polysaccharide able to act as a drug carrier with additional antibacterial and antifungal activity. Loaded with gentamycin, it seems to be a clinically useful method for the treatment of bone and soft tissue infections, delivering effective concentrations of antibiotics for about 8 weeks. Further, developed as a composite carrier together with nanohydroxyapatite and ethyl cellulose microspheres, it has been demonstrated to elute gentamycin above the minimum inhibitory concentration for 45 days [4].

Pexiganan is a synthetic analogue of the natural antimicrobial peptide magainin II. It is a broad-spectrum agent active against most of the microorganisms isolated in diabetic foot infections, including MR *Staphylococcus aureus* and multidrug-resistant gram-negative strains [9].

Advantages: Gradual resorb, so that they can act as a matrix for new bone growth; with degradation, additional release of antibiotics occurs, prolonging their action and preventing biofilm formation on their surface. No need for surgical removal.

Disadvantages: Calcium sulfate seems to be unable to provide significant long-term mechanical support, or to act as a scaffold for tissue regeneration, since it dissolves relatively quickly [10].

Evidences for the use of local antibiotics in diabetic foot infections

Once again, most of the experiences acquired in this field come from orthopedics, via the use of numerous antibiotic vehicles in the treatment of osteomyelitis, as well as for minimizing the risk of postoperative infections [11].

The available data concerning the use of local antibiotics in the treatment of diabetic foot infections are limited, and a robust body of evidence is missing; in particular, is still to be clearly defined whether local antibiotics are to be used alone or in conjunction with systemic antibiotics and/or surgical interventions.

From 2000 onward, some case reports, case series, randomized controlled multicenter trials, and retrospective cohort study have been published.

Case report and case series

The use of PMMA beads impregnated with either tobramycin or vancomycin in three diabetic patients with foot osteomyelitis (including a case of calcaneal osteomyelitis by MR *Staphylococcus aureus*), together with local resection of infected bone and systemic antibiotical infusion had success, and the three patients healed [20].

Other authors have reported the successful case of a forefoot osteomyelitis and septic arthritis treated with tobramycin-impregnated calcium sulfate pellets inserted in a bone cavity beneath the foot ulcer, in addition to oral antibiotic treatment, in a patient declining the ray excision [12].

The local application of calcium sulfate and hydroxyapatite beads loaded with vancomycin with bone resection in a patient with forefoot osteomyelitis was effective in healing osteomyelitis without requiring further surgery or removal of beads [13].

More recently, the same author has published a case series of 143 lower extremity osteomyelitis (foot and ankle) in 125 patients (104 of whom had diabetes) treated with a percutaneous antibiotic delivery technique delivering intraosseous calcium sulfate and hydroxyapatite loaded with vancomycin, followed by either oral or intravenous antibiotics for 4 weeks according to bone cultures. There was no recurrence of osteomyelitis in more than 96% of the treated patients [23].

The same result in a similar situation has been achieved after excision of infected bone and local application of calcium sulfate beads impregnated with vancomycin ad gentamycin combined with systemic antibiotics [14].

Some authors have presented the use of tobramycin-impregnated PMMA beads as adjunctive treatment after resection of distal fibula for Charcot ankle osteomyelitis in conjunction with systemic antibiotic treatment, the result was effective, and the patient had no recurrence of osteomyelitis [15].

In a series of 20 diabetic patients with diabetic foot ulcers and underlying osteomyelitis of the forefoot, after the failure of standard care through wound debridement, systemic antibiotics, and offloading, the application of a biodegradable highly purified synthetic calcium sulfate pellets loaded with vancomycin and gentamicin after the excision of bone sequestrate was once again effective: All patients healed; no recurrence was reported after a follow-up of 12 months [16].

Similar successful outcomes have been reported in a series of eight diabetic patients affected by chronic metatarsal or calcaneal osteomyelitis and treated with the use of PMMA beads or bone graft substitutes loaded with gentamicin in conjunction with systemic antibiotics and minor surgery. The osteomyelitis was successfully treated in all cases [3].

Another retrospective review of 70 patients with DFU (Texas Classification grade 3B and 3D) and osteomyelitis treated by debridement followed by adjuvant local antibiotic (an antibiotic-loaded absorbable calcium sulfate and hydroxyapatite biocomposite were used) was performed. Patients were followed until infection eradication or ulcer healing for a mean follow-up of 10 months.

Authors concluded that adjuvant, local antibiotic therapy with an absorbable biocomposite can help achieve up to 90% cure rates in DFUs with osteomyelitis, preventing reinfections and the need for multiple surgical procedures [17].

Retrospective comparative studies

Only a few studies present comparative data between outcomes, with or without the addition of local antibiotics to standard treatment, or between local versus systemic treatment.

A retrospective comparative study assessed the effectiveness of local application of bioabsorbable, tobramycin-impregnated calcium sulfate beads in addition to standard treatment after transmetatarsal amputation in diabetic patients with nonhealing forefoot full-thickness ulcerations with osteomyelitis or skin necrosis.

In total, data from 65 cases of amputations were considered, including 49 cases in the "beads group" and 16 cases without beads. The wound breakdown rate following transmetatarsal amputation was significantly lower in the "beads group." Also a difference favoring the "beads group" was observed regarding the time to wound healing, but this did not reach the statistical significance. The length of hospital stay and the proportion of patients who required conversion to transtibial amputation did not significantly differ between groups [18].

Some authors have retrospectively compared the outcomes of two treatment methods (infected bone resection combined with adjuvant calcium sulfate impregnated with vancomycin and/or gentamycin; infected bone resection alone) in 48 diabetic patients with foot osteomyelitis; systemic antibiotics, postoperative wound care, and offloading were continued to be applied following surgery in both groups.

There was no difference between the two groups in mean time to healing, healing rate, and postoperative amputation rate, while osteomyelitis recurrence rate was significantly lower (0% vs. 36.4%) with the use of antibiotic-impregnated calcium sulfate [2].

Randomized trials

Some authors investigated the effectiveness of a bioabsorbable, gentamicin-impregnated collagen sponge application into wounds after minor amputation for nonhealing ulcer with osteomyelitis.

Fifty diabetic patients were randomized to have or not to have the gentamicin sponge applied. All patients received systemic antibiotics according to the antibiogram profile.

The application of a gentamicin sponge significantly shortened wound healing duration by almost 2 weeks. No effect was observed on the length of hospital stay or any difference in the number of revisions for wound breakdown or consequent amputations between groups [19].

In another randomized controlled multicenter trial, 56 diabetic patients with moderately infected foot ulcers were randomized for the use or for the nonuse of a gentamicin collagen sponge in addition to standard care. Significantly higher rates of clinical cure and eradication of baseline pathogens were achieved in the group treated with the gentamicin collagen sponge [30].

A large, randomized controlled double-blind trial compared the topical application of the investigational antimicrobial peptide pexiganan versus oral ofloxacin.

About 835 diabetic patients with a mild infection of a diabetic foot ulcer were randomized to receive pexiganan or oral antibiotic, plus a respective inactive placebo.

Similar clinical improvement rates, microbiological eradication rates, and wound healing rates were achieved with both active treatments; no significant resistance to pexiganan has been reported, while bacterial resistance to ofloxacin was noted in some of the patients [20].

Summary

In conclusion, limited data exist in the field of diabetic foot infections (mostly case reports and case series), and a robust body of evidence is missing, in so far as whether local antibiotics are to be used alone or in conjunction with systemic antibiotics and/or surgical intervention.

Local delivery of antibiotics appears to be a promising option and an effective adjuvant treatment in cases of surgically treated osteomyelitis, and there also appears to be potential for soft tissue infection management in cases of infected deep ulceration.

References

[1] Lázaro Martínez JL, Álvarez YG, Tardáguila-García A, Morales EG. Optimal management of diabetic foot osteomyelitis: challenges and solutions. Diabetes Metab Syndrome Obes Targets Ther 2019;12: 947−59. https://doi.org/10.2147/DMSO.S181198.

[2] Armstrong DG, Boulton AJM, Bus SA. Diabetic foot ulcers and their recurrence. N Engl J Med 2017; 76(24):2367−75. https://doi.org/10.1056/NEJMra1615439.

[3] Boulton AJM, Vileikyte L, Ragnarson-Tennvall G, Apelqvist J. The global burden of diabetic foot disease. Lancet 2005;366(9498):1719−24. https://doi.org/10.1016/S0140-6736(05)67698-2.

[4] Schofield CJ, Libby G, Brennan GM, MacAlpine RR, Morris AD, Leese GP. Mortality and hospitalization in patients after amputation: a comparison between patients with and without diabetes. Diabet Care 2006;29(10):2252−6. https://doi.org/10.2337/dc06-0926.

[5] Lipsky BA, Berendt AR, Cornia PB, Pile JC, Peters EJG, Armstrong DG, Deery HG, Embil JM, Joseph WS, Karchmer AW, Pinzur MS, Senneville E. Infectious diseases society of america clinical practice guideline for the diagnosis and treatment of diabetic foot infections. J Am Podiatr Med Assoc 2013;103(1):2−7. https://doi.org/10.7547/1030002.2012.

[6] Pecoraro RE, Ahroni JH, Boyko EJ, Stensel VL. Chronology and determinants of tissue repair in diabetic lower-extremity ulcers. Diabetes 1991;40(10):1305−13. https://doi.org/10.2337/diab.40.10.1305.

[7] Qin CH, Zhou CH, Song HJ, Cheng GY, Zhang HA, Fang J, Tao R. Infected bone resection plus adjuvant antibiotic-impregnated calcium sulfate versus infected bone resection alone in the treatment of diabetic forefoot osteomyelitis. BMC Muscoskel Disord 2019;20(1):1−8. https://doi.org/10.1186/s12891-019-2635-8.

[8] Lavery LA, Armstrong DG, Wunderlich RP, Mohler MJ, Wendel CS, Lipsky BA. Risk factors for foot infections in individuals with diabetes. Diabet Care 2006;29(6):1288−93. https://doi.org/10.2337/dc05-2425.

[9] Lipsky BA, Aragón-Sánchez J, Diggle M, Embil J, Kono S, Lavery L, Senneville É, Urbančič-Rovan V, Van Asten S. IWGDF guidance on the diagnosis and management of foot infections in persons with diabetes. Diabet Metab Res Rev 2016;32(Suppl 1):45−74. https://doi.org/10.1002/dmrr.2699.

[10] Mutluoglu M, Lipsky BA. Non-surgical treatment of diabetic foot osteomyelitis. Lancet Diabet Endocrinol 2017;5(8):668. https://doi.org/10.1016/S2213-8587(16)30141-3.

[11] Panagopoulos P, Drosos G, Maltezos E, Papanas N. Local antibiotic delivery systems in diabetic foot osteomyelitis: time for one step beyond? Int J Lower Extrem Wounds 2015;14(1):87−91. https://doi.org/10.1177/1534734614566937. SAGE Publications Inc.

[12] Markakis K, Faris AR, Sharaf H, Faris B, Rees S, Bowling FL. Local antibiotic delivery systems: current and future applications for diabetic foot infections. Int J Low Extrem Wounds 2018;17(1):14−21. https://doi.org/10.1177/1534734618757532.

[13] Wininger DA, Fass RJ. Antibiotic-impregnated cement and beads for orthopedic infections. Antimicrob Agents Chemother 1996;40(12):2675−9. https://doi.org/10.1128/aac.40.12.2675.

[14] Aiola S, Amico G, Battaglia P, Battistelli E. The large-scale polarization explorer (LSPE). Clin Microbiol Rev 2002;15(2):167−93. https://doi.org/10.1128/CMR.15.2.167.

[15] Wichelhaus TA. Elution characteristics of vancomycin, teicoplanin, gentamicin and clindamycin from calcium sulphate beads. J Antimicrob Chemother 2001;48(1):117−9. https://doi.org/10.1093/jac/48.1.117.
[16] Maxson, Mitchell. 乳鼠心肌提取 HHS public access. Physiol Behav 2016;176(1):139−48. https://doi.org/10.1016/j.physbeh.2017.03.040.
[17] Flamm RK, Rhomberg PR, Simpson KM, Farrell DJ, Sader HS, Jones RN. In vitro spectrum of pexiganan activity when tested against pathogens from diabetic foot infections and with selected resistance mechanisms. Antimicrob Agents Chemother 2015;59(3):1751−4. https://doi.org/10.1128/AAC.04773-14.
[18] Inzana JA, Schwarz EM, Kates SL, Awad HA. Biomaterials approaches to treating implant-associated osteomyelitis. Biomaterials March 2016;81:58−71. https://doi.org/10.1016/j.biomaterials.2015.12.012.
[19] Van Vugt TAG, Geurts J, Arts JJ. Clinical application of antimicrobial bone graft substitute in osteomyelitis treatment: a systematic review of different bone graft substitutes available in clinical treatment of osteomyelitis. BioMed Res Int 2016;2016. https://doi.org/10.1155/2016/6984656.
[20] Roeder B, Van Gils CC, Maling S. Antibiotic beads in the treatment of diabetic pedal osteomyelitis. J Foot Ankle Surg 2000;39(2):124−30. https://doi.org/10.1016/s1067-2516(00)80037-x.
[21] Bokhari S, Israelian Z, Schmidt J, Brinton E, Meyer C. Effects of angiotensin II type 1 receptor blockade on β-cell function in humans. Diabetes Care 2007;30(1):181. https://doi.org/10.2337/dc06-1745.
[22] Karr JC. Management in the wound-care center outpatient setting of a diabetic patient with forefoot osteomyelitis using cerament bone void filler impregnated with vancomycin: off-label use. J Am Podiatr Med Assoc 2011;101(3):259−64. https://doi.org/10.7547/1010259.
[23] Karr JC. Lower-Extremity Osteomyelitis Treatment Using Calcium Sulfate/Hydroxyapatite Bone Void Filler with Antibiotics: Seven-Year Retrospective Study. J Am Podiatr Med Assoc 2018;108(3):210−4. https://doi.org/10.7547/16-096.
[24] Morley R, Lopez F, Webb F. Calcium sulphate as a drug delivery system in a deep diabetic foot infection. Foot 2016;27:36−40. https://doi.org/10.1016/j.foot.2015.07.002.
[25] Ramanujam CL, Zgonis T. Antibiotic-loaded cement beads for Charcot ankle osteomyelitis. Foot Ankle Spec 2010;3(5):274−7. https://doi.org/10.1177/1938640010376563.
[26] Jogia RM, Modha DE, Nisal K, Berrington R, Kong MF. Use of highly purified synthetic calcium sulfate impregnated with antibiotics for the management of diabetic foot ulcers complicated by osteomyelitis. Diabetes Care 2015;38(5):e79−80. https://doi.org/10.2337/dc14-3100.
[27] Niazi NS, Drampalos E, Morrissey N, Jahangir N, Wee A, Pillai A. Adjuvant antibiotic loaded bio composite in the management of diabetic foot osteomyelitis — a multicentre study. Foot 2019;39:22−7. https://doi.org/10.1016/j.foot.2019.01.005.
[28] Krause FG, DeVries G, Meakin C, Kalla TP, Younger ASE. Outcome of transmetatarsal amputations in diabetics using antibiotic beads. Foot Ankle Int 2009;30(6):486−93. https://doi.org/10.3113/FAI.2009.0486.
[29] Varga M, Sixta B, Bem R, Matia I, Jirkovska A, Adamec M. Application of gentamicin-collagen sponge shortened wound healing time after minor amputations in diabetic patients - a prospective, randomised trial. Arch Med Sci 2014;10(2):283−7. https://doi.org/10.5114/aoms.2014.42580.
[30] Lipsky BA, Kuss M, Edmonds M, Reyzelman A, Sigal F. Topical application of a gentamicin-collagen sponge combined with systemic antibiotic therapy for the treatment of diabetic foot infections of moderate severity: A randomized, controlled, multicenter clinical trial. J Am Podiatr Med Assoc 2012;102(3):223−32. https://doi.org/10.7547/1020223.
[31] Lipsky BA, Holroyd KJ, Zasloff M. Topical versus systemic antimicrobial therapy for treating mildly infected diabetic foot ulcers: a randomized, controlled, double-blinded, multicenter trial of pexiganan cream. Clin Infect Dis 2008;47(12):1537−45. https://doi.org/10.1086/593185.
[32] International Diabetes Federation. IDF Diabetes Atlas, 9th edn, Brussels, Belgium; 2019. Available at: https://www.diabetesatlas.org.

CHAPTER 17

Vascular mechanobiology and metabolism

Sarah Basehore[1,2], Jonathan Garcia[1,2], Alisa Morss Clyne[1,2]
[1]Drexel University, Philadelphia, PA, United States; [2]University of Maryland, College Park, MD, United States

Introduction

People with diabetes suffer from early, accelerated atherosclerosis, which then leads to myocardial infarction, stroke, and peripheral vascular disease [1,2]. While diabetes is an established cardiovascular risk factor, the precise mechanism by which diabetes accelerates atherosclerotic plaque development remains unknown. Glucose fluctuations characteristic of both type 1 and type 2 diabetes have been implicated in accelerated atherosclerosis, since tight glycemic control delayed the onset and decreased progression of diabetic vascular complications, including reducing the risk of myocardial infarction and stroke in people with diabetes by more than 50% [3–5].

Atherosclerosis is thought to initiate with endothelial cell (EC) dysfunction [6]. ECs are uniquely positioned at the fluid–tissue interface, where they interact with flowing blood at their apical surface and vascular wall proteins and cells at their basolateral surface. Healthy ECs express an **atheroprotective phenotype**, maintaining vascular homeostasis through tight control of permeability, inflammation, vascular tone, and injury repair [7]. However, when ECs are exposed to cardiovascular disease risk factors, they become dysfunctional and express an **atheroprone phenotype.** EC dysfunction is characterized by impaired endothelium-dependent vasodilation due to reduced nitric oxide bioavailability [8], as well as increased inflammatory adhesion molecules, permeability, and low-density lipoprotein oxidation [9].

ECs in high glucose express an atheroprone phenotype. They are highly permeable, which allows macromolecules to pass into and through the vascular wall [10]; express increased adhesion molecules [11] and produce less nitric oxide [12], which recruits more inflammatory cells and reduces vasodilation; and display diminished migration [13] and proliferation [14], which inhibits angiogenesis in response to injury and ischemia [15]. While most research focuses on high glucose, hypoglycemia similarly induces EC dysfunction [16]. Acute blood glucose changes are thought to induce EC dysfunction via mitochondrial superoxide overproduction, as well as increased polyol and hexosamine pathway flux, advanced glycation end products, and protein kinase C (PKC) [17–20].

ECs are also dysfunctional when subjected to disturbed rather than laminar flow [21,22]. In straight arterial sections, blood flow is primarily laminar to produce high, directed shear stress (10–70 dyn/cm^2) at the endothelial surface [23]. Under these conditions, ECs align, elongate, organize long actin stress fibers parallel to the flow direction, and express an atheroprotective phenotype [24,25]. At arterial branches and bends (and in some surgical intervention sites), ECs are exposed to disturbed flow, a general term which includes flow recirculation, separation, and reversal [26]. Disturbed flow is associated with low or reciprocating shear stress (in space and/or time) [21,27,28]. ECs exposed to disturbed flow express an atheroprone phenotype and are round with random, short actin filaments concentrated in the cell periphery [29,30]; proliferate more rapidly [30,31]; increase their permeability [32–34]; and promote monocyte adhesion via inflammatory adhesion protein expression (e.g., ICAM-1, E-selectin) [35,36]. These functional effects occur through changes in transcription factors (e.g., NF-kB), signaling molecule activation (e.g., JNK), or gene and protein expression (e.g., ICAM-1) [37–39].

Altered blood glucose and flow both contribute to EC dysfunction and are extensively studied individually. Several studies have started to address interactions between metabolic and mechanical changes in the EC environments, including our own prior work showing that both low and high glucose cultures disturb EC response to laminar flow [40–42]. Understanding the integrated effects of glucose metabolism and mechanobiology is critical to understand cardiovascular disease, especially in people with metabolic diseases such as diabetes. In this chapter, we describe current knowledge in endothelial metabolism, mechanobiology, and their interactions, including important tools used in their study.

Endothelial cell metabolism

Although blood flow provides a constant oxygen supply, ECs primarily use anaerobic glycolysis for adenosine triphosphate (ATP) production [43,44]. In fact, less than 1% of glucose-derived pyruvate undergoes oxidative metabolism in endothelial mitochondria [44]. Glycolysis converts glucose into pyruvate yielding two ATP molecules through a sequence of 10 enzymatic reactions. Glycolysis inhibition caused reactive oxygen species–triggered autophagy in ECs, demonstrating the importance of glycolysis in EC viability [45].

Glucose primarily enters ECs through the high affinity glucose transporter GLUT1 [46]. Glycolysis is then regulated by three rate-limiting enzymes: hexokinase, phosphofructokinase, and pyruvate kinase [47]. Hexokinase catalyzes the first step of glycolysis, converting glucose into glucose 6-phosphate (G6P) [47] (Fig. 17.1). Hexokinase activity is inhibited by high G6P levels [47]. After G6P is converted to fructose-6-phosphate (F6P), 6-phosphofructo-1-kinase (PFK-1) catalyzes the irreversible conversion of F6P to fructose-1,6-bisphosphate (FBP) [48]. PFK-1 is allosterically regulated by fructose-2,6-bisphosphate (F26BP) to increase activity and high ATP to decrease activity

Vascular mechanobiology and metabolism 293

```
Polyol pathway  ←——————  glucose
                 hexokinase ↓
               glucose 6-phosphate  ——→  Pentose phosphate pathway
                 phosphoglucose
                    isomerase ↓
Hexosamine  ←—  fructose 6-phosphate
biosynthesis    phosphofructokinase-1
pathway             (PFK-1) ↓
               fructose 1,6-bisphosphate
                            ↓
               glyceraldehyde 3-phosphate
Methylglyoxal  glyceraldehyde 3-phosphate
pathway            dehydrogenase ↓
               1,3-bisphosphoglycerate
                 phosphoglycerate
                    kinase ↓
               3-phosphoglycerate
                 phosphoglyceromutase ↓
               2-phosphoglycerate  ——→  One carbon metabolism
                    enolase ↓
               phosphoenolpyruvate
                 pyruvate kinase ↓
                    pyruvate
                   ↙      ↘
               lactate    TCA Cycle
```

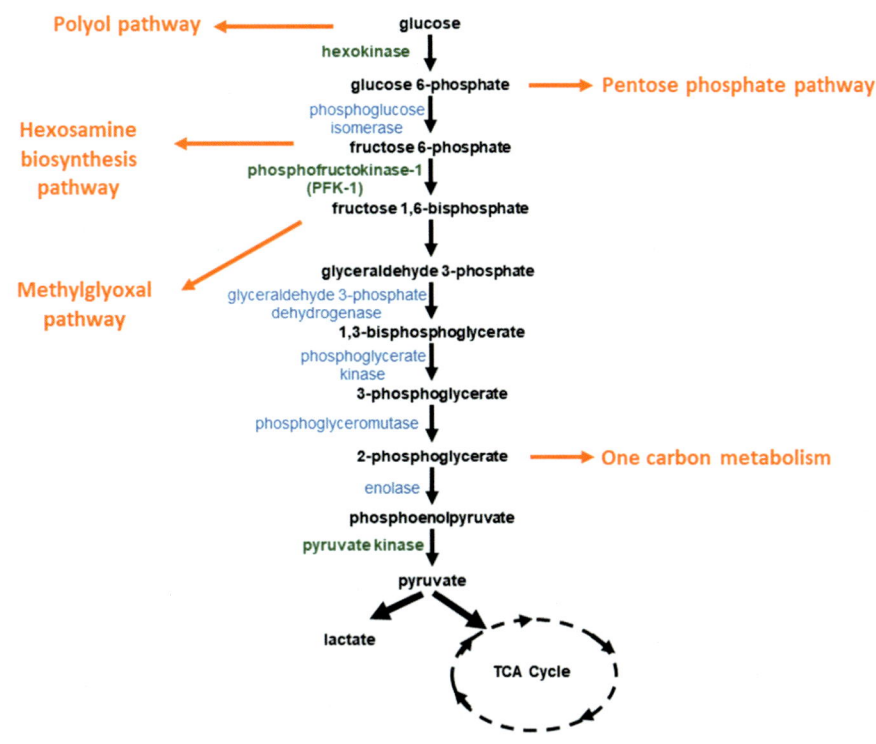

Figure 17.1 Metabolic pathways of importance in endothelial cells and diabetes. In endothelial cells, glucose enters through the GLUT1 transporter after which it is converted into glucose-6-phosphate (G6P). G6P further metabolizes and ultimately results in pyruvate. Pyruvate either turns to lactate and exits the cell or continues into the mitochondria for oxidative phosphorylation. The polyol pathway shunts glucose metabolic flux immediately upon entering the cell, while the pentose phosphate pathway (PPP) and hexosamine biosynthetic pathway (HBP) branch off at G6P and fructose-6-phosphate (F6P), respectively. The methylglyoxal pathway branches off at fructose-1,6-bisphosphate while one carbon metabolism branches at 2-phosphoglycerate.

[47,49]. FBP ultimately breaks down into phosphoenolpyruvate, which is converted to pyruvate by pyruvate kinase. FBP is an allosteric effector of pyruvate kinase [47]. Pyruvate is then either converted to lactate to exit the cell or shuttled into the mitochondria for oxidative metabolism in the tricarboxylic acid (TCA) cycle.

Glycolytic intermediates are also used in metabolic pathways that branch off glycolysis [50]. The polyol pathway, PPP, and HBP are of particular relevance in diabetes. The polyol pathway rarely metabolizes glucose in normal conditions. However, when hexokinase becomes saturated in hyperglycemia, as much as 33% of glucose is reduced through the polyol pathway. Glucose is first converted to sorbitol by aldose reductase in a reaction which also oxidizes NADPH to NADP+. Sorbitol is then converted to fructose by sorbitol dehydrogenase, a reaction which produces NADH from NAD+. Thus, enhanced polyol pathway flux enhances sorbitol accumulation and reduces

NADPH availability. While the mechanism by which increased polyol flux leads to EC damage is unclear, aldose reductase inhibitor clinical trials have proven effective in decreasing diabetic neuropathy [51]. However, the effects on retinopathy and capillary basement membrane thickening are inconsistent [52].

The PPP converts glucose 6-phosphate into the nucleotide synthesis precursor ribose 5-phosphate (R5P) [53]. Further down the pathway, erythrose 4-phosphate is used to synthesize aromatic amino acids [54]. The PPP is also the primary generator of NADPH, which is used in reductive biosynthetic reactions and reducing reactive oxygen species and therefore may be protective in diabetes [54,55]. Bovine aortic ECs induced to overexpress glucose 6-phosphate dehydrogenase (G6PD), the PPP rate-limiting enzyme, showed decreased reactive oxygen species and increased nitric oxide synthase activity when exposed to hydrogen peroxide [56]. NADH further serves as a donor for glutathione reductases that reduce oxidized glutathione and enhance catalase activity to convert hydrogen peroxide to water and oxygen [55,57]. PPP metabolites then reenter glycolysis at F6P and glyceraldehyde 3-phosphate (G3P) [47,55].

About 2%−5% of fructose-6-phosphate is shunted from glycolysis into the HBP [54]. In the first step of the HBP, fructose-6-phosphate (F6P) is converted to glucosamine-6-phosphate (Glc6P) by the rate-limiting enzyme glutamine:fructose 6-phosphate amidotransferase (GFAT). Alternatively, exogenous glucosamine can enter the HBP at this step, after conversion to Glc6P. Three subsequent reactions use acetyl-CoA and UTP to form UDP-N-acetylglucosamine (UDP-GlcNAc). UDP-GlcNAc then serves as a substrate for the *O*-linked β-N-acetylglucosamine (*O*-GlcNAc) modification of protein serine/threonine residues [58]. Two enzymes are responsible for *O*-GlcNAcylation: OGT catalyzes *O*-GlcNAc transfer onto proteins [59,60], while OGA catalyzes *O*-GlcNAc removal [61]. *O*-GlcNAcylation regulates both nuclear and cytoplasmic protein functions, often through an inverse functional relationship with phosphorylation at the same serine/threonine residues [62]. Of particular interest, endothelial nitric oxide synthase (eNOS) is activated by AKT-mediated phosphorylation at Ser1177, but eNOS was O-GlcNAcylated in vessels from diabetic animals, which left eNOS in an inactivated state [63]. The HBP is considered a "nutrient-sensing pathway" because it requires glucose, glutamine, acetyl-CoA, and ATP to form the final product [54,64].

Although ECs produce most of their glucose-derived ATP through glycolysis, they have fully functional mitochondria and can also use the TCA cycle and oxidative metabolism to produce ATP [50,54]. There are several possible reasons why ECs would reduce their glucose flux into the TCA cycle, including decreased reactive oxygen species production, reduced oxygen dependence to grow into hypoxic environments, and rapid ATP production at the cell periphery (e.g., lamellopodia). To supplement TCA cycle flux, ECs also preferentially take up glutamine via the solute carrier family 1 member 5 (SLC1A5) transporter [65]. Glutamine primarily provides carbon and nitrogen for nucleotide, hexosamine, glutathione, and nonessential amino acid synthesis [66]. Kim et al. used ^{13}C labeled glutamine to show that glutamine supplies the majority of TCA carbons [67]. Inhibition or deletion of glutaminase prevented EC proliferation but did not stop migration which relied on glycolysis [67]. These findings agree with Huang et al., who also showed impaired TCA anaplerosis when glutaminase was inhibited [68].

Endothelial cell metabolism impacts function

Endothelial metabolism has recently emerged as a powerful tool to regulate the vasculature. Metabolic pathways, such as glycolysis, fatty acid oxidation, and glutamine metabolism, have essential roles during vessel formation. Both cancer and ECs generate the majority of their energy through glycolysis even in the presence of oxygen (the Warburg effect) [69–71]. Although ECs stay mostly quiescent, they retain the capacity to rapidly initiate new vessel formation in response to injury or in pathological conditions. Glycolytic inhibitors have been shown to decrease new vessel formation both in vitro and in vivo. For example, phosphofructokinase-2/fructose-2,6-bisphosphatase-3 (PFKFB3) activity increases PFK-1 activity to increase glycolytic flux production [72]. When EC glucose metabolism was decreased by PFKFB3 blockade, EC proliferation, angiogenic sprouting, and cancer metastasis were inhibited both in vitro and in vivo [49,73,74].

Protein O-GlcNAcylation, which is a function of HBP flux, is important in physiologic cell processes such as transcription, stress response, calcium cycling, and signaling [75,76]. However, protein O-GlcNAcylation also contributes to vascular pathology. O-GlcNAcylation increases in a high glucose environment due to increased UDP-GlcNAc production [77,78]. Du et al. showed that eNOS O-GlcNAcylation increases in diabetic animals, which inhibits eNOS phosphorylation and NO production [79]. Aortic rings from high fat-fed mice with elevated O-GlcNAc levels had less vascular sprouting as compared to controls, which was abrogated by O-GlcNAc removal [80]. In ECs, elevated glucose and HBP flux increased eNOS O-GlcNAcylation twofold and thereby decreased eNOS phosphorylation at serine 1177. NO bioavailability decreased at baseline and during insulin and blood flow stimulation [79,81]. This effect was reversed by GFAT inhibition [79]. Recently, Barnes et al. reported increased O- GlcNAc in human pulmonary artery hypertension, a vascular disease associated with increased glycolytic flux [82].

Tools to measure endothelial cell metabolism

Techniques to measure EC glucose metabolism include YSI analysis, fluorescent glucose analogs, Seahorse assay, and mass spectrometry (MS). YSI analyzers measure nutrients and metabolites including glucose, lactate, glutamine, glutamate, etc., in biospecimens including blood, plasma, serum, and cerebrospinal fluid. Therefore, this technique is extensively used for diabetes and cancer [83]. The device uses biosensors to quantify the sample substrate reaction with an enzyme, which produces a current. The immobilized enzymes present on the membrane tip of the amperometric electrodes are highly specified for their target analyte, such as glucose or lactate, which allows for fast and accurate measurements to be made [84]. While this technique provides overall glycolytic flux analysis, the major limitation is that only glycolytic sources (e.g., glucose and glutamine) and by-products are measured. Additionally, it does not provide a dynamic analysis, but rather a static analysis after sample collection. Also, to ensure accurate readings, samples need to produce metabolite concentrations within the specific analyte measurement range. For example, the glucose measurement range for the YSI 2950 is 0.05—25 g/L, lactate is 0.05—2.70 g/L, glutamate is 15—1460 mg/L, and glutamine is 30—1169 mg/L. Therefore, for in vitro cell studies, large measurable changes in analytes may not occur for hours to days depending on the cell number and sample volume.

Fluorescent glucose derivatives are another technique to measure glucose uptake [85]. In 1996, researchers developed a novel fluorescent product, 2-(N-(7-nitrobenz-2-oxa-1,3-diazol-4-yl)amino)-2-deoxyglucose (2-NBDG) and used it to assess glucose uptake of *Escherichia coli* [86]. 2-NBDG is a fluorescent 2-deoxyglucose (2-DG) glucose analog. Cellular 2-DG uptake leads to partial trapping of 2-DG inside cells because it cannot be fully used in glycolysis. Once inside cells, 2-DG analogs, including 2-NBDG, are phosphorylated by hexokinase at the sixth carbon position. Because of the 2-deoxyglucose modification, the 2-DG 6-phosphate form cannot be further metabolized by glucose-6-isomerase, the next step in glycolysis. Therefore, the 2-DG 6-phosphate form accumulates inside the cell, and the nitrobenzoxadiazole (NBD) fluorescent form of 2-DG can be visualized using fluorescein optical filters [85]. While this is a qualitative technique to measure glucose uptake, live cell imaging provides a real-time look at how cells are using glucose for energy. However, 2-NBDG is larger than glucose; therefore it may not be transported in the same way and hence may not be an accurate indicator of glucose transporter activity. Additionally, 2-NBDG only indicates cellular glucose uptake differences but does not provide any information on how the glucose is being used.

Another live cell technique to measure changes in glucose metabolism is the Seahorse XF Analyzer. XF analyzers measure oxygen consumption rate (OCR) and extracellular acidification rate (ECAR) of live cells in a multiwell plate. OCR is a mitochondrial respiration indicator and ECAR is a glycolysis indicator. A sensor probe detects changes in dissolved oxygen and free proton concentration, yielding measurable rates of change for oxidative phosphorylation and glycolysis, respectively [87]. This technique provides real-time measurements; however, an XF analyzer requires the XF analyzer multiwell plate. Therefore, cells would either need to be trypsinized and plated into the XF well plates following shear stress exposure. Alternatively, ECs would have to be exposed to flow directly in XF well plates using an appropriately sized multiwell cone-and-plate viscometer. The biggest drawback to Seahorse XF Analyzers is that essentially no information is discovered regarding how the metabolic changes are occurring, including changes in side branch pathways that could affect the ECAR. For example, extracellular acid produced by cells is derived from both lactate, produced by anaerobic glycolysis, and CO_2, produced in the citric acid cycle during respiration. The proportions of glycolytic and respiratory acidification depend on the experimental conditions, including cell type or flow condition. The technique would not necessarily reveal that cells exposed to laminar flow may produce more lactate whereas cells exposed to disturbed flow may produce more CO_2 from the TCA cycle. Therefore, this technique does not provide an adequately in-depth analysis of glucose metabolic changes.

Recently, MS analysis has emerged as a novel technique to measure changes in glucose metabolism. MS measures the mass-to-charge ratio of ions to identify and quantify molecules. All mass spectrometers have an ion source, a mass analyzer, and an ion detector. Samples are loaded into the mass spectrometer in liquid, gas, or dried form and then vaporized and ionized by the ion source. Ionization allows users to obtain mass "fingerprints" that can be matched to proteins or other molecules to help identify known and unknown targets [88]. For glucose, metabolic intermediates can be identified based on the changing mass-to-charge ratio as glucose is metabolized. Isotopic tagging methods allow the user to quantify target molecules both in relative and absolute quantities. One of the most common techniques to measure changes in relative flux is using stable isotope labeling (e.g., ^{13}C glucose). The stable isotopes give a distinct mass shift over unlabeled metabolites and this mass difference can be detected by MS. The ratio of unlabeled-to-labeled metabolite levels can be quantified. The change in percent labeled for metabolic intermediates provides insight into metabolic flux. MS technology provides measurement of total metabolite levels for all detectable intermediates as well as changes in metabolic flux through heavy isotope labeling, thus delivering a multifaceted analysis of glycolytic changes in cells.

Endothelial cell response to flow

ECs are constantly exposed to shear stress, a frictional force between the flowing blood and the vessel surface [21]. ECs sense shear stress and transmit it through multiple signal transduction pathways, including ion channels, caveolae, the cytoskeleton, tyrosine kinase receptors, the glycocalyx, and primary cilia [41,89−93]. For example, Tzima et al. showed that 12 dyn/cm^2 shear stress activated a mechanosensory complex composed of platelet endothelial cell adhesion molecule (PECAM)-1, vascular endothelial growth factor receptor-2 (VEGFR2), and vascular endothelial cadherin (VE-cadherin) [94].

EC response to shear stress is thought to play a critical role in a variety of endothelial functions, including angiogenesis, vascular remodeling, and atherosclerosis [95−97]. Arterial shear stress magnitude depends on flow rate, vessel radius, and blood viscosity [98]. Straight arterial sections are exposed to laminar flow and high, unidirectional shear stress (10–70 dyn/cm² shear stress). ECs in these arterial regions are spindle shaped and aligned with their long axis and actin fibers arranged parallel to the blood flow direction [99–101]. Steady laminar flow is associated with atherogenic gene downregulation and antioxidant and growth-arrest gene upregulation [93,102,103]. Additionally, ECs in laminar flow release NO, a vasodilator that is also important in preventing vascular disease through regulation of thrombosis, inflammation, dilation, and remodeling [104].

Arterial branches and curvatures create areas of disturbed flow, and ECs in these areas experience a lower net shear stress (<4 dyn/cm²) [98]. Disturbed flow patterns include recirculation eddies and flows that change direction with space and time [96,105,106]. ECs in disturbed flow regions have a rounder morphology without a uniform orientation [99,100]. Recent studies indicate that disturbed flow and the associated low and oscillatory shear stress induce sustained atherogenic gene activation in ECs. For example, monocyte chemotactic protein-1 (MCP-1) [107–109], which increases monocyte binding, and fibroblast growth factor-2 (FGF-2) and platelet-derived growth factors (PDGFs) [110,111], which enhance EC turnover, are increased in ECs exposed to disturbed flow. Additionally, ECs in disturbed flow areas have impaired NO production [112,113]. Therefore, disturbed flow is typically linked to EC dysfunction and diseases such as atherosclerosis [113,114].

ECs are often experimentally exposed to flow in vitro in a parallel plate flow chamber, cone-and-plate viscometer, parallel disk viscometer, orbital shaker, or capillary tube [21]. With each device, shear stress magnitude, flow direction, and flow pattern (e.g., steady or pulsatile) can be changed. The parallel disk viscometer and orbital shaker do not yield a uniform shear stress across the entire monolayer, and capillary tubes do not yield sufficient cell numbers for some bioassays. Therefore, the parallel-plate flow chamber and cone-and-plate viscometer are the best choices for studying endothelial metabolism in flow. Both systems produce a wide range of uniform shear stress magnitudes and provide enough cells for common metabolic assays.

In a parallel plate flow chamber, fluid enters the flow chamber through a tube and then flows between two parallel plates, passing over cells cultured on the lower plate [115]. The fluid friction against the cells creates a shear stress, and the small gap between the cells and the top plate creates laminar flow. The applied shear stress (τ), which is calculated using the Navier Stokes equations applied to two infinite parallel plates, can be controlled by manipulating the volume flow rate, fluid properties, or chamber geometry.

$$\tau = \frac{6Q\mu}{bh^2}$$

Either pressure or gravity can create the required pressure differential to drive flow [116]. The parallel plate flow chamber has multiple advantages, including excellent control of shear stress magnitude, as well as the ability to easily image the sample on a microscope stage. Newly published disturbed flow gaskets can also create larger areas of ECs exposed to disturbed flow [117]. However, the parallel plate setup requires vacuum sealing and bubble traps, making it impractical for acute studies and high-throughput applications.

A cone-and-plate system is commonly used to research laminar and disturbed flow effects on ECs in vitro. Since the flow is not pressure driven, the cone and plate device can provide a better model for isolating and studying shear stress effects without altered pressure [118]. The cone-and-plate system generates flow by rotating a cone around an axis perpendicular to the plate [119]. Shear stress is produced between the stationary plate and rotating cone. The angled cone geometry allows the shear stress to remain constant across the surface since the distance between the cone surface and the cells at the device center is reduced [120]. The ratio of centrifugal to viscous forces is given by a modified Reynolds number

$$Re = \frac{r^2 \omega \alpha^2}{12\nu}$$

where r is the radial distance from the cone apex, ω is the cone angular velocity, α is the cone angle, and ν is the fluid kinematic viscosity. If $Re \ll 1$, then the centrifugal forces are so small that the fluid velocity is purely tangential ($v_\theta = \omega r$; $v_r = v_z = 0$). The shear stress can then be determined from the equation

$$\tau_w = \mu \frac{dv_\theta}{dz} = \mu \frac{v_\theta}{z} = \frac{\mu \omega}{rz} = \frac{\mu \omega}{\alpha}$$

where μ is the fluid dynamic viscosity and z is the vertical distance between the cone and plate [121,122]. This equation relies on the small angle assumption for the cone angle, $\sin(\alpha) = z/r = \alpha$.

In arterial curvatures and bifurcations, disturbed flow is present, characterized by low overall shear stress and potential flow reversal. To produce this flow type, the cone can be programmed to rotate in an oscillatory manner [121]. The oscillation is defined by the sine wave function

$$(3) \quad \omega(t) = B + A\sin(2\pi f t)$$

where f is frequency, t is time, A is the sine wave amplitude, and B is the amplitude shift.

Many endothelial signaling pathways are activated by shear stress [123]. A DNA microarray gene expression analysis conducted in human umbilical vein endothelial cells (HUVECs) and coronary artery endothelial cells (HCAECs) showed that 3% of 5600 genes examined responded to shear stress [124]. While HUVEC exposed to steady laminar flow (15 dyn/cm^2 shear stress) decreased gene expression involved in DNA synthesis and the cell cycle, disturbed flow (1.5 dyn/cm^2 shear stress) affected gene expression involved in vascular remodeling, including endothelin-1 (ET-1) and transforming growth factor beta (TGF-β) [124]. NO production is another commonly measured endothelial shear stress response. ECs exposed to laminar flow release increased NO, whereas EC exposed to disturbed flow have reduced NO bioavailability, which contributes to EC dysfunction and cardiovascular disease [125–127]. In bovine aortic ECs, shear stress of 15 dyn/cm^2 for 3–24 h resulted in a two- to threefold increase of eNOS mRNA as a result of potassium channel openings [128]. In contrast, eNOS gene expression (NOS3) and bradykinin-induced vasorelaxation was drastically decreased in porcine carotid segments exposed to oscillatory shear stress (0.3 ± 3 dyn/cm^2) compared with unidirectional shear stress (6 ± 3 dyn/cm^2), suggesting disturbed flow patterns impair endothelial NO production [129].

Endothelial cell metabolism in varied flow conditions

ECs have long been known to be dysfunctional in high glucose conditions, and ECs cultured in both low and high glucose are not able to sense and adapt to steady laminar flow and therefore show decreased eNOS phosphorylation and NO production [40,42,130–132]. More recent studies suggest that endothelial glucose metabolism may in itself be critical to vascular disease [50,133]. In the microvasculature, glycolysis is thought to be essential to angiogenesis. When 6-phosphofructo-2-kinase/fructose-2,6-biphosphatase 3 (PFKFB3), a rate-limiting glycolytic enzyme, was inhibited either in vitro or in vivo, angiogenesis decreased dramatically [49,73]. In a murine tumor model, PFKFB3 inhibition led to restoration of tumor vessel barrier function, which decreased tumor cell metastasis [74].

Glycolysis and PFKFB3 have similarly been studied in ECs exposed to flow. Doddaballapur et al. showed that chronic laminar flow (72 h) decreases endothelial metabolism in HUVECs by inducing expression of the flow-sensitive transcription factor Kruppel-like factor 2 (KLF2). Shear stress caused KLF2 to bind to the PFKFB3 promoter, which then decreased PFKFB3 expression resulting in reduced glycolytic flux. This then put ECs in a quiescent state and decreased angiogenesis [134]. EC glycolytic flux was also acutely increased by shear stress in explanted guinea pig hearts. The ascending aorta was cannulated and connected to a nonrecirculating perfusion system at constant flow rate (3, 6, 20, and 25 mL/min) for 30–60 min. Glycolytic flux, determined from ^3H_2O content analysis, increased with shear stress exposure time [135].

Despite these recent studies, little is known about how oscillating disturbed flow impacts endothelial glucose metabolism or how flow impacts glycolytic side branches, including those that are important in diabetic vascular complications. Thus we are only beginning to understand the mechanisms by which shear stress affects endothelial glucose metabolism, its downstream pathways, and endothelial dysfunction.

Computational models of vascular mechanobiology and metabolism

EC mechanobiology and metabolic networks are complex and interrelated. While many techniques have been developed to examine these processes in vitro and in vivo, experimental studies are limited to a small number of input and outcome variables. Computational models of ECs provide valuable insight by more fully evaluating the experimental space and highlighting interactions among the various signaling and metabolic networks.

Models of endothelial cells in flow

ECs are thought to convert flow to a biochemical response through multiple steps. First, the cell glycocalyx, cilia, and other cell surface components are deformed by the shear stress, and these deformations enable transmission of the shear stress into the cell. The internalized force is then converted into a biochemical signal, which activates downstream biochemical signaling pathways. In this decentralized mechanotransduction model, multiple cell components work individually and together to convert shear stress into changes in cell function [136].

No computational models fully describe cellular mechanotransduction. However, several models have been developed to explore distinct aspects of this process. Most models examine fluid shear stress sensing and amplification at the cell surface (mechanotransmission) or shear stress activation of biochemical pathways, including changes in cellular signaling; gene, protein, or metabolite levels; and biochemical fluxes.

Dabagh et al. were the first to incorporate the glycocalyx [137—140], focal adhesions [141,142], nucleus [143,144], cytoskeleton [143,145—147], cilia [148—150], and adherens junctions [141] into a mechanotransmission model [151]. Using the finite element method, they generated a model to simulate an in vitro EC monolayer exposed to steady shear stress at $10-20$ dyne/cm^2. Cells were defined to have a hexagonal shape. The model predicted more than 100-fold amplification of stresses at adherens junctions and focal adhesions over stresses at the apical surface. Their results suggest that force-induced mechanotransmission by a single molecule is possible, supporting the decentralized mechanotransduction model. However, since this model did not account for cell elongation in the flow direction, it could not predict cell response after early exposure to flow.

Several models also predicted EC surface signaling molecule binding kinetics with fluid-induced shear stress [152—156]. Zhao et al. developed a computational model based on FGF2 binding to an experimental synthetic hollow fiber tube bioreactor lined with bovine aortic ECs under pulsatile flow. Their model and experimental validation concluded that there is significant FGF2 capture under low flow conditions but FGF2 binding decreases with increasing shear stress. They also found that loss of heparan sulfate proteoglycans—a major glycocalyx component—further decreased FGF2 binding to the cell surface. Patel et al. introduced a model of FGF2 binding in a parallel plate flow chamber examining continuous and bolus FGF2 delivery at various flow rates. This model predicted integrated effects based on delivery mode, fluid flow rate, and cell surface—binding site density. While these models provided insight into binding dynamics in fluid flow conditions, they did not take into account changes in the cell surface as ECs adapt to flow. For example, after flow adaptation ECs upregulate heparan sulfate proteoglycan mRNA, decrease cell surface receptors, and decrease permeability. Garcia et al. therefore expanded this model to incorporate cell surface changes with flow adaptation. Experimentally, ECs adapted to 24 h of flow demonstrated biphasic FGF2 binding to heparan sulfate proteoglycans, complexes increasing up to 20 dyn/cm^2 shear stress and then decreasing at higher shear stresses. The model predicted that adaptive EC surface remodeling in response to shear stress affects FGF2 binding through changes in cell surface receptor quantity, availability, and binding kinetics [157].

Models of endothelial cell metabolism

Many experimental studies of endothelial metabolism focus on a single metabolite and its direct metabolic pathways [158]. However, ECs metabolize a variety of nutrients and often shift their metabolic pathways based on environmental cues [67,158–160]. In addition, both glycolysis and the TCA cycle are connected to many branch pathways, which are essential to cell survival. Unfortunately, it is not possible to simultaneously probe all these metabolites and pathways in either in vitro systems or in vivo animal models. Therefore, computational metabolic models are a powerful supplement to experimental studies, enabling us to predict what may happen across the cellular metabolic network [161].

Since metabolic kinetic parameters are largely unknown, stoichiometric models are generally used to describe metabolic systems because they require only the reaction stoichiometry. Stoichiometric models are static in time and can only describe the network steady state [162]. Stoichiometric models are described by an m × r sparse matrix S where m is the metabolite number and r is the reaction number in the model. This matrix contains positive stoichiometric coefficients for products and negative coefficients for substrates. Most of the entries are zero because metabolic reactions use only a small subset of metabolites (Fig. 17.2). Stoichiometric models require less detail for each reaction, which allows genome-scale models which include all known reactions for a given organism to be constructed [162]. Stoichiometric models are solved as a system of linear equations which are more efficient to solve computationally. The primary disadvantage of stoichiometric models is that they cannot be used to simulate changes in time or predict metabolite concentrations.

Most early large-scale stoichiometric metabolic models were developed for unicellular organisms, particularly *E. coli* [163–168]. The first genome-scale stoichiometric *E. coli* model was published shortly after the annotated genome sequence was available. This model consisted of 627 unique reactions and due to the long history of *E. coli* research, every model enzymatic reaction was supported by biochemical and genetic evidence [169,170]. Each reaction's upper and lower flux bounds were set to $\pm \infty$ for bidirectional reactions and zero to $+\infty$ for unidirectional reactions. The authors were interested in

A

A →(v1)→ B →(v3)→ D →(v5)→ E
B →(v2)→ C
C →(v4)→ D

B

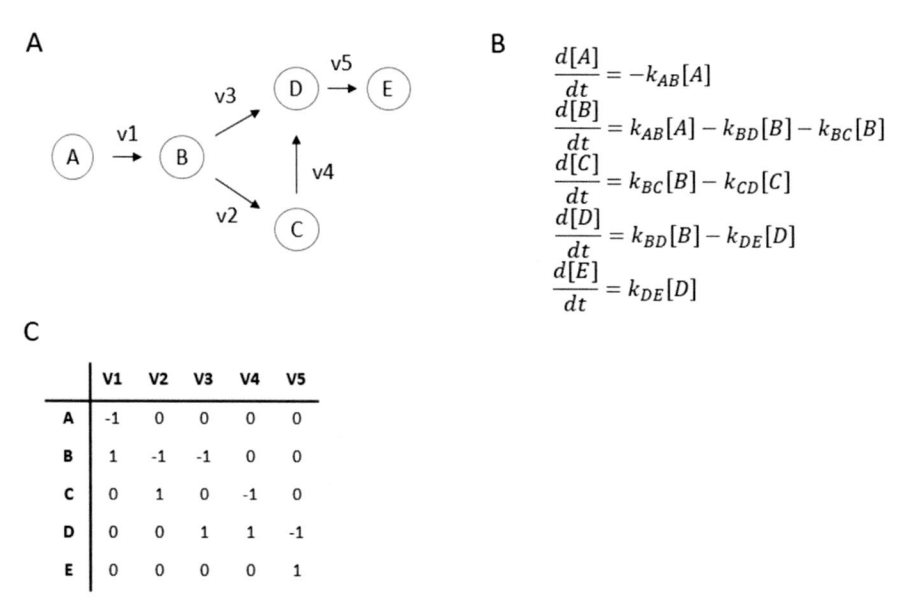

$$\frac{d[A]}{dt} = -k_{AB}[A]$$

$$\frac{d[B]}{dt} = k_{AB}[A] - k_{BD}[B] - k_{BC}[B]$$

$$\frac{d[C]}{dt} = k_{BC}[B] - k_{CD}[C]$$

$$\frac{d[D]}{dt} = k_{BD}[B] - k_{DE}[D]$$

$$\frac{d[E]}{dt} = k_{DE}[D]$$

C

	V1	V2	V3	V4	V5
A	-1	0	0	0	0
B	1	-1	-1	0	0
C	0	1	0	-1	0
D	0	0	1	1	-1
E	0	0	0	0	1

Figure 17.2 Metabolic network conversion to a stoichiometric model. (A) A simple metabolic reaction network which takes in metabolite A and produces a product E through a sequence of reactions involving intermediates B,C,D. (B) A kinetic model describes the simple metabolic network using differential equations where the population of each metabolite is tracked with respect to time, t. The rate of change per unit time is proportional to the metabolite concentration and the kinetic constant. (C) A stoichiometric model representation of the same metabolic reaction network is described by a sparse matrix. Each reaction is represented as column in the stoichiometric model. There is a row for each metabolite in the network.

determining the capabilities of the *E. coli* model to simulate experimental conditions. They generated 79 theoretical gene knockout strains by restricting flux through the corresponding reaction to zero. Simulations were run for each knockout strain to determine the maximal growth rate under aerobic conditions with minimal glucose. The predicted growth from the knockout strains was qualitatively compared to published experimental results (growth vs. no growth). Of 79 predictions, 68 (86%) were consistent with experimental observations.

Since then, several genome-scale stoichiometric models have been constructed for mammalian cells, including ECs [171,172]. Patella et al. developed a model to describe the metabolic state of HUVECs during tube formation based on a generic genome-scale human metabolism model (Recon 1) [173]. HUVECs were cultured on Matrigel, and intracellular proteins were quantified with mass spectrometry. Intracellular metabolic fluxes were then estimated with the iMAT algorithm, which determined the likelihood of enzyme activity based on protein levels. They also simulated the effect of CPT1A (fatty acid oxidation rate-limiting enzyme) inhibition by setting all the related reaction fluxes to zero. CPT1A inhibition significantly decreased oxygen consumption and increased monolayer permeability, which was verified in vitro using HUVEC monolayers. In vivo, CPT1A knockout led to blood vessel leakage suggesting that fatty acid metabolism is correlated with vessel permeability. The model by Patella et al. highlighted stoichiometric model advantages, which include simulating gene knockout effect and incorporating experimental data [162]. However, protein levels have limited accuracy in predicting intracellular fluxes since they do not always reflect enzyme activity. A second model limitation is that it is based on the entire annotated human genome, which includes reactions and pathways that are not active or expressed in ECs [55].

More recently, another EC genome-scale model was developed [174]. The goal was to create a model that would predict sepsis patient survival based on endothelial markers in the plasma metabolome. Similar to Patella et al., this model used Recon 1 as a starting point but leveraged available EC transcriptomic data to generate several cell specific models (HUVECs, microvascular ECs, and pulmonary artery ECs). The three models were quite similar in their metabolic gene expression, suggesting similar metabolism at least in vitro. The cells were then challenged with lipopolysaccharide (LPS) to simulate inflammation in sepsis, and the metabolic changes associated with increased permeability and glycocalyx loss were defined through metabolomics. When these data were used to parameterize the model, glycan production and fatty acid metabolism were associated with the LPS-induced cell changes. Finally, when the plasma metabolome was input into the model, glycan metabolism appeared to correspond with sepsis nonsurvivors in a way that was similar to the in vitro endothelial response. While the models generated were specific to ECs, the metabolic pathways were not adequately explored and the LPS effects were not quantified in terms of reaction rates. Simulations were also run with the assumption that ECs were trying to maximize proliferation, although confluent monolayers are quiescent.

While these past models provided insight into endothelial metabolic functions, more accurate models are needed. Genome-scale EC models should only include reactions that are known to be expressed within ECs. Intracellular reaction rates, or fluxes, should be estimated using isotope labeling experiments, in which heavy isotopes are introduced at known quantities. This is considered the gold standard for determining intracellular fluxes [175,176]. Finally, simulating the endothelial metabolic steady state requires the correct assumption to be made concerning the cell's objective function [177]. In the future, data-driven methods can be used to create models of endothelial metabolism in response to fluid-induced shear stress.

Translational potential and future work

Vascular mechanobiology and metabolism are both tightly linked to EC function. Yet few studies, either experimental or computational, have examined how mechanobiology and metabolism interact in the vasculature. Future work should focus on how mechanical stimuli affect EC metabolism, and in turn how these metabolic shifts impact EC function. In addition, it is equally important to address the question of how EC metabolism impacts mechanosensing and mechanotransmission. Finally, we must understand how these interrelated networks change in altered metabolic environments, for example, in people with diabetes. Since the mechanobiology and metabolic networks are so complicated, it is essential that these problems be addressed through both experimental and computational mechanisms. With deeper knowledge in the cell- and system-level integration of mechanobiology and metabolism, improved metabolic therapeutics can be developed to ameliorate cardiovascular disease in people with diabetes.

References

[1] Nicholls SJ, et al. Effect of diabetes on progression of coronary atherosclerosis and arterial remodeling: a pooled analysis of 5 intravascular ultrasound trials. J Am Coll Cardiol 2008;52(4):255—62.
[2] Wagenknecht LE, et al. Diabetes and progression of carotid atherosclerosis. Arterioscler Thromb Vasc Biol 2003;23(6):1035—41.
[3] Nathan D, et al. Intensive diabetes treatment and cardiovascular disease in patients with type 1 diabetes. N Engl J Med 2005;353(25):2643—53.
[4] Sustained effect of intensive treatment of type 1 diabetes mellitus on development and progression of diabetic nephropathy. Epidemiol Diabetes Interv & Complicat. (EDIC) Study JAMA 2003;290(16): 2159—67.
[5] Intensive blood glucose control and vascular outcomes in patients with type 2 diabetes. N Engl J Med 2008;358(24):2560—72.
[6] Libby P, Ridker PM, Maseri A. Inflammation and atherosclerosis. Circulation 2002;105(9):1135—43.
[7] Michiels C. Endothelial cell function. J Cell Physiol 2003;196:430—43.
[8] Lerman A, Burnett Jr JC. Intact and altered endothelium in regulation of vasomotion. Circulation 1992;86(6 Suppl.). III12—19.
[9] Reddy KG, et al. Evidence that selective endothelial dysfunction may occur in the absence of angiographic or ultrasound atherosclerosis in patients with risk factors for atherosclerosis. J Am Coll Cardiol 1994;23(4):833—43.
[10] Hempel A, et al. High glucose concentrations increase endothelial cell permeability via activation of protein kinase Cα. Circ Res 1997;81(3):363—71.
[11] Richardson M, et al. Increased expression in vivo of VCAM-1 and E-selectin by the aortic endothelium of normolipemic and hyperlipemic diabetic rabbits. Arterioscler Thromb Vasc Biol 1994;14(5): 760—9.
[12] Chakravarthy U, et al. Constitutive nitric oxide synthase expression in retinal vascular endothelial cells is suppressed by high glucose and advanced glycation end products. Diabetes 1998;47(6): 945—52.
[13] Hamuro M, et al. High glucose induced nuclear factor kappa B mediated inhibition of endothelial cell migration. Atherosclerosis 2002;162(2):277—87.
[14] Lorenzi M, Cagliero E, Toledo S. Glucose toxicity for human endothelial cells in culture. Delayed replication, disturbed cell cycle, and accelerated death. Diabetes 1985;34(7):621—7.
[15] Martin A, Komada MR, Sane DC. Abnormal angiogenesis in diabetes mellitus. Med Res Rev 2003; 23(2):117—45.
[16] Wang J, et al. Acute exposure to low glucose rapidly induces endothelial dysfunction and mitochondrial oxidative stress. Arterioscler Thromb Vasc Biol 2012;32:712—20.
[17] Lee TS, et al. Activation of protein kinase C by elevation of glucose concentration: proposal for a mechanism in the development of diabetic vascular complications. Proc Natl Acad Sci U S A 1989;86(13):5141—5.

[18] Brownlee MD, Michael. Advanced protein glycosylation in diabetes and aging. Annu Rev Med 1995;46(1):223–34.
[19] Hamada Y, et al. Rapid formation of advanced glycation end products by intermediate metabolites of glycolytic pathway and polyol pathway. Biochem Biophys Res Commun 1996;228(2):539–43.
[20] Du X-L, et al. Hyperglycemia-induced mitochondrial superoxide overproduction activates the hexosamine pathway and induces plasminogen activator inhibitor-1 expression by increasing Sp1 glycosylation. Proc Natl Acad Sci U S A 2000;97(22):12222–6.
[21] Chiu JJ, Chien S. Effects of disturbed flow on vascular endothelium: pathophysiological basis and clinical perspectives. Physiol Rev 2011;91(1):327–87.
[22] Giacco F, Brownlee M. Oxidative stress and diabetic complications. Circ Res 2010;107(9):1058–70.
[23] Malek AM, Alper SL, Izumo S. Hemodynamic shear stress and its role in atherosclerosis. J Am Med Assoc 1999;282(21):2035–42.
[24] Dewey C, et al. The dynamic response of vascular endothelial cells to fluid shear stress. J Biomech Eng 1981;103(3):177–85.
[25] Fisslthaler B, et al. Phosphorylation and activation of the endothelial nitric oxide synthase by fluid shear stress. Acta Physiol Scand 2000;168(1):81–8.
[26] Duraiswamy N, et al. Stented artery flow patterns and their effects on the artery wall. In: Annual review of fluid mechanics; 2007. p. 357–82.
[27] Frangos SG, Gahtan V, Sumpio B. Localization of atherosclerosis: role of hemodynamics. Arch Surg 1999;134(10):1142–9.
[28] Caro CG, Fitz-Gerald JM, Schroter RC. Arterial wall shear and distribution of early atheroma in man. Nature 1969;223(5211):1159–60.
[29] Reidy MA, Langille BL. The effect of local blood flow patterns on endothelial cell morphology. Exp Mol Pathol 1980;32(3):276–89.
[30] Chiu JJ, et al. Effects of disturbed flow on endothelial cells. J Biomech Eng 1998;120(1):2–8.
[31] Wright HP. Mitosis patterns in aortic endothelium. Atherosclerosis 1972;15(1):93–100.
[32] Chuang PT, et al. Macromolecular transport across arterial and venous endothelium in rats. Studies with evans blue-albumin and horseradish peroxidase. Arteriosclerosis 1990;10(2):188–97.
[33] Himburg HA, et al. Spatial comparison between wall shear stress measures and porcine arterial endothelial permeability. Am J Physiol Heart Circ Physiol 2004;286(5):H1916–22.
[34] Phelps JE, DePaola N. Spatial variations in endothelial barrier function in disturbed flows in vitro. Am J Physiol Heart Circ Physiol 2000;278(2):H469–76.
[35] Barber KM, Pinero A, Truskey GA. Effects of recirculating flow on U-937 cell adhesion to human umbilical vein endothelial cells. Am J Physiol 1998;275(2 Pt 2):H591–9.
[36] Chiu JJ, et al. Analysis of the effect of disturbed flow on monocytic adhesion to endothelial cells. J Biomech 2003;36(12):1883–95.
[37] Nagel T, et al. Vascular endothelial cells respond to spatial gradients in fluid shear stress by enhanced activation of transcription factors. Arterioscler Thromb Vasc Biol 1999;19(8):1825–34.
[38] Hahn C, et al. The subendothelial extracellular matrix modulates JNK activation by flow. Circ Res 2009;104(8):995–1003.
[39] Davies PF, et al. Hemodynamics and the focal origin of atherosclerosis: a spatial approach to endothelial structure, gene expression, and function. Ann N Y Acad Sci 2001;947:7–16. Discussion 16–7.
[40] Brower JB, et al. High glucose-mediated loss of cell surface heparan sulfate proteoglycan impairs the endothelial shear stress response. Cytoskeleton (Hoboken) 2010;67(3):135–41.
[41] Tarbell JM, Pahakis MY. Mechanotransduction and the glycocalyx. J Intern Med 2006;259(4):339–50.
[42] Kemeny SF, Figueroa DS, Clyne AM. Hypo- and hyperglycemia impair endothelial cell actin alignment and nitric oxide synthase activation in response to shear stress. PLoS One 2013;8(6):e66176.
[43] Krützfeldt A, et al. Metabolism of exogenous substrates by coronary endothelial cells in culture. J Mol Cell Cardiol 1990;22(12):1393–404.
[44] De Bock K, Georgiadou M, Carmeliet P. Role of endothelial cell metabolism in vessel sprouting. Cell Metabol 2013;18(5):634–47.

[45] Wang Q, et al. 2-Deoxy-D-glucose treatment of endothelial cells induces autophagy by reactive oxygen species-mediated activation of the AMP-activated protein kinase. PLoS One 2011;6(2):e17234.

[46] Mann GE, Yudilevich DL, Sobrevia L. Regulation of amino acid and glucose transporters in endothelial and smooth muscle cells. Physiol Rev 2003;83(1):183−252.

[47] Berg J, Tymoczko J, Stryer L. Biochemistry. 5th ed. PubMed; 2002. ISBN-10: 0-7167-3051-0 Search.

[48] Ros S, Schulze A. Balancing glycolytic flux: the role of 6-phosphofructo-2-kinase/fructose 2,6-bisphosphatases in cancer metabolism. Canc Metabol 2013;1(1):8.

[49] De Bock K, et al. Role of PFKFB3-driven glycolysis in vessel sprouting. Cell 2013;154(3):651−63.

[50] Goveia J, Stapor P, Carmeliet P. Principles of targeting endothelial cell metabolism to treat angiogenesis and endothelial cell dysfunction in disease. EMBO Mol Med 2014;6(9):1105−20.

[51] Greene DA, Arezzo JC, Brown MB. Effect of aldose reductase inhibition on nerve conduction and morphometry in diabetic neuropathy. Neurology 1999;53(3):580−91.

[52] Engerman RL, Kern TS, Larson ME. Nerve-conduction and aldose reductase inhibition during 5 years of diabetes or galactosemia in dogs. Diabetologia 1994;37(2):141−4.

[53] Kruger NJ, von Schaewen A. The oxidative pentose phosphate pathway: structure and organisation. Curr Opin Plant Biol 2003;6(3):236−46.

[54] Eelen G, et al. Endothelial cell metabolism. Physiol Rev 2017;98(1):3−58.

[55] Duarte NC, et al. Global reconstruction of the human metabolic network based on genomic and bibliomic data. Proc Natl Acad Sci U S A 2007;104(6):1777−82.

[56] Leopold JA, et al. Glucose-6-phosphate dehydrogenase overexpression decreases endothelial cell oxidant stress and increases bioavailable nitric oxide. Arterioscler Thromb Vasc Biol 2003;23(3):411−7.

[57] Kirkman HN, et al. Mechanisms of protection of catalase by NADPH kinetics and stoichiometry. J Biol Chem 1999;274(20):13908−14.

[58] Hanover JA, Krause MW, Love DC. Bittersweet memories: linking metabolism to epigenetics through O-GlcNAcylation. Nat Rev Mol Cell Biol 2012;13(5):312−21.

[59] Kreppel LK, Blomberg MA, Hart GW. Dynamic glycosylation of nuclear and cytosolic proteins. Cloning and characterization of a unique O-GlcNAc transferase with multiple tetratricopeptide repeats. J Biol Chem 1997;272(14):9308−15.

[60] Lubas WA, et al. O-Linked GlcNAc transferase is a conserved nucleocytoplasmic protein containing tetratricopeptide repeats. J Biol Chem 1997;272(14):9316−24.

[61] Wells L, et al. Dynamic O-glycosylation of nuclear and cytosolic proteins: further characterization of the nucleocytoplasmic beta-N-acetylglucosaminidase, O-GlcNAcase. J Biol Chem 2002;277(3):1755−61.

[62] Hart GW, Housley MP, Slawson C. Cycling of O-linked beta-N-acetylglucosamine on nucleocytoplasmic proteins. Nature 2007;446(7139):1017−22.

[63] Du XL, et al. Hyperglycemia inhibits endothelial nitric oxide synthase activity by posttranslational modification at the Akt site. J Clin Invest 2001;108(9):1341−8.

[64] Banerjee PS, Lagerlöf O, Hart GW. Roles of O-GlcNAc in chronic diseases of aging. Mol Aspect Med 2016;51:1−15.

[65] Herskowitz K, et al. Characterization of L-glutamine transport by pulmonary artery endothelial cells. Am J Physiol 1991;260(4 Pt 1):L241−6.

[66] DeBerardinis RJ, Cheng T. Q's next: the diverse functions of glutamine in metabolism, cell biology and cancer. Oncogene 2010;29(3):313−24.

[67] Kim B, et al. Glutamine fuels proliferation but not migration of endothelial cells. EMBO J 2017;36(16):2321.

[68] Huang H, et al. Role of glutamine and interlinked asparagine metabolism in vessel formation. EMBO J 2017;36(16):2334−52.

[69] DeBerardinis RJ, et al. The biology of cancer: metabolic reprogramming fuels cell growth and proliferation. Cell Metabol 2008;7(1):11−20.

[70] Warburg O. On the origin of cancer cells. Science 1956;123(3191):309−14.

[71] Verdegem D, et al. Endothelial cell metabolism: parallels and divergences with cancer cell metabolism. Canc Metabol 2014;2:19.
[72] Yalcin A, et al. Regulation of glucose metabolism by 6-phosphofructo-2-kinase/fructose-2,6-bisphosphatases in cancer. Exp Mol Pathol 2009;86(3):174—9.
[73] Schoors S, et al. Partial and transient reduction of glycolysis by PFKFB3 blockade reduces pathological angiogenesis. Cell Metabol 2014;19(1):37—48.
[74] Cantelmo AR, et al. Inhibition of the glycolytic activator PFKFB3 in endothelium induces tumor vessel normalization, impairs metastasis, and improves chemotherapy. Canc Cell 2016;30(6):968—85.
[75] Bond MR, Hanover JA. A little sugar goes a long way: the cell biology of O-GlcNAc. J Cell Biol 2015;208(7):869—80.
[76] Marsh SA, Collins HE, Chatham JC. Protein O-GlcNAcylation and cardiovascular (patho) physiology. J Biol Chem 2014;289(50):34449—56.
[77] Wu G, et al. Presence of glutamine:fructose-6-phosphate amidotransferase for glucosamine-6-phosphate synthesis in endothelial cells: effects of hyperglycaemia and glutamine. Diabetologia 2001;44(2):196—202.
[78] Brownlee M. The pathobiology of diabetic complications: a unifying mechanism. Diabetes 2005;54(6):1615—25.
[79] Du XL, et al. Hyperglycemia inhibits endothelial nitric oxide synthase activity by posttranslational modification at the Akt site. J Clin Invest 2001;108(9):1341—8.
[80] Luo B, Soesanto Y, McClain DA. Protein modification by O-linked GlcNAc reduces angiogenesis by inhibiting Akt activity in endothelial cells. Arterioscler Thromb Vasc Biol 2008;28(4):651—7.
[81] Beleznai T, Bagi Z. Activation of hexosamine pathway impairs nitric oxide (NO)-dependent arteriolar dilations by increased protein O-GlcNAcylation. Vasc Pharmacol 2012;56(3—4):115—21.
[82] Barnes JW, et al. O-linked beta-N-acetylglucosamine transferase directs cell proliferation in idiopathic pulmonary arterial hypertension. Circulation 2015;131(14):1260—8.
[83] Astles JR, Sedor FA, Toffaletti JG. Evaluation of the YSI 2300 glucose analyzer: algorithm-corrected results are accurate and specific. Clin Biochem 1996;29(1):27—31.
[84] Kurtz C, Smelko J. Which factors to consider when selecting an analytical method for cell culture fermentation: a comparison of four different metabolic analyzer instruments. Bioprocess J 2014;13(3):12—31.
[85] Yoshioka K, et al. A novel fluorescent derivative of glucose applicable to the assessment of glucose uptake activity of *Escherichia coli*. Biochim Biophys Acta Gen Subj 1996;1289(1):5—9.
[86] Yoshioka K, et al. Evaluation of 2-[N-(7-nitrobenz-2-oxa-1,3-diazol-4-yl)amino]-2-deoxy-D-glucose, a new fluorescent derivative of glucose, for viability assessment of yeast Candida albicans. Appl Microbiol Biotechnol 1996;46(4):400—4.
[87] Divakaruni Ajit S, Rogers George W, Murphy Anne N. Measuring mitochondrial function in permeabilized cells using the Seahorse XF analyzer or a clark-type oxygen electrode. Curr Protoc Toxicol 2014;60(1):25.2.1—25.2.16.
[88] Aebersold R, Mann M. Mass spectrometry-based proteomics. Nature 2003;422:198.
[89] Florian JA, et al. Heparan sulfate proteoglycan is a mechanosensor on endothelial cells. Circ Res 2003;93(10):e136—42.
[90] Nauli SM, et al. Endothelial cilia are fluid shear sensors that regulate calcium signaling and nitric oxide production through polycystin-1. Circulation 2008;117(9):1161—71.
[91] Chachisvilis M, Zhang Y-L, Frangos JA. G protein-coupled receptors sense fluid shear stress in endothelial cells. Proc Natl Acad Sci U S A 2006;103(42):15463.
[92] Fisher AB, et al. Endothelial cellular response to altered shear stress. Am J Physiol Lung Cell Mol Physiol 2001;281(3):L529—33.
[93] Li YS, Haga JH, Chien S. Molecular basis of the effects of shear stress on vascular endothelial cells. J Biomech 2005;38(10):1949—71.
[94] Tzima E, et al. A mechanosensory complex that mediates the endothelial cell response to fluid shear stress. Nature 2005;437(7057):426—31.
[95] Kamiya A, Togawa T. Adaptive regulation of wall shear stress to flow change in the canine carotid artery. Am J Physiol 1980;239(1):H14—21.

[96] Cunningham KS, Gotlieb AI. The role of shear stress in the pathogenesis of atherosclerosis. Lab Invest 2005;85(1):9–23.
[97] Lancerotto L, Orgill DP. Mechanoregulation of angiogenesis in wound healing. Adv Wound Care 2014;3(10):626–34.
[98] Adel M, Malek M, Alper SL, Izumo S. Hemodynamic shear stress and its role in atherosclerosis. J Am Med Assoc 1999;282:2035–42.
[99] Langille BL, Adamson SL. Relationship between blood flow direction and endothelial cell orientation at arterial branch sites in rabbits and mice. Circ Res 1981;48(4):481–8.
[100] Nerem RM, Levesque MJ, Cornhill JF. Vascular endothelial morphology as an indicator of the pattern of blood flow. J Biomech Eng 1981;103(3):172–6.
[101] Wong AJ, Pollard TD, Herman IM. Actin filament stress fibers in vascular endothelial cells in vivo. Science 1983;219(4586):867–9.
[102] Berk BC. Atheroprotective signaling mechanisms activated by steady laminar flow in endothelial cells. Circulation 2008;117(8):1082–9.
[103] Traub O, Berk BC. Laminar shear stress: mechanisms by which endothelial cells transduce an atheroprotective force. Arterioscler Thromb Vasc Biol 1998;18(5):677–85.
[104] Federici M, et al. Insulin-dependent activation of endothelial nitric oxide synthase is impaired by O-linked glycosylation modification of signaling proteins in human coronary endothelial cells. Circulation 2002;106(4):466–72.
[105] Asakura T, Karino T. Flow patterns and spatial distribution of atherosclerotic lesions in human coronary arteries. Circ Res 1990;66(4):1045–66.
[106] Bharadvaj BK, Mabon RF, Giddens DP. Steady flow in a model of the human carotid bifurcation. Part I–flow visualization. J Biomech 1982;15(5):349–62.
[107] Hsiai TK, et al. Pulsatile flow regulates monocyte adhesion to oxidized lipid-induced endothelial cells. Arterioscler Thromb Vasc Biol 2001;21(11):1770–6.
[108] Hsiai TK, et al. Monocyte recruitment to endothelial cells in response to oscillatory shear stress. FASEB J 2003;17(12):1648–57.
[109] Hwang J, et al. Oscillatory shear stress stimulates endothelial production of O_2- from p47phox-dependent NAD(P)H oxidases, leading to monocyte adhesion. J Biol Chem 2003;278(47):47291–8.
[110] Malek AM, et al. Fluid shear stress differentially modulates expression of genes encoding basic fibroblast growth factor and platelet-derived growth factor B chain in vascular endothelium. J Clin Invest 1993;92(4):2013–21.
[111] Wilcox JN, et al. Platelet-derived growth factor mRNA detection in human atherosclerotic plaques by in situ hybridization. J Clin Invest 1988;82(3):1134–43.
[112] Marina Noris MM, Donadelli R, Aiello S, Foppolo M, Marta Todeschini, Orisio S, Remuzzi G, Remuzzi A. Nitric oxide synthesis by cultured endothelial cells is modulated by flow conditions. Circ Res 1995;76:536–43.
[113] Yetik-Anacak G, Catravas JD. Nitric oxide and the endothelium: history and impact on cardiovascular disease. Vasc Pharmacol 2006;45(5):268–76.
[114] Ku DN, et al. Pulsatile flow and atherosclerosis in the human carotid bifurcation. Positive correlation between plaque location and low oscillating shear stress. Arteriosclerosis 1985;5(3):293–302.
[115] Chiu J-J, Chien S. Effects of disturbed flow on vascular endothelium: pathophysiological basis and clinical perspectives. Physiol Rev 2011;91(1):327–87.
[116] Davis CA, et al. Device-based in vitro techniques for mechanical stimulation of vascular cells: a review. J Biomech Eng 2015;137(4):040801.
[117] Sedlak JM, Clyne AM. A modified parallel plate flow chamber to study local endothelial response to recirculating disturbed flow. ASME J Biomech Eng 2020;142(4):041003. https://doi.org/10.1115/1.4044899.
[118] Davis R, et al. Theories of behaviour and behaviour change across the social and behavioural sciences: a scoping review. Health Psychol Rev 2015;9(3):323–44.
[119] Brown TD. Techniques for mechanical stimulation of cells in vitro. J Biomech 2000;33:3–14.
[120] Malek AM, et al. A cone-plate apparatus for the in vitro biochemical and molecular analysis of the effect of shear stress on adherent cells. Methods Cell Sci 1995;17(3):165–76.

[121] Buschmann MH, et al. Analysis of flow in a cone-and-plate apparatus with respect to spatial and temporal effects on endothelial cells. Biotechnol Bioeng 2005;89(5):493–502.
[122] Bussolari SR. Apparatus for subjecting living cells to fluid shear stress. Rev Sci Instrum 1982;53(12):1851.
[123] Davies PF. Hemodynamic shear stress and the endothelium in cardiovascular pathophysiology. Nat Clin Pract Cardiovasc Med 2009;6(1):16–26.
[124] Ohura N, et al. Global analysis of shear stress-responsive genes in vascular endothelial cells. J Atherosclerosis Thromb 2003;10(5):304–13.
[125] Buga GM, et al. Shear stress-induced release of nitric oxide from endothelial cells grown on beads. Hypertension 1991;17(2):187–93.
[126] Korenaga R, et al. Laminar flow stimulates ATP- and shear stress-dependent nitric oxide production in cultured bovine endothelial cells. Biochem Biophys Res Commun 1994;198(1):213–9.
[127] Cooke JP, Dzau VJ. Nitric oxide synthase: role in the genesis of vascular disease. Annu Rev Med 1997;48:489–509.
[128] Uematsu M, et al. Regulation of endothelial cell nitric oxide synthase mRNA expression by shear stress. Am J Physiol 1995;269(6 Pt 1):C1371–8.
[129] Gambillara V, et al. Plaque-prone hemodynamics impair endothelial function in pig carotid arteries. Am J Physiol Heart Circ Physiol 2006;290(6):H2320–8.
[130] Kemeny SF, et al. Glycated collagen and altered glucose increase endothelial cell adhesion strength. J Cell Physiol 2013;228(8):1727–36.
[131] Kemeny SF, et al. Glycated collagen alters endothelial cell actin alignment and nitric oxide release in response to fluid shear stress. J Biomech 2011;44(10):1927–35.
[132] Lopez-Quintero SV, et al. High glucose attenuates shear-induced changes in endothelial hydraulic conductivity by degrading the glycocalyx. PLoS One 2013;8(11):e78954.
[133] Eelen G, et al. Endothelial cell metabolism in normal and diseased vasculature. Circ Res 2015;116(7):1231–44.
[134] Doddaballapur A, et al. Laminar shear stress inhibits endothelial cell metabolism via KLF2-mediated repression of PFKFB3. Arterioscler Thromb Vasc Biol 2015;35(1):137–45.
[135] Suarez J, Rubio R. Regulation of glycolytic flux by coronary flow in Guinea pig heart. Role of vascular endothelial cell glycocalyx. AJP Heart & Circ Physiol 1991;261(6):H1994–2000.
[136] Davies PF. Flow-mediated endothelial mechanotransduction. Physiol Rev 1995;75(3):519–60.
[137] Kabedev A, Lobaskin V. Structure and elasticity of bush and brush-like models of the endothelial glycocalyx. Sci Rep 2018;8(1):240.
[138] Cruz-Chu ER, et al. Structure and response to flow of the glycocalyx layer. Biophys J 2014;106(1):232–43.
[139] Jiang XZ, Feng M, Luo KH, Ventikos Y, et al. Large-scale molecular dynamics simulation of flow under complex structure of endothelial glycocalyx. Comput Fluid 2018;173:140–6. https://doi.org/10.1016/j.compfluid.2018.03.014.
[140] Jiang XZ, et al. Large-scale molecular dynamics simulation of coupled dynamics of flow and glycocalyx: towards understanding atomic events on an endothelial cell surface. J R Soc Interface 2017;14(137):20170780.
[141] Biton Y, Safran S. Theory of the mechanical response of focal adhesions to shear flow. J Phys Condens Matter 2010;22(19):194111.
[142] Ferko MC, et al. Finite-element stress analysis of a multicomponent model of sheared and focally-adhered endothelial cells. Ann Biomed Eng 2007;35(2):208–23.
[143] Jean RP, Chen CS, Spector AA. Finite-element analysis of the adhesion-cytoskeleton-nucleus mechanotransduction pathway during endothelial cell rounding: axisymmetric model. J Biomech Eng 2005;127(4):594–600.
[144] Zeng Y, et al. A three-dimensional random network model of the cytoskeleton and its role in mechanotransduction and nucleus deformation. Biomech Model Mechanobiol 2012;11(1–2):49–59.
[145] Luo T, et al. Molecular mechanisms of cellular mechanosensing. Nat Mater 2013;12(11):1064.
[146] Shafrir Y, Forgacs G. Mechanotransduction through the cytoskeleton. Am J Physiol Cell Physiol 2002;282(3):C479–86.

[147] Palmer JS, Boyce MC. Constitutive modeling of the stress—strain behavior of F-actin filament networks. Acta Biomater 2008;4(3):597—612.
[148] Lim YC, Cooling MT, Long DS. Computational models of the primary cilium and endothelial mechanotransmission. Biomech Model Mechanobiol 2015;14(3):665—78.
[149] Lim YC, et al. Mechanical models of endothelial mechanotransmission based on a population of cells. In: Computational biomechanics for medicine. Springer; 2016. p. 63—73.
[150] Downs ME, et al. An experimental and computational analysis of primary cilia deflection under fluid flow. Comput Methods Biomech Biomed Eng 2014;17(1):2—10.
[151] Dabagh M, et al. Shear-induced force transmission in a multicomponent, multicell model of the endothelium. J R Soc Interface 2014;11(98):20140431.
[152] Zhao B, et al. Endothelial cell capture of heparin-binding growth factors under flow. PLoS Comput Biol 2010;6(10):e1000971.
[153] Patel NS, Reisig KV, Clyne AM. A computational model of fibroblast growth factor-2 binding to endothelial cells under fluid flow. Ann Biomed Eng 2013;41(1):154—71.
[154] Shen W, et al. A computational model of FGF-2 binding and HSPG regulation under flow. IEEE (Inst Electr Electron Eng) Trans Biomed Eng 2009;56(9):2147—55.
[155] Cappadona C, et al. Phenotype dictates the growth response of vascular smooth muscle cells to pulse pressure in vitro. Exp Cell Res 1999;250(1):174—86.
[156] Zhang C, et al. A numerical study of pulsatile flow through a hollow fiber cartridge: growth factor-receptor binding and dissociation analysis, In: Bioinformatics, systems biology and intelligent computing, 2009. IJCBS'09. International joint conference on. 2009; n.d. IEEE.
[157] Garcia J, et al. Fibroblast growth factor-2 binding to heparan sulfate proteoglycans varies with shear stress in flow-adapted cells. Ann Biomed Eng 2019;47(4):1078—93.
[158] McGarrity S, et al. Understanding the causes and implications of endothelial metabolic variation in cardiovascular disease through genome-scale metabolic modeling. Front Cardiovasc Med 2016;3:10.
[159] Coloff JL, et al. Differential glutamate metabolism in proliferating and quiescent mammary epithelial cells. Cell Metabol 2016;23(5):867—80.
[160] Patella F, et al. Proteomics-based metabolic modeling reveals that fatty acid oxidation (FAO) controls endothelial cell (EC) permeability. Mol Cell Proteomics 2015;14(3):621—34.
[161] Palsson BØ. Systems biology: constraint-based reconstruction and analysis. Cambridge University Press; 2015.
[162] Palsson B. Systems biology. Cambridge university press; 2015.
[163] Majewski R, Domach M. Simple constrained-optimization view of acetate overflow in *E. coli*. Biotechnol Bioeng 1990;35(7):732—8.
[164] Varma A, Boesch BW, Palsson BO. Biochemical production capabilities of *Escherichia coli*. Biotechnol Bioeng 1993;42(1):59—73.
[165] Varma A, Boesch BW, Palsson BO. Stoichiometric interpretation of *Escherichia coli* glucose catabolism under various oxygenation rates. Appl Environ Microbiol 1993;59(8):2465—73.
[166] Varma A, Palsson BO. Stoichiometric flux balance models quantitatively predict growth and metabolic by-product secretion in wild-type *Escherichia coli* W3110. Appl Environ Microbiol 1994;60(10):3724—31.
[167] Varma A, Palsson BO. Metabolic capabilities of *Escherichia coli*: I. Synthesis of biosynthetic precursors and cofactors. J Theor Biol 1993;165(4):477—502.
[168] Covert MW, Palsson BØ. Transcriptional regulation in constraints-based metabolic models of *Escherichia coli*. J Biol Chem 2002;277(31):28058—64.
[169] Edwards J, Palsson B. The *Escherichia coli* MG1655 in silico metabolic genotype: its definition, characteristics, and capabilities. Proc Natl Acad Sci U S A 2000;97(10):5528—33.
[170] Blattner FR, et al. The complete genome sequence of *Escherichia coli* K-12. Science 1997;277(5331):1453—62.
[171] Agren R, et al. Reconstruction of genome-scale active metabolic networks for 69 human cell types and 16 cancer types using INIT. PLoS Comput Biol 2012;8(5):e1002518.

[172] Fouladiha H, Marashi SA, Shokrgozar M. Reconstruction and validation of a constraint-based metabolic network model for bone marrow-derived mesenchymal stem cells. Cell Prolif 2015;48(4): 475—85.
[173] Patella F, Schug ZT, Persi E, Neilson LJ, Erami Z, Avanzato D, Maione F, Hernandez-Fernaud JR, Mackay G, Zheng L, Reid S, Frezza C, Giraudo E, Fiorio Pla A, Anderson K, Ruppin E, Gottlieb E, Zanivan S, et al. Proteomics-based metabolic modelling reveals that fatty acid oxidation controls endothelial cell permeability. Mol Cell Proteom 2015;14(3):621—34. https://doi.org/10.1074/mcp.M114.045575. mcp. M114. 045575.
[174] McGarrity S, et al. Metabolic systems analysis of LPS induced endothelial dysfunction applied to sepsis patient stratification. Sci Rep 2018;8(1):6811.
[175] Nöh K, Droste P, Wiechert W. Visual workflows for 13 C-metabolic flux analysis. Bioinformatics 2014;31(3):346—54.
[176] Wiechert W, Niedenführ S, Nöh K. A primer to 13C metabolic flux analysis. Fundamental Bioengineering; 2015. p. 97—142.
[177] Feist AM, Palsson BO. The biomass objective function. Curr Opin Microbiol 2010;13(3):344—9.

CHAPTER 18

Effect of type 2 diabetes on bone cell behavior

Rachana Vaidya, Anna Church, Lamya Karim
Department of Bioengineering, University of Massachusetts Dartmouth, Dartmouth, MA, United States

Introduction

Diabetes mellitus is a disease with an increasing impact on healthcare worldwide and is quickly becoming the world's most significant cause of morbidity and mortality [1,2]. With an estimated 463 million people with diabetes, it is now the seventh leading cause of death worldwide [2]. Diabetes is characterized by an elevation in blood glucose known as hyperglycemia, which has detrimental effects on the heart, kidney, retinal, and vascular tissues. Hyperglycemia can lead to many diabetic complications such as myocardial infarction, kidney failure, blindness, neuropathy, and infections that can result in limb amputations [3]. In addition, it has recently been identified that diabetic patients also have an increased risk of bone fracture that is independent of their bone mineral density (BMD).

While patients with type 1 diabetes (T1D) have low BMD and a six- to sevenfold higher risk for bone fractures, patients with type 2 diabetes (T2D) have normal or slightly higher BMD and up to threefold higher fracture risk [4–7]. Despite the similarity of chronic hyperglycemia, T1D and T2D have distinct pathophysiological mechanisms, which may differently affect bone metabolism. In both cases, the underlying mechanisms of poor bone strength are not well understood.

Since T2D accounts for more than 90% of diabetes cases, it is crucial to understand the mechanisms that contribute to bone fragility in T2D. However, the explanation for higher fracture risk yet normal or increased BMD in patients with T2D is seemingly more complicated. Several reports suggest that diabetic complications, duration of T2D, older age, poor balance, poor glycemic control, hypoglycemia, and certain diabetic medications (thiazolidinediones) further increase the fracture risk of patients in T2D, independent of the BMD [8,9]. Patients with T2D have normal BMD for a given femoral neck BMD T-score and age compared to nondiabetic controls as assessed by dual-energy X-ray absorptiometry (DXA) [10,11]. This has led to the suggestion that diabetes-related alterations in skeletal properties may result in fragility fractures in T2D, not captured by DXA [12,13]. Because estimation of fracture risk by DXA-derived BMD is unreliable for diabetic subjects, poor bone quality is the most appropriate and explicable cause for higher fracture risk in these populations.

Bone quality is composed of characteristics that define the overall quality of the tissue matrix itself. This collection of variables consists of tissue microarchitecture, turnover, microdamage, and tissue composition (e.g., extent of mineralization, mineral stoichiometry, collagen, other matrix constituents, and water content) (Fig. 18.1). These measures, in addition to BMD, contribute to bone strength [14]. Investigating specific factors causing deterioration of bone quality, which may be associated with the risk of fracture independent of BMD, may serve as a powerful clue for elucidating the pathology of bone fragility in diabetic patients. Many reports suggest hyperglycemia promotes excessive glycosylation of collagen, resulting in the formation of cross-links that are responsible for changing the material properties of bone and maybe plausible causes for poor bone quality in diabetic patients [15–18]. Also, various factors involved in the regulation of bone turnover are reported to be involved in fracture risk, independent of BMD. The collagen products and factors for mineralization along with bone formation and resorption markers are indices of bone turnover. Parathyroid hormone (PTH) secretion, bone formation markers such as procollagen type 1 N-terminal propeptide (P1NP), osteocalcin (OCN), bone-specific alkaline phosphatases (BAPs), as well as bone resorption markers such as C-terminal telopeptide (Ctx) are significantly lower in T2D patients compared to nondiabetic subjects, indicating a suppressed bone turnover in these patients. The decrease in these bone remodeling markers appears to be predictive of fracture risk regardless of BMD [19–22].

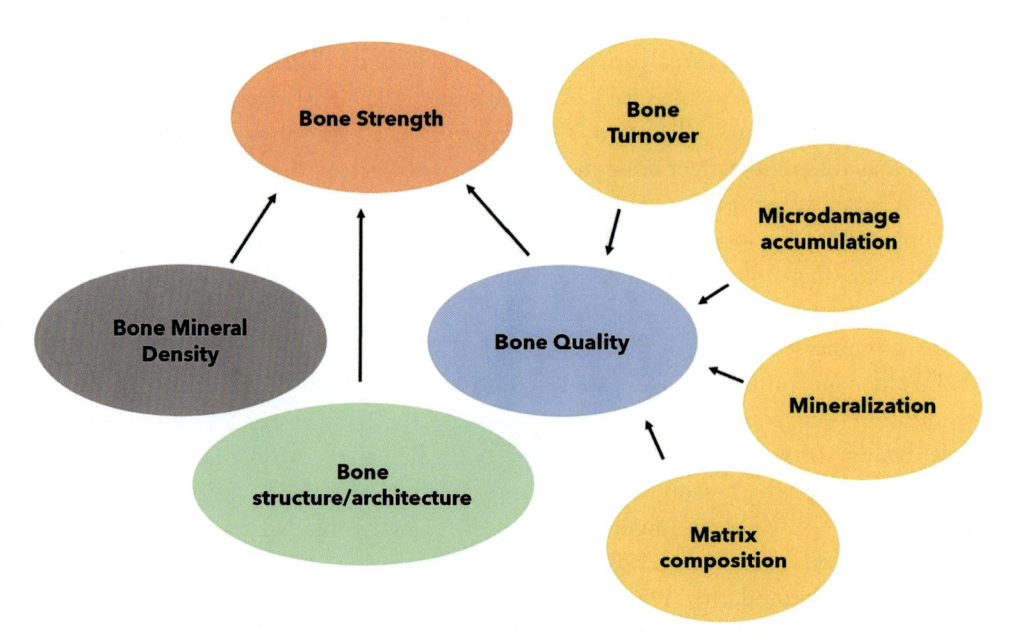

Figure 18.1 Bone strength is determined by bone mineral density, bone structure and architecture, and bone quality.

Bone cells

Bone contains four different types of cells. Osteoblasts, osteoclasts, and bone lining cells are present on bone surfaces, whereas osteocytes are found embedded in the mineralized matrix. Osteoblasts, osteocytes, and bone lining cells originate from pluripotent osteoprogenitor cells, which can also give rise to adipocytes and chondrocytes [23]. Osteoclasts originate from hematopoietic progenitors in the bone marrow, which can also give rise to monocytes and macrophages [23]. Although inert in appearance, bone is a highly dynamic organ that is continuously resorbed by osteoclasts and formed by osteoblasts. Osteocytes act as mechanosensors and orchestrate this bone remodeling process [23]. The function of bone lining cells is unclear, but these cells seem to play a role in coupling bone resorption to bone formation [24].

Osteoblasts

Osteoblasts are cuboidal cells found on the bone surface and are responsible for bone formation. Mesenchymal stromal cells (MSCs) located on periosteal surfaces and within bone marrow provide a source of osteoblasts that act in concert with osteoclasts to model bone during growth and maintain bone architecture during adulthood [25]. The commitment of MSCs toward osteoblast lineage requires expressions of Runt-related transcription factor 2 (Runx2), distal-less homeobox 5 (Dlx5), and osterix (Osx), which are crucial for osteoblast differentiation. Runx2 is a master gene of osteoblast differentiation [26,27]. As osteoblasts differentiate from their precursors, they begin to secrete bone matrix proteins such as type 1 collagen, osteocalcin (OCN), and bone sialoprotein (BSP1/11). Type I collagen is the major protein in the bone matrix, representing about 90% of the organic matrix. The network of type I collagen fibers provides the structure on which bone mineral is deposited [25]. Osteoblasts also secrete noncollagenous proteins, including proteoglycans, glycoproteins, and γ-carboxylated (Gla) proteins, which play regulatory roles in bone cell adhesion, migration, proliferation, and differentiation [25,26]. Osteoblasts not only secrete the organic components of bone matrix but are also indirectly responsible for the mineralization of osteoid via osteoblast-derived bone sialoprotein (BSP) and alkaline phosphatase (ALP) [25].

Osteoclasts

Osteoclasts are terminally differentiated multinucleated cells originated from mononuclear cells of the hematopoietic stem cell lineage, under the influence of several factors. They are derived from precursors in the myeloid/monocyte lineage that circulate in the blood after their formation in the bone marrow. These osteoclast precursors (OCPs) are attracted to sites on bone surfaces destined for resorption and fuse with one another to form the multinucleated cells that resorb calcified matrices under the influence of osteoblastic cells in bone marrow [27,28]. Several factors such as macrophage colony-stimulating factor (M-CSF), secreted by osteoprogenitor mesenchymal cells and osteoblasts [29], and receptor activator of nuclear kappa β ligand (RANKL), secreted by osteoblasts, osteocytes, and stromal cells, promote the activation of transcription factors and gene expression in osteoclasts [30].

In osteoclast precursors, binding of M-CSF to its receptor (cFMS) stimulates their proliferation and inhibits their apoptosis [29]. RANKL plays a crucial role in osteoclastogenesis by inducing osteoclast formation [30]. On the contrary, another factor called osteoprotegerin (OPG), which is produced by a wide range of cells including osteoblasts and stromal cells, binds to RANKL, preventing the RANK/RANKL interaction and, consequently, inhibiting the osteoclastogenesis [31]. Thus, the RANKL/RANK/OPG system is an essential mediator of osteoclastogenesis [30,31].

Osteocytes

Osteocytes compose 90%−95% of all bone cells in adult bone. They are the most abundant and long-lived cells, with a life span of up to 25 years [32]. During bone formation, a subset of osteoblasts embed themselves into the bone matrix they produced and differentiate into mechanosensing osteocytes [32]. Once the mature osteocyte is entrapped within the mineralized bone matrix, several of the previously expressed osteoblast markers such as OCN, BSPII, type I collagen, and ALP are downregulated. On the other hand, osteocyte markers, including dentine matrix protein 1 (DMP1) and sclerostin (SOST), are highly expressed [33−36].

The osteocyte cell body is located inside the lacuna, its cytoplasmic processes (up to 50 per each cell) cross tiny tunnels (259 ± 129 nm in diameter) [37] that originate from the lacuna space called canaliculi, forming the osteocyte lacunocanalicular system [38]. These cytoplasmic processes are connected to other neighboring osteocytes processes by gap junctions, as well as to cytoplasmic processes of osteoblasts and bone lining cells on the bone surface, facilitating the intercellular transport of small signaling molecules such as prostaglandins and nitric oxide among these cells [39]. Besides, the osteocyte lacunocanalicular system is in close proximity to the vascular supply for the exchange of oxygen and nutrients [40]. The cell−cell communication is also achieved by interstitial fluid that flows between the osteocytes processes and canaliculi [41].

By the lacunocanalicular system, the osteocytes act as mechanosensors as their interconnected network has the capacity to detect mechanical pressures and loads, thereby helping the adaptation of bone to daily mechanical forces [41]. In this way, the osteocytes seem to act as orchestrators of bone remodeling through regulation of osteoblast and osteoclast activities [42,43]. For example, osteocytes can regulate osteoclastogenesis by expressing RANKL and its decoy receptor OPG. Osteocytes can also regulate osteoblast differentiation by secreting SOST, an inhibitor of Wnt signaling [44−47]. Moreover, osteocyte apoptosis is recognized as a chemotactic signal to osteoclastic bone resorption [48]. In agreement, it has been shown that during bone resorption, apoptotic osteocytes are engulfed by osteoclasts [49,50].

Bone modeling and remodeling

Our skeleton is a dynamic, metabolically active, and functionally diverse organ. It plays a role in both mineral metabolism via calcium and phosphate homeostasis. It also provides levers for the muscle to allow locomotion, supports and protects vital organs, and is the site of hematopoietic marrow [23]. It may also have additional important endocrine roles in fertility, glucose metabolism, appetite regulation, and muscle function [51–54].

Throughout life, the skeleton is dynamically "constructed" and "reconstructed" by two processes: bone modeling and remodeling. During early childhood, both bone modeling (formation and shaping) and bone remodeling (replacing or renewing) occurs, whereas in adulthood, bone remodeling is the predominant process to maintain skeletal integrity, except for massive increases in bone formation that occur after a fracture [55].

Bone modeling begins early in skeletal development whereby the size and shape of the bone is modified. This process involves bone formation and resorption by osteoblasts and osteoclasts in an uncoupled manner, anatomically and temporally. Abnormalities in bone modeling cause skeletal deformities [56]. The majority of bone modeling is completed by skeletal maturity, but modeling can still occur, even in adulthood, such as in an adaptive response to mechanical loading and exercise. For example, in tennis players, the arm used for tennis has a higher bone mass than the other arm [43].

Bone remodeling occurs continuously throughout life to repair skeletal damage, prevent accumulation of hypermineralized bone, and maintain mineral homeostasis by providing rapid access to stores of calcium and phosphorous [57,58]. Small regions of bone are resorbed by osteoclasts and replaced by osteoblasts; this close coordination between resorption and formation ensures that structural integrity is maintained while allowing up to 10% of the skeleton to be replaced each year. In contrast to bone modeling, the formation and resorption processes are tightly coupled both spatially and temporally, such that the overall volume and structure of bone remains unchanged [40].

Anatomically the cycle takes place within a basic multicellular unit (BMU), which is composed of osteoclasts, osteoblasts, and a capillary blood supply [59]. The BMU lasts longer than the life span of the osteoblasts and osteoclasts within it, and so requires constant replenishment of these cells, which is critically controlled by the osteocyte [60]. The remodeling cycle occurs in a highly regulated and stereotyped fashion with five overlapping steps of activation, resorption, reversal, formation, and termination occurring over a period of 120–200 days in both cortical and trabecular bone [61]. Osteocytes orchestrate bone remodeling by regulating the differentiation and, thus, the activity of both osteoclasts and osteoblasts.

Remodeling is a process characterized by five phases. In the activation phase, the osteoclasts are attracted to remodeling sites. In the resorption phase, the osteoclasts resorb bone to release calcium. In the reversal phase, the osteoclasts undergo apoptosis, and the osteoblasts are formed and differentiated from the mesenchymal stem cells. In the formation phase, the osteoblasts lay down a new organic bone matrix that subsequently

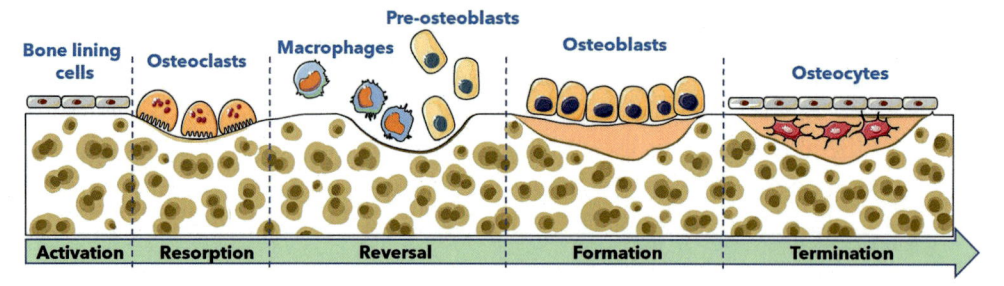

Figure 18.2 The bone remodeling cycle contains activation, resorption, reversal, formation, and termination phases that involve interaction between key bone cells. *(Image modified from Servier Medical Art under Creative Commons Attribution 3.0 Unreported license).*

mineralizes; and the termination phase, where the osteoblasts undergo apoptosis, change into bone lining cells or become entombed within the bone matrix and terminally differentiate into osteocytes [56,61] (Fig. 18.2). Osteocytes play a crucial role in signaling the end of remodeling via secretion of antagonists to osteogenesis, specifically antagonists of the Wnt signaling pathway such as SOST [33].

After the attainment of peak bone mass, bone remodeling is balanced, and bone mass is stable for a decade or two until age-related bone loss begins. Age-related bone loss is caused by increases in resorptive activity and reduced bone formation [62]. Abnormalities in bone remodeling cause bone loss or bone gain and are the basis of low and high bone mass syndromes [56,63].

Effect of T2D on bone cells

The pathophysiological link between diabetes and bone fragility is not well understood. Bone homeostasis may be influenced by several mechanisms which impair the function of osteoblast, osteoclast, and osteocytes and by changing the structural properties of the bone tissue (Fig. 18.3).

T2D and osteoblasts

T2D individuals have high serum glucose concentrations, insulin resistance, and a high body mass index [64]. In nondiabetic conditions, insulin promotes osteoblast differentiation leading to an increased expression of carboxylated form of osteocalcin. Osteocalcin plays a role in regulating glucose metabolism and improves glucose handling by promoting insulin secretion by β-cells as well as by favoring insulin sensitivity [65,66]. Glucose serves as an essential energy source for the production of collagen fibers in osteoblasts [67]. High glucose levels in T2D suppress osteoblast differentiation [68]. Serum concentrations of carboxylated osteocalcin are also reduced and found to be inversely associated with fasting glucose levels and insulin resistance [69]. Serum bone formation markers

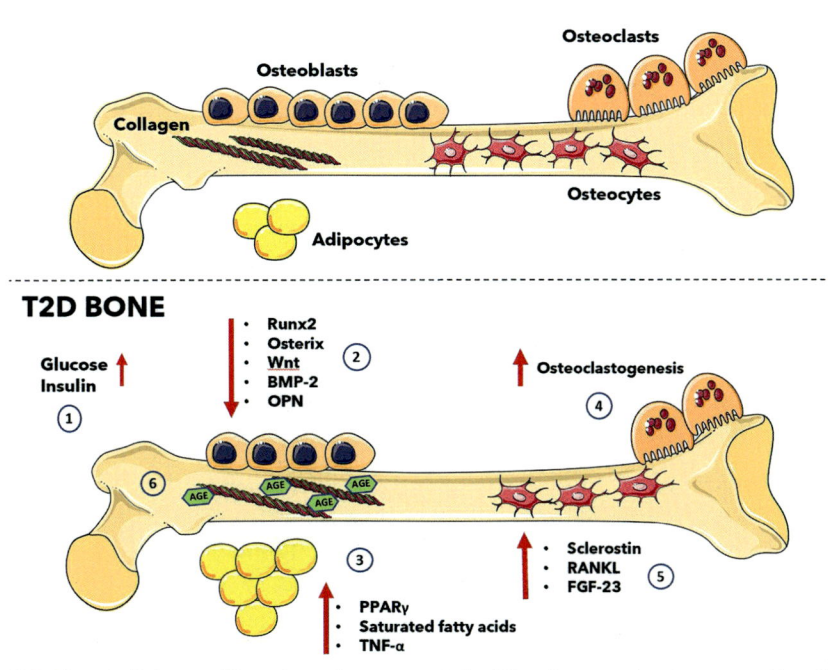

Figure 18.3 Type 2 diabetes affects bone homeostasis by (1) an increase in glucose and insulin levels, (2) a decrease in osteoblast function and expression of osteoblast marker genes, (3) differentiation of mesenchymal cells in bone marrow into adipocytes instead of osteoblasts to increase bone marrow adiposity, (4) a reduction in osteoclast function that negatively affects bone turnover, (5) reduction in osteocytes due to increased apoptosis, and (6) an increase in formation of advanced glycation end products. *(Image modified from Servier Medical Art under Creative Commons Attribution 3.0 Unreported license).*

such as P1NP and ALP have been reported to be unaltered or reduced in T2D, although increased levels of ALP have also been reported [70–73].

High glucose concentrations lead to the formation of advanced glycation end products (AGEs) in the bone matrix, which has a detrimental effect on bone quality. High glucose and AGE treatment of human osteoblasts show a reduced expression of transcription factors such as Runx2 and Osterix, which direct preosteoblasts into immature osteoblasts that express bone matrix genes [74–76]. More severely, AGEs increase the rate of apoptosis of osteoblasts and its precursor cells [77,78].

Bone morphogenetic pathway (BMP) and Wnt signaling are critical for osteoblast differentiation. In the presence of high glucose concentrations, osteogenic cell lines show decreased expression of BMP-2 and reduced Wnt activity, which is associated with reduced osteogenic differentiation [79,80]. Wnt signaling pathway is also one of the critical regulators of osteoblast versus adipocyte fate decisions of MSCs.

Peroxisome proliferator—activated receptor γ (PPARγ) is a master regulator of adipogenesis, whose activity is at least partially dependent on Wnt signaling. PPARγ, when activated, converts cells of osteoblast lineage to adipocytes [81]. In T2D, osteogenesis is reduced while adipogenesis is increased, resulting in bone marrow adiposity due to increased PPARγ signaling [82,83].

Osteopontin is another bone matrix protein that is highly expressed in mature osteoblasts at the sites of bone remodeling [84]. Decreased expression of osteopontin alongside reduced Wnt signaling was also found in osteoblasts obtained from type 2 diabetic rats [85], suggesting suppressed osteoblast differentiation. In accordance with reduced Wnt signaling in vitro, serum concentrations of the Wnt inhibitors, SOST, and Dickoppf-1 (DKK-1) are increased in T2D [86,87]. SOST inhibits bone formation, while DKK-1 promotes bone resorption [88]. Culturing preosteoblasts in this T2D serum leads to reduced ALP activity and diminished matrix mineralization [74,86]. Thus increased serum concentrations of these Wnt inhibitors provide potential clues to increased bone fragility in T2D patients.

In addition, fatty acid composition has a significant impact on osteoblast function. T2D individuals have an increased amount of saturated compared to monounsaturated fatty acids, which leads to suppressed osteoblast differentiation and mineralization capacity as well as increased apoptosis rate due to their lipotoxic effect [89,90]. Collectively, T2D exerts direct adverse effects on osteoblasts via several molecular mechanisms. Furthermore, it favors the fate decision of MSCs to turn into adipocytes, which further impairs osteoblast function, bone formation, and bone mass.

T2D and osteoclasts

In a healthy bone, bone formation by osteoblasts and bone resorption by osteoclasts are balanced. In T2D, osteoblast function is disturbed, and osteoclast activity is altered, leading to impaired bone remodeling. However, the literature on the effects of T2D on osteoclasts is debatable.

Serum concentrations of the bone resorption marker collagen type I C-terminal telopeptide (CTX) are reported to have either increased or decreased in T2D patients. A meta-analysis of 66 studies revealed an overall low bone turnover with low levels of CTX in diabetic patients [91]. However, in type 2 diabetic rodents (i.e., TallyHo mice, ZDF rats), bone resorption parameters are mostly increased (i.e., serum CTX, TRAP, histological numbers of osteoclasts) [85,92–94].

Osteoclast-like Raw264.7 cells cultured in high glucose concentrations show reduced expression of osteoclast-specific genes such as the nuclear factor of activated T-cells, cytoplasmic 1 (NFATC1), tartrate-resistant acid phosphatase (TRAP), and osteoclast-associated receptor (OSCAR). Moreover, high glucose decreases cell proliferation and cell size by suppressing the formation of osteoclast-specific actin ring [80].

When mimicking hyperglycemia and hyperinsulinemia in combination, osteoclast differentiation and expression of osteoclast-specific marker genes are downregulated [95].

RANKL must activate its receptor RANK present on preosteoclasts to initiate osteoclastogenesis. OPG acts as a decoy receptor of RANKL, thus inhibiting osteoclast differentiation. Osteoblasts, as well as osteocytes, highly express both RANKL and OPG. Culturing osteoblasts in high glucose concentration increases both RANKL and OPG expression [80,96] while the direct effect of RANKL on osteoclastogenesis is reduced [95]. Also, culturing osteocyte-like MLO-Y4-A2 cells in high glucose concentration and AGEs highly increases RANKL expression [97].

In addition to osteogenic cells, other cells also contribute to RANKL and OPG production under certain inflammatory conditions. T2D patients suffer from body and bone marrow adiposity, which is associated with increased tumor necrosis factor α (TNFα) serum level [98]. Human bone marrow adipocytes, when additionally treated with TNFα, express more RANKL and less OPG, resulting in an increased resorption capacity of osteoclasts [99]. Moreover, TNFα can induce osteoclastogenesis in combination with M-CSF [100] and also potently increases osteoclastogenesis when low RANKL concentrations are present [101].

Finally, T2D is associated with a higher amount of saturated fatty acids that stimulate osteoclastogenesis and enhance osteoclast survival by inhibiting apoptosis of mature osteoclasts [102,103]. Altogether, several diabetes-associated factors have an impact on osteoclast differentiation and survival, yet, sometimes in contrasting ways. Thus, based on current data, it is difficult to form a general statement on the role of T2D on osteoclasts.

T2D and osteocytes

The osteocytic network within the bone matrix is impaired in both T2D and high-fat diet conditions [104]. There is a decrease in osteocyte density resulting from osteocyte apoptosis under hyperglycemic conditions. This results in an increased number of empty lacunae, in which the osteocytes reside, thus leading to a suppressed mechanosensory response to a mechanical and chemical stimulus in the bone [97]. Increased osteocyte apoptosis leads to a significant decrease in activity in osteoclasts and osteoblasts due to a lack of stimulus to do so.

Osteocytes play a role in inducing bone resorption via their osteoclast activating factor RANKL and in inhibiting bone formation by secreting Wnt inhibitor sclerostin, encoded by the SOST gene [105,106]. In vitro, incubation with a high concentration of glucose and AGEs increases both SOST and RANKL expression [97]. In T2D patients, SOST serum levels are elevated and associated with glycated hemoglobin levels and insulin resistance [107,108].

Osteocytes regulate phosphate homeostasis by expression of fibroblast growth factor-23 (FGF23), which acts in the kidney and inhibits renal phosphate reabsorption.

FGF23 is also involved in the progression of atherosclerosis through its effects on endothelial cell function and is a predictor of cardiovascular disease risk [109,110]. Accordingly, FGF23 serum concentrations are increased in T2D patients that have a high risk of developing cardiovascular diseases [95].

Conclusion

Research over past years has highlighted the deleterious effect of T2D on bone strength and bone quality. Patients with T2D have a higher risk of fragility fractures, not predictable by BMD measurements. This higher risk is likely multifactorial, and at least in part, due to altered bone remodeling and bone cell function in T2D. In hyperglycemic patients, circulating biochemical markers of bone formation, including P1NP, osteocalcin, and BAP, are found to be decreased. A decrease in the serum concentration of bone resorption marker CTX in T2D has also been reported, along with an increase in serum SOST levels. Collectively, these biochemical markers may be predictive of fracture risk independent of BMD. Overall bone formation decreases and all mechanisms described so far contribute to the more deficient bone formation and quality, increasing fracture risk. Other factors such as bone marrow adiposity and fat saturation, AGE accumulation, and microarchitectural changes might also impact bone cell function and increase fracture risk in diabetes. Despite these features, there is a lack of current recommendations for routine screening or initiation of preventive medications for the treatment of fragility fractures in patients with T2D. Although the benefits of improvement and maintenance of glycemic control have been established for the prevention of many diabetes complications, its role in diabetes-related fracture risk is still controversial. Thus, considering the increasing prevalence of T2D, more basic and translational research needs to be undertaken to provide insights required to maintain good bone quality in T2D patients.

References

[1] Whiting DR, et al. IDF diabetes atlas: global estimates of the prevalence of diabetes for 2011 and 2030. Diabetes Res Clin Pract 2011;94(3):311—21.
[2] Organisation WH. The top 10 causes of death worldwide [Fact sheet] [cited 2020 4/1/2020]; Available from: https://www.who.int/news-room/fact-sheets/detail/the-top-10-causes-of-death.
[3] Trikkalinou A, Papazafiropoulou AK, Melidonis A. Type 2 diabetes and quality of life. World J Diabetes 2017;8(4):120—9.
[4] Janghorbani M, et al. Systematic review of type 1 and type 2 diabetes mellitus and risk of fracture. Am J Epidemiol 2007;166(5):495—505.
[5] Schwartz AV, et al. Older women with diabetes have an increased risk of fracture: a prospective study. J Clin Endocrinol Metab 2001;86(1):32—8.
[6] Bonds DE, et al. Risk of fracture in women with type 2 diabetes: the women's health initiative observational study. J Clin Endocrinol Metab 2006;91(9):3404—10.
[7] Hofbauer LC, et al. Osteoporosis in patients with diabetes mellitus. J Bone Miner Res 2007;22(9): 1317—28.

[8] Schwartz AV, et al. Diabetes-related complications, glycemic control, and falls in older adults. Diabetes Care 2008;31(3):391—6.
[9] Shanbhogue VV, et al. Type 2 diabetes and the skeleton: new insights into sweet bones. Lancet Diabetes Endocrinol 2016;4(2):159—73.
[10] Farr JN, Khosla S. Determinants of bone strength and quality in diabetes mellitus in humans. Bone 2016;82:28—34.
[11] Vestergaard P. Discrepancies in bone mineral density and fracture risk in patients with type 1 and type 2 diabetes–a meta-analysis. Osteoporos Int 2007;18(4):427—44.
[12] Schwartz AV, et al. Association of BMD and FRAX score with risk of fracture in older adults with type 2 diabetes. J Am Med Assoc 2011;305(21):2184—92.
[13] Giangregorio LM, et al. FRAX underestimates fracture risk in patients with diabetes. J Bone Miner Res 2012;27(2):301—8.
[14] Boskey AL, Imbert L. Bone quality changes associated with aging and disease: a review. Ann N Y Acad Sci 2017;1410(1):93—106.
[15] Saito M, Fujii K, Marumo K. Degree of mineralization-related collagen crosslinking in the femoral neck cancellous bone in cases of hip fracture and controls. Calcif Tissue Int 2006;79(3):160—8.
[16] Saito M, et al. Reductions in degree of mineralization and enzymatic collagen cross-links and increases in glycation-induced pentosidine in the femoral neck cortex in cases of femoral neck fracture. Osteoporos Int 2006;17(7):986—95.
[17] Odetti P, et al. Advanced glycation end products and bone loss during aging. Ann N Y Acad Sci 2005;1043:710—7.
[18] Yamamoto M, et al. Serum pentosidine levels are positively associated with the presence of vertebral fractures in postmenopausal women with type 2 diabetes. J Clin Endocrinol Metab 2008;93(3):1013—9.
[19] Yamamoto M, et al. Decreased PTH levels accompanied by low bone formation are associated with vertebral fractures in postmenopausal women with type 2 diabetes. J Clin Endocrinol Metab 2012;97(4):1277—84.
[20] Yamamoto M, Yamauchi M, Sugimoto T. Elevated sclerostin levels are associated with vertebral fractures in patients with type 2 diabetes mellitus. J Clin Endocrinol Metab 2013;98(10):4030—7.
[21] Dobnig H, et al. Type 2 diabetes mellitus in nursing home patients: effects on bone turnover, bone mass, and fracture risk. J Clin Endocrinol Metab 2006;91(9):3355—63.
[22] Wu Y, et al. Upregulated serum sclerostin level in the T2DM patients with femur fracture inhibits the expression of bone formation/remodeling-associated biomarkers via antagonizing Wnt signaling. Eur Rev Med Pharmacol Sci 2017;21(3):470—8.
[23] Clarke B. Normal bone anatomy and physiology. Clin J Am Soc Nephrol 2008;3(Suppl. 3):S131—9.
[24] de Baat P, Heijboer MP, de Baat C. [Development, physiology, and cell activity of bone]. Ned Tijdschr Tandheelkd 2005;112(7):258—63.
[25] Shapiro F. Bone development and its relation to fracture repair. The role of mesenchymal osteoblasts and surface osteoblasts. Eur Cell Mater 2008;15:53—76.
[26] Capulli M, Paone R, Rucci N. Osteoblast and osteocyte: games without frontiers. Arch Biochem Biophys 2014;561:3—12.
[27] Fakhry M, et al. Molecular mechanisms of mesenchymal stem cell differentiation towards osteoblasts. World J Stem Cell 2013;5(4):136—48.
[28] Roodman GD. Advances in bone biology: the osteoclast. Endocr Rev 1996;17(4):308—32.
[29] Boyce BF. Advances in the regulation of osteoclasts and osteoclast functions. J Dent Res 2013;92(10):860—7.
[30] Boyce BF, et al. Recent advances in bone biology provide insight into the pathogenesis of bone diseases. Lab Invest 1999;79(2):83—94.
[31] Yavropoulou MP, Yovos JG. Osteoclastogenesis–current knowledge and future perspectives. J Musculoskelet Neuronal Interact 2008;8(3):204—16.
[32] Boyce BF, Xing L. Functions of RANKL/RANK/OPG in bone modeling and remodeling. Arch Biochem Biophys 2008;473(2):139—46.
[33] Bonewald LF. The amazing osteocyte. J Bone Miner Res 2011;26(2):229—38.

[34] Charles JF, Aliprantis AO. Osteoclasts: more than 'bone eaters'. Trends Mol Med 2014;20(8): 449—59.
[35] Franz-Odendaal TA, Hall BK, Witten PE. Buried alive: how osteoblasts become osteocytes. Dev Dynam 2006;235(1):176—90.
[36] Mikuni-Takagaki Y, et al. Matrix mineralization and the differentiation of osteocyte-like cells in culture. J Bone Miner Res 1995;10(2):231—42.
[37] You LD, et al. Ultrastructure of the osteocyte process and its pericellular matrix. Anat Rec A Discov Mol Cell Evol Biol 2004;278(2):505—13.
[38] Poole KE, et al. Sclerostin is a delayed secreted product of osteocytes that inhibits bone formation. Faseb J 2005;19(13):1842—4.
[39] Mc Garrigle MJ, et al. Osteocyte differentiation and the formation of an interconnected cellular network in vitro. Eur Cell Mater 2016;31:323—40.
[40] Manolagas SC. Birth and death of bone cells: basic regulatory mechanisms and implications for the pathogenesis and treatment of osteoporosis. Endocr Rev 2000;21(2):115—37.
[41] Bellido T. Osteocyte-driven bone remodeling. Calcif Tissue Int 2014;94(1):25—34.
[42] Napoli N, et al. Fracture risk in diabetic elderly men: the MrOS study. Diabetologia 2014;57(10): 2057—65.
[43] Kontulainen S, et al. Effect of long-term impact-loading on mass, size, and estimated strength of humerus and radius of female racquet-sports players: a peripheral quantitative computed tomography study between young and old starters and controls. J Bone Miner Res 2003;18(2):352—9.
[44] Buenzli PR, Sims NA. Quantifying the osteocyte network in the human skeleton. Bone 2015;75: 144—50.
[45] Kobayashi Y, et al. Regulation of bone metabolism by Wnt signals. J Biochem 2016;159(4):387—92.
[46] Tu X, et al. Osteocytes mediate the anabolic actions of canonical Wnt/beta-catenin signaling in bone. Proc Natl Acad Sci U S A 2015;112(5):E478—86.
[47] Xiong J, et al. Osteocytes, not osteoblasts or lining cells, are the main source of the RANKL required for osteoclast formation in remodeling bone. PloS One 2015;10(9):e0138189.
[48] Plotkin LI. Apoptotic osteocytes and the control of targeted bone resorption. Curr Osteoporos Rep 2014;12(1):121—6.
[49] Boabaid F, Cerri PS, Katchburian E. Apoptotic bone cells may be engulfed by osteoclasts during alveolar bone resorption in young rats. Tissue Cell 2001;33(4):318—25.
[50] Cerri PS, Boabaid F, Katchburian E. Combined TUNEL and TRAP methods suggest that apoptotic bone cells are inside vacuoles of alveolar bone osteoclasts in young rats. J Periodontal Res 2003;38(2): 223—6.
[51] DiGirolamo DJ, Clemens TL, Kousteni S. The skeleton as an endocrine organ. Nat Rev Rheumatol 2012;8(11):674—83.
[52] Mera P, et al. Osteocalcin signaling in myofibers is necessary and sufficient for optimum adaptation to exercise. Cell Metabol 2016;23(6):1078—92.
[53] Mosialou I, et al. MC4R-dependent suppression of appetite by bone-derived lipocalin 2. Nature 2017;543(7645):385—90.
[54] Oldknow KJ, MacRae VE, Farquharson C. Endocrine role of bone: recent and emerging perspectives beyond osteocalcin. J Endocrinol 2015;225(1):R1—19.
[55] Seeman E, Delmas PD. Bone quality—the material and structural basis of bone strength and fragility. N Engl J Med 2006;354(21):2250—61.
[56] Langdahl B, Ferrari S, Dempster DW. Bone modeling and remodeling: potential as therapeutic targets for the treatment of osteoporosis. Ther Adv Musculoskelet Dis 2016;8(6):225—35.
[57] Bentolila V, et al. Intracortical remodeling in adult rat long bones after fatigue loading. Bone 1998; 23(3):275—81.
[58] Mori S, Burr DB. Increased intracortical remodeling following fatigue damage. Bone 1993;14(2): 103—9.
[59] Frost HM. Skeletal structural adaptations to mechanical usage (SATMU): 2. Redefining Wolff's law: the remodeling problem. Anat Rec 1990;226(4):414—22.
[60] Katsimbri P. The biology of normal bone remodelling. Eur J Canc Care 2017;26(6).

[61] Kenkre JS, Bassett J. The bone remodelling cycle. Ann Clin Biochem 2018;55(3):308—27.
[62] Raisz LG. Pathogenesis of osteoporosis: concepts, conflicts, and prospects. J Clin Invest 2005;115(12): 3318—25.
[63] Brunkow ME, et al. Bone dysplasia sclerosteosis results from loss of the SOST gene product, a novel cystine knot-containing protein. Am J Hum Genet 2001;68(3):577—89.
[64] Stolk RP, et al. Hyperinsulinemia and bone mineral density in an elderly population: the Rotterdam Study. Bone 1996;18(6):545—9.
[65] Ferron M, Lacombe J. Regulation of energy metabolism by the skeleton: osteocalcin and beyond. Arch Biochem Biophys 2014;561:137—46.
[66] Karsenty G, Ferron M. The contribution of bone to whole-organism physiology. Nature 2012; 481(7381):314—20.
[67] Wei J, et al. Glucose uptake and Runx2 synergize to orchestrate osteoblast differentiation and bone formation. Cell 2015;161(7):1576—91.
[68] Napoli N, et al. The alliance of mesenchymal stem cells, bone, and diabetes. Internet J Endocrinol 2014;2014:690783.
[69] Starup-Linde J, Vestergaard P. Biochemical bone turnover markers in diabetes mellitus - a systematic review. Bone 2016;82:69—78.
[70] Capoglu I, et al. Bone turnover markers in patients with type 2 diabetes and their correlation with glycosylated haemoglobin levels. J Int Med Res 2008;36(6):1392—8.
[71] Jiffri EH, Dahr MSA. Impact of non-insulin dependent diabetes mellitus on bone structure biomarkers in postmenopausal obese women. Advances in Obesity, Weight Manage Control 2017;7(Issue 1).
[72] Kulkarni SV, et al. Association of glycemic status with bone turnover markers in type 2 diabetes mellitus. Int J Appl Basic Med Res 2017;7(4):247—51.
[73] Maghbooli Z, et al. The association between bone turnover markers and microvascular complications of type 2 diabetes. J Diabetes Metab Disord 2016;15:51.
[74] Ehnert S, et al. Factors circulating in the blood of type 2 diabetes mellitus patients affect osteoblast maturation - description of a novel in vitro model. Exp Cell Res 2015;332(2):247—58.
[75] Komori T. Regulation of osteoblast differentiation by transcription factors. J Cell Biochem 2006; 99(5):1233—9.
[76] Miranda C, et al. Influence of high glucose and advanced glycation end-products (ages) levels in human osteoblast-like cells gene expression. BMC Muscoskel Disord 2016;17:377.
[77] Alikhani M, et al. Advanced glycation end products stimulate osteoblast apoptosis via the MAP kinase and cytosolic apoptotic pathways. Bone 2007;40(2):345—53.
[78] Kume S, et al. Advanced glycation end-products attenuate human mesenchymal stem cells and prevent cognate differentiation into adipose tissue, cartilage, and bone. J Bone Miner Res 2005; 20(9):1647—58.
[79] Lopez-Herradon A, et al. Inhibition of the canonical Wnt pathway by high glucose can be reversed by parathyroid hormone-related protein in osteoblastic cells. J Cell Biochem 2013;114(8):1908—16.
[80] Picke AK, et al. Sulfated hyaluronan improves bone regeneration of diabetic rats by binding sclerostin and enhancing osteoblast function. Biomaterials 2016;96:11—23.
[81] Ma X, et al. Deciphering the roles of PPARgamma in adipocytes via dynamic change of transcription complex. Front Endocrinol 2018;9:473.
[82] Kim TY, Schafer AL. Diabetes and bone marrow adiposity. Curr Osteoporos Rep 2016;14(6): 337—44.
[83] Sheu Y, et al. Vertebral bone marrow fat, bone mineral density and diabetes: the Osteoporotic Fractures in Men (MrOS) study. Bone 2017;97:299—305.
[84] Sodek J, et al. Regulation of osteopontin expression in osteoblasts. Ann N Y Acad Sci 1995;760: 223—41.
[85] Hamann C, et al. Delayed bone regeneration and low bone mass in a rat model of insulin-resistant type 2 diabetes mellitus is due to impaired osteoblast function. Am J Physiol Endocrinol Metab 2011; 301(6):E1220—8.

[86] Kim S, et al. Stat1 functions as a cytoplasmic attenuator of Runx2 in the transcriptional program of osteoblast differentiation. Genes Dev 2003;17(16):1979—91.

[87] Napoli N, et al. Serum sclerostin and bone turnover in latent autoimmune diabetes in adults. J Clin Endocrinol Metab 2018;103(5):1921—8.

[88] Wang N, et al. Role of sclerostin and dkk1 in bone remodeling in type 2 diabetic patients. Endocr Res 2018;43(1):29—38.

[89] Hardouin P, Rharass T, Lucas S. Bone marrow adipose tissue: to Be or not to Be a typical adipose tissue? Front Endocrinol 2016;7:85.

[90] Khan MP, et al. Pathophysiological mechanism of bone loss in type 2 diabetes involves inverse regulation of osteoblast function by PGC-1alpha and skeletal muscle atrogenes: AdipoR1 as a potential target for reversing diabetes-induced osteopenia. Diabetes 2015;64(7):2609—23.

[91] Hygum K, et al. Mechanisms IN endocrinology: diabetes mellitus, a state of low bone turnover - a systematic review and meta-analysis. Eur J Endocrinol 2017;176(3):R137—57.

[92] Devlin MJ, et al. Early-onset type 2 diabetes impairs skeletal acquisition in the male TALLYHO/JngJ mouse. Endocrinology 2014;155(10):3806—16.

[93] Tamasi JA, et al. Characterization of bone structure in leptin receptor-deficient Zucker (fa/fa) rats. J Bone Miner Res 2003;18(9):1605—11.

[94] Won HY, et al. Prominent bone loss mediated by RANKL and IL-17 produced by CD4+ T cells in TallyHo/JngJ mice. PloS One 2011;6(3):e18168.

[95] Arnlov J, et al. Serum FGF23 and risk of cardiovascular events in relation to mineral metabolism and cardiovascular pathology. Clin J Am Soc Nephrol 2013;8(5):781—6.

[96] Cunha JS, et al. Effects of high glucose and high insulin concentrations on osteoblast function in vitro. Cell Tissue Res 2014;358(1):249—56.

[97] Tanaka K, et al. Effects of high glucose and advanced glycation end products on the expressions of sclerostin and RANKL as well as apoptosis in osteocyte-like MLO-Y4-A2 cells. Biochem Biophys Res Commun 2015;461(2):193—9.

[98] Xu J, et al. High glucose inhibits receptor activator of nuclear factorkappaB ligand-induced osteoclast differentiation via downregulation of vATPase V0 subunit d2 and dendritic cellspecific transmembrane protein. Mol Med Rep 2015;11(2):865—70.

[99] Goto H, et al. Primary human bone marrow adipocytes support TNF-alpha-induced osteoclast differentiation and function through RANKL expression. Cytokine 2011;56(3):662—8.

[100] Kobayashi K, et al. Tumor necrosis factor alpha stimulates osteoclast differentiation by a mechanism independent of the ODF/RANKL-RANK interaction. J Exp Med 2000;191(2):275—86.

[101] Fuller K, et al. TNFalpha potently activates osteoclasts, through a direct action independent of and strongly synergistic with RANKL. Endocrinology 2002;143(3):1108—18.

[102] Cornish J, et al. Modulation of osteoclastogenesis by fatty acids. Endocrinology 2008;149(11):5688—95.

[103] Oh SR, et al. Saturated fatty acids enhance osteoclast survival. J Lipid Res 2010;51(5):892—9.

[104] Mabilleau G, et al. High fat-fed diabetic mice present with profound alterations of the osteocyte network. Bone 2016;90:99—106.

[105] Winkler DG, et al. Osteocyte control of bone formation via sclerostin, a novel BMP antagonist. EMBO J 2003;22(23):6267—76.

[106] Xiong J, O'Brien CA. Osteocyte RANKL: new insights into the control of bone remodeling. J Bone Miner Res 2012;27(3):499—505.

[107] Catalfamo DL, et al. Hyperglycemia induced and intrinsic alterations in type 2 diabetes-derived osteoclast function. Oral Dis 2013;19(3):303—12.

[108] Garcia-Martin A, et al. Circulating levels of sclerostin are increased in patients with type 2 diabetes mellitus. J Clin Endocrinol Metab 2012;97(1):234—41.

[109] Hu X, et al. Elevation in fibroblast growth factor 23 and its value for identifying subclinical atherosclerosis in first-degree relatives of patients with diabetes. Sci Rep 2016;6:34696.

[110] Silswal N, et al. FGF23 directly impairs endothelium-dependent vasorelaxation by increasing superoxide levels and reducing nitric oxide bioavailability. Am J Physiol Endocrinol Metab 2014;307(5):E426—36.

CHAPTER 19

What makes a good device for the diabetic foot

Evan Call[1], Darren F. Groberg[2], Nick Santamaria[3,4]

[1]Department of Microbiology, Weber State University, Ogden, UT, United States; [2]DPM, Utah Musculoskeletal Specialists, Salt Lake City, UT, United States; [3]Faculty of Medicine Dentistry and Health Sciences, University Of Melbourne Victoria, Melbourne, Australia; [4]Visiting Professorial Fellow, Cardiff University, Wales, United Kingdom

Introduction (Daren Groberg, DPM)

Not a day goes by in my clinic that I'm not confronted with a puzzling diabetic patient who has one or multiple manifestations of their diabetes in the lower extremity that I am failing at treating. The reality is that every patient presents a unique twist or complication that does not fit the standard mold for care. As I consider their unique challenges I consider what modification I can make to our current plan or the device I could use to turn the tables and see progress. I have dreamed of something just like this …

Imagine a pair of socks (or any device for that matter) that could sense and respond to the needs of its wearer so their skin was never too warm and never too cold. What if it could sense pressure or friction and it was able to cool the warm spots and reduce that pressure before it progressed to soft tissue breakdown. This pair of socks would be able to sense if any area of the body with which it was in contact, if it was wet, it would have the ability to wick away that moisture. If there was already a wound or a wound developed it would be able to assess the environment of the wound bed and keep it appropriately wet or dry. It would be able to adjust to any bacterial presence and limit its ability to thrive and proliferate. It would be able to change its shape in response to activity and stiffen with weight bearing to offload pressure points or soften when appropriate to avoid collateral damage. It would have stretch receptors that respond to a change in lower extremity swelling and it would respond by applying perfect graded pressure to resist the effects of gravity and the venous dysfunction in the person wearing it. It would be able to stimulate small sympathetic nerve fibers to mitigate the effects of peripheral neuropathy and resist peripheral vascular disease by increasing tissue perfusion and vascular collateralization. If this device existed it would be the perfect device for a complicated diabetic patient because these are many of the major issues under consideration with each medical device developed for care in the lower extremity.

The foot, by nature of its anatomy and function, poses the most difficult challenge for prevention and treatment of injury especially when considering the diagnosis or comorbidity of diabetes. With the significance of the challenge of bearing the weight of the body in quiet standing and in locomotion in every form from gentle walking to high exertion sports, it is no wonder that the foot is required to be the most complex load bearing structure in the human body, to meet these challenges. With over 100 muscles, tendons, and ligaments, 26 separate bones, and 33 joints it is as specialized as the thumb and fingers for fine manual control [1].

When all of this is taken into consideration we have to add, that people prefer to walk, even when told to remain off their feet following injury or surgery. They generally make that one trip to the bathroom, or one quick trip to answer the phone or door, or perhaps to get snacks or a drink. Whatever the justification, people prefer to walk. Factoring this basic human need for mobility into care requires that we plan for the extreme need for offloading, shear prevention, and moisture removal. After all, a sense of hope in future well-being is essential to a mindset that is compatible with healing.

Important functions of the foot that likewise drive the demands made of devices include:

1. The Human foot is uniquely stiff so that it can contribute to propulsion, yet elastic enough to provide energy storage in propulsion and locomotion. The foot possesses springlike qualities allowing it to store energy and return it during locomotion. New evidence is that the central nervous system activates muscles in the storage of energy and then returns the energy in the subsequent stride. The foot contributes up to 17% of the energy to power a stride [2].
2. Energy recycling, returning energy during gait.
3. Energy sink in gait and jumping/landing.
4. Posture and comfort in quiet standing.
5. Flexure of foot and flexure of the shoe generate different bending radii; this generates shear and uneven loading.
6. The flight of the foot and resulting impact governs the location of damage to the foot, or defines the site of greatest deformation.

Background

The complex interactions of daily life and the function of the lower extremities make the changes induced by diabetes critical. Without proper management of these risks and challenges the patient is moving toward breakdown, loss of function, or amputation. The speed of the progression toward these fateful outcomes is greatly influenced by disease severity and duration, access to high-quality assessment, and treatment by a specialized, multidisciplinary diabetic foot unit and by the individual's ability to adhere to preventative and treatment interventions.

Diabetes-related challenges or risks that must be considered are:
1. Neuropathy
2. Compromised arterial blood flow
3. Proprioception

4. Fractures and deformities [3]
5. Stress fractures
6. Callus formation
7. Altered pressure distribution
    a. From deformities
    b. Callus
    c. Altered gait
    d. Blood pressure
    e. Weight gain
    f. Swelling and edema
8. Infection and inflammation
9. Tissue elasticity
10. Weight gain
11. Sedentary life style
12. Amputations
13. Peripheral vascular disease
14. Microvascular disease
15. Changing conditions in use
    a. Arch drop with loading, 5 mm in longitudinal arch in young healthy volunteers, expected to be worse in aging, overweight diabetics [1]
    b. 1/1.5 deviation from vertical, changes the ankle, but also skin compression on the plantar surface [1]

What constitutes a good device

A good device must be useable, easily applied by the medical professional, a caregiver, or the patient themselves. If it is difficult to apply, the potential benefit is lost to frustration and lack of patience in a world where patient visits are cut to 10—15 min and this is expected to cover device application time.

A good device must be effective. If we apply the device it should accomplish the stated intention for use, i.e., reduction of edema, additional resistance to strain or injury, support that extends the useful day, improved function, improved gait, etc. If there is no perception of improvement, then the motivation to continue using the device wains. This is of course a significant problem, often even when there is evidence of strong functional effect [3].

While a good device must provide visible and measurable results it must also be cost-effective. I was once told that a good skin care product should cost $9—10 per 8 oz bottle, but if the device works or has a visible function, it should sell for $39.99 a bottle. While capitalism supports the reward for performance process, the user must recognize the potential cost advantage. Insurance and government pay programs tend to degrade this measure [4]. One patient told me that his insurance company pays $1700.00 per case of catheters, which lasts him 1 month. But the 1 month that he ran out, he would have paid that much for one.

A good device for diabetic feet must perform effectively throughout a reasonable life. Now, life of medical devices is relative to the job it performs. The reasonable life of an insulin syringe is a single use so about 3 min. While a compression wrap or a CROW Boot must last 3 weeks to 2 years, respectively. This implies cleanability, reuse in a reasonable fashion, guides to use and application that allow it to be applied and reapplied without compromising the function, extension beyond normal elastic limits, and must perform properly on day 1 as well as day 730. Many medical devices have reimbursement limits or windows. For example, custom diabetic shoes are considered durable medical equipment in the United States. This means that the device or shoe is expected to last 5 years with reasonable maintenance. Sadly, I have never had any foot wear last 5 years except my fishing waders, which last that long due to disuse. Please note that these are in my mind a medical device because of the critical role they play in maintaining good mental health. Of course, their survivability is based in intermittent use. Something that I insist on working aggressively in the near future.

Benefits a device must provide

1. Protect from Neuropathy. Ultimately neuropathy compromises function, sensation, and proprioception. Wearing protective foot coverings prevents accidental impact with damaging objects like bed posts or door jambs, stepping on sharp objects without realizing it; we have even seen a needle broken off in the foot that the patient did not know was there.
2. Compression Wraps and Hosiery. With the reduction of blood flow and loss of the typical vascular response to dilation and constriction signals from the nervous system, blood and lymph accumulate in the lower extremities. This can cause the dilation of veins to the point that the bicuspid reflux valves can no longer function and exacerbate the problem. Compression from wraps and to a lesser effect hosiery can reduce the venous volume to the point that the valves can regain function; this provides proper return flow and proper balance between superficial and deep veins [5].
3. Protect from Proprioception Errors. Impact with surrounding objects and falls can be due to loss of proprioception in cases of severe neuropathy. The base of diabetic shoes is intentionally oversized in comparison to the sole of a sport training shoe, the reason for this is to increase the area of contact and cause the shoe base to contact the ground early in the gait cycle so as to inform the cerebellum as to the foots location enhancing the balance response to uneven surfaces or impact with trip objects [6].

The rocker shoe design is intended to reduce flexure of joints in the foot reducing the focusing of loads on the heads of the bones in the flexed joints. This design has been shown to reduce pressure on joints and reduce diabetic foot ulcers [7].

4. Force Redistribution. Since the force of gravity acting on a person puts pressure or force on the skin of the feet which cannot be "relieved" without some sort of antigravity machine, we are stuck with redistributing the force. This can be a complete redistribution like lying supine to remove the force, or load, from the feet. Most often though, we cannot just have patients lay down, due to the requirement of ambulation and mobility to carry on with life's activities. This means that force redistribution in the foot has to be much less than complete and is usually accomplished by providing a small cushioning device, for example, that removes the force from the joint at the base of the hallux to the surrounding tissue. This might be accomplished by a felt pad being cut to remove material from the area in contact with the joint and then making this pad large enough to redistribute this force to the adjacent soft tissue. This approach can provide offloading of an ulcer, or blister, or surgery site.
5. Shear Reduction. Shear is one of those ever-present forces. Even in direct vertical compression the force on the skin causes it to spread away from the source of compression but also inducing shear and causing the tissue to displace laterally. Other forms of shear force are more obvious. For example, when the striding foot strikes the ground and the force transfers to the foot in forward motion, the loading and unloading of the foot introduce shear first away from the heel and then toward the heal. The cyclic two-way shear observed in walking is one of the reasons that blisters arise [8].
6. Friction Management. Friction occurs whenever two surfaces come in contact with each other, and there is a lateral force applied to one of the surfaces but not the other. This may be a simple single direction force, or a multidirectional force that might be seen in the example of walking described above. In this case when walking over an uneven surface, a foot strike may add lateral forces that cause friction to be experienced in multiple directions at one time, like when your foot falls on a rock and turns your foot. This friction is the force that is responsible for the formation of shear. Without friction, shear would not be present. Of course, we need to keep in mind that without friction, the foot would slide to the toe or heel of the shoe, and instead of the potential for blisters, we would see impact injury that might cause the loss of toe nails, or the creation of a heel ulcer [8].
7. Patient Compliance. Compliance is a complex behavioral process comprised, in part, of knowledge, understanding, and skills. This is to say that the patient needs know the goal of prevention or treatment as it relates to their disease and the specific problem. They need to understand the relationship between the protective/treatment intervention and the problem being managed. Finally, the patient may need the capacity to perform a specific set of actions such as inspection of the foot, the proper application of an offloading device, and the care of the device to ensure its ongoing performance. Obviously, compliance to preventative or treatment interventions may be affected by the individual's cognitive capacity, emotional state, and social and cultural background. Clinicians need to conduct careful assessment of all the above

elements to best assist the individual in adhering to preventative or treatment interventions. An example of this is the use of commitment devices [3].
8. Restoration of Functionality. When a load-bearing structure, skin, ligament, tendon, etc., has been damaged, certain forms of restraint or offloading can be used to restore function. This intervention is short term and intended to provide the patient with marginal function during the healing process. Examples of this might be an ankle brace, an ace bandage, a CROW boot, or total contact cast that allows function without further damage.
9. Extend Functionality Through the Course of the Day. In some cases, function exists, but due to fatigue, lack of endurance, or during regaining of function following injury or surgery, function can be extended with the use of wraps, cushions, braces, etc.

Drawbacks of devices

Knowing that placing any device in contact with the body creates a potential site of loading, irritation, or fulcrum for injury, caution must be exercised. All devices applied to the body must be evaluated for the potential of inducing further harm. If such potential harm exists then the clinician must mitigate that harm prior to discharge of the patient.
1. Improper introduction of friction by a loose or oversized shoe resulting in superficial or progressively worse injury.
2. Improper shear relief, constraint of the foot with high top ankle in a shoe does not stop the migration of the foot in a pattern that would introduce shear and could be very damaging.
3. Hot and Sweaty, Microclimate. Any time a layer of material is placed on some body structure, that layer becomes a resistor to heat and moisture escape. Unavoidably included with that introduction is then the insulative nature of the device in contact with the skin and the moisture trapping nature of the material. Heat and moisture are constantly transpiring from the skin, to stop that process places additional metabolic and structural strain on the skin. Excess moisture softens skin and increases susceptibility to friction blisters and excoriation. Excess heat reduces the tissues ability to cope with metabolic strains of exertion and shear and compressive loading [9].
4. Load Focusing. Ischemic injury due to sliding/bunching of compression hosiery or slipping wraps has been described [10]. Prevention of the sliding down of compression hosiery and compression wraps can become a critical issue. The taper's nature of the limb combined with the compressive force of the hosiery device can generate a sliding of the device, if it bunches in multiple layers or wrinkles, the combined force of multiple layers can induce an ischemic injury [10].
5. Improper Sizing.
6. Shoe Inserts That are Ill Fitting. Shoe inserts are generally manufactured to fit a given shoe; commonly users of these devices may move them from one to another shoe.

When the fit is not as prescribed, a very undesirable misuse condition can exist. The insert that is too long will roll up in the toe of the shoe and effectively shorten the shoe; this has been seen to cause impact injury on the toes contained in the now smaller shoe. If the shoe is too long and the insert leaves a shelf at the end of the insert, this becomes a high pressure, friction, and shear location and has been seen to cause blisters, abrasions, and ulceration. Similar lateral issues exist with ill-fitting inserts. Lastly, an insert that introduces a forward slope causes the foot to slide forward creating space where sliding friction and shear can occur introducing toe impact injury.
7. Create Need for Microclimate Intervention. Heel boots and other devices applied to cushion and offload the foot may inadvertently trap heat that becomes adequate to cause sweating and potentially maceration between toes and on surfaces exposed to friction and shear.
8. Positioning of Foot Influence on Stance, Posture, and Lumbar Curvature. Many patients complain of hip and back pain that is ultimately resolved with a new pair of shoes. Awareness of this on the part of the caregiver and proper education of the patient can prevent debilitating pain.
9. Amputees. When conditions are dire, and a lower extremity is amputated, all of the issues discussed above can apply to the socket liner. Particularly the accumulation of sweat and moisture and the accompanying friction and shear.

Importance of interventions

No matter what we do or how we try, we cannot take the professional out of healthcare. Working with a large and prominent healthcare chain on a process of selection of devices to provide the best outcome for their patients, we generated an elaborate patient care model considering 22 different nursing functions and gauged how they interact with the devices used to care for the patients. This was combined with a rigorous algorithm to rank products based on the predicted performance of eight devices using the defined nursing functions. The end result of this process was that no matter how we stacked up the devices, the greatest impact on the outcome was the nursing functions. The devices in the comparison included the current device in use, and seven competitive products. While we were able to define that the current device was not the one that the process would predict as the top performer, none of the devices generated as much impact on predicted outcome as the nursing practice did. This is despite the fact that cost was not considered, and the costs of the highest performing products were significantly higher than products that had been used successfully in the past. Nursing practice was still more influential in predicting a positive outcome than the devices.

We might ask why do we bother with the new, more expensive devices then. The basic reason is that if we can determine a better way, that gives a patient a better chance of recovery, or a better quality of life, or less pain and suffering in the end of life, we owe it to the patient to do it. To do it better today than yesterday. And, though it is not today, the day will come that we fully understand the processes of the diabetic complications, allowing us to preserve life and limb. Maintaining functional ambulation is critical in considering what makes a device a Good Device.

Offloading in prevention

Offloading support surfaces

For bed-bound patients, the heel is at the greatest risk, this is especially true for diabetic patients [11]. The prominence of the calcaneus due to its extension posterior to the calf makes it the structure that falls in contact with the support surface when the patient is lying supine. This can be exacerbated by the natural inclination of the feet to external rotation while lying in bed, where the feet fall outward placing the load on the outside radii of the calcaneus. This has the effect of further focusing the load of the foot and calf being pulled downward on the very small surface of the lateral radius of the calcaneus. In addition, to the reductions in blood flow and stiffening of tissues seen with diabetes, it is critical to be able to move damaging forces away from sites of critical concern, i.e., wounds or bony prominences. Support surfaces frequently provide a heel slope, designed for the average person's calf to strike the surface as a fulcrum, so that the attachment at the knee raises the heel like a lever would. This is very effective in offloading the heel. It should be noted that this can be overdone, in which case the heel is held high off the support surface. This is undesirable because it generates discomfort in the knee due to the lifting force this creates, like the "anterior drawer test" used to diagnose torn ligaments [12].

In prevention successful, offloading will prevent the buildup of callus which alters force distribution and focuses the load of the body that is applied to the foot to the bony prominence that the callus is forming over. This load focusing is one of the major forces (along with shear and friction) to damage tissue [13–16].

Offloading in treatment

In treatment, the issues remain the same but the mechanism is slightly different. Loading of a wound interferes with the repair mechanisms. Friction delivers shear that disrupts the migration of fibroblasts and keratinocytes or even disruption of the formation of the delicate sensing filopodia and the migration driving lamellipodia in the migrating cells thus preventing attachment and migration of cells attempting to close the wound [17].

Compression further alters the healing environment. Compression can be properly applied to the edematous limb to improve blood flow as well as lymph drainage. This has been shown to dramatically improve healing [18]. At the same time compression can dramatically impair healing as discussed below. Therefore, offloading of the wound in the face of continued ambulation becomes critical. Construction of offloading dressings, where felt donuts or 2-piece crescents are used to move the compression and shear forces to the surrounding area, which is ideally larger than the wound area, so that the surrounding tissue bears the load that would impair healing, thus protecting the wound from undesirable mechanical forces. While these interventions are essential, the role of the wound dressing cannot be compromised in this process. Proper moisture balance for wound closure and protection of the wound from invading microorganisms remains critical.

Science of tissue damage

Compressive tissue damage is well documented. The application of force through multiple mechanisms including compression, shear, and ischemia induces significant damage to tissues resulting in pressure ulcers or injury. Often the forces responsible for this damage are lumped into the phrase "tissue distortion." This is a useful term because it induces the vision of changing the shape of the cells that are challenged by either thinning under compression, thinning under shear, or displacement under both [19,28].

These forces ultimately result in the destruction of cells when they reach the limit of their ability to resist deformation and die. We have previously referred to this as elastic limit [20]. Realistically, the forces, stresses, strains, and conditions of patients in bed are much more complicated. The typical stresses of rolling shear displacement, cross-sectional bulk shearing, shear of dressing layers, dressing elastomeric shear, displacement of adhered skin, volumetric containment, and pressure redistribution and especially body location like the foot all affect the elastic limit of cells in different ways. The patient's hydration, tissue turgor, age, collagen levels, interstitial fluids, race, comorbidity, and genetics also affect the elasticity of the skin. The authors propose that clinical interactions, further testing, and research on the elastic limit of skin should be performed to determine more precisely the interactions between these factors and the contribution of individual factors [20].

As force is applied to a material like steel as in Fig. 19.1, the strain, or the amount of elongation or deflection of the material, rises. Stress/strain curves show the material's strength, where "stress" is the force divided by the cross-sectional area, and "strain" is the amount of elongation or deflection of the material (length divided by the original length). Fig. 19.1 is a typical stress/strain curve of a structural steel.

The elastic region is the region of the stress/strain curve in which the material will return to its original shape when the stress is removed. Once a material has reached its elastic limit, or yield point, it enters into the plastic region. Once here, the material is permanently deformed. Engineers and architects commonly design structures to work in the elastic region to ensure that a structure retains its size and shape.

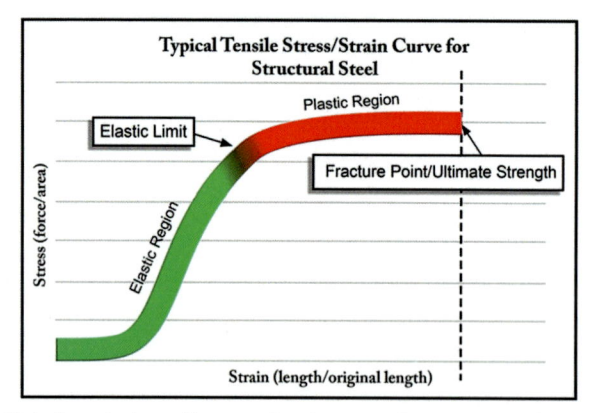

Figure 19.1 A typical tensile stress/strain curve for a structural steel [20,21].

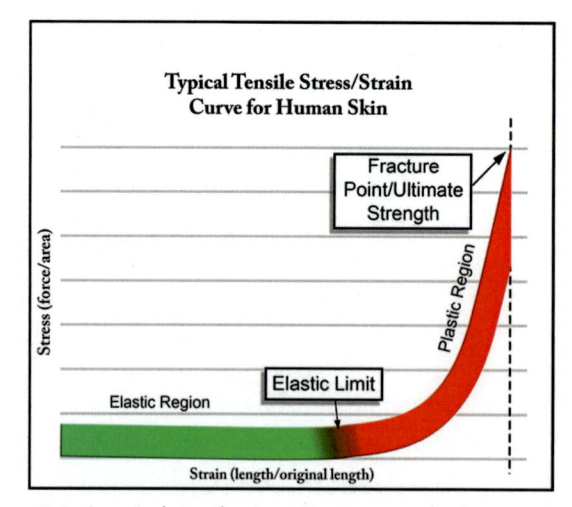

Figure 19.2 A typical tensile stress/strain curve for human skin [20].

Human skin has a similar ability to stretch to a certain point and then return to its natural form. Traditional tests on the elasticity of human skin, usually involving tensile samples taken from cadavers, are depicted in stress/strain curves similar to those of structural materials. Fig. 19.2 is the authors' proposed graph to show a typical stress/strain curve for human skin [20].

The skin's stress/strain curve is opposite to the structural steel's curve. The slopes of the stress/strain curves are especially notable, as they show the "stiffness" of the material. The "stiffness" is the amount of strain or elongation for a given stress. In stress/strain curves, the stiffness is shown as the slope of the line in the plastic region; a higher stiffness is represented by a more vertical line, while a lower stiffness has a more horizontal line. A more horizontal slope indicates that a small force will cause a large strain, and a more vertical slope indicates that it will take a large force to cause a small strain [20].

The structural steel in Fig. 19.1 starts out with a high stiffness, or a steep slope, of the elastic curve, meaning it takes a relatively large stress to deflect or elongate the object while it is in the elastic region. But in Fig. 19.2, the slope of the skin starts out very horizontal, indicating that the stiffness is low and it takes very small stress to cause damage to the skin even before it reaches its technical elastic limit. Because it takes such a small amount of force to deflect the skin, a relatively small strain easily damages cells, and at the deep tissue level this damage often results in pressure injuries [20].

Because steel is inanimate and skin is biologic, the comparison of the two breaks down quickly. The elasticity, or modulus, is what defines the difference between skin and steel. While steel's molecules do not change under stress until the material reaches its elastic limit, skin cells under stress are deformed and killed well before skin reaches its technically defined elastic limit (the point at which the skin is unable to return to its former shape). Therefore, the authors define the elastic limit of the skin as the point when individual skin cells deform and die due to either prolonged or a high amount of stress. While this definition applies to a single cell, the elastic limit of the cellular aggregate helps us understand how the skin handles stress [20].

Studies have indicated that prolonged external pressure causes deep tissue damage, which is the main cause of pressure injuries [22–25]. In 2009, A. Gefen wrote, "To understand the etiology of deep tissue injury, health professionals should be able to predict whether or not a certain state of internal mechanical loads in deep tissues, such as tissue deformations and forces per unit area of tissue, would lead to localized irreversible cell damage" [26]. The concept of the cell's elastic limit provides an increased understanding of the high occurrence of pressure injuries for long-term hospital patients, especially when a patient's mobility is restricted due to medical complications.

As previously mentioned, several variables contribute to the skin's ability to withstand pressure, including the type of stress, the condition of the skin, and the amount of time the stress is applied. An example of a typical stress that an immobile patient encounters is when the head of the patient's bed is up at an angle, and he or she slides down the bed. Patients sitting up at 30 in bed slid down 29.34 (±5.53 cm) over 2 h, which is a significant amount of shear for fragile skin to handle [27]. The combined force and shear are more than enough stress to stretch the cells to their elastic limit [20].

For example, when a patient with fragile skin loses his or her balance and a caregiver reacts by grabbing the patient by the arm, the skin will often tear. The most frequently reported wound found in the aged care sector is the skin tear. This wound is most commonly found on the arm or leg and is often associated with patient/resident repositioning. This tear is a result of the skin exceeding its elastic limit very quickly and with great force. A similar reaction in the skin happens with supine patients who develop pressure injuries. The shear is not applied as rapidly or with as high of a force as a skin tear, but over time causes displacement wrinkles and deforms the skin, the magnitude of which kills the individual cells that proportionally have reached the elastic limit. This shear damage occurs when the elastic limit is reached [20].

Gefen proposes that cells begin to die at approximately 50% strain [24] and the authors propose that the elastic limit of cells is in that range. Skin cells are easily deformed, but, borrowing from the engineer analogy, when a material's area is doubled the amount of stress it can withstand is doubled (stress = force/area). The same concept applies to skin. Bandages and prophylactic dressings reinforce the skin so that it can withstand a greater amount of force before reaching its elastic limit. Essentially, dressings increase the skin's stiffness by reducing the forces and distortion that are applied to the skin cells [20].

Specific devices

The prescription of any specific device may be based on a number of intentions on the part of the physician. The measure of fitness for the intention is up to the prescriber. However, we can guide the prescription selection with the use of testing that outlines the performance of the various devices in the conditions required to meet the complex needs of the patient. The challenge for any prescribing care provider is to be aware of the myriad of devices available and relative performances of each in multiple conditions. This complex need drives the current effort being considered in the world of dressings for prevention. The NPIAP and the EPUAP are considering the formation of a joint committee that would work toward the production of standards that drive simplification of device selection process by making it easy for a caregiver to identify the critical performance elements of multiple devices so that the most important requirements are met with the highest performance device. The elements that this group of standards should cover are the critical performance elements that provide protection from the risks and challenges discussed above, such as pressure or force redistribution, shear reduction, friction between the device and the body as well as between the device and the support surface, microclimate, usability, and product life. It is particularly important for such an undertaking to include all of the potential tools available to optimize the process. This should include the use of finite element analysis and modeling with inputs that can provide manipulation of the performance data to make the caregiver capable of predicting positive intervening effect. This seems to be the most advantageous path forward in a world where the impact of implementation of the high-level tools that have been developed will be immense in the form of prevention of injury as well as treatment. While this requires moving from the theoretical to clinical, the potential benefits certainly justify the effort.

Heel lift boots

Heel boots are used with the supine or bed-bound patient to protect the heels from pressure/force on the heels caused by gravity acting on the leg and foot of the patient.

This mechanism is well documented and the prevention of pressure injuries generally requires the elevation of the foot from the support surface (NPIAP Guidelines 2019). The value of these products is in the successful removal of all pressure from the heels and has been demonstrated in RCT research [28].

The number one complaint with heel boots is the heat that accumulates inside the boot. Bench tests for the microclimate generated by the use of a heel boot indicate that heat is generally trapped explaining why the patients tend to kick them off or refuse them due to discomfort. Efforts at improving the heat and the microclimate generated by heel boots include; the inclusion of breather holes, sections that can be opened, use of gel modified foam, and opening soles of the boot to allow air exchange.

Total contact cast

The goal of containing a foot and lower leg in a completely enclosed cast is to entirely immobilize the limb and foot to accelerate healing and maximize offloading. Given the process of application of the cast offloading is usually easily accomplished with the force of the body being pulled by gravity toward the foot, being nicely redistributed to both the soft tissue surrounding the location of concern and the lower calf and gastrocnemius. The tapered nature of the gastrocnemius muscle and the load bearing created by a very slight dorsal flection while applying the cast allows the body weight of the patient to be borne by these structures both above the foot and the soft tissues surrounding the wounds and sites of concern [3,29].

Compression systems, wraps, and garments

The science of the compression wrap is well published and highly effective in restoring circulation to patients with venous insufficiency. This is accomplished by the compression of the soft tissues to allow damaged venous reflux valves to become functional [30]. Compression wraps are one item where good guidelines and performance definitions are available, making the selection of which product to use, much more diagnostic [31]. This is also true of compression garments. The level of compression required defines which product should be chosen. Class I through IV in the STRIDE document by Bjork and Ehmann are compared for three national standards: British, French, and German. A formal document does exist in Europe under CEN ENV 12718 for medical compression hosiery [32]. While there are no formal Standards in the United States, values provided in the STRIDE document are reported presumably as expert opinion and taken from clinical practice [18]. These values are similar, yet overlapping for each class and national standard, for example, in Class I they range from 10 to 21 mmHg, for Class II they range from 15 to 32 mmHg, for Class III they range from 20 to 40 mmHg, and for Class IV 36—50 mmHg.

Static stiffness index

Two of the most knowledgeable experts in the field of compression wrap therapy state that "the failures of compression therapy are not caused by poor compression material but due to poor knowledge and application techniques of the care providers." This seems to say that even given a poor device a person with excellent knowledge can make it work [33].

While it is not our intent to argue the difference between devices in performance, when considering the use of compression hosiery versus compression wraps, it is reported that patients using compression wraps are two times more likely to heal than those using hosiery [34]. This brings us back to the reliance on the medical professional and their ability to apply the proper device to the diagnosis because hosiery is also reported to be successful in reducing healing time and aiding progression toward healing [18].

Dressings for treatment

Use of dressings in ambulation is rough

The goals of treatment of a diabetic foot wound from the perspective of the dressing needs to be based on the specific characteristics of the wound and the patient. Thorough and skilled initial and ongoing wound assessment are the critical steps in achieving and effective treatment plan. The challenges of providing an effective foot wound treatment dressing in an ambulant patient can be significant. Focusing on the wound specifically, dressing selection must be based on the characteristics of the presenting wound such as tissue appearance, presence of slough, necrosis, signs of infection, odor, exudate type and amount, presence of undermining, cavity or tunneling, and visibility of tendon or bone. Remembering that in the presence of significant neuropathy, the patient may have an insensate foot and therefore not be able to provide good reports of level of pain. Similarly, diabetic disease changes inflammatory responses to infection and places even greater importance on high-quality wound assessment [35].

Ultimately the wound dressing needs to provide a primary component in contact with the wound surface that does not cause further trauma to healing tissue, does not adhere to the tissue, and absorbs and transfers excess wound exudate away from the wound bed. In wounds that are colonized or locally infected, topical antimicrobial dressings such as silver- or iodine-impregnated dressings may be appropriate. In cavity wounds, gelling fiber dressings that absorb and transfer wound exudate would be a better choice due to their ability to conform to the shape of the cavity. Once again, the presence of local infection may require an antimicrobial gelling fiber dressing. Importantly the gelling fiber dressing should not form a "plug" trapping exudate and should not leave any components or fibers within the wound cavity once removed. This is of particular importance as any dressing residue left in the wound may form the basis of further subsequent inflammation. The secondary dressing should be chosen based on the initial wound assessment and would be based on the ability of the secondary dressing to absorb wound exudate transferred from the primary dressing, the ability to manage wound microclimate via the ability to transpire moisture vapor while at the same time provide an antimicrobial barrier. Additionally, the secondary dressing would provide protection to the wound from pressure, friction, and shear.

The presence of foot deformities such as Charcot foot adds further complexity to dressing selection if we are to achieve both an effective wound dressing but also one able to conform to the anatomical changes in the foot while providing protection to the wound from pressure, friction, and shear forces and at the same time be able to tolerate the forces exerted on the dressing by ambulation.

We report the use of elastic tube products to maintain dressings on the heel on bed-bound patients. However, ambulation and weight bearing have a tendency to overcome the adhesive nature of the dressing.

Dressings for prevention

The use of soft silicone multilayer foam dressings in the prevention of sacral and heel pressure injuries is well established in terms of clinical and cost-effectiveness evidence [36–41]. However, the majority of these studies have been carried out in the acute in-patient settings of ICU, general wards, and in aged care. Similarly, the use of dressings to prevent medical device–related pressure injuries is also based in the acute setting. The science underpinning the effectiveness of these prophylactic has been well elucidated [42,43], for the bed-bound patient at high risk of developing a pressure injury. There is, however, a significant need for research to investigate if this class of dressings may offer the diabetic patient similar protection as there is currently limited evidence for clinical efficacy in the ambulant diabetic patient with a foot wound.

Device life

A good device for prevention and treatment of the diabetic foot must demonstrate therapeutic effect throughout the life of the product. The natural degradation of performance that occurs through use of the device should not reduce performance below the therapeutic level. This is again one of those points where cost, availability, and payer policy can adversely impact quality of care. For example, if the level of compression required to generate limb reduction using an elastic compressive wrap requires a high level of tension be applied in use, then the wrap will presumably fail earlier than a device that is applied with a lower level of tension. If the life of the product is shortened because of this demanding use environment, the product may yield prior to the patient qualifying for a replacement device. Testing to determine the effective life should be required to ensure that the product functions properly through its typical use period.

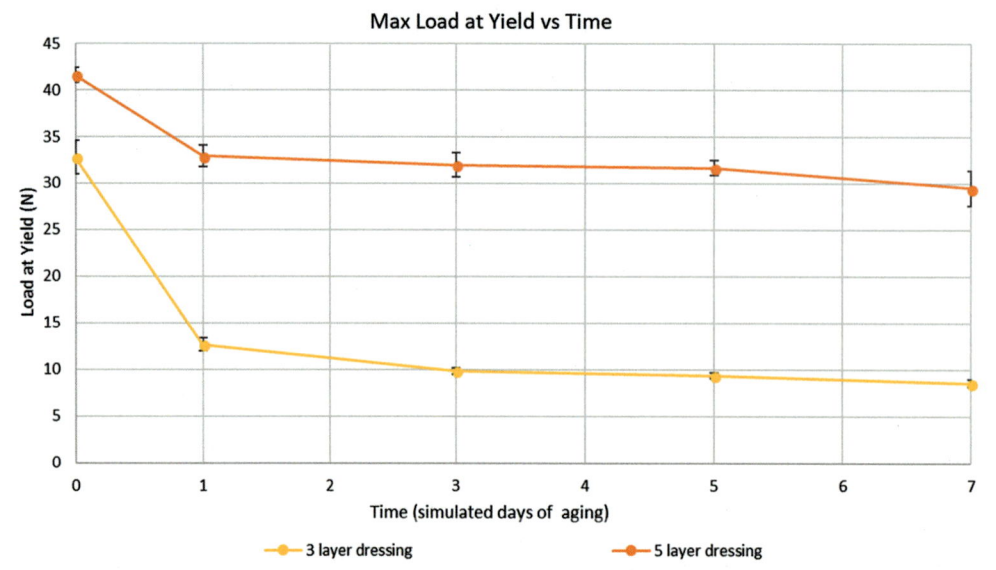

Figure 19.3 Reduction in tensile loading in N observed over the days of use in sacral dressings (taken from testing in our laboratory) [44].

The life of dressings was recently estimated in our lab, using a shear method representing sliding in bed on the sacrum, not unlike the wear observed on the foot in walking [44]. While this test is more appropriate for the sacrum than the heel, the concept remains the same; the application of shear displacement on the surface of the dressing breaks down the normal performance of the dressing. In the example test we demonstrated that the tensile or pulling force required to cause a dressing to fail drops with successive shear excursions. The dressings would release from the test surface and roll, much the way unrestrained dressings do on the foot of an ambulatory patient (Fig. 19.3).

Similar performance requirements should be made of all cushioning devices, temporary offloading aids and devices interfacing with the body to ensure that the therapeutic benefit exceeds the typical use period.

Conclusions

The rigorous demands placed on devices interfacing with the diabetic foot are unique to the protection and healing of the diabetic foot. Nowhere else on the body is the same level of compressive loading from standing, shear loading from walking, and microclimate challenges due to the presence of protective boots and shoes seen.

Design of device use on the diabetic foot must consider the exceptional environment in which they must perform, and the protective and therapeutic effect of the devices must demonstrate proper function in initial use and at the end of the life of the device.

The function of the diabetic foot intervention device must also be able to be applied by a physician or a layman. The learning curve in training a family member to apply the device for the patient must be simple enough to be carried out correctly and repeatedly in the home environment.

The benefits of the device must be weighed against the potential complications of its improper use to prevent the formation of adverse events like compression wrap injury due to rolling or sliding of the wrap inducing pressure ulcer/injury. The foresight and compassion required of the clinical staff in dealing with the diabetic foot is very important, particularly where a positive attitude on the part of the caregiver can impart hope in the patient's mind and thus encourage compliance with device use instructions. Finally, when considering using a device in contact with the foot of a diabetic patient, exercise patience, caution, high level of education, and empathetic encouragement to generate compliance and healing.

References

[1] Wright WG, Ivanenko YP, Gurfinkel VS. Foot anatomy specialization for postural sensation and control. J Neurophysiol December 7, 2011;107(107):1513—21.
[2] Hao XY, Li HL, Su H, Cai H, Guo TK, Liu R, Jiang L, Shen YF. Topical phyenytoin for treating pressure ulcers (review) the cochrane collaboration. 2017. p. 3—35.
[3] Jarl G. Commitment devices in the treatment of diabetic foot ulcers. J Foot Ankle Res 2019;44:1—4.
[4] GAO, Medicare. Need to overhaul costly payment system for medical equipment and supplies. 1998. GAO/GEGS-98-102.
[5] Lim CS, Davies AH. Graduated compression stockings. Can Med Assoc J 2014;186(10):E391—8.
[6] Paton JS. Does footwear affect balance. J Am Podiatr Med Assoc 2013;103(6):508—15.
[7] Chapman JD, Preece SJ, Braunstein B, Hone A, Nester CJ. Effect of rocker shoe design features on forefoot plantar pressures in people with and without diabetes. Clin Biomech 2003;28(6):679—85. https://doi.org/10.1016/j.clinbiomech. 2013.05.005.
[8] Sulzberger, Marion B. Studies on blisters produced by friction. I. Results of linear rubbing and twisting technics. J Invest Dermatol 1966;47(5):456—65.
[9] Gefen A. How do microclimate factors affect the risk for superficial pressure ulcers: a mathematical modeling study. J Tissue Viability 2011;20:81—8.
[10] Rathfore FA, Ahmad F, Khan OJ. Compression stockings and pressure ulcers: case series of a neglected issue. Cureus 2017:e1763. https://doi.org/10.7759/cureus.1763.
[11] 3777 European Pressure Ulcer Advisory Panel. National pressure injury advisory panel and Pan Pacific pressure injury alliance. Prevention and treatment of pressure ulcers/injuries: clinical practice guideline. In: Emily H, editor. The international guideline. EPUAP/NPIAP/PPPIA; 2019.
[12] Butler DL, Noyes FR, Grood ES. Ligamentous restraints to anterior-posterior drawer in the human knee. J Bone Joint Surg Am 1980;62:259—70.
[13] Gefen A. Plantar soft tissue loading under the medial metatarsals in the standing diabetic foot. Med Eng Phys 2003;25:491—9.
[14] Amit G. Reswick and Rogers pressure-time curve for pressure ulcer risk. Part 2. Nurs Stand 2009;23: 40—4.
[15] Gefen A. How much time does it take to get a pressure ulcer? Integrated evidence from human, animal, and in vitro studies. Ostomy Wound Manag 2008;54(10):26—35.
[16] Amit G. Reswick and Rogers pressure-time curve for pressure ulcer risk. Part 1. Nurs Stand 2009;23: 64—74.

[17] Lauffenburger DA, Horwitz AF. Cell migration: a physically integrated molecular process. Cell 1996; 84:359–69.
[18] Bjork R, Ehmann S. S.T.R.I.D.E. Professional guide to compression garment selection for the lower extremity. J Wound Care 2019;28(6 Suppl. 1):1–44.
[19] Linder-Ganz E, Gefen A. Mechanical compression-induced pressure sores in rat hindlimb: muscle stiffness, histology, and computational models. J Appl Physiol 2004;96:2034–49.
[20] Evan C, Randy J, Katherine DM, Joshua NB, Susan J, Allyn B, Craig O. "The elastic limit": introducing a novel concept in communicating excessive shear and tissue deformation. WCET J 2017; 37(4):16–20.
[21] Boresi AP, Schmidt RJ. Stress-strain relations. Advanced mechanics of materials. 6th ed. Hoboken, USA: John Wiley & Sons, Inc.; 2003. p. 10. Ch. 1, Sec. 3.
[22] Zahouani H, et al. Assessment of the elasticity and tactile properties of the human skin surface by tribological tests. In: Proceedings of the 22nd IFSCC congress, edinburgh; 2002.
[23] Dick JC. The tension and resistance to stretching of human skin and other membranes, with results from a series of normal and oedematous cases. J Physiol 1951;112:102–13.
[24] Gawlitta D, et al. The relative contributions of compression and hypoxia to development of muscle tissue damage: an in vitro study. Ann Biomed Eng 2007;35:273–84.
[25] Berlowitz DR, Brienza DM. Are all pressure ulcers the result of deep tissue injury? A review of the literature. Ostomy Wound Manag 2007;53:34–8.
[26] Gefen A, van Nierop B, Bader DL, Oomens CWJ. Strain-time cell-death threshold for skeletal muscle in a tissue-engineered model system for deep tissue injury. J Biomech 2008;41:2003–12.
[27] Oomens CWJ, Bader DL, Loerakker S, Baaijens F. Pressure induced deep tissue injury explained. Ann Biomed Eng 2015;43:297–305.
[28] Donnely J, Winder J, Kernohan WG, Stevenson M. An RCT to determine the effect of a heel elevation device in pressure ulcer prevention post-hip fracture. J Wound Care 2011;20(7):309–18.
[29] Lavery, Lawrence A. 5 questions - and answers - about off-loading. Adv Skin Wound Care 2003; 16(5):231–4.
[30] Partsch H, Mosti G. Unexpected venous diameter reduction by compression stocking of deep but not of superficial veins. Veins & Lymphat 2012;1(3):7–9.
[31] Milic DJ, Zivic SS, Bogdanovic DC, Jovanovic MM, Jankovic RJ, Milosevic ZD, Stamenkovic DM, Trenkic MS. The influence of different sub-bandage pressure values on venous leg ulcers healing when treated with compression therapy. J Vasc Surg n.d.;51(3):655-660.
[32] Medical Compression Hosiery. Prestandard ONORM ENV 12718 European standard. 2001.
[33] Partsch H, Mortimer P. Compression for leg wounds. Br J Dermatol 2015;173:359–69.
[34] 2799 Finlayson K, Courtney M, Gibb M, O'Brien J, Parker C, Edwards H. The effectiveness of a four-layer compression bandage system in comparison to class 3 compression hosiery on healing and quality of life for patients with venous leg ulcers: a randomised controlled trial. Int Wound J 2014;11(1):21–7.
[35] Kravitz SR, McGuire J, Shanahan SD. Physical assessment of the diabetic foot. Adv Skin Wound Care 2003;16(2):68–75.
[36] Santamaria N, Gerdtz M, Sage S, McCann J, Freeman A, Vassiliou T, et al. A randomised controlled trial of the effectiveness of soft silicone multi-layered foam dressings in the prevention of sacral and heel pressure ulcers in trauma and critically ill patients: the border trial. Int Wound J 2013. https://doi.org/10.1111/iwj12101.
[37] Santamaria N, Liu W, Gerdtz M, Sage S, McCann J, Freeman A, et al. The cost-benefit of using soft silicone multilayered foam dressings to prevent sacral and heel pressure ulcers in trauma and critically ill patients: a within-trial analysis of the border trial. Int Wound J 2015;12(3):344.
[38] Kalowes P. Five-layered soft silicone foam dressing to prevent pressure ulcers in the Intensive Care Unit. Am J Crit Care 2016;25(6):E108–19.
[39] Clark M, Black J, Alves P, Brindle C, Call E, Dealey C, et al. Systematic review of the use of prophylactic dressings in the prevention of pressure ulcers. Int Wound J 2014;11(5):460.
[40] Santamaria N, Gerdtz M, Kapp S, Wilson L, Gefen A. A randomised controlled trial of the clinical effectiveness of multi-layer silicone foam dressings for the prevention of pressure injuries in high-risk aged care residents: the border III trial. Int Wound J 2017. https://doi.org/10.1111/iwj.12891.

[41] Padula WV, Chen YH, Santamaria N. Five-layer border dressings as part of a quality improvement bundle to prevent pressure injuries in US skilled nursing facilities and Australian nursing homes: a cost-effectiveness analysis. Int Wound J 2019. https://doi.org/10.1111/iwj.13174.

[42] Levy A, Gefen A. The biomechanical efficacy of dressings in preventing heel ulcers. J Tissue Viability 2015;24(1):1—11. https://doi.org/10.1016/j.jtv.2015.01.001.

[43] Black J. Prevention and management of pressure injury to the heel. Wounds Int 2018;9(2):43—9.

[44] Burton J, Fredrickson A, Capunay C, Tanner L, Oberg C, Santamaria N, Gefen A, Call E. New clinically relevant method to evaluate the life span of prophylactic sacral dressings. Adv Skin Wound Care July 2019:1—7.

CHAPTER 20

Allostasis: a conceptual framework to better understand and prevent diabetic foot ulcers

Laurel Tanner[1], Craig Oberg[2], Evan Call[2]
[1]EC-Service, Centerville, UT, United States; [2]Department of Microbiology, Weber State University, Ogden UT, United States

Introduction

In a hospital setting, the heel is a common site for pressure injury, as are other bony prominences such as the sacrum, head, ischial tuberosities, and shoulders. For a person with diabetes, the foot is at elevated risk of tissue damage. These risks include the mechanical load brought on during normal activities such as walking, excessive pressure from shoes that don't fit well, unrelieved pressure from altered gait, or even through lack of sufficient movement. With thin skin and a modified foot pad, the diabetic foot is at higher risk of tissue damage, too often leading to diabetic foot ulcers (DFUs). DFUs are an ongoing problem in the healthcare field, in spite of many people working to treat and to prevent them.

Known complications from diabetes include peripheral artery disease (PAD) and peripheral neuropathy, which concerns present complications for the diabetic foot especially. The resulting loss of protective sensation affects not just pain, but also a patient's balance, alignment, and proper motion, or discomfort due to increased pressure, friction, and shear forces placed on muscles, soft tissue, and bone. Skin integrity is also at risk from excess heat and moisture as well as from dry or cracked skin, leading to blisters and possible infection [1]. Both decreased sensation due to nerve damage and the lack of oxygen/nutrients due to vascular disease increase a patient's risk of tissue damage, potentially leading to DFU. The cost of DFU to patients is high in physical complications and mental and emotional distress. The annual incidence of DFU hovers near 6% depending on patient ethnicity and location [2]. Approximately 15% of patients with diabetes develop a foot ulcer Table 20.1 in their lifetime, which is 2.4 million patients [3].

Preventing and treating DFU are the focus of numerous studies by experts around the world. These studies examine a wide range of approaches, including clinical prevalence and incidence, cadaver histology, finite element analysis, cell culture, tissue distortions, use of preventative wound dressings, and innovations in support surface interventions to help with pressure offloading, to name a few. Many studies focus on interventions. However, since the primary etiology of DFU is still being elucidated, then the interventions themselves are subject to discussion on which is most suited to address the underlying causality. Because the issue of DFU is complex and multifaceted, and thus difficult to fully alleviate, it remains a persistent problem.

The continuing rate of DFUs would indicate a need for continued innovation both in intervention and in clinical care. Additionally, this indicates a need for better understanding of the physiological process, and how to intervene to provide clinician support and device integration to meet the continued prevalence and incidence of DFU. Given the sheer volume of patents extended and the studies on just diabetes care, and still the ongoing DFU incidence, it would seem that not just new devices and new interventions would be called for, but a new underlying conceptual approach.

One innovative approach to address this healthcare problem is through application of a new treatment paradigm. It is well recognized that the body adapts to input [4—9]. For example, a transcutaneous electrical net stimulation, or a TENS unit, is a battery-operated device that some clinicians use to treat pain. The unit works by delivering small electrical impulses through adhesive pads attached to a person's skin. However, common treatment protocol includes roughly 10 min of initial treatment, followed by the clinician resetting the level of stimulation, because the patient has adapted to this input, and thus the input needs to be adjusted to be effective again. This is an example of the body adapting to input, although the body's adapting can be to internal or external stimuli.

Although it is well recognized that the body adapts to input, this idea has not been incorporated into treatment paradigms or methodologies. Instead, standard wound care seeks to apply the principles of homeostasis to the treatment of DFU. However, this line of thinking and treatment is not inclusive enough. While homeostasis is helpful, there is more to this issue than previously identified: This chapter will examine the concept of *allostasis*, a more-nuanced concept which addresses the limitations of the homeostasis concept by providing a more resolved method of looking at the problem of patient treatment and prevention of DFU.

Homeostasis emphasizes a return to a fixed state, and treats any variation from that fixed state as a deviation from normal and therefore something to be managed [10]. However, variability is vital to health [6,9,10]. The human body is constantly subjected to a wide variety of stimuli both internal and external, and the body is in a constant state of adaption to stimuli. Nevertheless, all physiological systems have limits, and thus the very absence of adaption itself may indicate damage or disease [4,9,10] and may indicate a need for increased clinical intervention. The concept of allostasis better explains a range of individual adaptions as well as the absence of adaption in patients with compromised health.

We need to apply the principles of allostasis and create a paradigm shift in wound care, moving from a reactionary focus to an anticipatory plan that can more effectively treat as well as prevent DFU in wound care.[1] Our aim in this chapter is to bring the concept of allostasis to the forefront of understanding and discussion, in the context of physiological response to loading and potential interventions, in order to better prevent DFUs and resulting negative outcomes.

In order to illustrate how the concept of allostasis can help improve treatment for DFUs, in this chapter we will discuss:

I The difference between homeostasis and allostasis
II Key components of the allostasis approach
III Physical response to tissue loading and tissue injury risks
IV Potential application/integration of allostasis in treatment approaches
V Conclusion and key takeaways

Homeostasis versus allostasis

Whenever a patient is placed on a support surface, there is a physiological response to the forces applied to their tissues. One method to understand this response is through the familiar clinical term homeostasis, which is consistently defined as the state of equilibrium in the body as the body responds to various functions and the composition of fluids and tissues [11]. The Greek roots are *homeo* (same/similar) and *stasis* (state), with primary emphasis on the return to the "same state" after the body responds to changing conditions (Table 20.1).

The homeostasis approach to care considers each patient need separately and in sequence. While there is some logic to this approach, allostasis acknowledges that not only do the physical needs which the body must meet happen simultaneously, but that the body's systems work interactively rather than sequentially. Homeostasis is usually considered in terms of a single parameter, and a specific clinical response, and then a return to equilibrium [9,10,12]. For instance, exercise would produce excessive heat. This increased heat would elicit the response of increased sweat output, until a return to an appropriate temperature level. Under the homeostasis model, physicians try to restore each parameter to what is considered the appropriate fixed level, i.e., equilibrium [9]. The homeostatic response to a patient presenting with fever would be to treat with acetaminophen in order to lower the temperature (or, return the body to normal state).

Homeostasis is a concept most clinicians are familiar with: patients return to normal, and "normal" is a specific set point, such as "normal" blood pressure, "normal" rate of breathing, even a "normal" body temperature. In contrast, allostasis includes the concept that *normal* itself will fluctuate, as the body responds to stimuli both internal and external.

Homeostasis has been a good initial description of how a highly complex, organic system works, and the concept of adaption and then a return to normal state certainly

---

[1] Darren Groberg, DPT, email communication, April 2019.

Table 20.1 Glossary of terms.

Homeostasis	Maintaining stability through simple response to negative feedback; return to equilibrium. Ignores or eradicates variance.
Allostasis	Maintaining survivability by adapting to fluctuating needs through reallocating interactive but finite resources. Assumes variance is normal and seeks to support this adaptability.
Reallocation/Borrowing	Redistribution of unstorable resources between systems or organs to fill short-term needs in order to meet a more urgent need.
Decompensation	Failure to meet physiological demands when at the end of resources, whereby tissue damage and injury can occur.
Skin ulcer	Localized damage to the skin and/or underlying soft tissue. Usually over a bony prominence such as the heel, as a result of intense and/or prolonged pressure, friction, shear, or microclimate. Soft tissue tolerance may be affected by internal and external factors in addition to pressure, etc.
Acclimatization	When the body has fully adjusted to internal or external environmental factors, and no longer continues efforts to adapt. May happen in advance/predictively.

has its merits. However, while the familiar concept of homeostasis can provide some insight into the body's response to loading, the concept of homeostasis has some inherent limitations which the concept of **allostasis** can better address. The Greek roots are *allo* (different) and *stasis* (state), thus referring to changing states. This newer concept represents a shift away from a return to equilibrium or stability, with instead an emphasis more on meeting demands that fluctuate as needs change Table 20.1.

Schulkin explains that the main difference between these two body regulation models is that one emphasizes maintaining normal balance while the other emphasizes "heightened activity with an accelerated rate of defense of bodily viability." Thus, when faced with "unusually heavy demands, ordinary homeostasis is not enough" [12].

Homeostasis is based on a fundamental assumption that individual variation in any system is just statistical noise, with an average of all data as the best representation of how a human body functions, rather than allowing for individual adaptation as a normal response to physiological demands stemming from internal or external input. The complex systems of the body, such as respiratory or cardiovascular systems, have been described in a statistical approach in order to make sense of a highly complex system. But as a direct result of the statistical approach, homeostasis disregards important considerations as irrelevant details, and as such, will have inherent limitations in how well it will capture both the full concept of how the organs and systems interact to meet various demands, and also where the danger threshold will be for tissue damage which requires intervention to prevent + [6,10,13]. As a result of this system of clinical thinking, the body's ability to adapt to simultaneous and interconnected needs is viewed as something to address and to correct rather than something to expect and to support.

If treating allostatically instead of homeostatically, then that same patient presenting with a fever may have an infection and thus need antibiotics, or may need better nutrition to address an imbalance, in addition to needing acetaminophen to address high temperature. Deviating from the normal set point is widely considered an indication of disease or distress under the homeostasis model, rather than a dynamic system responding to a wide range of demands which are simultaneous and interconnected. Allostasis is a dynamic model rather than one with discrete parameters to be addressed sequentially. As such, it allows for finely nuanced treatment which can be better adapted to individual patient needs.

Key components of the allostatic approach
Resolution of observation
When looking at how a body functions allostatically, we observe that functions like body temperature are not held perfectly at 37°C but vary with activity and time of day. The diurnal temperature cycle is a good example of this. With this and the natural variation that occurs in blood pressure in similar fashion we begin to understand that these seemingly small variations do play a role in overall health, healing, and typical body function. Knowing the allostatic nature of the body suggests that we will want to begin to acknowledge that there are times of day that the foot will experience greater heat load in the shoe and as a result will have to deal with additional moisture at these times of day. The effects of this fluctuation suggest the need for either excess capacity in terms of handling this moisture or a change in daily care practices that address heavier late afternoon sweating in the diabetic foot. Through an allostatic approach to treatment, this higher resolution in observation and response will enable the additional skin protection provided by more refined responses to tissue requirements [13].

Reallocation
Necessarily, the expanded concept of allostasis entails understanding the allostatic process, and as such, how complex and detailed that process is. Given that a body has simultaneous demands to be met, the allostatic process necessarily involves more complex regulation than the simpler homeostatic model would be able to allow for. Allostasis involves the whole brain and body rather than the very localized, individual response of homeostasis. Thus, allostasis is a coordinated variation, meaning the responses involve complex interplay between systems. This complex coordinated response is a means to optimize overall performance at the least physiological cost [12].

The body adapts to multiple needs, which necessitates a constant give and take, and thus the body must prioritize which need is more urgent as the body reallocates finite resources. Rather than supplying each need through a dedicated reserve for each individual system, the body draws on resources in reserve from other systems. For instance, muscles at rest require much lower blood flow, while muscles during heavy exercise will require triple the blood flow. To accommodate this increased demand during peak requirements, almost 10% of the total blood flow to muscles is borrowed [9].

Fluctuating demand is part of the allostasis model. In order to meet constant but varied demand from all organs and systems, the body must be able to draw on increased resources, which means it must either draw on a cache of reserves or find the resources elsewhere. For efficiency (in terms of storage capacity, for instance, and cost of untapped resources), the body will reallocate resources between various body systems. An inherent part of the allostasis model is the concept of borrowing resources from other systems in order to adapt to a more urgent need. This means, for example, that the body will divert blood flow from skin to sustain lungs or other organs, as more acute needs take priority [9]. In short term, for healthy people, this reallocation Table 20.1 process typically poses no difficulty. For others, the cost of borrowing resources can cause damage, even in very short term, when the body does not have sufficient resources to meet the volume of demand.

Poor perfusion will negatively affect patients with diabetes especially, as diabetes reduces the ability of blood vessels to contract and dilate, and therefore will impede the ability to borrow from other resources: Diabetes impairs the body's basic ability to adapt. Additionally, reduced ability to adapt will put other body sources at risk, as reallocating from lower priority sources risks taking from a system that was already inadequate but is still not the most urgent. And a system once-borrowed from may become the default to borrow from. This becomes a problem for those with diabetes, as at times the people with the most need for resources to borrow may be the very ones with less resources to work with. For these patients especially, it becomes important to think allostatically to recognize when and how to offer support and interventions. In terms of demands the body must meet, allostasis provides a higher resolution picture of how the body meets fluctuating but interactive needs than that allowed through a more simplified homeostatic approach. The details which homeostasis overlooks as unimportant or even needing corrected are the same details which allostasis incorporates to provide better individualized care for each patient across varying situations and needs.

Difference in response between homeostasis and allostasis

To illustrate the difference between homeostasis and allostasis, consider the everyday situation of blood pressure regulation. In the homeostasis model, the body has a "normal" level, or the most frequent value of 110/70. However, blood pressure fluctuates throughout a 24-hour period in response to various situations: heavy exercise, after eating, at moments of fright, or during sleep [9] (see Fig. 20.1).

Medication to medically clamp (or to restrict) blood pressure to a preestablished "normal" will mean that the body cannot adjust adequately, and patients can pass out because the body cannot adapt to increased demand during exercise, or even to the orthostatic fluctuations from simply standing up. As a result of the inability to adapt to fluctuating needs, patients have a high rate of discontinuance for blood pressure medication, or may

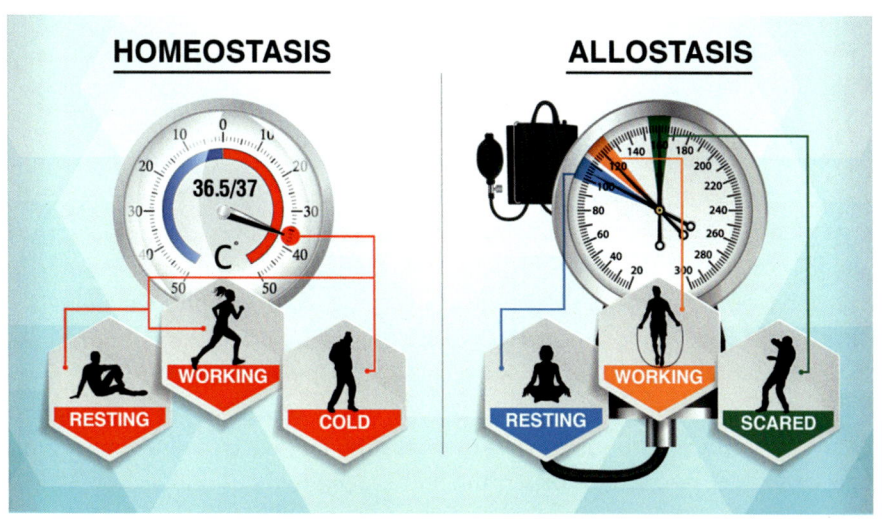

Figure 20.1 Homeostasis versus allostasis.

change medication within a short time [9]. Some indications are that less than 25% of hypertensive patients in the United States are controlled [9]. Patient compliance and education becomes a significant issue for people with diabetes, given that much of clinical intervention relies on what a patient understands and is willing or able to comply with for ongoing treatment and prevention. Using the allostatic treatment approach can help clinicians better educate patients about how devices and treatment will impact individual patient plans.

As the body responds to load, some of the interactive allostatic responses include modified blood flow, stress marker response, pain, inflammation, localized temperature response, moving in discomfort, as well as many others. Homeostasis considers each as a separate response. Allostasis considers these as complex, interacting responses. Garcia-Fernandez et al. list more than 80 factors which impact the creation of various skin lesions, broken down by dimension and ranking of importance [7]. (Their discussion is illuminating, but broader in scope than our current focus.) Understanding the interaction of the body systems and how patients meet these varying—and at times competing—physiological demands with finite resources can lead to better options for supporting treatment and methodologies.

Allostatic compensation

Any surface the human body is placed on for support will deliver mechanical stress to the body. This mechanical stress, also called loading, results from gravity trapping at risk tissue between the mass of the body and the support surface. Some of the forces that generate these allostatic responses include normal forces on the skin (pressure at 90 degrees); the radial shear that forms in the tissue due to the normal forces; additional shearing forces induced by friction and compression, and by heat and metabolic stresses; and the stress induced by comorbidities. All of these stresses seem to be most damaging when compression traps soft tissue between the support surface and the bony prominences, creating significant tissue distress than can result in various stages of tissue damage.

When loading occurs, the body responds through the process of allostatic compensation. In general, healthy and mobile individuals can compensate for these loading stresses via physiological responses, especially in short term. However, many patients cannot fully compensate for the inherent pressures and stress to tissues during loading, especially those patients who are elderly or unwell, physically unstable, or even, admittedly, those who are simply noncompliant. Additionally, even otherwise healthy and seemingly low risk patients have been known to develop ulcers, suggesting an inherent vulnerability in certain patients that would be hard to predict by usual methods of assessment [14].

Patients' ability to cope with increased health challenges is already at risk for patients with diabetes. For these patients, the more pertinent consideration is not returning to everyday, set-point equilibrium, but in survivability, i.e., what the body needs to address in order of strictest priority, no matter the higher cost, to focus on survival. The allostasis model better explains the range of physiological responses in patients, as well as the associated potential costs to addressing only urgent needs rather than ongoing considerations.

Physical response to tissue loading and risk of tissue injury

Understanding what is happening physiologically becomes critical to prevention and care, especially as DFU development is a multifactorial, complex problem [5,7,15]. Further, understanding how allostasis fits into the physiological process will allow improved prevention and more comprehensive patient care.

It has long been recognized that etiology for ulceration may subsume a number of causative pathways [16,17], although the two primary pathways involve issues with (1) blood flow and oxygenation and (2) compression and tissue distortion. While these are the two primary causative pathways, it does become clear that the complete process by which tissue damage occurs is not adequately understood, especially as it relates to loading and the diabetic foot [18].

Blood flow and oxygenation

Maintenance of good tissue health requires an adequate supply of nutrients from the blood and adequate tissue oxygenation [4]. Any time there is reduced blood flow, this can contribute to the development of DFU. Blood flow reduction is caused by "a loss of capillary network, failure of the skeletal muscle pump mechanism, muscle inactivity, drop in blood pressure, and occlusion of blood vessels" [19]. Resulting capillary occlusion results in tissue ischemia and thus hypoxia [8], but hypoxia is not the only adverse effect during ischemia: Research shows that reperfusion injury, lymphatic blockage, and cellular deformation each play a role in pressure injury development [8,15,20,21]. Tissue damage also occurs with pressure-induced vasodilation [8] with resultant turbulent blood flow [22], which causes endothelial damage [8,22] and interstitial fluid migration or edema [20].

Without adequate blood oxygenation and circulation, toxic metabolites accumulate in cells, accelerating the rate of cell death, which leads to ulceration and necrosis of skin and underlying tissue, and eventually to ulcer development [4]. Metabolite issues in ischemic tissues include glucose acidosis [8,22] and lactate build up, as well as calcium overload [8] and accumulation of free radicals [8,22].

Other factors such as inflammation [8,23], effects of general anesthesia [15] compromised skin [1,14,25,26], and even scar tissue [23] play important roles in tissue damage that can ultimately lead to ulcer formation [23]. Work in our own lab, as well as others' such as Oomens, Brienza, Gefen, etc., shows that while ischemia may not be the initial cause of tissue damage, it certainly contributes to tissue stress.

Risks: compression, deformation, and DTI

While tissue damage can occur on the skin's surface, deeper tissue is also subject to compression and damage. This deep tissue injury (DTI) occurs from unrelieved pressure or compression, where bony prominences intrude into muscle and soft tissue, occurring on the heel as well as throughout the bones and tissues of the foot. This tissue compression results in a DTI when deformation and time exposure exceed the body's tolerances [17,18,27]. Tissue distortion can cause near-immediate damage [27] yet may take up to several days before the damage is visible, and even then, the extent of damage can be difficult to fully assess.

Sustained tissue deformation is the direct cause of cell and tissue death [8,20]. In addition to the tissue damage from ischemia and hypoxia, tissue breakdown can also occur internally. Rather than study these etiologies as separate—or competing—issues, it is important to understand that tissue damage from both of these etiologies are interrelated: initial cell damage is dominated by deformation, and then aggravated by ischemia at later time points [8]. The core concept for allostasis includes the understanding that these physiological systems have limits [10,12,13] . Specifically, tissue which is either at the skin surface or deeper has an elastic limit based on its ability to withstand pressure, shear, and tissue distortion. When the tissue reaches its elastic limit, where it can no longer withstand loading forces of shear and distortion, it begins to break down [28].

When the body starts to reallocate resources from other areas, for example, when shifting blood flow from the skin to support healing or other functions such as brain or organ support, then, even though it takes some time for ischemia-based injury to occur, the tendency for the body to draw more and more on those same resources increases susceptibility for tissue deformation and the breakdown of tissue based on shear and pressure. As Coleman et al. indicate, there is strong scientific evidence that poor perfusion and status of skin and/or pressure ulcer reduce patient tolerance to pressure and increase the likelihood of ulcer development [5]. Poor perfusion from neuropathy affects the body's ability to borrow resources as the vessels do not dilate and contract as well.

Studies have been done on diabetic foot injury for decades, yet there is still work to be done to fully understand all the factors involved, and the interplay between these factors. However, DFU is hard to assess since the foot is in the shoe and it takes a bit of sleuthing to identify the cause of injury. Determining if the injury is device- or user-related, or perhaps single event driven, is essential to proper foot care. Researchers are using newer methods to more fully assess a wider range of physiological responses which include finite element modeling, interface pressure studies, tissue distortion, and cell culture, all to expand understanding of the interaction of the diabetic foot and the dressing or shoe. Approaches are still being refined, as well as what the results mean for pressure injury understanding and prevention [4]. The extent of tissue injury associated with mechanical forces (sheer and stress) has yet to be determined [15,23]. The concept of allostasis can bridge some of the current gap in understanding, and as such can guide research, treatment, prevention, and intervention.

When in allostasis the body responds to load: It adapts when it can, for as long as it can. When the body cannot sufficiently adapt, or can no longer sustain the responses, then allostatic compensation gives way to decompensation, and tissue damage occurs. This decompensation Table 20.1 leads to damage that can vary in level and intensity. The end result for many of these physiological responses is pressure injury. This would be seen as blisters, DFU, or diabetic foot infection, and require constant vigilance to prevent and treat.

Potential application/integration of allostasis in treatment approaches

Many devices currently exist that attempt to alleviate problems stemming from increased load to tissue. Devices range from preventative dressings, low-friction socks, specialty inserts, and custom shoes. As of yet, research has shown no stand-alone pressure therapy or single product that can solely prevent the occurrence of diabetic foot injury. Without adequate, proven products, the burden of prevention and treatment then falls primarily on the clinical staff. This would indicate a need for change in research focus until technology or experimental methods improve. The only proven method of DFU prevention is a collaborative effort that combines pressure therapy approaches from device support in combination with clinical best practices [29].

The Braden Scale for Predicting Pressure Sore Risk (Braden Scale) is a common assessment tool to establish a patient's risk for pressure injury [32]. (While assessment tools such as Norton and others are also in use, the Braden Scale is used very broadly, and has some advantages over the others [33].) Another way to view this assessment tool is to realize that the Braden Scale is assessing the extent of resources each individual patient has. An examination of some patient risk factors for DFU (influenced by Braden Scale subscore factors), in connection with potential clinical intervention and potential support surface intervention, illustrates the collaborative effort between caregivers and their use of tools to intervene for individual patient needs. As the risk factors for injury do not work in isolation, the combination of risk factors for each patient will influence the amount of intervention that may be necessary. A homeostatic approach to patient care would tend to address each category separately and sequentially. An allostatic approach will incorporate a multifactorial response to patient needs, as the needs will intersect (see Table 20.2).

Many of the interventions apply to both clinician and patient, so the process of how to incorporate prevention and treatment overlaps between intervention categories (Table 20.2). Allostasis lets us see how diverse and interactive the demands placed on a body are and helps to illustrate which types of intervention are most useful for each individual patient. Understanding allostasis will help to establish when additional intervention is needed for a patient whose resources are not adequate to fully prevent injury on their own.

Table 20.2 Allostasis: A multifactorial approach to addressing diabetic foot needs.

Patient risk factors	Possible patient intervention	Possible clinician intervention
Patient sensory perception	- Protection when walking - Drugs and topicals for neuropathy - Pressure redistribution shoes - Conformability and proper fit	- Provide walking aids, shoes, canes, and walkers - Assist with bed egress and bathroom trips - Foot inspections and education on foot inspections
Microclimate (moisture and temperature)	- Breathable socks and shoes - Use of protective open design shoes and house shoes - Proper use of skin care products - Injury prevention - Safe environment when shoeless	- Monitor for moisture accumulation and potential maceration - Monitor for dry or damaged skin - Monitor for patient dehydration - Emphasize clean socks daily - Clean/bathe patient, as needed - Teach proper skin care for product use, protection of toes, etc.
Patient movement	- Home preparation, including elimination of tripping hazards, i.e., rugs, exposed bed posts, etc. - Dressing aids to ensure proper sock and shoe use	- Education - Training on use of mobility aids - Occupational training as needed to increase mobility in different settings

Continued

Table 20.2 Allostasis: A multifactorial approach to addressing diabetic foot needs.—cont'd

Patient risk factors	Possible patient intervention	Possible clinician intervention
	- Cane and walker use to improve proprioception - Ambulation to improve circulation and reduce edema - Decreased fall risk through mobility aids	- Increase patient activity level, safe walking - Keep patient moving, to keep patient's body engaged in making needed adjustments - Orthotic and Pedorthist referrals
Nutrition, hydration, perfusion	- Proper nutrition and supplements - Proper hydration - Increased activity for perfusion	- Monitor intake - Monitor hydration and rehydration - Monitor medication which affects patient perfusion - Manage compliance - Refer to home dietitian consultation - Apply compression hosiery and compression wraps as needed for diabetic wounds and peripheral vascular disease
Friction and shear	- Proper fitting shoes and orthotics - Proper sock selection and use - Use of low-friction material, to allow for blister-free walking - Monitoring high-pressure points for tissue relief - Awareness of changes in gait, balance, extra pressure concentration points	- Refer for diabetic shoe fitting - Demonstrate proper sock selection - Train in foot inspection for blister and maceration - Moleskin, skin tougheners, tape, compression wrapping, etc.

Limitations

Attempting to lower incidence rate with interventions can become more challenging than helpful, as simply assessing which patients are at most risk is not always easy to do correctly, evidenced by continued incidence rates in spite of all current attempts to address this issue. Obviously, prevention is superior to treatment. However, to be most effective, the treatment approach needs to be broad enough to consider the intersection of feasibility and practicality as well as how any theoretical or intended use plays out in patient ability and compliance.

Conclusion

A new approach, to better address problem of DFU

Not everyone is willing to accept current statistics as simply the status quo. Many sources are working to change the numbers, and there is urgency to address the issue of DFU. Notwithstanding any basic unwillingness to address the urgency behind the statistic, part of the difficulty is a lack of data to work with and a lack of interventions that will clearly help. Solis et al. tell us that in the absence of a significant reduction in pressure injuries, new preventative measures are needed [8]. Similarly, part of the problem in addressing the persistent occurrence of DFU may not rest in the devices themselves, nor in caregiver interventions, but in the current plateau in how we approach the underlying problem.

The concept of allostasis provides a better explanation for both the interactions of the multiple potential risk factors as well as the body's response to the many factors. This in turn provides a more clear understanding of the interaction of the tissue stresses that result in pressure injury and DFU. Padula suggested that pressure injury is a result of risk factors overlapping, where the cumulative effect is greater than the sum of the many parts involved [31]. As we better understand the role of allostasis, and the process of the body adapting, then we will have considerably more options to allow us to treat this complex, multifactorial problem. The Goldilocks principle applies here, as we aim to address a dynamic problem with solutions that are neither oversimplified, nor overspecialized in their approach. We must be sure we are asking the right questions, and part of asking the right question is that we must make sure future supporting solutions will engage the body more effectively, without making things worse (actively causing harm) or untenable (such as needing training at all levels of care, or too time-intensive to individualize properly). Armed with a better understanding of physiological responses to loading, and strengthened by its connection with the concept of allostasis, we then have more-nuanced conceptual approach to address this persistent, multifaceted problem. As allostasis covers a wide and converging range of responses which tend to be studied, tested, treated, and funded individually, this new framework will provide better options to address a complex problem through a concept that brings higher resolution, in part because this concept will better address not just individual factors, but the interplay between them as well.

We need to address the current shortfalls in the industry. The older, more familiar concept of homeostasis is inadequate to address the broad spectrum of risk factors for DFUs as well as the more current understanding of etiologies, their reciprocity, and their complexities. The concept of allostasis will help us to frame new and better approaches as we actively engage the body's own allostatic process in assisting in treatment, rather than overlooking this valuable concept and the better options for care that this new understanding can create.

Key takeaways

- We know that the body adapts but we are not examining the full range of allostatic responses, nor the interplay between those responses, nor the cost of borrowing finite resources.
- The body's ability to adapt to simultaneous and interconnected needs is something to expect and to support.
- *Allostasis* more fully explains the body's interconnected responses to the varied demands in care situations, and the cost to the body of meeting fluctuating demands with finite resources available.
- We need to apply the principles of allostasis to create a paradigm shift in wound care, moving from a reactionary focus to an anticipatory plan that can more effectively treat as well as prevent DFU in wound care.
- *Allostasis* provides a better understanding of how and when to intervene to better support and engage the body in its efforts to prevent tissue damage. The new allostasis paradigm lets us better prevent and treat DFU by directing research and clinical intervention strategies toward more integrative solutions.

Conflict of interest

This project has been supported in part by an unrestricted grant from Encompass Group which provided funding for this chapter.

References

[1] Phipps L, Gray M, Call E. Time of onset to changes in skin conditioning during exposure to synthetic urine: a prospective study. J Wound Ostomy Continence Nurs 2019;46(6):315−20.
[2] Data points No.2: incidence of diabetic foot ulcer and lower extremity amputation among medicare beneficiaries, 2006-2008. Effective health care program. AHRQ research report. US Dept of HHS; 17 Feb. 2011. p. 1−7.
[3] Diabetic wound care: foot health. American Podiatric Medical Assoc; 2020. www.apma.org/diabeticwoundcare.
[4] Kim J, Wang X, Ho C, Bogie K. Physiological measurements of tissue health: implications for clinical practice. Int Wound J 2012;9:656−64.
[5] Coleman S, Nixon J, Keen J, et al. A new pressure ulcer conceptual framework. J Adv Nurs 2014: 2222−34.
[6] Fossion R, Rivera AL, Estanol B. A physicists's view of homeostasis: how time series of continuous monitoring reflect the function of physiological variables in regulatory mechanisms. Physiol Meas 2018;39:084007. https://doi.org/10.1088/1361-6579/aad8db. 10 pages.
[7] Garcia-Fernandez FP, Agreda JJS, Verdu J, Pancorbo-Hidalgo PL. A new theoretical model for the development of pressure ulcers and other dependence-related lesions. J Nurs Scholarsh 2014;46(1): 28−38.
[8] Gawlitta D, Oomens C, Bader D, Baaijens P, Bouten C. Temporal differences in the influence of ischemic factors and deformation on the metabolism of engineered skeletal muscle. J Appl Physiol 2007;103:464−73.

[9] Sterling P. Principles of allostasis: optimal design, predictive regulation, pathophysiology, and rational therapeutics. In: Allostasis, homeostasis and costs of physiological adaptation, Schulkin. NY: Cambridge UP; 2004. p. 17–58.
[10] West BJ. Homeostasis and Gauss statistics: barriers to understanding natural variability. J Eval Clin Pract 2010;16:403–8.
[11] Stedman's medical dictionary illustrated in color. homeostasis. In: Schulkin J, editor. Allostasis, homeostasis and costs of physiological adaptation. Cambridge UP: NY, 2004. 27th ed. Baltimore, MD: LWW; 2000. pg 827.
[12] Schulkin J, editor. Allostasis, homeostasis and costs of physiological adaptation. NY: Cambridge UP; 2004.
[13] Tanner L, Rappl L, Oberg C, Call E. Keeping patients under the damage threshold for pressure injury. JNCQ [e-version]. June 2020. p. 1–7.
[14] Bergstrand S, Kallman U, Ek A-C, Engstrom M, Lindgren M. Microcirculatory responses of sacral tissue in healthy individuals and inpatients on different pressure-redistribution mattresses. J Wound Care Aug 2015;24(8):346–58.
[15] Prevention and treatment of pressure ulcers/injuries: clinical practice Guideline European Pressure Ulcer Advisory Panel (EPUAP), National Pressure Injury Advisory Panel (NPIAP) and the Pan Pacific Pressure Injury Alliance (PPPIA). 3rd ed. 2019.
[16] Gray R, Voegeli D, Bader D. Features of lymphatic dysfunction in compressed skin tissues–Implications in pressure aetiology. J Tiss Via 2016;25:26–31.
[17] Gefen A, et al. The etiology of pressure injuries. In: Prevention and treatment of pressure ulcers/injuries: clinical practice guideline European Pressure Ulcer Advisory Panel (EPUAP), National Pressure injury Advisory Panel (NPIAP) and the Pan Pacific Pressure Injury Alliance (PPPIA). 3rd ed. 2019.
[18] Gefen A. How medical engineering has changed our understanding of chronic wounds and future prospects. Med Eng Phys 2019;72:13–8.
[19] van Londen A, Herwegh M, van der Zee M, et al. The effect of surface electrical stimulation of the gluteal muscles on the interface pressure in seated people with spinal cord injury. Arch Phys Med Rehabil 2008;89:1724–32.
[20] Demarre L, Beeckman D, Vanderwee K, Defloor T, Grypdonck M, Verhaeghe S. Multi-stage versus single-stage inflation and deflation cycle for alternating low pressure air mattresses to prevent pressure ulcers in hospitalised patients: a randomised-controlled clinical trial. Int J Nurs Stud 2012;49:416–26.
[21] Rithalia S. Assessment of patient support surfaces: principle, practice and limitations. J Med Eng Tech July 2005;29(4):163–9.
[22] Spahn J, Duncan C, Butts L. Effects of a support surface on homeostasis. Keep it simply scientific. EHOB; 2000.
[23] Solis L, Hallihan D, Uwiera RR, Thompson RB, Pehowich ED, Mushahwar V. Prevention of pressure-induced deep tissue injury using intermittent electrical stimulation. J Appl Physiol 2007; 102:1992–2001.
[24] Black JM, Edsberg LE, Baharestani MM, et al. Pressure ulcers: avoidable or unavoidable? Results of the national pressure ulcer advisory panel consensus conference. Ostomy Wound Manage February 2011; 57(2):24–37.
[25] Levine JM. Skin failure: an emerging concept. J Am Med Dir Assoc 2016;17:666–9.
[26] Blume-Peytavi U, Kottner J, Sterry W, et al. Age-associated skin conditions and diseases: current perspectives and future options. Gerontologist 2016;56(S2):S230–42. https://doi.org/10.1093/geront/gnw003.
[27] Bader D, Brienza D, Oomens C, Hammond AW, Gefen A. Improving patient outcomes: bridging the gap between science and efficacy. In: Presentations: 2015 EPUAP conference. Ghent University; 16 Sept. 2015.
[28] Call E, Jones R, DeMonja K, et al. "The elastic limit": introducing a novel concept in communicating excessive shear and tissue deformation. WCET J 2017;37(4):16–20.
[29] Niederhauser A, Lukas CV, Parker V, Ayello EA, Zulkowski K, Berlowitz D. Comprehensive programs for preventing pressure ulcers: a review of the literature. Adv Skin Wound Care 2012;25(4): 167–88. https://doi.org/10.1097/01.ASW.0000413598.97566.d7.

[30] Soban LM, Hempel S, Munjas BA, Miles J, Rubenstein L. Preventing pressure ulcers in hospitals: a systematic review of nurse-focused quality improvement interventions. Joint Comm J Qual Patient Safety 2011;37(6):245—52. https://doi.org/10.1016/S1553-7250(11)37032-8. AP1-AP16.

[31] Padula W. Gateway to regulatory compliance: through the looking glass. Presentation at. St. Louis, MO: NPUAP Annual Conference; 2019. March 2, 2019.

[32] Braden Scale for Predicting Pressure Sore Risk ©. Prevention Plus, home of the Braden Scale website. [Accessed January 6, 2020]. http://bradenscale.com/images/bradenscale.pdf. Copyright, Barbara Braden and Nancy Bergstrom, 1988. Reprinted with permission. All rights reserved. doi:10.1097/00006199-198809000-00014.

[33] Pancorbo-Hidalgo PL, Garcia-Fernandez FP, Lopez-Medina I, Alvarez-Nieto C. Risk assessment scales for PU prevention: a systematic review. J Adv Nurs 2006;54(1):94—100. https://doi.org/10.1111/j.1365-2648.2006.03794.x.

CHAPTER 21

Footwear for persons with diabetes at high risk for foot ulceration: offloading, effectiveness, and costs

Sicco A. Bus
Associate Professor and Head Gait Lab, Amsterdam UMC, University of Amsterdam, Department of Rehabilitation Medicine, Amsterdam, Netherlands

Introduction

People with diabetic foot disease who are at risk of developing a foot ulcer are commonly prescribed with special footwear to help prevent such an ulcer. One of the main mechanisms by which this goal is achieved is redistribution of mechanical stress or pressure applied to the foot. In this chapter, these biomechanical mechanisms will be discussed, the special features of the footwear will be described, the effectiveness of such footwear in preventing foot ulcers will be highlighted, with an important role for adherence, and the costs involved will be discussed.

The person with diabetes at high risk

The International Working Group on the Diabetic foot (IWGDF) is tasked with the development of clinical evidence—based guidelines for prevention and treatment of diabetic foot disease [1,2]. The IWGDF stratifies people with diabetes according to their risk of developing a foot ulcer. At very low risk, people have diabetes but no other complications; at low risk, they have peripheral neuropathy or artery disease as complication; at moderate risk, they have both these factors or either one in addition to having foot deformity; and at high risk, they went through an episode of foot ulceration. The annual risk of developing a foot ulcer increases from about 2% for those at very low risk to 7% for those at low risk to almost 40% for those at high risk [3]. As a result, most attention and interventions toward preventing foot ulcers have focused on the high-risk patient who is considered to be "in remission" by already undergoing an episode of ulceration [3,4]. Footwear as intervention is not an exception to that and the footwear recommended for these different risk categories ranges from a good advice for appropriate footwear for the low-risk patient to that of custom-made footwear with a demonstrated pressure-relieving effect for the high-risk patient who healed from a plantar ulcer [4].

Offloading

Approximately 50% of people with diabetes will eventually have loss of protective sensation (LOPS) in their feet due to peripheral neuropathy [5]. LOPS is sufficient to allow these patients to injure their feet without noticing it [3]. High local mechanical stress, measured as elevated plantar peak pressure during ambulation, is in the presence of LOPS a causative factor in the development of foot ulcers and also in the recurrence thereof [6,7]. These elevated plantar foot pressures are generally caused by changes in foot structure, such as deformity, limited joint mobility, and loss of fat pad quality [8].

Reducing mechanical stress or plantar peak pressure is called "offloading." If the foot is not offloaded adequately, ulcers may develop or present ulcers may not heal. Therefore, offloading is a cornerstone of treatment for and against plantar foot ulcers in people with diabetes [9]. With a risk of ulcer recurrence up to 40% in the first 12 months, and 60%—65% after 3 years [3], the need for continuous offloading in these high-risk patients is evident. Offloading is measured by using a plantar pressure measurement system, which is an insole consisting of an array of sensors that is inserted inside the shoe and measures the plantar pressure distribution. From these measurements, the peak value of the pressure during the stance phase of walking is the most common parameter used, and an alternative is the pressure—time integral. Pressure measurement is becoming more used in clinical footwear practice, mainly through the effect of several studies demonstrating the benefit of such measurements [10,11]. A detailed discussion of mechanical stress in diabetic foot disease can be found elsewhere [12].

The offloading effect of footwear

Therapeutic footwear deals with shoes and insoles (i.e., orthoses) that aim for a therapeutic effect, and in people with diabetic foot diseases this is mostly the prevention of foot ulcers. Fig. 21.1 shows several examples of custom-made therapeutic footwear. The mechanism through witch this aim is achieved is the provision of sufficient space and pressure relief at high-risk areas. Such footwear is built up from different design features that have a pressure-relieving or offloading effect on the foot [13]. A rocker-bottom outsole configuration is very effective in relieving plantar pressure in the forefoot, with up to 52% of pressure relief found compared to a shoe that does not have a rocker outsole [14—16]. Custom-made insoles are effective in relieving peak pressures compared to more flat standard insoles (Fig. 21.2) [17]. Furthermore, specific insole design elements such as metatarsal pad or bars or a medial arch support are effective in relieving pressure in regions where foot ulcers in people with diabetes most often occur, the metatarsal heads [18,19]. Additionally, the top layer of an insole has significant pressure-relieving effects on single locations but also over the entire plantar surface of the foot. More local modifications such as removing and softening insole materials at at-risk regions are also

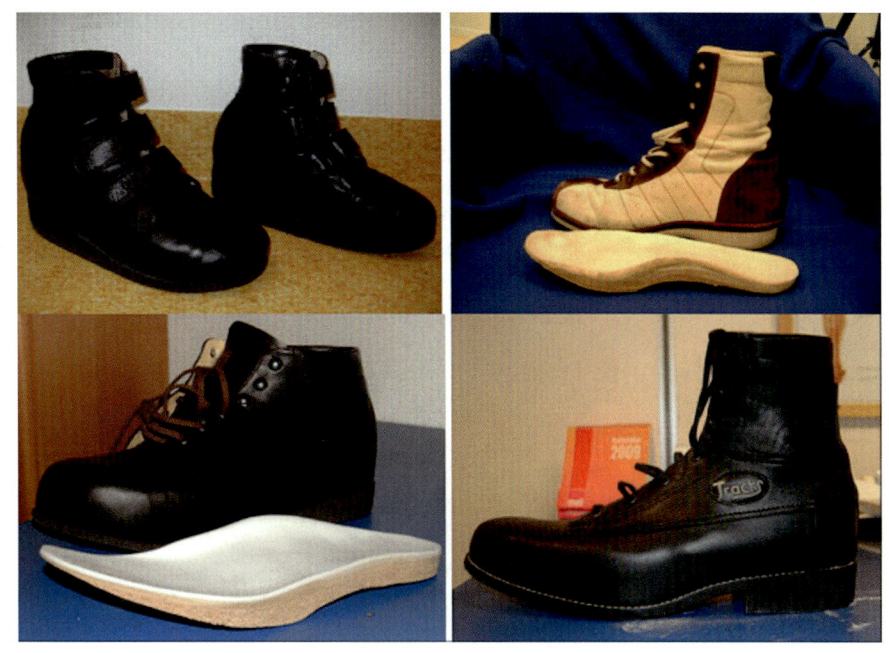

Figure 21.1 Examples of therapeutic shoes and insoles that are prescribed to people with diabetes who are at high risk of developing a foot ulcer.

effective in offloading [19]. The use of a combination of these design features has an additive offloading effect. One should consider, however, that while such insole design elements can effectively offload plantar foot regions, the specific form, materials used and placement of these elements is critical in the final pressure outcome; incorrect placement can result in pressure increase and not decrease [20]. The use of barefoot or in-shoe plantar pressure measurement is a valuable method to optimize the placement of these insole design elements [21,22].

Effectiveness of footwear for ulcer prevention

Many studies have been performed on the effect of therapeutic footwear to help prevent foot ulcers from occurring. Most of these studies have focused on the prevention of ulcer recurrence as patients are prescribed most of the time with such footwear when a foot problem like an ulcer has already occurred. Only few studies exist on the prevention of a first-ever foot ulcer, mainly because the risk of developing a foot ulcer is lower and footwear for people at risk who have not yet had an ulcer is less specified. Generally, the prospective trials that have been conducted have shown a beneficial effect of the use of therapeutic footwear compared to standard footwear in preventing ulcer recurrence, although variation exists [23]. The contrasting results that are present can be attributed to

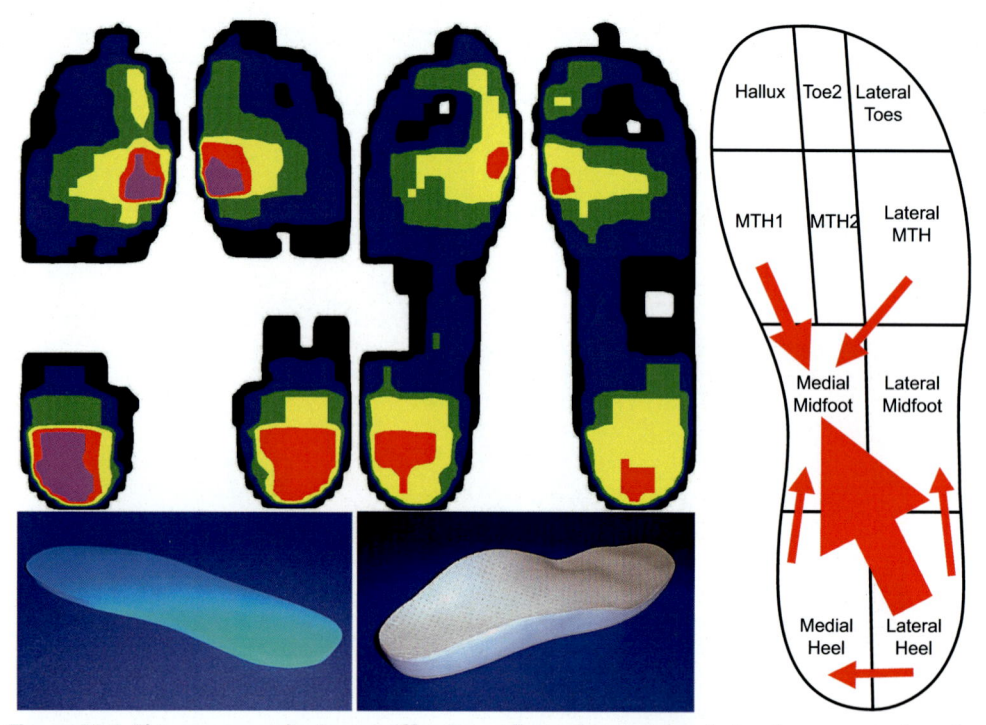

Figure 21.2 The pressure-relieving or offloading effect of custom-made insoles compared to flat insoles. Through the total contact principle and the incorporation of a medial arch support and metatarsal bar, load is redistributed away from the forefoot and heel regions to the midfoot area, giving peak pressure relief at the metatarsal heads and increase in pressure at the medial midfoot.

the diversity of intervention and control conditions tested and, quite importantly, to the lack of information about whether the footwear tested was effective in offloading. This complicates the comparison of studies in this area [24].

More recently, two multicenter RCTs have further improved our understanding of how offloading in therapeutic footwear helps to prevent plantar foot ulcer recurrence [10,11]. In one trial, in-shoe plantar pressure analysis was used as an evaluation tool to guide modifications to custom-made footwear that was prescribed and delivered to patients based on the expertise and skills of the clinical team. By identifying high-pressure regions and modifying the footwear to target these regions, this approach significantly improved the pressure-relieving properties of the custom-made footwear with an average 30% peak pressure relief. However, this approach of pressure optimization showed only a nonsignificant 11% reduction in the incidence of ulcer recurrence after 18 months follow-up compared to custom-made footwear that did not undergo such modifications and improvement based on pressure guidance [10]. But when only the patients were analyzed who, with objective measures [25], were shown to be adherent to wearing their prescribed footwear, pressure-improved footwear showed a significant 46% lower plantar foot ulcer incidence rate compared to the shoes that did not undergo this improvement. In the other multicenter RCT, custom-made insoles were designed and manufactured based on barefoot plantar pressure distribution and 3D foot shape data. These insoles, worn in extra-depth shoes, were shown in proof-of-principle studies to relieve peak pressure at the metatarsal heads about 30% more than traditional shape-based custom-made insoles. In the RCT on the efficacy of these custom-made insoles to prevent plantar forefoot ulcer recurrence with follow-up of 15 months, the pressure-based insole reduced risk of ulcer recurrence at the metatarsal heads by 63% compared to the traditionally shaped insoles [11]. These data-driven footwear concepts demonstrate that it is the combination of adequate pressure relief through pressure-based design and adequate adherence to wearing the footwear that leads to the best clinical outcome.

What adequate pressure relief means quantitatively is not exactly known, and will differ from patient to patient but also depends on the type of in-shoe plantar pressure measurement system used. Nevertheless, some useful indications for a target pressure threshold from comparative and prospective studies exist. One study examined patients who had remained healed after plantar foot ulceration with wearing custom-made footwear and found a mean pressure of approximately 200 kPa at the prior site of ulceration [26]. Data from the abovementioned RCT using in-shoe pressure analysis showed that when peak pressures were >200 kPa, these pressures could be effectively reduced, but when peak pressures were <200 kPa, further adaptation of the footwear proved futile [22]. Additionally, an analysis of risk factors of pressure-related plantar foot ulcer recurrence from the same trial showed that when peak pressure is below 200 kPa and adherence is above 80%, the risk of plantar foot ulcer recurrence is 60% lower compared to when these conditions are not met [7]. Even though a pressure threshold for foot ulceration is likely unique to each individual, we now have indications that the 200 kPa value can serve as a useful target for plantar offloading in footwear prescription [22]. This is under the condition that the pressures are measured with a validated and calibrated system with a spatial resolution of at least 1 sensor per 2 cm^2 [4]. These studies illustrate that diabetic footwear provision for ulcer prevention is moving from a skills and experience-based method to a more systematic, scientific, and data-driven approach. The effects of such an approach have been studied recently by comparing several data-driven scientific-based footwear design methods on plantar pressure relief and shows that more effective pressure relief and less need for modification of footwear after in-shoe pressure measurement is provided by these data-driven footwear designs [27].

Adherence to wearing diabetic footwear

Regardless of their pressure-relieving capacity, therapeutic footwear can only be effective when it is worn by the patient. Therefore, the effect of these shoes must be judged both by the capacity to relieve peak pressure and by the patient's adherence to wear the shoes. Studies show that of patients who are at high risk of developing a foot ulcer, based on presence of peripheral neuropathy, a foot ulcer history, and foot deformity, only 50% may wear their prescribed custom-made shoes for more than 80% of the steps they take. The same study showed that nonadherence is particularly a problem when patients are inside their home, with adherence being on average 67% compared to 87% outdoors. And this while they take more steps inside the house than when outside [28], amplifying the problem of nonadherence. This contributes to the lower effectiveness of footwear in ulcer prevention and stresses the importance of continued pressure relief to prevent foot ulcers [13,29].

Despite the important role of adherence, studies on how to improve footwear adherence are scarce [13,28]. One small explorative study on the effect of motivational interviewing as behavioral intervention to improve adherence shows variable effects between patients and some effect only in the short term [30,31]. More recently, offloading footwear specifically for indoor use has been developed. Such footwear should have similar offloading properties as the custom-made shoes the patient already has, but stimulate the use of protective footwear when inside the house, by making the footwear more lightweight, easy to don and doff, and comfortable in use. A study on the effect of these specially designed and manufactured shoes for indoor use showed significantly improved adherence indoors, both short and long term after prescribing such footwear next to the custom-made shoes the patient already has [32]. Patients seem to be satisfied with having such a special shoe for indoors. More development of methods and tools that can improve adherence and studies testing these methods is needed.

Recommendations from international guidelines

Being at increased risk for ulceration, it is important that the footwear of people with diabetes fits, protects, and accommodates the shape of their feet; this includes having adequate length, width, and depth. When a foot deformity or preulcerative sign is present, it becomes even more important to change foot biomechanics and reduce plantar pressure on at-risk locations. This may require custom-made footwear, custom-made insoles, or toe orthoses. For people who have healed from a plantar foot ulcer, the necessity to redistribute and reduce peak pressures under the foot becomes even more important. This importance of footwear in the prevention of foot ulceration in diabetes is supported by recommendations on its use in international guidelines from the IWGDF [4]. These recommendations are as follows:

> **Recommendation:** *Instruct a person with diabetes who is at moderate risk for foot ulceration (IWGDF risk 2) or who has healed from a non-plantar foot ulcer (IWGDF risk 3) to wear therapeutic footwear that accommodates the shape of the feet and that fits properly, to reduce plantar pressure, and help prevent a foot ulcer. When a foot deformity or a preulcerative sign is present, consider prescribing custom-made footwear, custom-made insoles, or toe orthoses (Strong recommendation; Low level of evidence).*

Recommendation: *Consider prescribing orthotic interventions, such as toe silicone or (semi-)rigid orthotic devices, to help reduce abundant callus in a person with diabetes who is at risk for foot ulceration (IWGDF risk 1-3) (Weak recommendation; Low level of evidence).*

Recommendation: *In a person with diabetes who has a healed plantar foot ulcer (IWGDF risk 3), prescribe therapeutic footwear that has a demonstrated plantar pressure relieving effect during walking, to help prevent a recurrent plantar foot ulcer; furthermore, encourage the patient to consistently wear this footwear (Strong recommendation; Moderate level of evidence).*

A "demonstrated plantar pressure-relieving effect" means that at high-pressure locations, there should be either a ≥30% reduction in the peak pressure during walking (compared with the current (therapeutic) footwear), or a measured peak pressure < 200 kPa (if measured with a validated and calibrated pressure measuring system with sensor size of 2 cm^2). The way to achieve such a pressure relief or level is described above and constitutes applying available state-of-the-art scientific knowledge on footwear designs that effectively offload the foot [19,23,27].

Costs related to the use of footwear for ulcer prevention

No studies are available that have investigated the cost-effectiveness of therapeutic footwear in preventing foot ulceration in people with diabetes. One trial has attempted to do a cost analysis for data-driven footwear design techniques in association with the risk of developing recurrent plantar foot ulcer, using data from abovementioned footwear trial [10]. In this trial 85 patients wore pressure-improved footwear that used the Pedar-X system for the evaluation of in-shoe plantar pressure and guidance for footwear modifications based on the peak pressure distribution found. The costs involved include write-off costs and maintenance of the system, training of technicians to perform pressure measurements, the time spent on conducting in-shoe pressure measurements, and the time spent on modifying the shoe based on the pressures found.

In this trial, the total sum of these costs was calculated to amount to €38.500. The costs for ulcer treatment were not directly assessed in the trial, and so were based on literature data that quite consistently show that foot ulcers treated in multidisciplinary foot clinics cost approximately €10.000 per episode [33,34]. Using these data and the difference in number of ulcers that developed in 18 months in the intervention group compared to the control group that did not undergo such modification and pressure optimization to the footwear, a small cost saving is achieved with the intervention when considering a group of about 85 patients. If footwear adherence can be assured, a much larger reduction in risk of recurrence is achieved by the pressure-improved footwear, and cost saving of almost €100.000 in about 40 patients can be achieved.

These data are useful when it comes to decision-making toward the implementation of data-driven footwear design and evaluation approaches and methods with which footwear adherence can be increased in people at high risk of developing plantar foot ulcers.

The remaining issues of who will pay for data-driven approaches (the footwear companies, insurance companies, or patients) are important ones that will greatly affect implementation in clinical footwear practice.

A few considerations

In quantifying offloading, we normally refer to the normal component of pressure, not to the shear component, as this cannot be measured with the currently available systems. In ulcer development, shear is suggested to play a contributing role. And hyperkeratotic tissue or callus, under which foot ulcers often occur, is considered to be the result of shear acting on the foot [3]. Given the lack of available shear measuring options, this biomechanical aspect of ulcer development remains underinvestigated and in need of technological advances in the field. Foot temperature has been opted as a possible surrogate for shear [35], which, if more clearly demonstrated, would open up options for interpretation of shear inside footwear in people with diabetes.

Despite the value of using plantar pressure measurement in the design and evaluation of diabetic footwear, and its more widespread use in clinical practice, it is not yet a standard in footwear design. A major advance would be the requirement that measurable and effective pressure reduction should result from any prescribed interventions to a high-risk foot [36]. Such requirements for demonstrated efficacy in pressure relief for prescribed custom-made shoes are effective in some places or countries, but not on a large scale. And these measurements may not be possible at every treatment location because of available resources, but the cost analysis shown above does suggest the cost-benefit of implementing such pressure measurements in clinical footwear practice [10].

Future research

Several aspects related to offloading the foot of people with diabetic foot disease require further research. More research and development is needed in measuring shear stress as a likely component of ulcer development in people with diabetes and neuropathy. More development is also needed in methods and tools that can improve adherence to wearing prescribed footwear, and studies on their efficacy are urgently needed. Finally, future research should define success and failure in implementing data-driven footwear design protocols and in-shoe plantar pressure measurement in footwear prescription practice so to improve state-of-the-art knowledge and tools and with that improve footwear for the person with diabetes who is at risk of ulceration.

Conclusions

This chapter discussed several aspects of footwear for people with diabetes who are at high risk of developing a foot ulcer: offloading, effectiveness, and costs. Footwear with a demonstrated pressure-relieving effect has shown to be effective in reducing risk of plantar foot ulcer recurrence in people with diabetes, under the condition that the footwear is worn. The recommendation for such pressure-relieving footwear is now integrated in international guidelines on the diabetic foot, and is likely to be cost-effective in ulcer prevention. Custom-made footwear design protocols and the implementation of data-driven approaches such as plantar pressure measurement help in translating these guidelines to clinical footwear practice. With that, footwear design and provision for the high-risk diabetic foot moves from a more traditional approach to a more modern scientific field that uses a systematic, science-based, and data-driven approach that clinically benefits the high-risk person with diabetes.

References

[1] Bus SA, van Netten JJ, Monteiro-Soares M, Lipsky BA, Schaper NC. Diabetic foot disease: "The Times They are A Changin". Diabetes Metab Res Rev 2020;36(Suppl. 1):e3249.

[2] Schaper NC, van Netten JJ, Apelqvist J, Bus SA, Hinchliffe RJ, Lipsky BA, et al. Practical Guidelines on the prevention and management of diabetic foot disease (IWGDF 2019 update). Diabetes Metab Res Rev 2020;36(Suppl. 1):e3266.

[3] Armstrong DG, Boulton AJM, Bus SA. Diabetic foot ulcers and their recurrence. N Engl J Med 2017; 376(24):2367–75.

[4] Bus SA, Lavery LA, Monteiro-Soares M, Rasmussen A, Raspovic A, Sacco ICN, et al. Guidelines on the prevention of foot ulcers in persons with diabetes (IWGDF 2019 update). Diabetes Metab Res Rev 2020;36(Suppl. 1):e3269.

[5] Boulton AJ, Vileikyte L, Ragnarson-Tennvall G, Apelqvist J. The global burden of diabetic foot disease. Lancet 2005;366(9498):1719–24.

[6] Monteiro-Soares M, Boyko EJ, Ribeiro J, Ribeiro I, Dinis-Ribeiro M. Predictive factors for diabetic foot ulceration: a systematic review. Diabetes Metab Res Rev 2012;28(7):574–600.

[7] Waaijman R, de Haart M, Arts ML, Wever D, Verlouw AJ, Nollet F, et al. Risk factors for plantar foot ulcer recurrence in neuropathic diabetic patients. Diabetes Care 2014;37(6):1697–705.

[8] Barn R, Waaijman R, Nollet F, Woodburn J, Bus SA. Predictors of barefoot plantar pressure during walking in patients with diabetes, peripheral neuropathy and a history of ulceration. PLoS One 2015; 10(2):e0117443.

[9] Wu SC, Crews RT, Armstrong DG. The pivotal role of offloading in the management of neuropathic foot ulceration. Curr Diabetes Rep 2005;5(6):423–9.

[10] Bus SA, Waaijman R, Arts M, de Haart M, Busch-Westbroek T, van Baal J, et al. Effect of custom-made footwear on foot ulcer recurrence in diabetes: a multicenter randomized controlled trial. Diabetes Care 2013;36(12):4109–16.

[11] Ulbrecht JS, Hurley T, Mauger DT, Cavanagh PR. Prevention of recurrent foot ulcers with plantar pressure-based in-shoe orthoses: the CareFUL prevention multicenter randomized controlled trial. Diabetes Care 2014;37(7):1982–9.

[12] Lazzarini PA, Crews RT, van Netten JJ, Bus SA, Fernando ME, Chadwick PJ, et al. Measuring plantar tissue stress in people with diabetic peripheral neuropathy: a critical concept in diabetic foot management. J Diabetes Sci Technol 2019;13(5):869–80.

[13] Bus SA, van Deursen RW, Armstrong DG, Lewis JE, Caravaggi CF, Cavanagh PR, et al. Footwear and offloading interventions to prevent and heal foot ulcers and reduce plantar pressure in patients with diabetes: a systematic review. Diabetes Metab Res Rev 2016;32(Suppl. 1):99–118.

[14] van Schie C, Ulbrecht JS, Becker MB, Cavanagh PR. Design criteria for rigid rocker shoes. Foot Ankle Int 2000;21(10):833–44.

[15] Praet SF, Louwerens JW. The influence of shoe design on plantar pressures in neuropathic feet. Diabetes Care 2003;26(2):441–5.
[16] Chapman JD, Preece S, Braunstein B, Höhne A, Nester CJ, Brueggemann P, et al. Effect of rocker shoe design features on forefoot plantar pressures in people with and without diabetes. Clin Biomech 2013;28(6):679–85.
[17] Bus SA, Ulbrecht JS, Cavanagh PR. Pressure relief and load redistribution by custom-made insoles in diabetic patients with neuropathy and foot deformity. Clin Biomech 2004;19(6):629–38.
[18] Guldemond NA, Leffers P, Schaper NC, Sanders AP, Nieman F, Willems P, et al. The effects of insole configurations on forefoot plantar pressure and walking convenience in diabetic patients with neuropathic feet. Clin Biomech 2007;22(1):81–7.
[19] Arts ML, de Haart M, Waaijman R, Dahmen R, Berendsen H, Nollet F, et al. Data-driven directions for effective footwear provision for the high-risk diabetic foot. Diabet Med 2015;32(6):790–7.
[20] Hastings MK, Mueller MJ, Pilgram TK, Lott DJ, Commean PK, Johnson JE. Effect of metatarsal pad placement on plantar pressure in people with diabetes mellitus and peripheral neuropathy. Foot Ankle Int 2007;28(1):84–8.
[21] Bus SA, Haspels R, Busch-Westbroek TE. Evaluation and optimization of therapeutic footwear for neuropathic diabetic foot patients using in-shoe plantar pressure analysis. Diabetes Care 2011;34(7):1595–600.
[22] Waaijman R, Arts ML, Haspels R, Busch-Westbroek TE, Nollet F, Bus SA. Pressure-reduction and preservation in custom-made footwear of patients with diabetes and a history of plantar ulceration. Diabet Med 2012;29(12):1542–9.
[23] van Netten JJ, Raspovic A, Lavery LA, Monteiro-Soares M, Rasmussen A, Sacco ICN, et al. Prevention of foot ulcers in the at-risk patient with diabetes: a systematic review. Diabetes Metab Res Rev 2020;36(Suppl. 1):e3270.
[24] Cavanagh PR, Bus SA. Off-loading the diabetic foot for ulcer prevention and healing. J Am Podiatr Med Assoc 2010;100(5):360–8.
[25] Bus SA, Waaijman R, Nollet F. New monitoring technology to objectively assess adherence to prescribed footwear and assistive devices during ambulatory activity. Arch Phys Med Rehabil 2012;93(11):2075–9.
[26] Owings TM, Apelqvist J, Stenstrom A, Becker M, Bus SA, Kalpen A, et al. Plantar pressures in diabetic patients with foot ulcers which have remained healed. Diabet Med 2009;26(11):1141–6.
[27] Zwaferink JBJ, Custers W, Paardekooper I, Berendsen HA, Bus SA. Optimizing footwear for the diabetic foot: data-driven custom-made footwear concepts and their effect on pressure relief to prevent diabetic foot ulceration. PLoS One 2020;15(4):e0224010.
[28] Waaijman R, Keukenkamp R, de haart M, Polomski WP, Nollet F, Bus SA. Adherence to wearing prescription custom-made footwear in patients with diabetes at high risk for plantar foot ulceration. Diabetes Care 2013;36(6):1613–8.
[29] Crews RT, Shen BJ, Campbell L, Lamont PJ, Boulton AJ, Peyrot M, et al. Role and determinants of adherence to off-loading in diabetic foot ulcer healing: a prospective investigation. Diabetes Care 2016;39(8):1371–7.
[30] Keukenkamp R, Merkx MJ, Busch-Westbroek TE, Bus SA. An explorative study on the efficacy and feasibility of the use of motivational interviewing to improve footwear adherence in persons with diabetes at high risk for foot ulceration. J Am Podiatr Med Assoc 2018;108(2):90–9.
[31] Binning J, Woodburn J, Bus SA, Barn R. Motivational interviewing to improve adherence behaviours for the prevention of diabetic foot ulceration. Diabetes Metab Res Rev 2019;35(2):e3105.
[32] Keukenkamp R, van Netten JJ, Busch-Westbroek T, Bus SA. Custom-made footwear for indoor use increases adherence in people at high-risk for ulceration. The Hague, Netherlands: Proceedings of the International Symposium Diabetic Foot; May 2019. p. 22–5.
[33] Rinkel WD, Luiten J, van Dongen J, Kuppens B, Van Neck JW, Polinder S, et al. In-hospital costs of diabetic foot disease treated by a multidisciplinary foot team. Diabetes Res Clin Pract 2017;132:68–78.

[34] Prompers L, Huijberts M, Schaper N, Apelqvist J, Bakker K, Edmonds M, et al. Resource utilisation and costs associated with the treatment of diabetic foot ulcers. Prospective data from the Eurodiale Study. Diabetologia 2008;51(10):1826—34.

[35] Yavuz M, Brem RW, Davis BL, Patel J, Osbourne A, Matassini MR, et al. Temperature as a predictive tool for plantar triaxial loading. J Biomech 2014;47(15):3767—70.

[36] Bus SA. The role of pressure offloading on diabetic foot ulcer healing and prevention of recurrence. Plast Reconstr Surg 2016;138(3 Suppl. l):179S—87S.

CHAPTER 22

Compounding effects of diabetes in vessel formation in microvessel fragment—based engineered constructs

Omar Mourad, Blessing Nkennor, Sara S. Nunes
Toronto General Hospital Research Institute, University Health Network, Toronto, ON, Canada

Introduction

Engineered tissues require adequate blood flow in order to facilitate their survival and function post-transplantation. While some engineered tissues have been successfully used in the clinic, including urethral [1], tracheal [2], vaginal [3], and nasal cartilage tissues [4], a major hurdle in the advancement of the field is the lack of effective vascularization [5]. Current studies on vascularization have used single cells, including endothelial cells and endothelial progenitor cells, or cocultures of different cell types [6]. Challenges to cell-based studies include technical issues involved in isolating and culturing single cells as well as their functional immaturity in contrast to native vessels such as is the case of human pluripotent stem cell—derived endothelial cells.

Another approach for encouraging angiogenesis in implanted tissue constructs is the addition of growth factors such as vascular endothelial growth factor (VEGF) and basic fibroblast growth factor (bFGF) [7–9]. Alternatively, engineered biomaterials have been used to stimulate angiogenesis in vascular cells [5,10,11]. For example, scaffold modifications, such as pore size or polymer composition, have been manufactured to encourage vascular ingrowth [12,13]. However, a major problem inherent in both approaches is the slow rate of vascularization in vivo, especially in the context of large engineered tissues, which become subject to ischemia caused by prolonged periods of hypoxia [5]. Moreover, these systems yield immature vessels that either become substituted by host cells or regress once the angiogenic stimulus ceases [14].

Isolated microvessel fragments

Isolated microvessel fragments (MFs) represent an alternative vascularization approach for tissue engineering. This system relies on the isolation of existing tissue microvessels using a limited collagenase digestion approach and a series of filtration steps to obtain ready-made MFs consisting predominantly of endothelial and perivascular cells in vessel form. They can be used to generate vascular networks within engineered tissues to enable rapid inosculation with the host microcirculation and perfusion with blood [5,15]. The isolation and suspension of MFs in 3D collagen type I gels was initially described by Hoying et al. in 1996. In brief, epididymal fat from retired breeder rats was surgically removed after euthanasia, minced, then subjected to a limited collagenase digestion. After centrifugation, buoyant adipocytes were discarded, and the resulting pellet was resuspended, washed, and subjected to sieving to remove single cells and to harvest highly pure intact vessel fragments (Fig. 22.1). These fragments can then be resuspended in collagen type I gels and grown in vitro where they will undergo angiogenic sprouting without the requirement of exogenous growth factors. MFs in 3D constructs can also be used for in vivo implantation [16]. It's been shown that MFs isolated from mice, rats, and humans generate effectively vascularized constructs in vivo [15–18].

MFs are an attractive vascularization strategy for regenerative medicine applications for several reasons. First, isolated MFs contain all cell types present in native vessels, including endothelial and perivascular cells; in addition, mesenchymal cells and endothelial progenitor cells from the adipose tissue are also present in isolated fragments [19]. Flow cytometry of dissociated mouse MFs revealed that they contain 21% CD73-positive cells and 5% CD117-positive cells, both of which are mesenchymal cell markers, in addition to 5% Sca-1/VEGFR-2-positive cells, which is indicative of endothelial progenitors [19]. Second, MFs can inosculate with host vessels and develop into perfused networks stabilized by perivascular cells [18,19]. These characteristics make them ideal for application in regenerative medicine in multiple diseases, including type 1 diabetes (T1D), which is known to cause severe damage to the microvasculature. For example, T1D significantly reduces vessel sprouting in peripheral blood vessels, with a notable

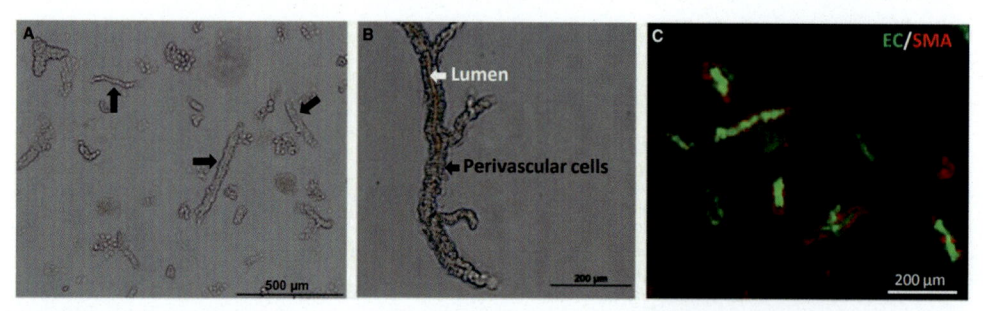

Figure 22.1 (A) Phase contrast micrograph of isolated microvessel fragments (MFs) from mouse brain cortex (*arrows*). (B) Magnified image of a brain MF showing a vessel lumen and perivascular cell coverage. (C) Brain MFs isolated from tie2-GFP transgenic mice. Only endothelial cells (ECs) express GFP; perivascular muscle cells are visualized via staining for α-smooth muscle actin (SMA—red). (*Reproduced from Nunes SS, Krishnan L, Gerard CS, Dale JR, Maddie MA, Benton RL, Hoying JB. Angiogenic potential of microvessel fragments is independent of the tissue of origin and can be influenced by the cellular composition of the implants. Microcirculation (New York, N.Y.: 1994) 2010;17(7):557–567. https://doi.org/10.1111/j.1549-8719.2010.00052.x with permission*).

exception being diabetic retinopathy, in which vessel overgrowth, disorganization, and pericyte dropout are observed [20,21]. Here we describe the vascularization potential of MFs and summarize the effects of diabetes in 3D grafts generated with MFs isolated from rat adipose tissue. We identify specific cellular and molecular contributors to reduced vascularization in both in vitro and in vivo models of T1D at different stages of vessel formation in engineered tissues.

Elements affecting microvessel fragment vascularization outcomes

Isolated MFs are composed of a heterogenous population of capillaries, arterioles, and venules varying in diameter and ranging mostly between 40 and 180 μm in length (measured immediately after isolation) [22,23]. MFs have α-smooth muscle actin-positive perivascular cells encircling them [16,18,19] and have been shown to secrete factors that improve their regenerative capacity such as VEGF, bFGF, and hepatocyte growth factor (HGF) [19,23–25]. Many different factors influence MF-derived angiogenesis and vascularization. Examples include: the extracellular matrix [26,27], donor age [28], biomechanics [29], and MF seeding density [30].

Two frequently used matrices in MF studies are porous synthetic scaffolds and collagen-based hydrogels. Typically, synthetic scaffolds rely on the body as a natural source of microvessels after implantation in highly vascularized regions [31]. Incorporating MFs from mouse adipose tissue into a porous polyurethane scaffold improved vascularization in mice postimplantation [19]. A major hindrance to the use of synthetic scaffolds is the risk of infiltration of inflammatory cells, subsequently leading to fibrosis surrounding the scaffold. Collagen-based matrices, on the other hand, offer a biocompatible approach to tissue vascularization. Collagen has been shown to support the organization of diverse vascular cell types into 3D structures, including endothelial cells and vessel fragments [16,32]. Additionally, MFs embedded in collagen have shown success in improving vascular supply in volumetric muscle loss injury [23] and full-thickness skin defects in mice [33–35]. Expanded polytetrafluoroethylene (ePTFE) disks embedded in an MF-containing collagen gel significantly reduced the presence of inflammatory cells in contrast to ePTFE embedded in collagen without MFs [36]. Collagen's relative biocompatibility and abundance within the body make it an ideal hydrogel substrate for MFs in tissue engineering applications [37].

Mechanisms controlling engineered vessel function and maturation

Several physiological, biological, and functional properties of MFs have been defined in recent literature, solidifying their potential as a novel vascularization method for applications in tissue engineering. In this section, we describe specific stages of MF growth and maturation in vivo, and in the following section, define how T1D affects specific stages.

Connecting the pieces: from isolated MFs to a functional neovasculature

Using MFs isolated from the adipose tissue of tie2:GFP^{Sato} mice [38], which are transgenic mice expressing GFP under the Tie2 promoter and thus mark endothelial cells, the stages of vascularization postimplantation of MFs can be easily monitored. During the first week postimplantation, MFs have been shown to undergo sprouting angiogenesis. This process was shown to be driven by HGF, which is intrinsically secreted by MFs in vitro [24], as HGF sequestration by anti-HGF antibodies or inhibition of the HGF receptor c-Met with selective small molecule inhibitors decreases MF angiogenic sprouting in a dose-dependent manner. MFs are also in a state of high proliferation, low apoptosis, and low perivascular cell coverage (Figs. 22.2 and 22.3A) [15]. During this time, the vessels have a relatively homogenous diameter and display high expression of angiogenesis-related genes, such as *MMP14*, and downregulation of antiangiogenic genes and vessel maturation genes, such as *THBS2* and *ANG1*, respectively [15]. Collectively, these data indicate that MFs are in a sprouting angiogenesis stage. In the second week postimplantation, MFs are in a state of moderate proliferation, high apoptosis, and show increased perivascular cell coverage (Fig. 22.3B) [15]. Vessel diameter is still relatively homogenous at this time, and MFs display lower expression of angiogenic genes, with high expression of destabilization genes such as *ANG2*. This stage is not simply an "intermediate" one, as different clusters of genes are differentially expressed specifically in this stage compared to either the preceding or following stages [15]. Of note is that this stage only begins once blood perfusion begins. It is thought that blood flow is an essential driver for MFs to enter this stage; in fact, the exact days that define these different vascularization

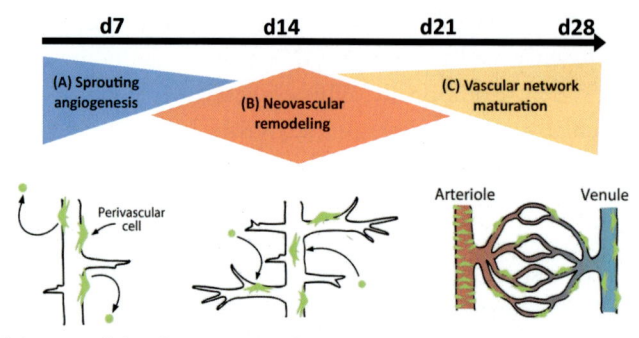

Figure 22.2 A schematic of the three stages of microvessel fragment–based engineered construct maturation in healthy conditions: sprouting angiogenesis, neovascular remodeling, and vascular maturation. (A) The sprouting angiogenesis stage exhibits high perivascular detachment, increased hepatocyte growth factor (HGF) secretion, high proliferation, low apoptosis, and perfusion. (B) The neovascular remodeling stage exhibits increased perivascular cell recruitment in a PDGF-dependent manner, perivascular immaturity, high proliferation, and perfusion. (C) The final stage, vascular maturation, is Notch3/Jagged1 dependent and exhibits a heterogeneous diameter in capillaries, as well as perivascular maturation and high perfusion.

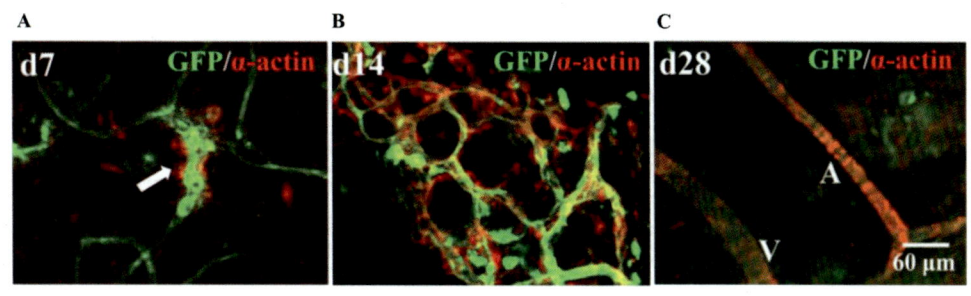

Figure 22.3 Neovascular stages in implanted 3D microvessel constructs. Confocal images of microvessel constructs at day 7 (A), day 14 (B), and day 28 (C) postimplantation show the distinct phases of neovascularization and their associated network architecture. Perivascular cells were stained for smooth muscle alpha-actin (red). Endogenous GFP fluorescence or Alexa 488—conjugated lectin GS-1 was used to stain endothelial cells in green. A, arteriole; V, venule. Arrow in day 7 panel indicates parent microvessel fragments. *(Reproduced from Nunes SS, Greer KA, Stiening CM, Chen HYS, Kidd KR, Schwartz MA, Sullivan CJ, Rekapally H, Hoying JB. Implanted microvessels progress through distinct neovascularization phenotypes. Microvasc Res 2010;79(1):10—20.https://doi.org/10.1016/j.mvr.2009.10.001 with permission).*

stages are not precise, and the appearance of blood is a more robust indicator of the termination of the angiogenic stage and the beginning of this "vascular remodeling" phase. In vitro, only the sprouting angiogenesis stage has been recapitulated, likely due to the absence of blood flow [15]. From around the start of the third week to the end of the fourth week postimplantation, MFs are in a state of low proliferation, low apoptosis, and a perivascular coverage indicative of mature vessels (Fig. 22.3C) [15]. The vessels now have heterogenous diameters, indicative of a hierarchical vasculature consisting of arterioles, venules, and capillaries. The expression levels of stabilization genes, such as *THBS2* and *ANG1*, are increased. In summary, MF constructs display three distinct vascular phenotypes post subcutaneous implantation in mice: sprouting angiogenesis, neovascular remodeling, and network maturation [15]. These different stages have been summarized in Fig. 22.2.

Characterization of vessel arterial—venous plasticity in MF-based constructs

One of the most striking characteristics of a mature vascular bed is the presence of a stereotypical tree structure with vessels of different arterial—venous (AV) identity: arterioles, venules, and capillaries. This is particularly relevant to neovascularization applications in which vessels need to adopt a mature and functional phenotype and appropriate arterial and venous identities for the generation of a functional vascular network. Isolated MFs from mouse brain and rat epididymal fat, which include arterioles, venules, and capillaries, undergo AV identity changes during neovascular sprouting angiogenesis [22]. For example, constructs have been generated with MFs exclusively of arterial identity from Efnb2-GFP reporter mice for implantation. Arterial MFs were isolated using magnetic beads, which were injected into the left carotid artery of mice where they become lodged in arterioles as the lumens become progressively smaller. These arterial microvessels were implanted into SCID mice and formed a vessel network composed of all microvessel types by 6-week postimplantation [22]. Also, using both brain and adipose-derived MFs, it was shown that during the sprouting angiogenesis phase, MFs show homogeneous diameter distribution and undergo identity changes characterized by Efnb2 (a marker of arterial identity) or Ephb4 (a marker of venule identity) reporter expression [22]. Overall, although there is an initial loss of AV identity around week 1 of neovascularization, as neovessels mature over time (week 6), they attain specific arterial or venous identities with specific arteriosclerosis—venous reporter marker expression. Therefore, AV identity during adult 3D-engineered tissue neovascularization using MFs displays phenotypic plasticity and is not necessarily predetermined.

Impact of diabetes on engineered vessel function

It is known that impaired angiogenesis is a common feature in diabetes, and that vascular complications account for a major subset of diabetes-related morbidities. Examples include diabetic retinopathy, impaired wound healing, microvascular rarefaction in the heart, and diabetic nephropathy, among many others [39]. Thus, the behavior and application of MF tissue constructs in diabetes are of significant interest, as they may be applicable in the abovementioned diseases as a regenerative medicine strategy. Early reports have used MFs to encourage vascularization of pancreatic islets for T1D applications. Hiscox et al. [40] developed a prevascularized pancreatic encapsulating device (PPED) to study the ability of prevascularized collagen constructs containing rat fat-derived MFs in preserving islet viability in vivo. To construct PPEDs, isolated MFs from rat epididymal fat were resuspended in type I collagen gels [16] then incubated in vitro for 7 days. Islets were isolated from rats and suspended in collagen type I. These gels were then sandwiched between two of the preincubated MF constructs to form the PPED and then implanted in immunodeficient mice [40]. PPEDs were able to increase islet survival for 4-week postimplantation compared to islets in collagen without MFs. In vitro analysis showed increased insulin secretion of PPEDs in response to glucose stimulation compared to free islets [40]. Although PPEDs show encouraging evidence for supporting islet function and survival, these findings need to be reexamined in the context of diabetes, i.e., using diabetic mice as hosts for implantation.

Type 1 diabetes delays sprouting in engineered grafts

To assess the potential effects of T1D in MF-based vascularization, the ability of MFs to form a mature vasculature upon implantation was assessed in vivo. MFs were isolated from nondiabetic (ND) mice and implanted into ND hosts as controls while microvessels from streptozotocin (STZ)-induced diabetic (D) mice were isolated and implanted into

D hosts (Fig. 22.4A). Between 6 and 9 days postimplantation, during the sprouting angiogenesis/early remodeling stages [15], MF constructs were removed, and RNA was extracted for qPCR analysis (Fig. 22.4B). Diabetic MF constructs showed reduced expression of KLF2, a flow-sensitive gene (Fig. 22.4C). This is indicative of delayed blood perfusion and therefore inosculation with host vessels [24]. Constructs explanted from D mice showed a significant reduction in HGF secretion, with no changes in c-MET, VEGF, and VEGF receptor 2 (VEGFR2) expression, indicating that T1D suppresses HGF secretion, which can impair angiogenic sprouting and therefore be responsible for the delay in inosculation with host vessels and in blood perfusion in the grafts (Fig. 22.4C) [24]. In vitro, cultivation of MFs in high glucose media significantly reduced angiogenic sprouting. This could be reversed by exogenous HGF supplementation in a dose-dependent manner [24]. Therefore, we can conclude that T1D delays the angiogenic growth of MFs, and that this effect is reversible [24].

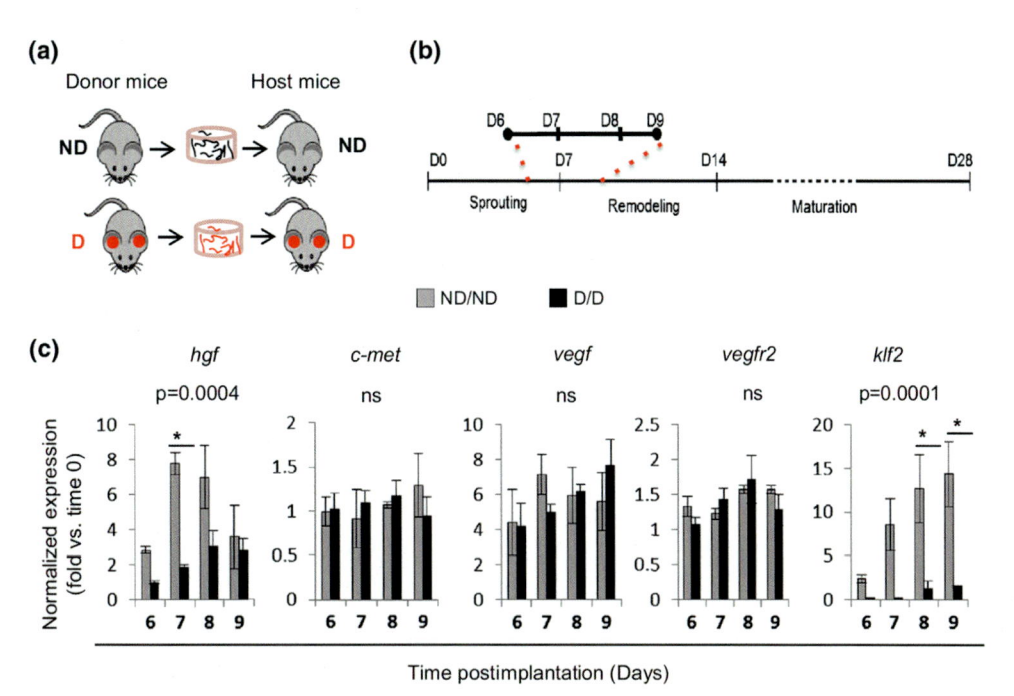

Figure 22.4 HGF secretion in prevascularized constructs implanted in vivo is suppressed in diabetic conditions. (A) An overview of vessel isolation. Microvessels were isolated from nondiabetic (ND) controls and implanted into ND hosts while microvessels from diabetic mice (D) were isolated and implanted into D hosts. (B) Experimental timeline. At 6–9 days postimplantation, explants were harvested, and RNA was extracted for analysis. (C) qPCR gene expression analysis during sprouting angiogenesis in both ND (gray) and D explants (black). *(Reproduced from Altalhi W, Hatkar R, Hoying JB, Aghazadeh Y, Nunes SS. Type I diabetes delays perfusion and engraftment of 3D constructs by impinging on angiogenesis; which can be rescued by hepatocyte growth factor supplementation. Cell Mol Bioeng 2019;12(5):443–454. https://doi.org/10.1007/s12195-019-00574-3 with permission).*

Type 1 diabetes affects vessel maturation in 3D-engineered constructs

Engineered constructs derived from EphrinB2-GFP reporter mice and implanted into ND control mice possess both EphrinB2 positive and negative vessels, which indicate the presence of arterioles and venules since EphrinB2 is an arterial marker (Fig. 22.5A). These vessels display blood perfusion and therefore anastomosis, visualized by fluorescent dextran perfusion after injection into host mice. Vessels isolated from EphrinB2-GFP reporter D mice and implanted into STZ-induced D mice display ubiquitous EphrinB2 expression 4 weeks after implantation. This ubiquitous expression pattern, along with the uniformly small vessel diameters, is indicative of immature vessels (Fig. 22.5A) [41]. In addition, MFs from D mice implanted into D mice display significantly reduced perivascular coverage, with perivascular cells loosely associated with vessels as shown by α-smooth muscle actin staining (Fig. 22.5B). In D mice, qPCR data of MFs postexplant show reduced expression of Jagged1 and Notch3 (Fig. 22.5C). Notch3 is expressed by perivascular cells, whereas Jagged1 is expressed in both endothelial and perivascular cells [41]. This provides evidence that perivascular cell interaction with endothelial cells is mediated by signaling between endothelial cell–derived Jagged1 and perivascular cell–derived Notch3 [42,43]. Furthermore, engineered constructs implanted in ND mice in which perivascular cell recruitment is blocked (by injecting anti-PDGFRb antibodies or a selective PDGFRb inhibitor) also display impaired AV specification and decreased perivascular cell coverage [41]. Therefore, T1D impinges on vessel maturation in MF constructs in vivo by decreasing perivascular cell coverage and impeding AV specification [41].

Future directions

Although we know and understand different processes that initiate or delay the neovascularization process in MF-based engineered constructs studied in healthy and diabetic conditions (Figs. 22.2 and 22.6), a number of questions remain unanswered. For instance, how do graft and host cells inosculate to form a single vasculature? What are the molecular cues that lead to effective graft–host connection? One key aspect to elucidate is how different tissue-specific parenchymal cells affect the vascularization potential and characteristics of MFs. For example, the inclusion of glial-restricted precursor cells in constructs containing adipose tissue–derived microvessel promotes morphological changes in vessels that lead to a brain-like phenotype. Specifically, they express the brain endothelial marker Glut-1 and exhibit reduced permeability [44]. Conversely, MFs isolated from rodent brain cortices are less permeable than those from adipose tissue, even in the absence of astrocytes, which are mediators of the blood–brain barrier [44]. Clearly, there are properties inherent to MFs that are retained postisolation as well as properties that depend on the fragments' microenvironment during implantation. These factors should be studied in detail and can inform the best tissue source for various tissue engineering applications, as well as which supporting cells to include.

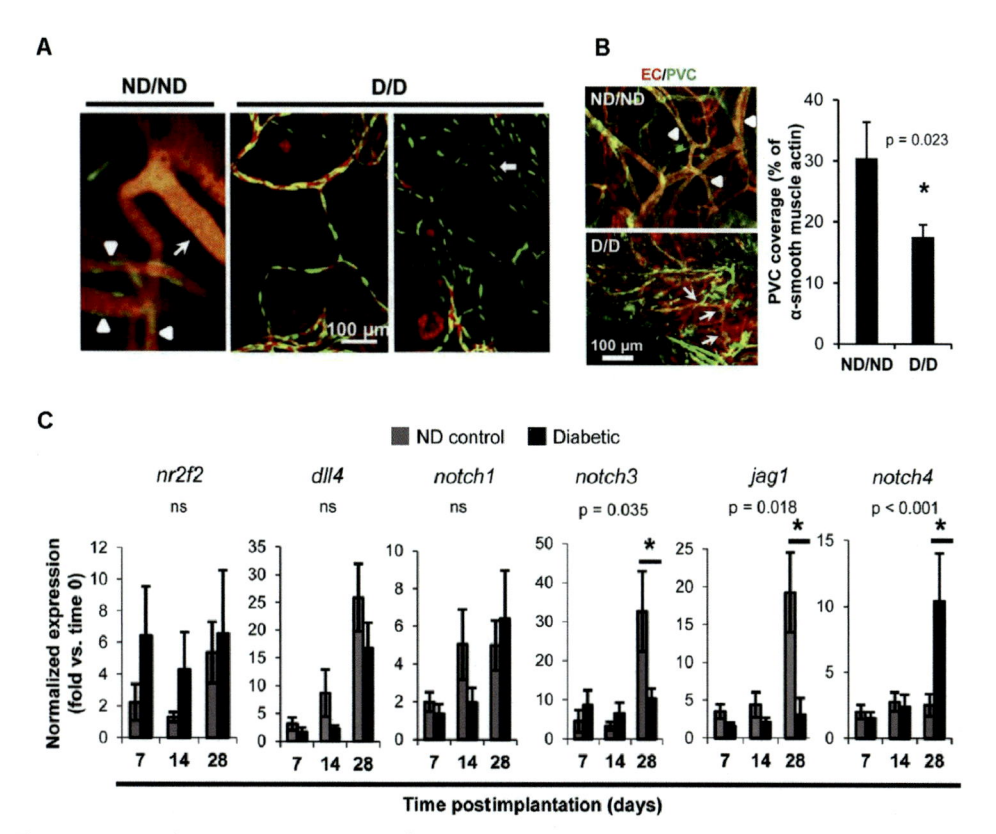

Figure 22.5 Diabetes impairs AV specification and downregulates the expression of endothelial-perivascular cell—cell interaction molecules Jagged1 and Notch3. (A) Engineered constructs derived from EphrinB2-GFP ND mice and implanted into ND mice display typical AV identities with EphrinB2 positive and negative vessels (*arrowheads* and *arrow*, respectively). Vessels isolated from EphrinB2-GFP reporter diabetic (D) mice and implanted into D mice display ubiquitous EphrinB2 expression. (B) Percent perivascular cell (green) coverage of microvessels (endothelial cells, red) 4 weeks after implantation shows lower perivascular coverage compared to nondiabetic mice conditions (*arrowheads*). (C) qPCR Analysis of the expression of AV signaling molecules throughout neovascularization. Explants were harvested at different time points for RNA analysis. All conditions were normalized to both TBP (a housekeeping gene) and to time zero expression. *(Reproduced from Altalhi W, Sun X, Sivak JM, Husain M, Nunes SS. Diabetes impairs arterio-venous specification in engineered vascular tissues in a perivascular cell recruitment- dependent manner. Biomaterials 2019;119:23—32, (2017) with permission).*

A key difference between typical angiogenesis in the body and the sprouting phenotype in MFs is blood flow. MFs do not experience blood flow until inosculation with the host vasculature. In contrast, sprouts from established host vasculature fill with blood as angiogenesis occurs [45]. Mechanical force, namely, shear stress due to blood flow, is an important regulator of microvascular function. Shear stress influences vascular remodeling and endothelial cell turnover, as well as nitric oxide production [46]. How these, as

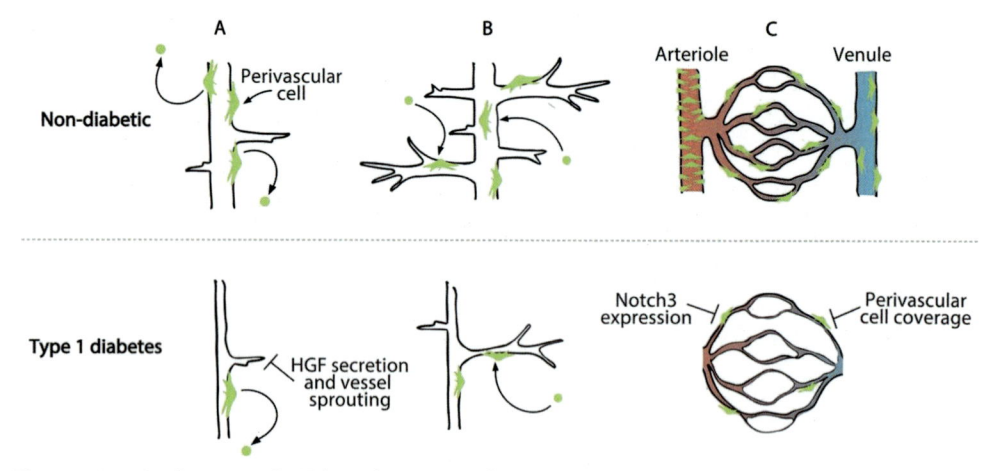

Figure 22.6 A schematic of MF-based engineered construct maturation in type 1 diabetes. (A) In the early stages of neovascularization, vessels exhibit reduced HGF secretion, sprouting, (B) low perivascular cell recruitment and delayed blood perfusion. (C) In the final stage, vessels exhibit a uniformly small diameter, immature AV identity, as well as low perivascular cell coverage evident through reduced Notch3 expression.

well as other molecular and cellular contributors to native microvascular homeostasis, might affect the future translational applications of MFs is not presently known. One positive effect of this phenomenon, however, is that we can study the discrete stages of vascularization of MFs in a more defined, compartmentalized manner.

Treatment of isolated MFs with various factors preimplantation has been studied in attempts to determine ways to improve vascularization. For example, subnormothermic short-term cultivation [47], high glucose exposure [48], insulin-like growth factor treatment [49], and erythropoietin treatment [50] of isolated MFs have all been shown to improve vascularization after they are implanted into mice. Interestingly, other findings have suggested that cultivating MFs in high glucose decreases their angiogenic potential, leading to delayed sprouting [24]. These constructs, when implanted into diabetic mice, show delayed perfusion [24]. It may be possible that specifically moving from a high glucose environment in vitro to a normoglycemic environment postimplantation has a beneficial effect. This has implications for tissue engineering applications in patients with diabetes; pretreatment of MFs with glucose may in fact be deleterious in this case. Additionally, it has been shown that MFs can be cryopreserved while maintaining their ability to form sprouts and connect with the host's vasculature in vivo [51]. Novel treatments, or combinations of treatments, may be beneficial in the applications of MFs in tissue engineering; it would be interesting to evaluate how these treatments and others may improve engineered tissue vascularization in patients with diabetes.

References

[1] Raya-Rivera A, Esquiliano DR, Yoo JJ, Lopez-Bayghen E, Soker S, Atala A. Tissue-engineered autologous urethras for patients who need reconstruction: an observational study. Lancet 2011;377: 1175–82.

[2] Elliott MJ, De Coppi P, Speggiorin S, Roebuck D, Butler CR, Samuel E, et al. Stem-cell-based, tissue engineered tracheal replacement in a child: a 2-year follow-up study. Lancet 2012;380:994–1000.

[3] Raya-Rivera AM, Esquiliano D, Fierro-Pastrana R, López-Bayghen E, Valencia P, Ordorica-Flores R, et al. Tissue-engineered autologous vaginal organs in patients: a pilot cohort study. Lancet 2014;384:329–36.

[4] Fulco I, Miot S, Haug MD, Barbero A, Wixmerten A, Feliciano S, et al. Engineered autologous cartilage tissue for nasal reconstruction after tumour resection: an observational first-in-human trial. Lancet 2014;384:337–46.

[5] Laschke MW, Menger MD. Prevascularization in tissue engineering: current concepts and future directions. Biotechnol Adv 2016;34(2):112–21. https://doi.org/10.1016/j.biotechadv.2015.12.004.

[6] Sun X, Altalhi W, Nunes SS. Vascularization strategies of engineered tissues and their application in cardiac regeneration. Adv Drug Deliv Rev 2016;96:183–94. https://doi.org/10.1016/j.addr.2015.06.001.

[7] Laschke MW, Rücker M, Jensen G, Carvalho C, Mülhaupt R, Gellrich NC, et al. Incorporation of growth factor containing Matrigel promotes vascularization of porous PLGA scaffolds. J Biomed Mater Res A 2008;85:397–407.

[8] Phelps EA, Garcia AJ. Update on therapeutic vascularization strategies. Regen Med 2009;4(1):65–80.

[9] Singh S, Wu BM, Dunn JC. Delivery of VEGF using collagen-coated polycaprolactone scaffolds stimulates angiogenesis. J Biomed Mater Res A 2012;100:720–7.

[10] Fischbach C, Mooney DJ. Polymers for pro- and anti-angiogenic therapy. Biomaterials 2007;28(12): 2069–76.

[11] Sun X, Nunes SS. Overview of hydrogel-based strategies for application in cardiac tissue regeneration. 2015.

[12] Choi SW, Zhang Y, Macewan MR, Xia Y. Neovascularization in biodegradable inverse opal scaffolds with uniform and precisely controlled pore sizes. Adv Healthcare Mater 2013;2:145–54.

[13] Rücker M, Laschke MW, Junker D, Carvalho C, Schramm A, Mülhaupt R, et al. Angiogenic and inflammatory response to biodegradable scaffolds in dorsal skinfold chambers of mice. Biomaterials 2006;27:5027–38.

[14] Sun X, Evren S, Nunes SS. Blood vessel maturation in health and disease and its implications for vascularization of engineered tissues. Crit Rev Biomed Eng 2015;43(5–6):433–54. https://doi.org/10.1615/CritRevBiomedEng.2016016063.

[15] Nunes SS, Greer KA, Stiening CM, Chen HYS, Kidd KR, Schwartz MA, Sullivan CJ, Rekapally H, Hoying JB. Implanted microvessels progress through distinct neovascularization phenotypes. Microvasc Res 2010;79(1):10–20. https://doi.org/10.1016/j.mvr.2009.10.001.

[16] Hoying JB, Boswell CA, Williams SK. Angiogenic potential of microvessel fragments established in three-dimensional collagen gels. In Vitro Cell Dev Biol Anim 1996;32(7):409–19. https://doi.org/10.1007/bf02723003.

[17] Krishnan L, Hoying JB, Nguyen H, Song H, Weiss JA. Interaction of angiogenic microvessels with the extracellular matrix. Am J Physiol Heart Circ Physiol 2007;293(6):H3650–8.

[18] Shepherd BR, Chen HY, Smith CM, Gruionu G, Williams SK, Hoying JB. Rapid perfusion and network remodeling in a microvascular construct after implantation. Arterioscler Thromb Vasc Biol 2004;24(5):898–904. https://doi.org/10.1161/01.atv.0000124103.86943.1e.

[19] Laschke M, Kleer S, Scheuer C, Schuler S, Garcia P, Eglin D, et al. Vascularisation of porous scaffolds is improved by incorporation of adipose tissue-derived microvascular fragments. Eur Cell Mater 2012; 24:266–77. https://doi.org/10.22203/ecm.v024a19.

[20] Mizutani M, Kern TS, Lorenzi M. Accelerated death of retinal microvascular cells in human and experimental diabetic retinopathy. J Clin Invest 1996;97:2883–90.

[21] Rask-Madsen C, King GL. Vascular complications of diabetes: mechanisms of injury and protective factors. Cell Metabol 2013;17:20−33. https://doi.org/10.1016/j.cmet.2012.11.012.
[22] Nunes SS, Rekapally HR, Chang CC, Hoying JB. Vessel arterial-venous plasticity in adult neovascularization. PLoS One 2011;6(11):e27332.
[23] Pilia M, et al. Transplantation and perfusion of microvascular fragments in a rodent model of volumetric muscle loss injury. Eur Cell Mater 2014;28:11−23.
[24] Altalhi W, Hatkar R, Hoying JB, Aghazadeh Y, Nunes SS. Type I diabetes delays perfusion and engraftment of 3D constructs by impinging on angiogenesis; which can be rescued by hepatocyte growth factor supplementation. Cell Mol Bioeng 2019;12(5):443−54. https://doi.org/10.1007/s12195-019-00574-3.
[25] McDaniel JS, et al. Characterization and multilineage potential of cells derived from isolated microvascular fragments. J Surg Res 2014;192:214−22.
[26] Edgar LT, Hoying JB, Utzinger U, Underwood CJ, Krishnan L, Baggett BK, Maas SA, Guilkey JE, Weiss JA. Mechanical interaction of angiogenic microvessels with the extracellular matrix. J Biomech Eng 2014;136(2):021001. https://doi.org/10.1115/1.4026471.
[27] Edgar LT, Underwood CJ, Guilkey JE, Hoying JB, Weiss JA. Extracellular matrix density regulates the rate of neovessel growth and branching in sprouting angiogenesis. PLoS One 2014;9(1):e85178. https://doi.org/10.1371/journal.pone.0085178.
[28] Laschke M, Grässer C, Kleer S, Scheuer C, Eglin D, Alini M, Menger M. Adipose tissue-derived microvascular fragments from aged donors exhibit an impaired vascularisation capacity. Eur Cell Mater 2014;28:287−98. https://doi.org/10.22203/ecm.v028a20.
[29] Chang CC, Krishnan L, Nunes SS, Church KH, Edgar LT, Boland ED, Weiss JA, Williams SK, Hoying JB. Determinants of microvascular network topologies in implanted neovasculatures. Arterioscler Thromb Vasc Biol 2012;32(1):5−14. https://doi.org/10.1161/ATVBAHA.111.238725.
[30] Später T, Körbel C, Frueh F, Nickels R, Menger M, Laschke M. Seeding density is a crucial determinant for the in vivo vascularisation capacity of adipose tissue-derived microvascular fragments. Eur Cell Mater 2017;34:55−69. https://doi.org/10.22203/eCM.v034a04.
[31] Laschke MW, Menger MD. Adipose tissue-derived microvascular fragments: natural vascularization units for regenerative medicine. Trends Biotechnol 2015;33(8):442−8. https://doi.org/10.1016/j.tibtech.2015.06.001.
[32] Montessano R, Orci L, Vassalli P. In vitro rapid organization of endothelial cells into capillary-like networks is promoted by collagen matrices. J Cell Biol 1983;97:1648−52.
[33] Frueh FS, Später T, Lindenblatt N, Calcagni M, Giovanoli P, Scheuer C, et al. Adipose tissue-derived microvascular fragments improve vascularization, lymphangiogenesis, and integration of dermal skin substitutes. J Invest Dermatol 2017;137(1):217−27. https://doi.org/10.1016/j.jid.2016.08.010.
[34] Frueh FS, Später T, Körbel C, Scheuer C, Simson AC, Lindenblatt N, Giovanoli P, Menger MD, Laschke MW. Prevascularization of dermal substitutes with adipose tissue-derived microvascular fragments enhances early skin grafting. Sci Rep 2018;8(1):10977. https://doi.org/10.1038/s41598-018-29252-6.
[35] Später T, Frueh FS, Nickels RM, Menger MD, Laschke MW. Prevascularization of collagen-glycosaminoglycan scaffolds: stromal vascular fraction versus adipose tissue-derived microvascular fragments. J Biol Eng 2018;12(1):24. https://doi.org/10.1186/s13036-018-0118-3.
[36] Gruionu G, et al. Encapsulation of ePTFE in prevascularized collagen leads to peri-implant vascularization with reduced inflammation. J Biomed Mater Res A 2010;95:811−8.
[37] Charriere G, Bejot M, Schnitzler L, Ville G, Hartmann DJ. Reactions to a bovine collagen implant. Clinical and immunologic study in 705 patients. J Am Acad Dermatol 1989;21:1203−8. https://doi.org/10.1016/S0190-9622(89)70330-3.
[38] Motoike T, Loughna S, Perens E, Roman BL, Liao W, Chau TC, Richardson CD, Kawate T, Kuno J, Weinstein BM, Stainier DY, Sato TN. Universal GFP reporter for the study of vascular development. Genesis 2000;28:75−81.
[39] Avogaro A, Fadini GP. Microvascular complications in diabetes: a growing concern for cardiologists. Int J Cardiol 2019;291:29−35. https://doi.org/10.1016/j.ijcard.2019.02.030.

[40] Hiscox AM, Stone AL, Limesand S, Hoying JB, Williams SK. An islet-stabilizing implant constructed using a preformed vasculature. Tissue Eng A 2008;14(3):433—40. https://doi.org/10.1089/tea.2007.0099.

[41] Altalhi W, Sun X, Sivak JM, Husain M, Nunes SS. Diabetes impairs arterio-venous specification in engineered vascular tissues in a perivascular cell recruitment- dependent manner. Biomaterials 2019;119:23—32. 2017.

[42] High FA, Lu MM, Pear WS, Loomes KM, Kaestner KH, Epstein JA. Endothelial expression of the Notch ligand Jagged1 is required for vascular smooth muscle development. Proc Natl Acad Sci U S A 2008;105(6). 1955e1959.

[43] Xia Y, Bhattacharyya A, Roszell EE, Sandig M, Mequanint K. The role of endothelial cell-bound Jagged1 in Notch3-induced human coronary artery smooth muscle cell differentiation. Biomaterials 2012;33(8). 2462e2472.

[44] Nunes SS, Krishnan L, Gerard CS, Dale JR, Maddie MA, Benton RL, Hoying JB. Angiogenic potential of microvessel fragments is independent of the tissue of origin and can be influenced by the cellular composition of the implants. Microcirculation 2010;17(7):557—67. https://doi.org/10.1111/j.1549-8719.2010.00052.x. New York, N.Y.: 1994.

[45] Carmeliet P. Mechanisms of angiogenesis and arteriogenesis. Nat Med 2000;6:389—95. https://doi.org/10.1038/74651.

[46] Xu S, Li X, LaPenna KB, Yokota SD, Huke S, He P. New insights into shear stress-induced endothelial signalling and barrier function: cell-free fluid versus blood flow. 2017. Retrieved from: https://academic.oup.com/cardiovascres/article/113/5/508/2965333.

[47] Laschke MW, Heß A, Scheuer C, Karschnia P, Menger MD. Subnormothermic short-term cultivation improves the vascularization capacity of adipose tissue-derived microvascular fragments. J Tissue Eng Regen Med 2019;13(2):131—42. https://doi.org/10.1002/term.2774.

[48] Laschke M, Seifert M, Scheuer C, Kontaxi E, Metzger W, Menger M. High glucose exposure promotes proliferation and in vivo network formation of adipose-tissue-derived microvascular fragments. Eur Cell Mater 2019;38:188—200. https://doi.org/10.22203/ecm.v038a13.

[49] Laschke MW, Kontaxi E, Scheuer C, Heß A, Karschnia P, Menger MD. Insulin-like growth factor 1 stimulates the angiogenic activity of adipose tissue—derived microvascular fragments. J Tissue Eng 2019;10. https://doi.org/10.1177/2041731419879837.

[50] Karschnia P, Scheuer C, Heß A, Später T, Menger M, Laschke M. Erythropoietin promotes network formation of transplanted adipose tissue-derived microvascular fragments. Eur Cell Mater 2018;35:268—80. https://doi.org/10.22203/ecm.v035a19.

[51] Laschke MW, Karschnia P, Scheuer C, Heß A, Metzger W, Menger MD. Effects of cryopreservation on adipose tissue-derived microvascular fragments. J Tissue Eng Regen Med 2017;12(4):1020—30. https://doi.org/10.1002/term.2591.

CHAPTER 23

Dressing selection challenges in diabetic foot local treatment

Paulo Alves[1,4], Tania Manuel[1,4], Nuno Mendes[2], Emília Ribeiro[2], Anabela Moura[3,4]

[1]Universidade Católica Portuguesa, Health Sciences Institute | Wounds Research Lab, Center for Interdisciplinary Research in Health (CIIS), Portugal; [2]Grupo Saúde Nuno Mendes, Portugal; [3]Centro Hospitalar de São João (CHSJ), Portugal; [4]Portuguese Wound Management Association (APTFeridas), Portugal

Therapeutic approach in DFUs

Diabetic foot ulcers (DFUs) management involves comprehensive multidisciplinary care. The National Institute for Health and Clinical Excellence [1] estimated that 10% of people with diabetes will develop a DFU during their life. Existing literature indicates that revascularization should always be considered in patients with DFUs and ischemia. However, it still remains unclear if such procedures have any added value for all patients including patients with mild-to-moderate perfusion deficits [2].

Due to the high cost associated with the treatment of these injuries, the impact of this problem is not limited only to the person, but to his family and the entire health system, so it is important that DFUs are treated promptly and properly to optimize healing conditions and minimize impact on patients' quality of life.

Although amputation is one of the most serious complications of DFUs, this can be reduced when multidisciplinary strategies are implemented in the area of prevention and treatment, from the balance of health status to the local treatment of the injury.

The treatment of DFUs has the ultimate goal of healing the lesion; however, it is not at all simple and wound dressings represent a part of the management of diabetic foot ulceration.

Treatment is a dynamic and complex process that implies the management of several more global factors of the patient, up to the implementation of an individualized treatment plan adapted to the needs of the user and its unique characteristics. In their global approach, healthcare professionals should account for the risk factors for foot ulceration, classification and grading of wounds, bacteriology, multidisciplinary team approach, types of debridement, and importance of offloading. We will focus our attention here on the wound care and choices based on the complexity of the wound and properties of the dressing regime in each category. Wound care plays a pivotal role in the management of DFU, which comprises cleaning the wound and promoting a moist wound-healing environment [3–5]. To add to this we should always keep in mind the words of Armstrong [6]: "It's not what you put on, but what you take off: techniques for debriding and off-loading;" the dressing alone does not fulfill all the requirements.

Local wound care, and in particular the selection of dressing materials, is of fundamental importance and the choice should be guided by the characteristics of the lesion, the necessities of the patient, and costs.

Wound dressings

Ulcer treatment is the most time-consuming activity for professionals in the field of diabetic foot treatment. However, local treatment of the ulcer is only part of the treatment, with health professionals aware that glycemic balance, infection, pain, oxygenation, and nutritional support are also essential factors to be corrected. Still, if there are stages of ulcer healing that is less demanding in the use of a dressing material, others are particularly demanding in this regard. The choice of dressing material suitable for the treatment of DFUs is a decisive factor for its success, as well as its continuous evaluation. There is no conventional guideline regarding the selection of wound care materials in diabetic foot wounds [5]; however, there is a huge variability and quantity of materials available on the world market, so your selection should take into account, namely, the location of the wound, the shape, the size, the depth, the type of tissue present, the quantity exudate, and the presence of signs of infection, without forgetting an economic component, with regard to its cost—benefit.

It has been recognized that a moist wound environment is optimal for wound healing, but dressings are not a substitute for sharp debridement, managing systemic infection, offloading devices, and control of the pathology. The dressing material should create optimal conditions for the wound to heal, hence a continuous assessment is essential, since the characteristics of the wound bed vary throughout the treatment, and it is necessary to adapt the treatment to the wound stage.

No single dressing fulfills all the needs of a diabetic patient; however, each category of dressings has particular characteristics that aid the selection. Table 23.1 summarizes the main groups of dressing materials simply adding the main aspects without considering effectiveness).

All dressings require frequent change for wound inspection, and heavily exuding ulcers require more frequent dressing changes to reduce maceration of surrounding skin.

The dressings are responsible to maintain a balanced wound environment that is not too moist or too dry, so they should help to manage wound exudates optimally and promote a balanced environment, as a key to improving outcomes.

There are some other problems regarding dressings for treating DFUs because the majority of the dressings are designed for nonfoot areas of the body and may be difficult to apply between or over the toes or plantar surface [7].

Table 23.1 Dressing materials and main characteristics.

Dressing materials (groups)	Description
Nonadhesive	Well tolerated, simple, and inexpensive.
	Fixing them becomes a more complicated task.
Foams	Highly absorbent and effective for heavily exuding wounds.
	Have a wide range of absorbency, provide thermal insulation, and are easily cut to shape.
Alginate	Highly absorbent and effective for heavily exuding wounds.
	May need wetting before removal.
Hydrogels	Facilitate autolysis and may be beneficial in managing ulcers containing necrotic tissue.
Iodine and silver	May aid in managing wound infection.
Hydrocolloids	Their use for highly exudative wounds can lead to maceration of the surrounding skin.
	Concerns persist regarding their use for infected wounds.

Since there is a lack of strong evidence of clinical or cost-effectiveness, healthcare professionals should use wound dressings that best match the clinical appearance and site of the wound, as well as patient preferences [1,7].

Before discussing how dressings influence the success of DFU treatments, we should take into account the global approach to the treatment, and the most recent list of recommendations from the 2019 International Working Group on the Diabetic Foot (IWGDF) Guidelines on the Prevention and Management of Diabetic Foot Disease [8] should be adopted (Table 23.2).

As previously said, the level of evidence on effectiveness of dressings is overall low and the table shows that when choosing a dressing material, three important factors must be taken into account: exudate control, comfort, and cost [8].

The dressing choice must begin with thorough patient and wound assessments, and the following factors should be considered: Location of the wound; Extent (size/depth) of the wound; The amount and type of exudate; The predominant tissue type on the wound surface; Condition of the periwound skin; Compatibility with other therapies (e.g., contact casts); Wound bioburden and risk of infection; Avoidance of pain and trauma at dressing changes; Quality of life and patient well-being [7].

It can be challenging to effectively dress wounds located on the foot and fingers, particularly in cases of deformity (Fig. 23.1) or disease such as Charcot foot, the dressings

Table 23.2 IWGDF Guideline on interventions to enhance healing of foot ulcers in persons with diabetes.

N	Recommendations	(GRADE strength of recommendation: Quality of evidence)
1	Remove slough, necrotic tissue, and surrounding callus of a diabetic foot ulcer with sharp debridement in preference to other methods, taking relative contraindications such as pain or severe ischemia into account.	(Strong; low)
2	Select dressings principally on the basis of exudate control, comfort, and cost.	(Strong; low)
3	Do not use dressings/applications containing surface antimicrobial agents with the sole aim of accelerating the healing of an ulcer.	(Strong; low)
4	Consider the use of the sucrose-octasulfate impregnated dressing in noninfected, neuroischemic diabetic foot ulcers that are difficult to heal despite best standard of care.	(Weak; moderate)
5	Consider the use of systemic hyperbaric oxygen therapy as an adjunctive treatment in nonhealing ischemic diabetic foot ulcers despite best standard of care.	(Weak; moderate)
6	We suggest not using topical oxygen therapy as a primary or adjunctive intervention in diabetic foot ulcers including those that are difficult to heal.	(Weak; low)
7	Consider the use of negative pressure wound therapy to reduce wound size, in addition to best standard of care, in patients with diabetes and a postoperative (surgical) wound on the foot.	(Weak; low)

Table 23.2 IWGDF Guideline on interventions to enhance healing of foot ulcers in persons with diabetes.—cont'd

N	Recommendations	(GRADE strength of recommendation: Quality of evidence)
8	As negative pressure wound therapy has not been shown to be superior to heal a nonsurgical diabetic foot ulcer, we suggest not using this in preference to best standard of care.	(Weak; low)
9	Consider the use of placental-derived products as an adjunctive treatment, in addition to best standard of care, when the latter alone has failed to reduce the size of the wound.	(Weak; low)
10	We suggest not using the following agents reported to improve wound healing by altering the wound biology: growth factors, autologous platelet gels, bioengineered skin products, ozone, topical carbon dioxide, and nitric oxide, in preference to best standard of care.	(Weak; low)
11	Consider the use of autologous combined leukocyte, platelet and fibrin as an adjunctive treatment, in addition to best standard of care, in noninfected diabetic foot ulcers that are difficult to heal.	(Weak, moderate)
12	Do not use agents reported to have an effect on wound healing through alteration of the physical environment including through the use of electricity, magnetism, ultrasound, and shockwaves, in preference to best standard of care.	(Strong; low)

Continued

Table 23.2 IWGDF Guideline on interventions to enhance healing of foot ulcers in persons with diabetes.—cont'd

N	Recommendations	(GRADE strength of recommendation: Quality of evidence)
13	Do not use interventions aimed at correcting the nutritional status (including supplementation of protein, vitamins, and trace elements, pharmacotherapy with agents promoting angiogenesis) of patients with a diabetic foot ulcer, with the aim of improving healing, in preference to best standard of care.	(Strong; low)

Font, IWGDF Guidelines on the Prevention and Management of Diabetic Foot Disease, 2019.

should not be too bulky as this will affect the fit of the shoe. To dress these wounds appropriately, healthcare professionals must often use nontraditional methods which limit or simplify the selected dressing. All DFUs have specific and singular characteristics that influence the dressing selection:

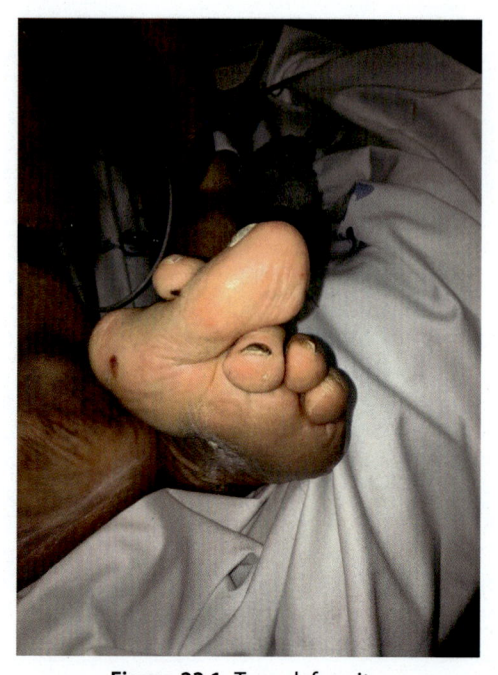

Figure 23.1 Toes deformity.

- Little size; these wounds may start as a small hole, in small fingers (Figs. 23.2 and 23.3)
- Deeper than extensive
- Located in areas of pressure
- Subject to peaks of pressure and forces shearing
- Anatomical location (e.g., fingers or in the sole of the foot)
- Convex areas

We need to also consider that a wound dressing that is in close contact with the wound bed and surrounding skin can disturb or damage the wound, particularly if it changes the optimal interface temperature, creates chemical imbalances and/or chemical stress, changes the optimal moisture balance, and does not solve the problem of adherence and mechanical stress, and if the dressings degrade and release debris or foreign bodies [9–11].

Regardless the local treatment and dressing selection, there are two main topics which are relevant, but not often addressed regarding these topics. These are: (1) ensuring routine wearing of appropriate footwear and (2) pressure offloading (Figs. 23.4 and 23.5).

The routine wearing of appropriate footwear is highly recommended since the patients are typically individuals with loss of protective sensation (LOPS), so, all footwear should be adapted to conform to any alteration in foot structure or foot biomechanics affecting the feet of the individual [12].

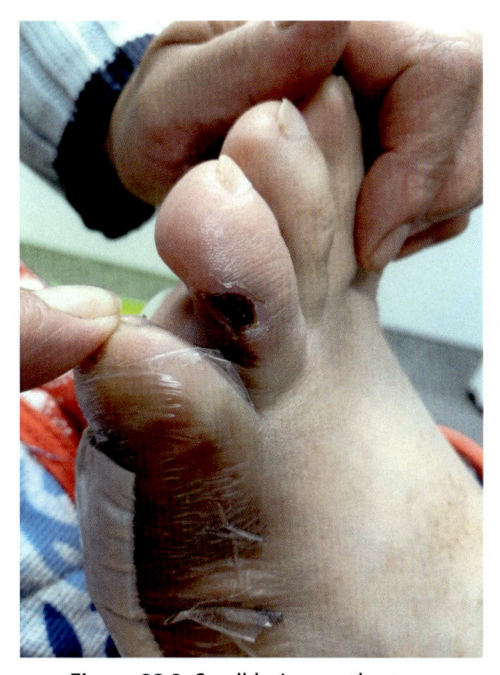

Figure 23.2 Small lesion on the toes.

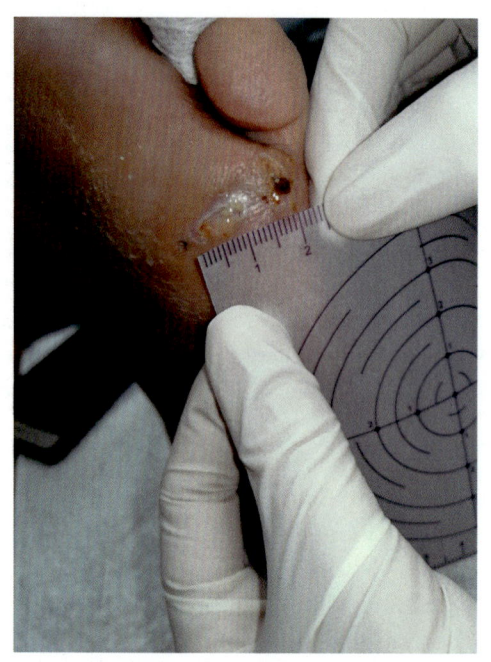

Figure 23.3 Small lesion on the plantar zone.

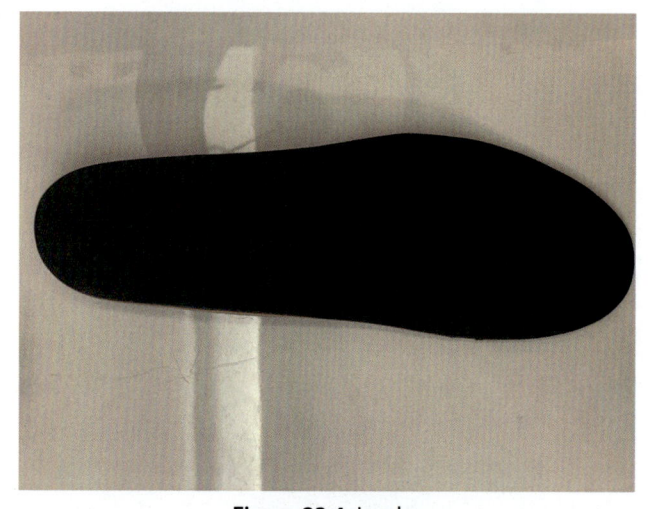

Figure 23.4 Insole.

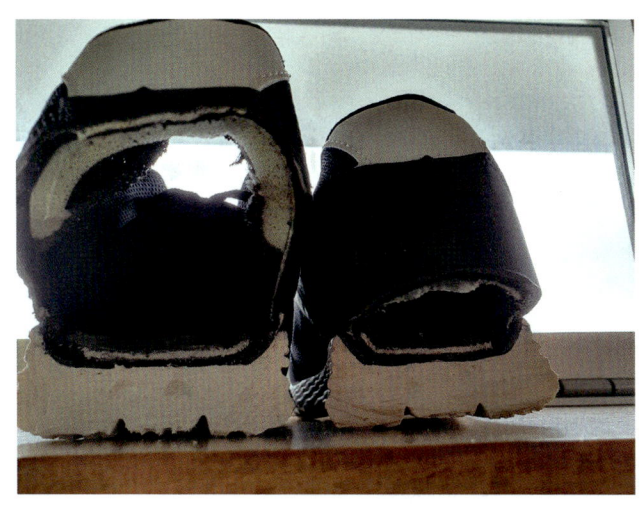

Figure 23.5 Artisanal adapted shoe.

Offloading is a cornerstone in treatment of ulcers that are caused by increased biomechanical stress, and the practical guidelines recommendation from the IWGDF says that the preferred offloading treatment for a neuropathic plantar ulcer is a nonremovable knee-high offloading device, either a total contact cast (TCC) or removable walker rendered irremovable [12]. The use of adapted shoes or medical devices adapted to relieve or redistribute pressures on the feet is recommended as an efficient measure, particularly in diabetic neuropathy; it significantly reduces the risk of injury to the feet [13], and consequently avoids deterioration/increase in pressure in the existing injury zones. Shoes and insoles, for example, for the diabetic should prevent discomfort and reduce the pressure area (Fig. 23.6).

The focus of this chapter is on the clinical and laboratory evaluations of the dressing selection on local treatments. Regarding the recommendations about the dressing selection of the IWGDF Guidelines [8]: Select dressings principally on the basis of exudate control, comfort, and cost; and do not use antimicrobial dressings with the goal of improving wound healing or preventing secondary infection, both should be taken in account, but it is the first recommendation that has been evaluated in clinical practice and in our laboratory work. Dressing selection takes into account three main characteristics: exudation management, dressing change frequency, and comfort.

Exudation, frequency and comfort

Before selecting the best dressing for a specific patient, with their unique wound characteristics, local care of the DFU is crucial. The regular inspection of the ulcer by a trained healthcare provider is indispensable, and its frequency depends on the local treatments

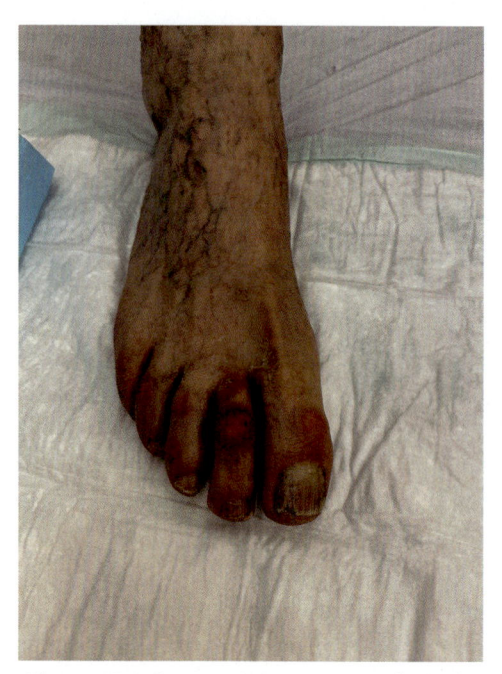

Figure 23.6 Pressure injury zone on the toes.

that are provided and several other factors such as the severity of the ulcer and/or underlying pathology, infection, and exudation. There are also other interventions that are recommended: for example, debriding the ulcer and removing surrounding callus; not soaking the feet as this may induce skin maceration and consideration of other therapies in the treatment of these lesions, such as negative pressure wound therapy to help heal postoperative wounds [14].

An effective dressing strategy is one of the best ways of managing a DFU and preventing infection. The previous variables should be taken into account and healthcare professionals should also consider if the dressing stays intact and remains in place throughout long periods; the dressing does prevent leakage between dressing changes; it does adequately manage the exudation without causing maceration/allergy or sensitivity; it does reduce pain and odor; and finally it does retain fluid and trap the exudate components [15].

One of the first descriptions about ideal dressing characteristics for DFUs was from Foster [16]: it does not take up too much room in the shoe, it performs well in an enclosed environment (shoe) and can withstand shear forces, it does not increase the risk of infection, it absorbs exudates and allows drainage, and it can be changed frequently and easily.

Exudation management

Although exudate formation is a normal part of the wound-healing process and an essential component of healing, excessive exudate that is not managed effectively can have a negative impact on the patient [17,18]. After cleaning the wound, it must be kept moist, but without excess exudate [19]. The ideal wound dressing should have optimal fluid handling ability (absorption and retention, even under pressure); limit leakage; limit the spread of exudate to the periwound area (thus reducing the risk of maceration); and act as a barrier to prevent bacterial ingress. If the wound is very exudative, the dressing material must be absorbent; on the opposing, if the wound is little exudative, the best option will be to hydrate the bed, promoting autolytic debridement in the case of the presence of dry and adherent necrosis [20]. Special care must be taken, with a dressing material that increases the possibility of potentiating the maceration, as well as with the nonrecommendation of using dressing materials that retain the exudate, in the presence of ischemia or dry necrosis [21,22] (Figs. 23.7 and 23.8).

Frequency of dressing change

The ability of a dressing to conform to body contours helps ensure optimal adhesion [20], and many dressings are not specifically designed for use on the foot and are consequently hard to apply over or between the toes and around the curvature of the heel, and this will

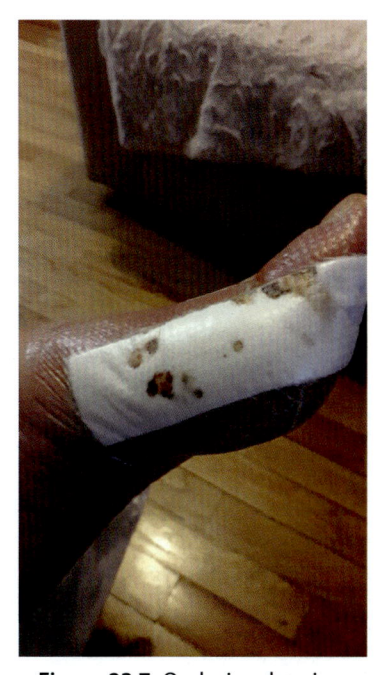

Figure 23.7 Occlusive dressing.

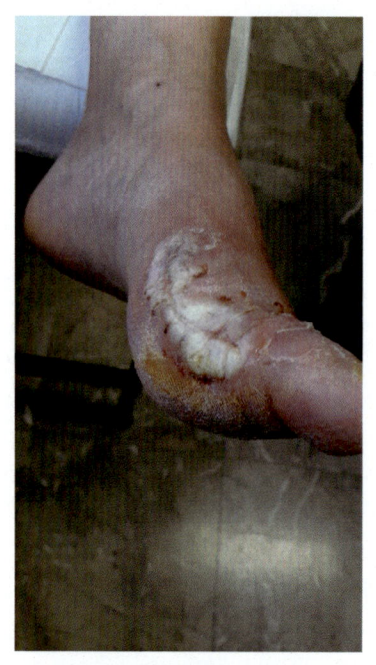

Figure 23.8 Maceration.

generate an impact on the frequency of dressing change. The adhesion of dressings is a factor that can affect dressing wear time. Additional dressing changes due to poor adherence are associated with increased costs associated with nursing time, materials, pain medication, and medication for dressing-related trauma. The dressing material should never make it impossible to use shoes; it must be functional and be in place without needing to be changed unnecessarily. The dressing material will be cost-effective if it is used to the maximum of its capacities, taking into account the maximum time limit of use, as well as its maximum absorption capacity.

Comfort

Flexibility and conformability of the dressing help prevent early detachment. Dressings should be selected that conform well to the wound, are comfortable to wear, are easy to use, minimize unnecessary wound disturbance, and are cost-effective [23]. There are some reports from many patients, that even if they have neuropathy or neuroischemia, they may feel pain due to the wound or the procedure [24]; it is important to adapt strategies in order to prevent trauma and minimize the pain experienced during treatment, more specifically the dressing material [15]. Choosing to use silicone dressing material, for example, and avoid unnecessary manipulating the wound are two strategies that can be adopted [24].

When choosing the dressing material, regardless of the brands and typology of the products, ease of use (patients and health professionals) must be taken into account; enhance self-care; level of independence of the user; and that the chosen dressings do not cause alteration of the self-image and self-esteem of the patient. Another feature, already mentioned, is the size of the dressing material, as large dressing materials can cause excessive pressure zones when the patient wears his shoes.

However, related to this last point, some doubts remain, namely: Did the mechanical properties of the dressing change after it has been saturated? Did its volume increase (thereby causing greater pressure in the places it should have protected)?

Ulcer protection and pressure relief

The multidisciplinary team should assess the need to use therapeutic shoes and/or discharge devices, in order to reduce pressure, while guaranteeing the ability to walk and simultaneously heal the injuries that are found on the foot. The understanding of biomechanical stress helps professionals to select among the various devices, but it must take into account the user's ability to comply with the suggested option, the available resources, and of course, the severity of the ulcer and the local treatment that it requires.

Biomechanical measurements of pressure on the plantar area of the foot have proven to be useful in the diagnosis and management of pressure-related foot problems [25].

Regarding these technologies and the three factors previously mentioned, we have conducted some biomechanical measurements in a convenience group of volunteers, simulating a lesion on the plantar area of the foot, next to hallux, and we then replicated four different situations to compare with the normal evaluation of the foot with regard to plantar pressures, discomfort, and adhesion of the dressing material.

It is already known that plantar pressure measurements without shoes may be of poor value because for diabetic patients it is unthinkable to walk barefoot, but we aimed at performing static and dynamic plantar pressure measurements to determine the loading, angle, and length of strides during walking in order to understand how different dressings may influence pressure distribution. Some peak pressure tests were carried out with the dressing dry and then saturated, to determine its mechanical effect on the injury. It was therefore possible to understand that rigidity of the dressing (lack of flexibility) is compromising the conditions of the feet and the gait of the patient. Shear forces in the diabetic foot are also highly relevant and have been implicated in the formation of callus [26,27]. It is well known that in the beginning, the body develops this callus for protection, but if the callus becomes excessive, it will contribute to higher pressures that will end as an ulcer.

The first peak pressure measurements attempted to evaluate the pressures on the foot, with and without a dressing on the hallux, and in all the volunteers there were localized pressure on the area of the dressing, but simultaneously, a redistribution of pressure emerged, related to the use of dressing (in Fig. 23.9 it is possible to see the difference in pressures on the same volunteer without and with a dressing on the hallux).

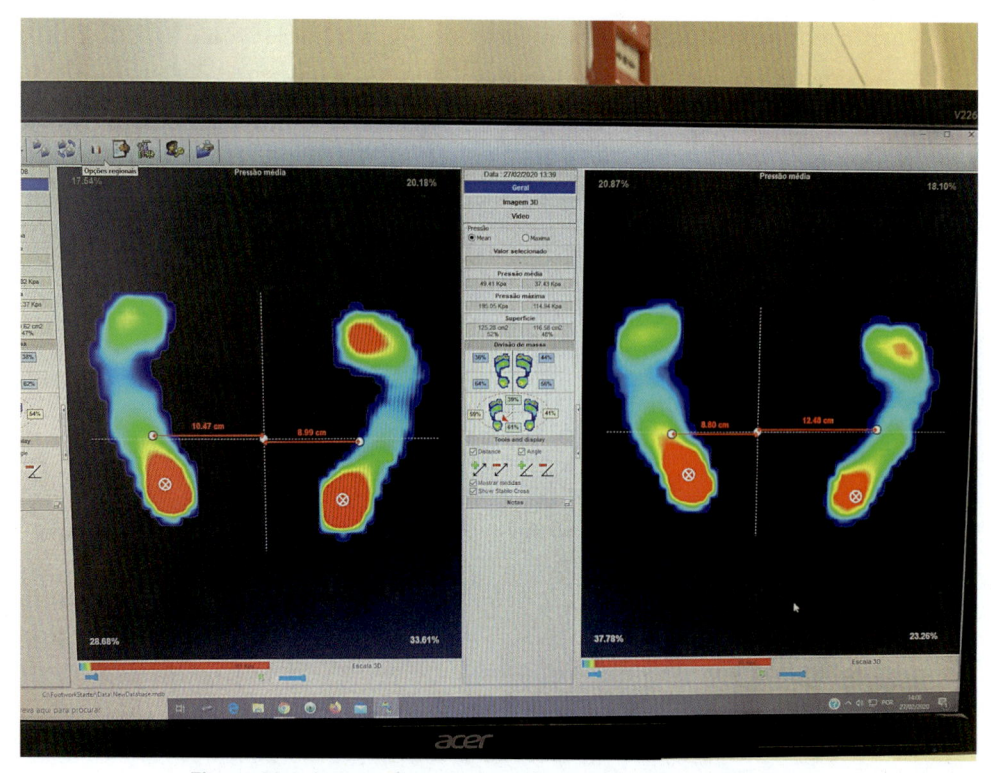

Figure 23.9 Image of pressures without and with the dressing.

With the use of the dressing the load increases on the foot that has the injury, decompensating the load distribution between the two lower members. After these first measurements, we applied an insole to the volunteers, adapted to their foot, and the redistribution of the pressure was similar to a foot without a dressing (Fig. 23.10).

The second variable assessed was the behavior of the dressing after saturation, and the results showed that there is not a significant difference between the dressings in their dry versus saturated conditions. It was possible to see that the silicone multilayer dressings were able to better control the exudation, prevent leakage, stay in place, and maintain good pressure redistribution. Contrarily, the single layer dressings had poor results in leakage and pressure redistribution (Fig. 23.11).

The multilayer dressings also had better behavior related to pressure redistribution, releasing pressure from the area of lesion and distributing the pressure to other areas of the foot.

Some of the problems identified were the low adhesion to the skin after walking and sweating; some dressings started to deteriorate and unstick, particularly the nonsilicone dressings.

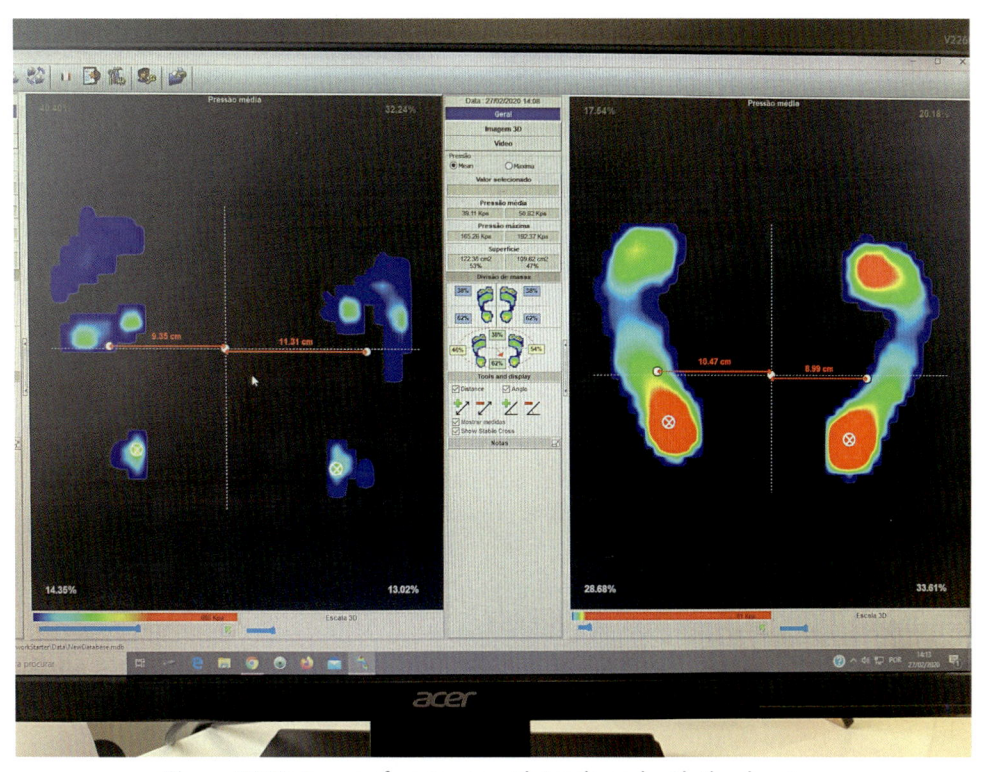

Dressing selection challenges in diabetic foot local treatment 403

Figure 23.10 Image of pressures with insole and with the dressing.

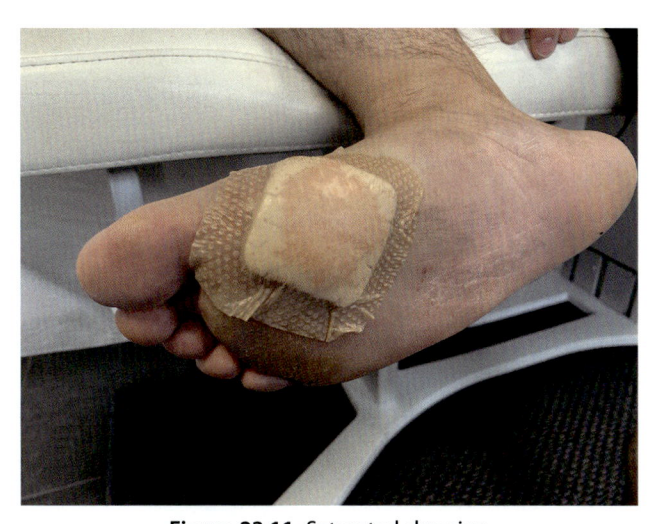

Figure 23.11 Saturated dressing.

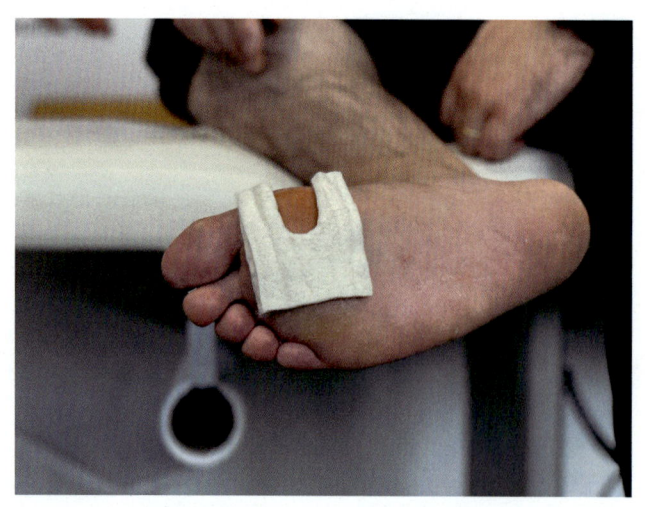

Figure 23.12 Felted foam dressing.

The last biomechanical test performed was with felted foam dressings which were the dressing that all volunteers documented as less comfortable. Considering all the dressings evaluated, the thinner dressing materials were associated with the highest level of stride symmetry, and in contrast thicker dressing materials had greater asymmetry. These tests that we share here reveal only a few concerns felt by clinicians in their daily practice, which is why testing with greater accuracy and a higher number of patients is recommended for future work (Fig. 23.12).

Importantly, when choosing the dressing material to treat a diabetic foot, one should take into account exudation management, the frequency of dressing changes and comfort, but also the cost-effectiveness factor, according to the rules established by the specific Health Institution where the treatment takes place. Ideally, dressings should alleviate the symptoms, provide protection to the wound, and encourage healing.

References

[1] National Institute for Health and Clinical Excellence. NICE NG19. Diabetic foot problems: prevention and management. London: NICE; 2015.
[2] Hinchliffe R, et al. Effectiveness of revascularization of the ulcerated foot in patients with diabetes and peripheral artery disease: a systematic review. Diabetes Metab Res Rev 2016;32(Suppl. 1):136–44.
[3] Queen D, Orsted H, Sanada H, Sussman G. A dressing history. Int Wound J 2004;1:59–77. https://doi.org/10.1111/j.1742-4801.2004.0009.x [PMID: 16722898].
[4] Sibbald RG, Torrance G, Hux M, Attard C, Milkovich N. Cost-effectiveness of becaplermin for nonhealing neuropathic diabetic foot ulcers. Ostomy/Wound Manag 2003;49:76–84 [PMID: 14652415].
[5] Kavitha KV, Tiwari S, Purandare VB, Khedkar S, Bhosale SS, Unnikrishnan AG. Choice of wound care in diabetic foot ulcer: a practical approach. World J Diabetes 2014;5(4):546–56. https://doi.org/10.4239/wjd.v5.i4.546.

[6] Armstrong DG, Lavery LA, Nixon BP, Boulton AJ. Clin Infect Dis August 1, 2004;39(Suppl. 2): S92—9.
[7] International Best Practice Guidelines. Wound management in diabetic foot ulcers. Wounds International; 2013.
[8] Rippon M, Davies P, White R. Taking the trauma out of wound care: the importance of undisturbed healing. J Wound Care 2012;21(8):359—68.
[9] International Working Group on the Diabetic Foot (IWGDF). IWGDF Practical guidelines on the prevention and management of diabetic foot disease. 2019.
[10] Messaoud M, Marsiquet C, Revol-Cavalier F, et al. Flexible sensors for real-time monitoring of moisture levels in wound dressings. J Wound Care 2018;27(6):385—91.
[11] Haycocks S, Chadwick P, Davies P. Case series: mepilex® Border Comfort in the treatment of diabetic foot ulcers with exudate. Diabetic Foot J 2018;21(4):265—71.
[12] Schaper, et al. Practical guidelines on the prevention and management of diabetic foot disease (IWGDF 2019 update). Diab Metab Res Rev 2020:e3266.
[13] Bus SA, Haspels R, Busch-Westbroek TE. Evaluation and optimization of therapeutic footwear for neuropathic diabetic foot patients using in-shoe plantar pressure analysis. Diabetes Care July 2011; 34(7):1595—600. https://doi.org/10.2337/dc10-2206. Epub 2011 May 24. PMID: 21610125; PMCID: PMC3120171.
[14] Rayman, et al. Guidelines on use of interventions to enhance healing of chronic foot ulcers in diabetes (IWGDF 2019 update). Diab Metab Res Rev 2020:e3283.
[15] World Union of Wound Healing Societies (WUWHS). Principles of best practice: wound exudate and the role of dressings. A consensus document. London: MEP Ltd; 2007.
[16] Foster AVM, Greenhill MT, Edmonds ME. Comparing two dressings in the treatment of diabetic foot ulcers. J Wound Care 1994;3:224—8.
[17] Tickle J. Wound exudate: a survey of current understanding and clinical competency. Br J Nurs 2016; 25(2):102—9.
[18] World Union of Wound Healing Societies (WUWHS) Consensus Document. Wound exudate: effective assessment and management Wounds International. 2019.
[19] Wu L, Norman G, Dumville JC, O'Meara S, Bell-Syer SE. Dressings for treating foot ulcers in people with diabetes: an overview of systematic reviews. Cochrane Database Syst Rev July 14, 2015;2015(7): CD010471. https://doi.org/10.1002/14651858.CD010471.pub2. PMID: 26171906; PMCID: PMC7083265.
[20] Rippon M, Waring M, Bielfeldt S. An evaluation of properties related to wear time of four dressings during a five-day period. Wounds U K 2015;11(1):45—54.
[21] Mulder G, e. a. Standart, appropriate, and advanced care and medical-legal considerations: part one - diabetic foot ulcerations Wounds 2003;15(4):92—106.
[22] Game F. The advantages and disadvantages of non-surgical management of the diabetic foot. Diabetes/Metab Res Rev 2008;24(1):S72—5.
[23] Chadwick P, McCardle J. Exudate management using a gelling fibre dressing. Diabetic Foot J 2015; 18(1):43—8.
[24] Baker N. Implications of dressing-related trauma and pain in patients with diabetes. Diabetic Foot J 2012;15(Suppl. l):S1—8.
[25] Mani R, Romanelli M, Skula V. Measurements in wound healing. London: Springer —Verlag; 2012. https://doi.org/10.1007/978-1-4471-2987-5.
[26] Yavuz M, Erdemir A, Botek G, Hischman GB, Bardsley L, Davis BL. Peak plantar pressure and shear locations. Diabetes Care 2007;30:2643—5.
[27] Yavuz M, Tajaddini A, Botek G, Davis BL. Temporal characteristics of plantar shear distribution: relevance to diabetic patients. J. Biomech 2008 2008;41(3):556—9.

Index

Note: 'Page numbers followed by "*f*" indicate figures and "*t*" indicate tables.'

A

Achilles tendon (AT), 36, 108, 201–202, 204
Activities of daily living (ADL), 13
Adenosine triphosphate (ATP) production, 292
Adherence, 367–368
Adipocytes, 143–144
  differentiation, 152
  mechanical stresses, 152–153
  mechanobiology of, 145f
    cell-level mechanics, 145
    migration ability, 145
    traction force microscopy, 146–147, 146f
  stem cells
    embryonic stem cells (ESCs), 147–148, 149f
    single preadipocytes, 149–152, 151f
  stiffness changes, 153–156, 154f–155f
Adipose-derived mesenchymal stem cells (ASCs), 174
Adipose tissue, 43, 44f, 45, 173
  and adipocytes, 143–144
Advanced glycation end products (AGEs), 66–68, 177–179, 201, 319
  collagenous tissues, 204–205
    clinical aspects, 207
    signaling role, 205
    structural consequences, 205, 206t
  mechanism of, 259, 259f
  tendons, 201–202
    microstructural interactions, 209
Adventitia, 72–73
Alkaline phosphatase (ALP), 176–177
Allostasis
  blood flow, 354–355
  components, 351–354
    compensation, 353–354
    reallocation, 351–352
    resolution of observation, 351
  compression, 355–356
  deep tissue injury (DTI), 355–356
  deformation, 355–356
  diabetic foot ulcers (DFUs), 347, 357t–358t
  homeostasis, 348–351
  limitations, 358
  multifactorial approach, 357t–358t
  oxygenation, 354–355
  peripheral artery disease (PAD), 347
  potential application/integration, 356–358
  tissue injury risk, 354–356
  tissue loading physical response, 354–356
  treatment approaches, 356–358
Angiosome theory, 23
Ankle brachial index (ABI), 15
Ankle brachial pressure index (ABPI), 3
Aortic stiffening, 64–66
Arterial branches, 298
Arterial stiffening, 64–66
Arterial–venous (AV) identity, 379–380
Artificial intelligence (AI), 115–117, 117f
Atheroprone phenotype, 291
Atheroprotective phenotype, 291
Atomic force microscopy (AFM), 69–73
Atopic dermatitis, 168, 170f
Atrophy, 108
Autoamputation, 20–21, 21f

B

Basic fibroblast growth factor (bFGF), 375
Basic multicellular unit (BMU), 317
Basic runs, 235–237, 235t, 237f–238f
Blood flow, 354–355
Body mass index (BMI), 64
Bone cell
  basic multicellular unit (BMU), 317
  bone modeling, 317–318
  bone remodeling, 317–318, 318f
  bone-specific alkaline phosphatases (BAPs), 314
  C-terminal telopeptide (Ctx), 314
  definition, 315–316
  dual-energy X-ray absorptiometry (DXA), 313
  osteoblasts, 315
  osteocalcin (OCN), 314
  osteoclasts, 315–316
  osteocytes, 316
  parathyroid hormone (PTH), 314
  quality, 314
  type 1 diabetes (T1D), 313

Bone cell (*Continued*)
   type 2 diabetes (T2D), 313, 318–322
      advanced glycation end products (AGEs), 319
      bone morphogenetic pathway (BMP), 319–320
      C-terminal telopeptide (CTX), 320
      fibroblast growth factor-23 (FGF23), 321–322
      osteoblasts, 318–320
      osteocalcin, 318–319
      osteoclasts, 320–321
      osteocytes, 321–322
      osteopontin, 320
      RANKL, 321
      tartrate-resistant acid phosphatase (TRAP), 320
Bone marrow–derived mesenchymal stem cells (BMSCs), 174, 179
Bone mass density (BMD), 174–175
Bone modeling, 317–318
Bone morphogenetic pathway (BMP), 177, 319–320
Bone remodeling, 317–318, 318f
Bone-specific alkaline phosphatases (BAPs), 314
Boussinesq solution, 146

C

Cadaveric dissection, 258–259
Calcification, 69, 108
Calcium sulfate, 285
Capillary loop depth (CLD), 168
Carbon paper method, 108–109
Cardiometabolic syndrome, 78–80, 79f
Cardiovascular disease (CVD), 64
Cell density, 228–229
Charcot ankle osteomyelitis, 287
Chitosan, 286
Chronic limb-threatening ischemia (CLTI), 14
Cleveland Clinic Foundation (CCF) device, 97–98, 100f
Cloud computing, 117–118, 118f
Collagen-based matrices, 377
Collagen fleece, 285–286
Collagen sliding, 207–209
Collagen type IV fibers, 43–44
Comfort, 397–401
Compensation, 353–354
Compressibility, 264–265, 265f, 271
Compression, 355–356
   garments, 339
   hosiery, 330
   systems, 339
   wraps, 330, 339

Compressive loading, 184–185
Computational mechanics, 227–228, 233–235
Computed topography, 36, 258–259
Core-binding factor alpha-1 (Cbfα1), 182
Critical limb ischemia (CLI), 14, 18f, 20–21, 20f
C-terminal telopeptide (Ctx), 314
Custom-made footwear, 366–367
Custom-made insoles, 364–365

D

Deep tissue injury (DTI), 355–356
Deformation, 355–356
Demonstrated plantar pressure-relieving effect, 369
DFI. *See* Diabetic foot infection (DFI)
DFO. *See* Diabetic foot osteomyelitis (DFO)
DFU. *See* Diabetic foot ulcers (DFU)
Diabetes mellitus (DM)
   homeostasis, 201–202
   mechanobiological changes
      cell-level changes, 248
      macro-level changes, 248
      molecular level, 249
   prevalence of, 247
   tendon dysfunction, 202
      Achilles tendon (AT), 204
      rotator cuff pathology, 203–204
      trigger finger, 202–203, 203f
Diabetic adipose-derived mesenchymal stem cells (dASCs), 174
Diabetic foot device
   benefits, 330–332
   compression
      garments, 339
      hosiery, 330
      systems, 339
      wraps, 330, 339
   drawbacks, 332–333
   extend functionality, 332
   force redistribution, 331
   friction management, 331
   functionality restoration, 332
   heel lift boots, 338–339
   interventions, 333–334
   life, 341–342
   neuropathy, 330
   offloading, 334
      prevention, 334
      support surfaces, 334
      tissue damage science, 335–338
      treatment, 334–335

patient compliance, 331–332
proprioception errors, 330
risks, 328–329
shear reduction, 331
specific devices, 338
static stiffness index, 340
total contact cast, 339
treatment, dressings for, 340–341
    ambulation, 340–341
    prevention, 341
Diabetic foot infection (DFI), 14
  local antibiotics, 286–288
Diabetic foot osteomyelitis (DFO), 283
  diagnosis, 283–284
  local antibiotic delivery systems, 285–286
  topical antibiotic therapy, 284–285
  treatment, 283–284
Diabetic foot ulcer (DFU), 247, 347
  computational modeling, 39f–40f
    boundary conditions, 38
    edge-effect, 35–36
    finite element (FE) analysis, 36
    geometry, 36, 37f
    mechanical properties, 36–38, 37t
    numerical method, 38
    outcome measures, 38–39
    pressure reduction, 35–36
  debridement and amputation, 21–22
  dressing selection, 389
  foot pressures and, 108–111, 109f
  global prevalence of, 13
  health and societal burden of, 107
  intelligent insole system, 118–119
    artificial intelligence (AI), 115–117, 117f
    cloud computing, 117–118, 118f
    computer vision, 115
    functions, 115
    and smartwatch, 113–115, 114f
  pathomechanics of
    biomechanics, 89
    ground reaction forces, 89, 90f–91f
    plantar callus and shear, 100–101
    plantar pressure, 92–96, 93f–94f
    pressure injuries (ulcers), 96
    shear forces, 89
    shear stress, 97–100, 98f–100f
    temperature increase and shear, 101–102
  practice pathway, 14, 15f
  prevention, 111
    continuous real-world pressure feedback technology, 111–113
    intelligent insole system, 113
  revascularization, 23–24
  risk factors for, 107–108
  soft tissue reconstruction, 25–29, 26f, 28f, 30f–31f
  tissue biomechanics, 124–127, 125f. *See also* Tissue biomechanics
  treatment, 20, 21f, 389
  wound assessment, 14–21, 16f–21f
  wound bed preparation (WBP), 21–22
  wound dressings, 390–397
    artisanal adapted shoe, 395, 397f
    comfort, 397–401
    dressing change frequency, 399–400
    dressing materials, 391t
    exudation, 397–401
    exudation management, 399
    footwear, 395
    frequency, 397–401
    insole, 395, 396f
    loss of protective sensation (LOPS), 395
    plantar zone small lesion, 395, 396f
    pressure of floading, 395
    pressure relief, 401–404
    toes deformity, 394f
    ulcer protection, 401–404
  wound healing (WH), 24–25
  wounds, 2–3
Diabetic neuropathy, 107–108
Diabetic peripheral sensorimotor neuropathy, 107–108
Diabetic wounds, 249–252
DM. *See* Diabetes mellitus (DM)
DNA fragmentation, 177–179
Dual-energy X-ray absorptiometry (DXA), 313

E

E-cadherin, 147–148
Edge-effect, 35–36
Elastic fiber degradation, 69
Elastic modulus, 266–267
Elastin, 210
Embryonic stem cells (ESCs), 147–148, 149f
Endothelial cell (EC)
  adenosine triphosphate (ATP) production, 292
  arterial branches, 298
  atheroprone phenotype, 291

Endothelial cell (EC) (*Continued*)
  atheroprotective phenotype, 291
  endothelial nitric oxide synthase (eNOS), 294
  extracellular acidification rate (ECAR), 296–297
  fingerprints, 297
  flow models, 301–302
  fluorescent glucose, 296
  future work, 304–305
  glucose 6-phosphate dehydrogenase (G6PD), 294
  glycolysis, 292–293
  glycolytic intermediates, 293–294
  human umbilical vein endothelial cells (HUVECs), 299–300
  metabolism, 292–295
    function, 295
    models, 302–304
    tools, 295–297
  monocyte chemotactic protein-1 (MCP-1), 298
  nitrobenzoxadiazole (NBD), 296
  oxygen consumption rate (OCR), 296–297
  protein kinase C (PKC), 291
  responses, 297–300
  ribose 5-phosphate (R5P), 294
  shear stress, 297–298
  translational potential, 304–305
  tricarboxylic acid (TCA), 292–293
  UDP-N-acetylglucosamine (UDP-GlcNAc), 294
  varied flow conditions, 300
  vascular endothelial growth factor receptor-2 (VEGFR2), 297
  vascular mechanobiology, 300–304
Endothelial cells (ECs), 70, 74f, 248
Endothelial Na^+ channel (EnNaC), 75–78
Endothelial nitric oxide synthase (eNOS), 294
Endovascular treatment (EVT), 16, 18f
EphrinB2, 382
Ethylene-vinyl acetate (EVA), 38
European Pressure Ulcer Advisory Panel (EPUAP), 96
Expanded polytetrafluoroethylene (ePTFE), 377
Extracellular acidification rate (ECAR), 296–297
Extracellular matrix (ECM), 43, 64–66, 181
  obesity and diabetes, 66–69
  stiffness, 248
  tensile/compressive stress, 248
Extracorporeal shock wave therapy (ESWT), 247, 251–252
Exudation, 397–401
  management, 399

F

Fat accumulation, 143
Fibroblasts, 227–228, 235–236
Fibrocytes, 249–251
Fibrosis, 43
Finger-pad interaction, 167, 167f
Fingerprints, 297
Finite element (FE) analysis modeling, 36, 45, 126–127
  application of, 134
  heel strike, 128, 129f
  indentation test, 133, 133f
  limitations, 58
Fluid shear, 185–187, 190–193, 192f
Fluorescent glucose, 296
Focal adhesion kinase (FAK), 182, 249
Foot deformities, 108
Foot pressure, 95, 108–111, 109f
FootSnap, 115, 118f
Footwear
  adherence, 367–368
  considerations, 370
  diabetes at high risk, person with, 363
  international guidelines recommendations, 368–369
  offloading, 364–365
  research, 370
  ulcer prevention, 365–367
    cost, 369–370
  wound dressings, 395
Foot wounds, 8, 8t
Force redistribution, 331
Fourier transformation, 230–231
Free flap transfer reconstruction, 27–29
Friction management, 331
Full-thickness skin grafts (FTSGs), 26–27
Functionality restoration, 332

G

Gentamycin, 287
Glucose 6-phosphate dehydrogenase (G6PD), 294
Glycolysis, 292–293
Green's function, 146
Ground reaction force (GRF), 38

H

Heel lift boots, 338–339
Heel pad, 257–259, 257f

Heel pressure ulcers (PUs), 3
Heel ulcers, 4—5
  in diabetes heal, 4
  diabetics and nondiabetics, 5—9, 7t—8t
Hepatocyte growth factor (HGF), 377
Homeostasis, 207—209, 348—351
Human bone marrow–derived mesenchymal stem cells (hBMSCs), 174
Human umbilical vein endothelial cells (HUVECs), 299—300
Hyperglycemia, 64, 143, 260
  mesenchymal stem cells (MSCs)
    adipogenesis, 179—181, 180f
    osteogenesis, 174—179, 178f
    proliferation, 174, 175f
Hyperkeratosis, 168
Hyperosmotic shock, 177—179
Hyperplasia, 44—45, 144
Hypertension, 64
Hypertrophy, 44—45

I

Immune cells, 71—72
Inflammatory cytokines, 71—72
Insole, 395, 396f
Insulin signaling, 75
Intelligent insole system, 113, 118—119
  artificial intelligence (AI), 115—117, 117f
  cloud computing, 117—118, 118f
  computer vision, 115
  functions, 115
  and smartwatch, 113—115, 114f
Interfascicular matrix (IFM), 207—208
Interlobular septa, 43
  thickening, 46f, 54f, 56—57
    biomechanical outcome measures, 48
    boundary conditions, 47
    effective stress distribution, 53, 53f
    geometry, 45—47
    mechanical properties, 47, 47t
    numerical method, 48
International Working Group on the Diabetic Foot (IWGDF), 363, 368—369, 391, 392t—394t
Ischemia, 14, 16f—18f
Isolated microvessel fragments (MFs), 375—377, 376f

K

Keratin, 255—256

L

Loss of protective sensation (LOPS), 364, 395
Lyapunov stability, 230
Lysyl oxidases (LOXs), 68—69

M

Magnetic resonance imaging (MRI), 129—130, 130f, 161, 207
Matlab detection algorithm, 163—166, 165f
Matrix metalloproteinase-13 (MMP13), 176—177
Matrix metalloproteinases (MMPs), 66—67
Mechanotherapy, 249—252
Mechanotransduction, 249
Mesenchymal stem cells (MSCs), 144, 195f
  differentiation, 173
  hyperglycemia
    adipogenesis, 179—181, 180f
    osteogenesis, 174—179, 178f
    proliferation, 174, 175f
  mechanical loading cues, 181
    adipogenesis, 187—193
    osteogenesis, 181—187
Metatarsal pads, 258—259
Microvessel fragment (MF)
  arterial—venous (AV) identity, 379—380
  basic fibroblast growth factor (bFGF), 375
  diabetes on, 380—382
  engineered vessel function, 377—380
  expanded polytetrafluoroethylene (ePTFE), 377
  functional neovasculature, 378—379
  future directions, 382—384
  isolated microvessel fragments (MFs), 375—377, 376f
  maturation, 377—380
  mechanisms, 377—380
  neovascular remodeling, 378—379, 378f
  sprouting angiogenesis, 378—379, 378f
  treatment, 384
  type 1 diabetes, 380—381
    3D-engineered constructs, vessel maturation in, 382
    engineered grafts, sprouting in, 380—381
  vascular endothelial growth factor (VEGF), 375
  vascularization outcomes, 377
  vascular maturation, 378—379, 378f

Microvessel fragment (MF) (*Continued*)
   vascular remodeling phase, 378–379
   vessel arterial–venous plasticity, 379–380
Mineralocorticoid receptor (MR), 75
Mitogen-activated protein kinases (MAPKs), 182
Modulus of elasticity, 124
Monocyte chemotactic protein-1 (MCP-1), 298
Monte Carlo simulations, 238–241, 239f–240f, 241t, 242f–243f
MSCs. *See* Mesenchymal stem cells (MSCs)
Multifactorial approach, 357t–358t
MyFootCare, 115
Myofibroblasts, 227–228

N
National Pressure Injury Advisory Panel (NPIAP), 96
Natural feedback mechanism, 111
Negative pressure wound therapy (NPWT), 24, 247, 249–251
Neovascular remodeling, 378–379, 378f
Neuropathy, 13, 127–128, 330
Newton's second law, 232
Nitric oxide synthase (NOS), 249–251
Nitrobenzoxadiazole (NBD), 296
Noncollagenous proteins
   elastin, 210
   proteoglycans, 210–211
   tendon fibroblasts, 211
Nondiabetic (ND), 380–381
Nondiabetic adipose-derived mesenchymal stem cells (nASCs), 174
Nonresorbable antibiotic delivery systems, 285

O
Obesity, 43
   biomechanical outcome measures, 52
   defined, 143
   extracellular matrix (ECM), 66–69
   hyperplasia, 44–45
   hypertrophy, 44–45
   interlobular septa thickening, 46f, 54f, 56–57
      biomechanical outcome measures, 48
      boundary conditions, 47
      effective stress distribution, 53, 53f
      geometry, 45–47
      mechanical properties, 47, 47t
      numerical method, 48
   mechanical loads, lean and obese tissues, 54–58, 55f
      boundary conditions, 48–49
      geometry, 48, 49f
      mechanical properties, 49–50, 51f
      numerical method and construction, 50–52, 52t
   prevalence of, 173
   vascular stiffening, 66–69
      cellular contributions, 69–73
      metabolic and endocrine activation, 73–80
OCT. *See* Optical coherence tomography (OCT)
Offloading, 334
   footwear, 364–365
   prevention, 334
   support surfaces, 334
   tissue damage science, 335–338
   treatment, 334–335
Ogden hyperelastic model, 126
One-dimensional model, 229–231
Optical coherence tomography (OCT)
   atopic dermatitis diagnosis, 168, 170f
   diabetes-related complications, 168
   fiber-based design, 162–163
   free-space design, 162–163
   for skin biomechanics measurements, 163–168, 164f–165f
   skin imaging, 161–163, 163t
   working principle of, 161, 162f
Osteoblasts, 315
Osteocalcin (OCN), 176–177, 314
Osteoclasts, 315–316
Osteocytes, 316
Osteopontin, 320
Oxygenation, 354–355
Oxygen consumption rate (OCR), 296–297

P
PAD. *See* Peripheral arterial disease (PAD)
Parathyroid hormone (PTH), 314
Pedar-X system, 369
Percentage energy loss, 269–270, 269f, 272–273, 273f
Peripheral arterial disease (PAD), 14, 347
Peripheral neuropathy, 367–368
Perivascular adipose tissue (PVAT), 72–73
Pexiganan, 286
Phenomenological model, 231–233, 233f
Plantar pressure measurement, 370
Plantar shear stress distribution, 97
Plantar soft tissue, 255
   biomechanics of
      compressibility, 271
      elasticity, 270–271

hardness, 271
hydrostatic structures, 261
loading and simulation, 273
measurement of, 262
percentage energy loss, 272–273, 273f
stiffness, 272
superficial tissue characterization, 262–264
viscoelastic structures, 261
whole tissue characterization, 264–270
diabetes, 259–260, 259f
plantar fat pads, 256–259
plantar skin, 255–256
Plastic deformations, 229
Poisson's ratio, 146–147
Polymethyl methacrylate (PMMA), 285
Preadipocytes, 149–152, 151f
Pressure measurement systems, 92, 364
Pressure relief, 401–404
Pressure ulcers (PUs), 96
  characteristics of, 6, 7t
  definitions of, 2
  impact of, 1
  wounds, 2–3
Prevascularized pancreatic encapsulating device (PPED), 380
Proprioception errors, 330
Protein kinase B (PKB), 177
Protein kinase C (PKC), 291
Proteoglycans, 210–211
Pulse wave velocity (PWV), 64
PUs. *See* Pressure ulcers (PUs)

R

Randomized controlled clinical trials, 5, 73–80, 110–111, 288, 366–367
Reactive oxygen species (ROS), 64–66, 177–179
Reallocation, 351–352
Reconstruction ladder, 25–26, 26f
Resolution of observation, 351
Resorbable antibiotic delivery systems, 285
Revascularization, 23–24
Ribose 5-phosphate (R5P), 294

S

Secant modulus, 268, 268f
Shear reduction, 331
Shear stress, 38, 97–100, 98f–100f, 297–298
Shear wave (SW) elastography, 134, 137–138, 137f

Silent information regulator type 1 (SIRT1), 182–183
Skin and soft tissue infection (SSTI), 17–20
Skin perfusion pressure (SPP), 15
Small-angle X-ray scattering (SAXS), 207–208
S-nitrosylation, 248–249
Spheroids, 147–148
Splitter, 161
Split-thickness skin graft (STSG), 26–27
Spring–dashpot system, 232, 233f
Static stiffness index, 340
Strain/real-time ultrasound elastography, 134–137, 135f
Strain, tissue biomechanics, 124, 125f
Stratum corneum (SC), 163, 255–256
Stratum corneum–stratum lucidum (SC-SL) junction, 163
Stress, tissue biomechanics, 124, 125f
Superficial plexus depth (SPD), 168
Superficial tissue characterization
  elasticity, 262–263, 263f
  hardness, 263–264

T

Tendon mechanics
  advanced glycation end products (AGEs), 207
    microstructural interactions, 209
  diabetes, 211, 212t–216t
  dysfunction, 202
    Achilles tendon (AT), 204
    rotator cuff pathology, 203–204
    trigger finger, 202–203, 203f
  microstructural mechanics, 207–209, 208f
  noncollagenous proteins, 210–211
Tensile loading, 181–184, 183f, 187–190, 189f, 191f
Thermal insulation, 43–44
Tissue biomechanics, 123
  internal tissue damage, 129–131, 130f
  shear wave (SW) elastography, 137–138, 137f
  strain/real-time ultrasound elastography, 134–137, 135f
  ulceration, 127–128, 129f
  ultrasound indentation, 131–134, 131f, 133f
Tissue inhibitors of matrix metalloproteinases (TIMPs), 66–67
Tissue injury, 124, 227, 354–356
Tissue loading physical response, 354–356
Tissue transglutaminase 2 (TG2), 68
Toes deformity, 394f
Total contact cast (TCC), 397

Total stress concentration exposure (TSCE), 38–39
Traction force microscopy, 146–147, 146f
Transcutaneous oxygen pressure (TcPO$_2$), 15
Tricarboxylic acid (TCA), 292–293
Trigger finger, 202–203, 203f
Type 1 diabetes (T1D), 143, 175–176, 313, 380–381
   3D-engineered constructs, vessel maturation in, 382
   engineered grafts, sprouting in, 380–381
Type 2 diabetes (T2D), 143, 175–176, 313
   advanced glycation end products (AGEs), 319
   bone morphogenetic pathway (BMP), 319–320
   C-terminal telopeptide (CTX), 320
   fibroblast growth factor-23 (FGF23), 321–322
   osteoblasts, 318–320
   osteocalcin, 318–319
   osteoclasts, 320–321
   osteocytes, 321–322
   osteopontin, 320
   RANKL, 321
   tartrate-resistant acid phosphatase (TRAP), 320
Type 2 diabetes mellitus (T2DM), 64, 66–67

U
UDP-N-acetylglucosamine (UDP-GlcNAc), 294
Ulcer prevention, 365–367
Ulcer protection, 401–404
Ultimate stress, 125
Ultrasound indentation test, 131–134, 131f, 133f

V
Vancomycin, 285
Varied flow conditions, 300
Vascular endothelial growth factor (VEGF), 176–177, 375
Vascular endothelial growth factor receptor-2 (VEGFR2), 297
Vascular maturation, 378–379, 378f
Vascular mechanobiology, 300–304
Vascular remodeling phase, 378–379
Vascular smooth muscle (VSMC), 70–71
Vascular stiffening
   mechanisms, 64–66, 65f
   obesity and diabetes
      cellular contributions, 69–73, 73f
      extracellular matrix (ECM), 66–69
      metabolic and endocrine activation, 73–80
Vascular stiffness, 64–66, 65f
Vessel arterial–venous plasticity, 379–380
VivoSight, 162–163, 163t, 166
Volumes of interest (VOIs), 38–39
von Mises stress, 38

W
Weight gain, 143
Whole in-shoe threshold, 110
Whole tissue characterization, 264f
   compressibility, 264–265, 265f
   percentage energy loss, 269–270, 269f
   stiffness, 265–269, 266f–268f
Wound bed preparation (WBP), 21–22
Wound dressings, 390–397
   artisanal adapted shoe, 395, 397f
   comfort, 397–401
   dressing change frequency, 399–400
   dressing materials, 391t
   exudation, 397–401
      management, 399
   footwear, 395
   frequency, 397–401
   insole, 395, 396f
   loss of protective sensation (LOPS), 395
   plantar zone small lesion, 395, 396f
   pressure of floading, 395
   pressure relief, 401–404
   toes deformity, 394f
   ulcer protection, 401–404
Wound healing (WH), 13
   diabetes mellitus (DM), 248–249
   treatments for, 24–25, 25f
Wound registry data, 5–9, 7t–8t, 9f

X
X-ray computed tomography, 161

Y
Yes-associated proteins (YAP), 190–191, 249
Young's modulus, 124, 146–147, 240f, 266

Z
Zero damping, 234
Zero inertia, 234
   with damping, 234–235